Computational Design and Robotic Fabrication

Series Editor

Philip F. Yuan, Tongji University, Shanghai, China

This open access book series includes compilations of selected papers from the International Conference on Computational Design and Robotic Fabrication. The books focus on novel techniques for computational design and robotic fabrication. It not only aims at the most recent research results from the key scholars in the computational design and robotic fabrication, but also offers an in-depth examination of intelligence in design and construction industry, referring to the invention and application of machine intelligence in architecture. The proceedings in this series are related with Sustainable Development Goals 11 (Sustainable Cities & Communities) and 9 (Industry Innovation & Infrastructure).

The contents make valuable contributions to academic researchers, designers, and engineers in the industry. As well, readers will encounter new ideas about understanding intelligence in architecture.

Philip F. Yuan · Hua Chai · Chao Yan · Keke Li ·
Tongyue Sun
Editors

Hybrid Intelligence

Proceedings of the 4th International Conference on Computational Design and Robotic Fabrication (CDRF 2022)

Springer

Editors
Philip F. Yuan
College of Architecture and Urban Planning
Tongji University
Shanghai, China

Chao Yan
College of Architecture and Urban Planning
Tongji University
Shanghai, China

Tongyue Sun
College of Architecture and Urban Planning
Tongji University
Shanghai, China

Hua Chai
College of Architecture and Urban Planning
Tongji University
Shanghai, China

Keke Li
College of Architecture and Urban Planning
Tongji University
Shanghai, China

ISSN 2731-9040 ISSN 2731-9059 (electronic)
Computational Design and Robotic Fabrication
ISBN 978-981-19-8639-0 ISBN 978-981-19-8637-6 (eBook)
https://doi.org/10.1007/978-981-19-8637-6

This Springer imprint is published by the registered company Springer Nature Singapore Pte Ltd.
The registered company address is: 152 Beach Road, #21-01/04 Gateway East, Singapore 189721, Singapore

Preface

DigitalFUTURES is an annual series of academic events, consisting of a conference, workshops and exhibition, hosted by the College of Architecture and Urban Planning, Tongji University, Shanghai. The aim of DigitalFUTURES is to encourage international collaboration and interaction and to promote theoretical and scientific research into computational design, robotic fabrication, and other areas of architectural intelligence. The "2022 DigitalFUTURES—The 4th International Conference on Computational Design and Robotic Fabrication (CDRF 2022)" provides an international platform for advanced researches addressing Hybrid Intelligence in architecture.

> What constitutes the head of the architect is neither its 'consciousness' nor its 'idea', but the tertiary retention that enables the heritage of circuits of transindividuation to which it is connected and which precede it as its preindividual milieu, supersaturated with potentials—which endows it with the power to dream and to realize its dreams.
>
> —-Bernard Stiegler, Translated by Daniel Ross

The past 30 years have witnessed the myriad ways in which digital technology not only improves the dynamicity of communication between human and machine but also provides entirely new directions for innovation and creation. In the digitally structured process of architectural design and construction, machines are not limited to visualizations and realizations of the conceived form by architects but are involved directly in the creation process. The essence of creativity is no longer limited to the mere informational landscape of the human mind. With shared tools, libraries, and procedures, it now includes an array of intelligent machines capable of extending human intentions outward, away from the individual, forming a hybrid intelligence. Architecture will become a collective endeavor, in which not only is architectural knowledge collectively produced but also design intent can be formed by decentralized networks of creativities. The digital will be not so much about expanding methods of generating and fabricating architectural objects as it is concerned with cultivating a new context of understanding the central nature of design, creativity, and even architecture itself.

Shanghai, China

Philip F. Yuan
Hua Chai
Chao Yan
Keke Li
Tongyue Sun

Organization

Committees

Honorary Advisors (in alphabetical order of last name)

Philippe Block, ETH Zurich, Switzerland
Jane Burry, Swinburne University of Technology, Australia
Mark Burry, Swinburne University of Technology, Australia
Ximing Chen, Nanyang Technological University, Singapore
Lieyun Ding, Huazhong University of Science and Technology, China
Jian Gong, Shanghai Construction Group (SCG), China
Guoqiang Li, Tongji University, China
Jiaping Liu, Xi'an University of Architecture and Technology, China
Areti Markopoulou, Institute for Advanced Architecture of Catalonia, Spain
Achim Menges, University of Stuttgart, Germany
Antoine Picon, GSD, USA
Patrik Schumacher, Zaha Hadid Architects (ZHA), UK
Mette Ramsgaard Thomsen, Royal Danish Academy, Denmark
Zhiqiang Wu, Tongji University, China
Yimin (Mike) Xie, RMIT University, Australia
Xianzhong Zhao, Tongji University, China

Organization Committees

Philip F. Yuan, Tongji University, China
Neil Leach, Tongji University, China

Scientific Committees (in alphabetical order of last name)

Felix Amtsberg, University of Stuttgart, Germany
Alisa Andrasek, RMIT University, Australia
Nic Bao, RMIT University, Australia
Thomas Bock, Technical University of Munich, Germany
Serban Bodea, University of Stuttgart, Germany
Biayna Bogisian, Florida International University in Miami, USA
Daniel Bolojan, Florida Atlantic University, USA
Matias Del Campo, University of Michigan, USA
Brad Cantrell, University of Virginia, USA
Tengwen Chang, National Yunlin University of Science and Technology, Taiwan, China
Kristof Crolla, University of Hong Kong, China
Benjamin Dillenburger, ETH Zurich, Switzerland
Marcus Farr, Tongji University, China
Melissa Goldman, University of Virginia, USA
Yunsong Han, Harbin Institute of Technology, China
Hua Hao, Southeast University, China
Wanyu He, Xkool, China
Tim Heath, University of Nottingham, UK
Alvin Huang, University of Southern California, USA
Weixin Huang, Tsinghua University, China
Guohua Ji, Nanjing University, China
Gene Ting-Chun Kao, ETH Zurich, Switzerland
Immanuel Koh, Singapore University of Technology and Design, Singapore
Neil Leach, Tongji University, China
Guan Lee, University College London, UK
Hyejin Lee, Tongji University, China
Biao Li, Southeast University, China
Linxue Li, Tongji University, China
Yujie Lu, Tongji University, China
Peng Luo, Tongji University, China
Andrea Macruz, Tongji University, China
Sandra Manninger, University of Michigan, USA
Wes McGee, University of Michigan, USA
Xianchuan Meng, Nanjing University, China
Virginia Melnyk, Tongji University, China/Clemson University, USA
Kris Mun, University of Minnesota, USA
Guvenc Ozel, University of California, Los Angeles, USA
Gilles Retsin, University College London, UK
Klaas de Rycke, Bollinger + Grohmann, Germany
Bob Sheil, University College London, UK
Xing Shi, Tongji University, China

Miroslaw J. Skibniewski, University of Maryland, USA
Roland Snooks, RMIT University, Australia
Satoru Sugihara, Architectural Technology Laboratorial Venture
Chengyu Sun, Tongji University, China
Kostas Terzidis, Tongji University, China
Oliver Tessmann, Technische Universität Darmstadt, Germany
Kathy Velikov, University of Michigan, USA
Tomas Vivanco, Tongji University, China
Xiang Wang, Tongji University, China
Makoto Sei Watanabe, Tokyo City University, Japan
Dylan Wood, University of Stuttgart, Germany
Jing Wu, Southern University of Science and Technology, China
Leiqing Xu, Tongji University, China
Weiguo Xu, Tsinghua University, China
Michael Weinstock, The Architectural Association, UK
Chao Yan, Tongji University, China
Feng Yang, Tongji University, China
Jiawei Yao, Tongji University, China
Kaiho Yu, University of Applied Arts Vienna, Austria
Philip F. Yuan, Tongji University, China
Xu Zhen, Tianjin University, China

Acknowledgements

This book is funded by National Science Foundation of China-shenzhen Joint Fund (No. U1913606), Science and Technology Innovation Plan of Shanghai Science and Technology Commission (No. 21DZ1204500).

Contents

Computation and Formation

Spatial Analysis of Villages in Jilin Province Based on Space Syntax and Machine Learning

Deli Liu(✉) and Keqi Wang(✉)

School of Architecture and Planning, Jilin Jianzhu University, Changchun, China
1661654037@qq.com, wkq0431@126.com

Abstract. The development of machine learning technology gives architects and urban planners a new tool that can be used for research and design. The topic of this paper is to analyze the rural space of Jilin Province with the machine learning algorithms and space syntax theory, and to obtain the inherent formation and development laws of rural spatial forms, which can be used as a reference and evaluation system for subsequent rural development, and also can emphasize the locality and continuity of rural development. First, based on geographic information data, researching the connection between the distribution of villages and geographic data at a macro level and to classify them. Then, from each category, selecting one township and use all villages in its area as samples for the more specific study. Spatial features of individual village are extracted based on space syntax theory, and representative spatial features which can as feature values for cluster analysis are selected through comparative analysis. Then classify villages from high-dimensional data and explore their type characteristics. Finally, we hope the result of this study can help provide useful theoretical references for rural construction and nature conservation in the future.

Keywords: Spatial characteristics · Village · Space syntax · Machine learning

1 Introduction

In China, most cities have reached the level of stock development, while the countryside is still lagging behind. More and more people are concentrating their attention on rural construction in the context of rural rejuvenation. However, the countryside, unlike the city, does not develop from the top down, but more from the bottom up, which is a very natural and slow development process. This means that it is difficult to have a very clear development pattern like the city. The characteristics of the space in village are very implicit and highly influenced by local environmental and cultural factors. As a result, a method distinct from urban design is required in dealing with rural concerns. Otherwise, there would be a slew of issues, including the adoption of a single paradigm in the face of significantly disparate villages, a failure to respect history and natural conditions and so on.

P. F. Yuan et al. (eds.), *Hybrid Intelligence*, Computational Design and Robotic Fabrication,
https://doi.org/10.1007/978-981-19-8637-6_1

In the process of building the countryside, it is important to follow the objective rules and propose development strategies with a targeted approach. So as a first step we need to have a more accurate understanding of the countryside and be able to grasp its objective laws. However, as mentioned in the previous part of this article, the development of the countryside is a natural process. So, it is difficult to find obvious patterns of the individual countryside, and we need to draw conclusions from a large amount of data in comparison and analysis. When dealing with such a large and complex amount of data, machines are able to be more efficient and can extract information from high-dimensional data in a way that the human brain cannot. In the process of feature extraction of individual villages, space syntax theory is introduced, which can be used to describe the spatial form, and complete the spatial analysis of individual village.

2 Related Work

The theory of space syntax was proposed by Bill Hillier, Julienne Hanson and colleagues at The Bartlett, University College London in the late 1970s to early 1980s. It's a theory about architecture and urban based on the theory of society and space in the earlier the Social Logic of Space [2]. Space syntax theory is mostly used in studies related to urban design [4–6], and its initial object of research is also the city, in order to reveal the'deep structure' in buildings and cities. It illustrates how urban spatial patterns are constituted through spatial laws that link the emergence of particular spatial patterns of the city to other factors [3]. While studying cities and buildings, human behavioral factors are often included in the study. Such as analyzing the connection between spatial structure and human behavior factors. Alabi [1] tried to use the space syntax as a research theory to find out how the urban form and socio-economic factors that influence human behavior patterns, and designers can use the findings to design successful transit and pedestrian-oriented developments. Montello [7] tried to combine space syntax and environmental psychology theories. Space syntax theory can provide a rich and diverse set of quantitative indices for characterizing places in many ways that are potentially relevant to a variety of psychological and behavioral responses.

In China, some scholars have also tried to do research with the Chinese villages as the object based on space syntax theory. Niu et al. [8] use four typical villages in Xiahuayuan District of Zhangjiakou city as samples to obtain the spatial structure characteristics based on space syntax theory, which could provide theoretical basis for the current village plan. Xu et al. [10], taking Jiangshan Hejia-Wujia village in Guchen township, Gaochun county, Nanjing as an example, in terms of Space Syntax, analyzes the spatial characteristics of traditional villages, and they hope to provide new thoughts for systematically protecting and healthily developing of the traditional villages.

However, the sample sizes of these studies are very small and can only reflect the spatial characteristics of individuals, and no common patterns can be derived. Therefore, in this study, we expand the sample size and expect to obtain some implicit common laws through further analysis of results of individual village analysis with machine learning algorithms, which can be used to provide some theoretical references for future rural development.

3 The Geographical Distribution Characteristics of Villages in Jilin Province

The geomorphic feature of Jilin Province varies greatly. We can obtain the contour data of Jilin Province from China National Catalogue Service For Geographic Information, and the elevation values were used as the population value for kernel density analysis. The analysis result shows that the topography of Jilin Province slopes from southeast to northwest, showing a clear feature of high southeast and low northwest (Fig. 1a). Bound by the central Big Black Mountain, it can be divided into two major landscapes: the southeastern mountains and the central and western plains. The water resources data of Jilin Province were also obtained from China National Catalogue Service For Geographic Information, and the area value of rivers was used as the population value for kernel density analysis (Fig. 1b), which shows that it is mostly concentrated in the central and northwestern regions.

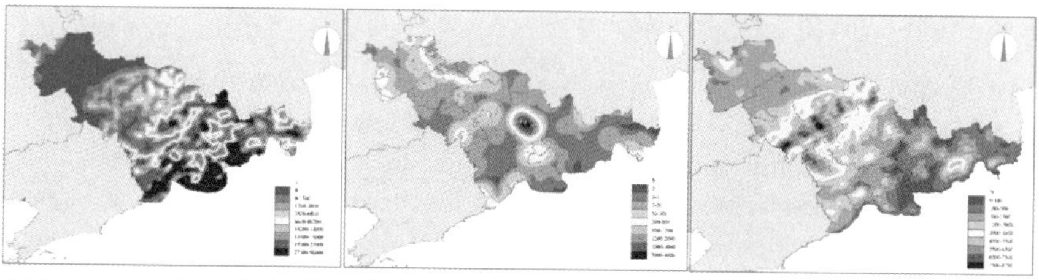

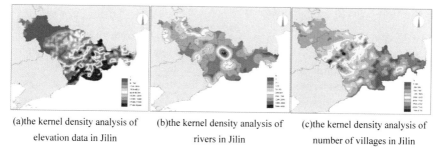

(a)the kernel density analysis of elevation data in Jilin (b)the kernel density analysis of rivers in Jilin (c)the kernel density analysis of number of villages in Jilin

Fig. 1. The kernel density analysis of elevation value, rivers and villages in Jilin

According to the kernel density analysis of the village POI data obtained from Amap (Fig. 1c), it can be seen that villages in Jilin Province are mostly concentrated in the central region, and its distribution is less in the southeast, which is strongly related to the topography of the mountain range. The raster data from the kernel density analysis of villages, elevation and water resources were summed up at the county level area of Jilin Province, and the results were mapped uniformly into the [0,10] value domain to draw a scatter plot (Fig. 2). Then the correlation coefficient between the elements is calculated according to Eq. (1). From Fig. 2 and correlation coefficient values, it is clear that the topography of the mountain ranges is more related to the distribution of villages, and the overall trend is that as the number of mountains increases, the distribution of villages decreases.

$$r = \frac{\sum_{i=1}^{n}(X_i - \bar{X})(Y_i - \bar{Y})}{\sqrt{\sum_{i=1}^{n}(X_i - \bar{X})^2}\sqrt{\sum_{i=1}^{n}(Y_i - \bar{Y})^2} \quad (1)$$

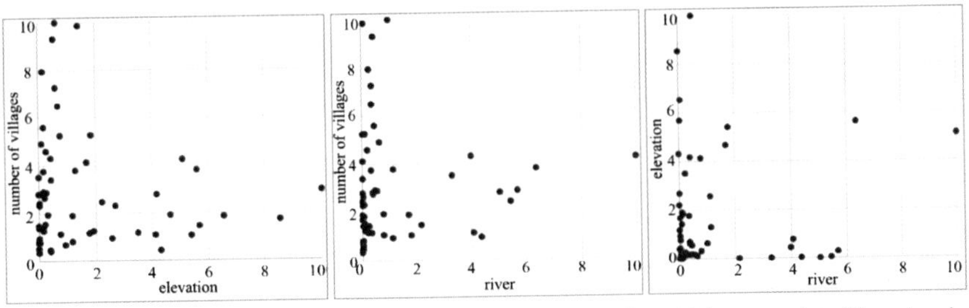

(a)the scatterplot of the value of villages and elevation value (r=-0.1207)

(b)the scatterplot of the value of villages and rivers(r=0.051)

(c)the scatterplot of the value elevation value and rivers(r=0.0883)

Fig. 2. The scatterplot of the value of rivers, number of villages and elevation data in Jilin

4 Village Spatial Feature Extraction Based on Space Syntax

4.1 Sample Selection for the Research

From the above correlation analysis, we know that the geographical distribution of villages is closely related to the elevation value, so the elevation value and the number of villages are used as influencing factors to complete the classification analysis, and then one township from each category is selected as the research object. The space syntax is used as the theory to extract the spatial characteristic values of individual village and use them for subsequent analysis. First, the number of villages and elevation data are used as feature values to cluster the county-level regions of Jilin Province, and the K-means algorithm of scikit-learn algorithm library is chosen for the clustering analysis, which can organize the data with similar feature values for classification. When selecting the number of clusters, the SSE value is used as a criterion to evaluate the good or bad clustering results, and different numbers of clusters and their corresponding SSE values are calculated according to Eq. (2) and plotted as a line graph.

$$\mathrm{SSE} = \sum_{i=1}^{k} \sum_{p \in C_i} |p - m_i|^2 \tag{2}$$

It can be found that the SSE value has a plummeting inflection point when the number of clusters is 2 and 3 (Fig. 3a), so it will be better to choose 2 or 3 for the number of clusters. However, since a representative from each cluster will be selected for specific analysis at a later stage, in order to increase the sample size, so 4 was chosen as the number of clusters for cluster analysis, and the clustering results are shown in Fig. 3b. One county-level region was randomly selected in each of the four clusters obtained, and one township-wide villages in each region was selected as the study sample. The four county-level regions selected were Jingyue District, Shulan City, Wangqing County, and Yushu City. The townships selected from these four county-level areas are Xinhu Township in Jingyue District, Xihe Township in Shulan City, Baichaogou Township in Wangqing County and Yumin Township in Yushu City. From their satellite maps

(Fig. 4), it can be seen that Baicaogou Townships is located in the mountainous area, Xihe Township is located in the hilly area, while Xinhu Township and Yumin Township are located in the plain.

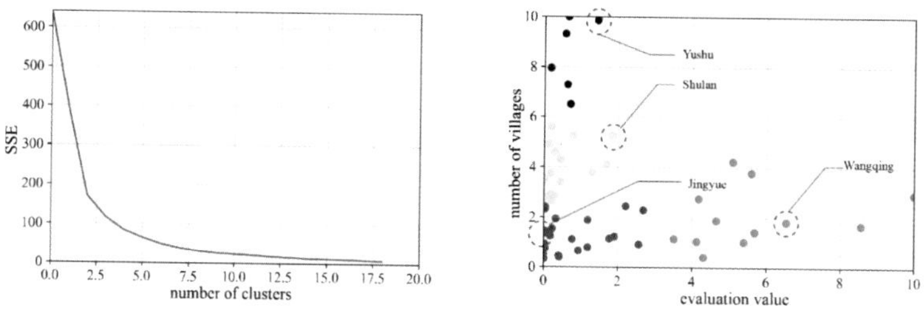

(a)SSE values corresponding to the number of different clusters (b)Clustering results when the number of clusters is 4

Fig. 3. Cluster analysis based on elevation value and the number of villages

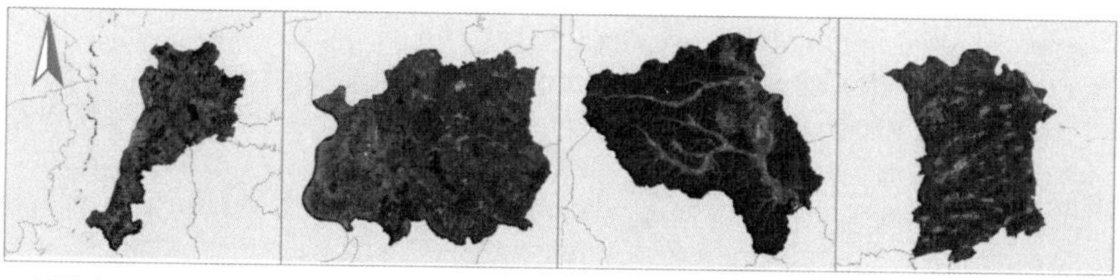

(a)Xinhu Township (b)Xihe Township (c)Baicaogou Township (d)Yumin Township

Fig. 4. The satellite map of four Townships

4.2 Spatial Analysis of the Individual Village

The analysis of axial maps in cities can reflect the external spatial structure and there are many studies to confirm this [9]. The axial maps were also used in the study about Chinese villages. Selecting connectivity, integration and intelligibility values as the values of the variables describing the spatial characteristics of the villages. The connectivity value can indicate the number of spaces that are directly connected to the space where that axis is located. The axis with higher integration means that the average topological depth value of that axis reaching all axes of the whole system is lower, indicating its higher accessibility. The intelligibility value can be represented by the R-squared value of the linear fit of the connectivity value of the axes and the integration value. The higher the R-squared value means that the better the connectivity value and the integration value are fitted, which means it is easier for people to infer the overall spatial structure from the local perception, and the intelligibility of the space is higher.

The R-squared value represents the linear correlation between the connectivity value and the integration value of all axes in a village, which is also a value that can be directly used as a feature value for the subsequent village clustering analysis. However, In the case of integration and connectivity analysis, each axis has a value attached to it, so it is difficult to describe the overall status of individual axial maps. In this case, mean, median, mode, standard deviation and range values are used as the statistical indicators of the list data. And among these representative values, only the mean and standard deviation give a better description of the overall condition of the data.

Four villages were randomly selected and then DepthmapX software was used to analyze them. From the result (Fig. 5), it can be seen that the mean value of connectivity of village-2 is 3.45, which is greater than the mean value of village-4, so the degree of spatial interconnection represented by its axes is higher than that of village-4.The standard deviation value reflects the dispersion of the data, although the average of the connectivity values of village-2 and village-3 are relatively close, the standard deviation value of village-3 is larger than that of village-2, indicating that its connectivity values are not evenly distributed. It is also obvious from the figure that the connectivity values of village-3 appear extreme, with a very obvious main axis, and its range values are also significantly larger than those of village-2. The integration can reflect the accessibility of the spaces where the axes of a village are located. The higher the value of integration, the easier it is to reach other spaces in the village. The standard deviation of the integration of all axes can reflect the difference of how easily to access to different space in the village. In terms of spatial intelligibility, we can find that villages whose R-squared values are closer to 1 have relatively simpler spatial structures. In the case of the four villages, they are sorted by their R-squared values close to 1, and their order is Village 4, Village 2, Village 3 and Village 1. From the axial maps, we can also clearly see that Village 4 has the simplest spatial structure, with a very obvious main axis. It is easier for people to perceive the whole external space in it. While village 1, which has the farthest R-squared difference value from 1, we can see that its space is more diffuse and the whole village shape is more organic.

From the above analysis, we can learn that when using the connectivity and integration values to describe the spatial characteristics, the standard deviation and the mean values are chosen to be more comprehensive because they can reflect all the data very well. The R-squared value, on the other hand, is already the result of the comparison of all the connectivity and integration values, and can also complete the description of the spatial characteristics very well.

5 Clustering of Villages Based on Spatial Features

There are four townships in this study, and axial maps of all villages are depicted by hand. There are 165 effective axial maps, including 27 axial maps in Baicaogou Township, 42 in Xihe Township, 54 in Xinhu Township, and 42 in Yumin Township. The selected spatial eigenvalues are the R-squared values, the mean and standard deviation of the connectivity values and the integration value. The clustering algorithm be selected is also the K-means algorithm, which is able to find classifications with similar values of each eigenvalue. Since all villages come from four different townships and it does not have a clear inflection point for the SSE analysis, the cluster number 4 was chosen as the cluster number, and the clustering results are shown in Fig. 6.

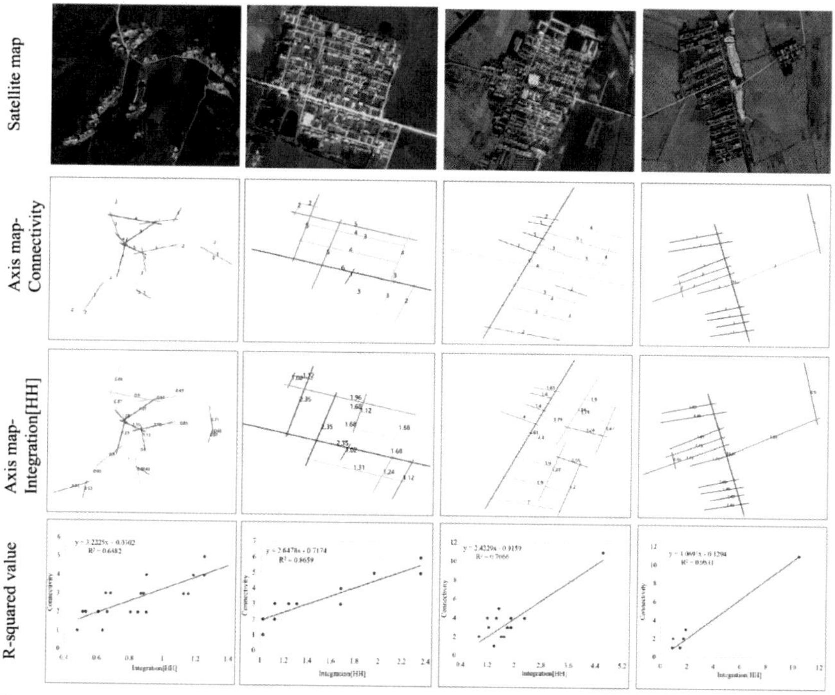

Connectivity Value

name	(a)Village-1 value	(b)Village-2 value	(c)Village-3 value	(d)Village-4 value
mean	2.72	3.45	3.33	2.14
standard deviation	1.09	1.36	2.18	2.53
range	4	5	10	10
mode	2	3	4	1

Integration[HH] Value

name	(a)Village-1 value	(b)Village-2 value	(c)Village-3 value	(d)Village-4 value
mean	0.83	1.58	1.75	2.13
standard deviation	0.28	0.48	0.75	2.31
range	0.97	1.33	3.66	9.5
mode	/	/	/	/

(a)Village-1 (b)Village-2 (c)Village-3 (d)Village-4

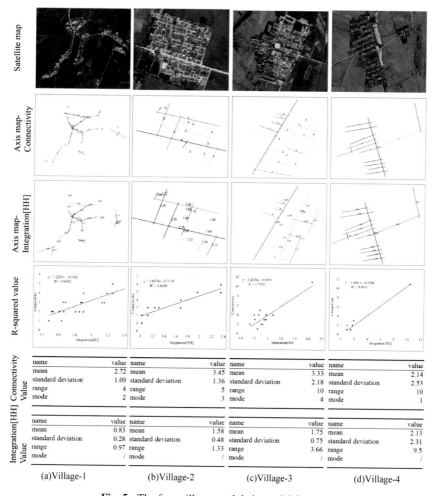

Fig. 5. The four villages and their spatial features

5.1 Spatial Characteristics of the Villages in Each Cluster

Table 1 shows the countryside feature parameters in each different set of clusters, and it can be seen that among the four clusters, the mean intelligibility in cluster-1 (red) and cluster-4 (blue) is higher than that in Cluster-2 (purple) and Cluster-3 (green). So, it can be known that the spatial structure of villages in cluster-1 (red) and cluster-4 (blue) is simpler than the other two, and it is easier for people to perceive the whole rural space in this cluster. In the eigenvalues of connectivity, there is no great variability in the mean values among the four clusters. In the eigenvalues of integration, Cluster-1 (red) has a higher average value of integration than the other three, indicating that the topological depth between its individual axes is relatively low and there are more connections between the individual spaces. The lowest mean value of integration in

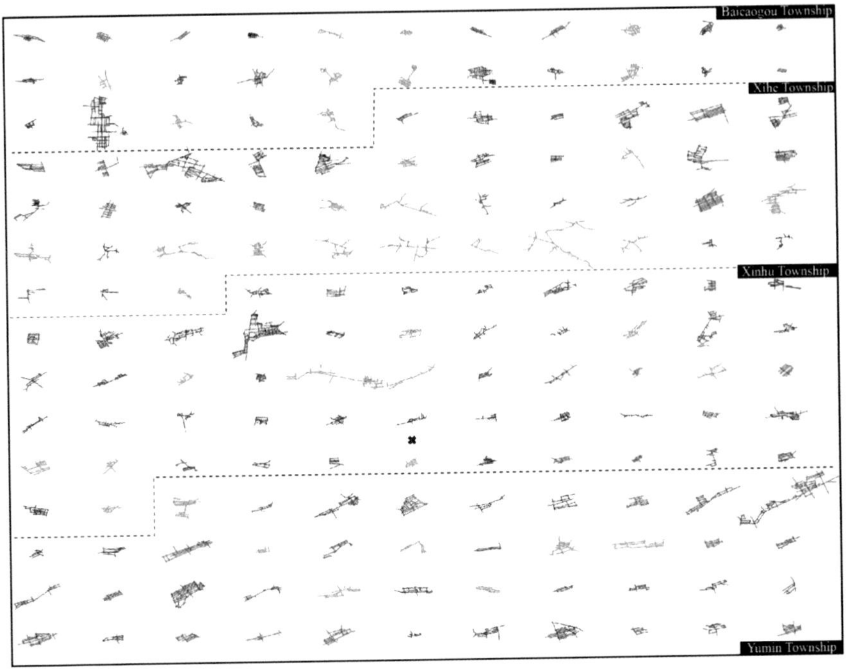

Fig. 6. The results of village clustering. This figure shows the villages sorted by belonging to the same township, with different colors representing the different clusters they belong to

Cluster-3(green) indicates that the spaces represented by each axis are less connected and more diffuse overall.

In summary, these four clustering results have the following characteristics. In Cluster-1 (red), they have high integration and connectivity values. So, the space where the axes are located in the village are more closely connected to each other, and all the spaces in the spatial system where the axes are located reach each other more easily. They also have high intelligibility value, which means the spatial structure of these villages is simpler, and it is easier for people to understand the structure of all the space system. In Cluster-2 (purple), the spatial intelligibility value is low, which means that the axial spatial structure is more complex. We can find that the central settlements in all four townships are in this classification, and their scales are larger compared with other villages. They have a higher integration values, it is means the space where the axes are also located in the village are more closely connected to each other.

In Cluster-3(green), they have the lowest intelligibility, integration and connectivity values. Therefore, the spatial structure of this type of countryside is the most complex and difficult to understand, as can be seen from Fig. 6, which shows that the axial maps of this type of countryside are more natural and disorderly compared to the other three classifications. The spatial intelligibility is higher in Cluster-4 (blue) and its spatial structure is relatively simple compared to Cluster-2 (purple) and Cluster-3 (green). Its connectivity value is not high, indicating that the direct connection between each axis is also weak, and it can be found that the number of axes in each axial map in this cluster

Table 1. The spatial feature values of the four clustering results

		Cluster-1 (red)	Cluster-2 (purple)	Cluster-3 (green)	Cluster-4 (blue)
Intelligibility	Range	[0.691,0.980]	[0.239, 0.632]	[0.116, 0.582]	[0.569, 0.931]
	Mean	0.836	0.459	0.398	0.764
Connectivity_ mean value	Range	[2, 4.385]	[2.750, 3.887]	[1.867, 3.415]	[1.867, 3.533]
	Mean	3.191	3.252	2.577	2.717
Integration[HH]_mean value	Range	[1.419, 2.1259]	[0.974, 1.671]	[0.45, 1.079]	[0.785, 1.442]
	Mean	1.67894	1.211644	0.824427	1.163224
Connectivity_ standard deviation	Range	[1.099, 2.941]	[1.187, 3.304]	[0.667, 1.762]	[0.786, 1.795]
	Mean	1.772424	1.879552	1.153918	1.300722
Integration[HH]_ standard deviation	Range	[0.371, 2.312]	[0.240, 0.431]	[0.117, 0.372]	[0.247, 0.571]
	Mean	0.716854	0.334469	0.222923	0.394921

is low. If we arrange these four clusters according to the degree of simpler or regular spatial structure and similar morphology of the villages in the cluster. Then the order should be Cluster-1 (red), Cluster-4 (blue), Cluster-2 (purple) and Cluster-3 (green).

5.2 Conclusion Based on Comparative Analysis of the Spatial Characteristics of Villages in the Four Townships

In these four townships, the ratio of these Cluster-1 (red),Cluster-2 (purple), Cluster-3 (green) and Cluster-4 (blue), four types of villages in Baicaogou Township is 0.297: 0.185: 0.333: 0.185, the ratio in Xihe Township is 0.214: 0.167: 0.333: 0.286, the ratio in Xinhu Township is 0.148: 0.185: 0.185: 0.482, and the ratio in Yumin Township is 0.238: 0.071: 0.167: 0.524. In all villages, the ratio of this four type villages is: 0.212: 0.152: 0.242: 0.394. In Baicaogou Township and Xihe Township, the proportion of villages of cluster-3 (green) type is higher, while the largest number of all villages in these four townships is the cluster-4 (blue) type. We know that the spatial feature of villages in cluster-3 (green) have the minimum mean value of Intelligibility and Integration[HH]_mean value, which means that these villages have a complex spatial structure, and they have a variety of spatial forms with natural patterns and fewer regularities.

In Xinhu Township and Yumin Township, the number of villages belong to cluster-4 (blue) is larger than the other types of villages, which means the plane of villages in this two cluster is more likely controllable and less affected by geography. In the previous part of this article we said Baicaogou Township is in the mountainous area, Xihe Township is located in the hilly area, while Xinhu Township and Yumin Township are located in the plain area. In fact, it is not difficult to imagine that in the mountainous and hilly areas, the plan of villages is more natural and the spatial structure is more disorderly due to the geographical topography. Villages located in the plain area, on the other hand, are relatively more regular in their planar texture, less constrained by geography and more controlled by human factors.

6 Conclusion

In this paper, we uses space syntax and machine learning algorithms to complete the analysis of the spatial characteristics of villages in Jilin Province, to find patterns and draw conclusions from the high-dimensional data, so that we can better understand the characteristics of rural spatial characteristics and apply them to the subsequent work on villages. It is worth noting that this analysis method does not yield unique results. For example, when changing the number of clusters in the final cluster or replacing the feature values, or weighting the feature values, the clustering results will become different, and different conclusions can be drawn based on the analysis of the clustering results. This is the characteristic of using machine learning algorithms to complete the analysis, it does not filter for us subjectively, while it just analyzes and outputs the results without any preference. When we change the input parameters, the output results will also be diverse. Algorithms and theories just help us to have more and deeper understanding of the object we are analyzing when we use them, and finally it is up to us to interpret and summarize the diverse results to get the information we need. This corresponds to the statement in the abstract that machine learning algorithms only provide a new analysis or design tool. They do not make decisions for us, but provide us with new references when we making decisions. Finally, this is also an attempt to apply space syntax theory and machine learning algorithms to the study of the countryside in question. Although there are still some shortcomings, however, we still hope it can give new reference and inspiration to researchers in the same field.

References

1. Alabi MO (2021) Space syntax: evaluating the influence of urban form and socio-economy on walking behaviour in neighbourhoods of Akure, Nigeria. Urban, Plan Transport Res 9(1):579–597
2. Hillier B, Hanson J (1989) The social logic of space. Cambridge University Press
3. Hillier B (2007) Space is the machine: a configurational theory of architecture. Space Syntax
4. Hillier B (1999) The hidden geometry of deformed grids: or, why space syntax works, when it looks as though it shouldn't. Environ Plann B Plann Des 26(2):169–191
5. Jiang B, Claramunt C, Klarqvist B (2000) Integration of space syntax into GIS for modelling urban spaces. Int J Appl Earth Obs Geoinf 2(3–4):161–171
6. Karimi K (2012) A configurational approach to analytical urban design:'Space syntax'methodology. Urban Design Int 17(4):297–318

7. Montello DR (2007) The contribution of space syntax to a comprehensive theory of environmental psychology. In: Proceedings of the 6th international space syntax symposium, pp 012.1–012.14
8. Niu MJ, Xu WG, Li YX (2020) Quantitative research and analysis of the rural settlement structure of Xianhuayuan district based on space syntax. Contemp Architect 2020(06):138–140
9. Penn A (2003) Space syntax and spatial cognition: or why the axial line? Environ Behav 35(1):30–65
10. Xu H, Zhao HS, Liu F (2016) Preliminary study in syntactical of the spatial form of traditional villages: a case study of Jiangshan Hejia-Wujia village in Guchen Town, Gaochun County, Nanjing. Modern Urban Res 2016(01):24–29

A Human–Machine Collaborative Building Spatial Layout Workflow Based on Spatial Adjacency Simulation

Ximing Zhong[1(✉)], Fujia Yu[2], and Beichen Xu[3]

[1] Aalto University, 02150 Espoo, Finland
ximing.zhong@aalto.fi
[2] Tampere University, 33100 Tampere, Finland
yufujia@tju.edu.cn
[3] University of Liverpool, L69 7ZN Liverpool, UK
B.Xu14@liverpool.ac.uk

Abstract. The space layout of a reasonable modular building prototype is a time consuming and complex process. Many studies have optimised automatic spatial layouts based on spatial adjacency simulation. Although machine-produced plans satisfy the adjacency and area constraints, people still need further manual modifications to meet other spatially complex design requirements. Motivated by this, we provide a human–machine collaborative design workflow that simulates the spatial adjacency relationship based on physical models. Compared with previous works, our workflow enhances the automated space layout process by allowing designers to use environment anchors to make decisions in automatic layout iterations. A case study is proposed to demonstrate that the solution generated by our workflow can initially complete different customised design tasks. The workflow combines the advantages of the designer's decision-making experience in manual modelling with the machine's ability in rapid automated layout. In the future, it has the potential to be developed into a designer-machine collaboration tool for completing complex building design tasks.

Keywords: Spatial adjacency simulation · Physical model · Responsive design process · Human–machine collaborative workflow · Real-time visualization

1 Introduction

The modular building spatial layout is a task involving many complex decision-making activities [1]. Designers need to spend a huge amount of time and make a great effort to consider the dimensions of the building modular and all possible relationships within and between the buildings. In addition, they need to dynamically modify the prototypes of individual modules and layout combinations to meet the user's changing needs and complex spatial adjacencies [2].

Since the 1960s, the rule-based approach has been widely used for automated spatial layout [3]. Several rule-based studies translate adjacency [4], density [5], and topological

P. F. Yuan et al. (eds.), *Hybrid Intelligence*, Computational Design and Robotic Fabrication,
https://doi.org/10.1007/978-981-19-8637-6_2

and geometric constraints [6] into mathematical constraint objectives to optimize layout solutions for spatial automation. Based on 'Rule-based', various classical studies treat automated building layout and prototyping as a multi-objective optimisation process. However, not all design requirements can be easily translated into explicit mathematical constraints [7]. Moreover, due to the partial errors in adjacencies sometimes arising and the weak interactivity of the layouts generated by this method. It is hard for designers to make decisions in automatic layout iterations [8].

Many studies regarding automatic layout based on the physical model show the potential to simulate the human–machine collaborative design process, thus resulting in the advantage of being highly interactive [9]. Based on these methods, some studies demonstrate the feasibility of simulating the spatial layout by analogy spring [10] and gravitational and repulsive forces [4] to satisfy complex adjacency demand constraints. However, the results of the physical model calculations still require a significant amount of manual modelling work and filtering to complete customized architectural design tasks. In their studies, designers are mainly limited to filtering and modifying machine-generated plans after simulating but not adjusting the results via directly changing the physical model during the machine iteration. Many studies on human–machine collaboration in recent years have demonstrated the great advantages of human–machine collaboration. Humans make decisions based on experience; machines can solve problems quickly and automatically, with each side playing to its strengths [1]. Impacted by this, the aim of our workflow is:

1. Utilizing physical models based on simulating spatial adjacency relationships to automate spatial layouts to meet customized design tasks' needs (adjacency, area, geometric relationships, etc.)
2. Enable designers to manually control the nodes interactively. Designers can set and move the environment anchor in the system to determine the final layout generated by machines. The workflow can solve different prototype design tasks (PDT) initially.

2 Related Works

In 1989, Baraff proposed an approach to the automatic layout of space by simulating and correcting physical collisions with physics, an interactive process [11]. In 2002, Arvin automated the conceptual design process by applying the physics of motion to the elements of spatial planning [10]. This approach provides a responsive design process. In 2010, Hao, H. and Ting-Li, J presented the optimization of adjacency relations for floating bubbles as an analogy to spatial layouts. By assigning two primary forces to agents: attraction and repulsion, it achieved a responsive layout of spaces with more complex adjacency requirements constraints [4]. While in 2016, ElSayed further proposed a theoretical approach to simulate the interactive layout process of collision and reorganization between the modified space and other prototype spaces that the designers provided in advance by simulating mechanical springs [8]. Furthermore, it demonstrated its highly feasible and interactive in the experiments used by Egyptian lay workers for custom modifications. However, 1. Conflicting adjacencies and geometric constraints sometimes make the system arise with conflicting partial layouts that need to be redone manually by architects. 2. Even though automated layouts based on physical layout models are more interactive than purely algorithmic simulation, it is still hard for designers to modify the physic model during the iteration.

Therefore, we are trying to explore a responsive building layout workflow based on space adjacency simulation with physical models by introducing the concept of environment anchors. Environment anchors in our paper are objects in modelling software that architects can define for different design tasks, such as sun, orientation, building main core, corridors, and public spaces.

3 Method

As shown in Fig. 1, the proposed workflow is divided into four major parts. I. User Input Component (UIC): Designers enter the Matrix of Space size and Matrix of Adjacent relationship (data of various spatial demands and complex adjacency requirements). II. Stimulating Component: Use physical models to achieve automatic layout by simulating the adjacency relationship between units. IV. Visualization Component: The computer provides real-time feedback with the results and evaluations (space adjacent, average distance) to designers. III. Designer Adjustment Component: Designers can interactively change the layout through moving environmental anchors, combing manual modifications with machine-generated solutions to generate an automated layout that satisfies the adjacency relationship.

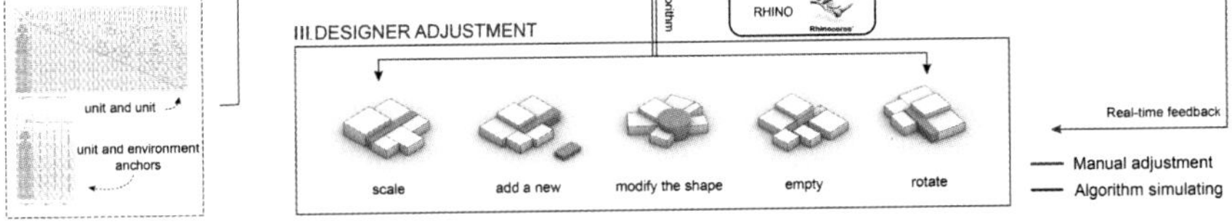

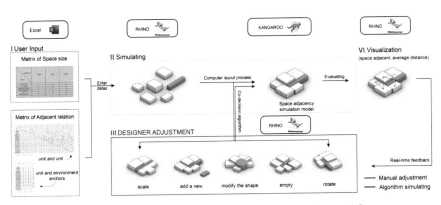

Fig. 1. The framework of human–machine collaborative workflow

3.1 UIC Component (Unit Size and Adjacency Matrix)

Unit Size Input: Designers record the size requirements for each unit by length, width, and height in an Excel file shown in Fig. 2a. These data are transferred to the Rhino modelling platform and then converted into a 3D physical model using Grasshopper. As Fig. 2b shows, each unit is generated automatically.

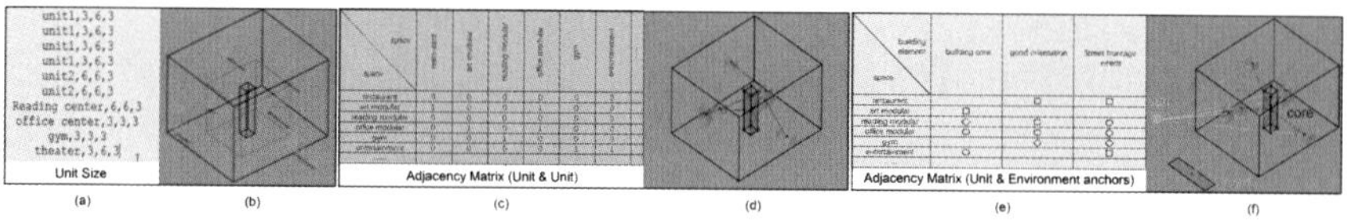

Fig. 2. UIC and preliminary results

Adjacency Matrix Input: Designers record different adjacency requirements in the adjacency matrix. Figure 2c shows the adjacency relationship between units, and Fig. 2e shows the relationship between units and environment anchors. Moreover, the different shapes represent w = 0.1, w = 0.5, w = 1. In our case, parameter w is a constant from 0 to 1, which represents the degree of adjacent requirements between the different modules.

Designers need to set these adjacencies manually. When the setting is completed, the system converts the adjacency into lines between spaces, as shown in Fig. 2d, f. The adjacency relationship lines will then be put into the modules to simulate forces. Different adjacency relationships correspond to different weights of attraction and repulsion. When the adjacent relationship becomes stronger, the more increasing corresponding force performs.

3.2 Simulating Component

3.2.1 Spatial Adjacency Relations Simulating by Physical Model

We first introduce the simulation of the spatial adjacency relationship between all units. Arvin. S and House. D present a prototype for simulating spatial layout [10]. Different axial gravitational or repulsive forces are exerted between the two units depending on their different adjacency requirements. For unit a, the force on it in the system is equal to the sum of the forces of all the objects that have an adjacency requirement in Fig. 3b. Specifically, the combined force of the unit a is equal to the gravitational force of each object multiplied by the weight w (Fig. 3a) and then summed. It is noted that the initial gravitational force values in the system are all 1 N. The designer can control which units previously had a stronger attraction by adjusting w to control the final layout. Furthermore, we use the Kangaroo plugin to automatically simulate the force and layout of units based on data we can read from the UIC.

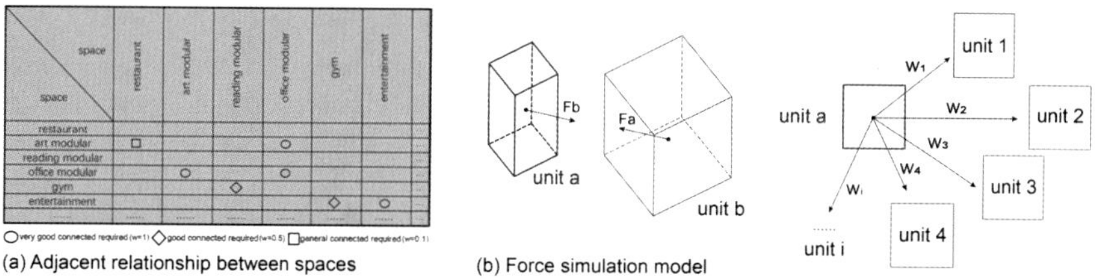

Fig. 3. Layout simulation based on adjacency between unit and unit by physical model

We then introduce the simulation of the adjacency relationship between units and environment anchors. In our workflow, environmental anchors can be defined as any design-related custom factor, such as the sun direction, the position of the main core, street frontage needs, main core, corridors, platforms where units are located and other factors that can influence the layout of the building. Meanwhile, we set these anchors as the 'object instance' in the system. Hence the designer can be manually moved, deformed, and turned (Fig. 4) by any modelling software. It provides a good basis for combining manual modelling and automated scheduling workflows. Designers try to solve layout problems with environment anchors. Then the machine will automatically fill in the solution based on the adjacency relationship.

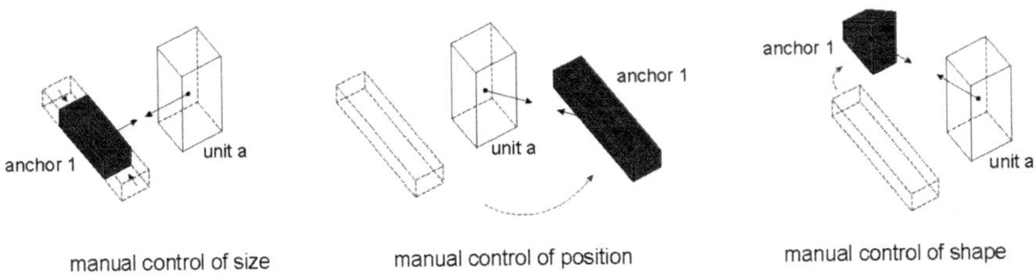

Fig. 4. Operability of environment anchors

Hao, H. and Ting-Li, J proposed a prototype of disjoint objects that generate repulsive forces when objects overlap and separate [4], ensuring that the space is geometrically constrained in the layout process. Based on their works, we use collision modules in Kangaroo to enable objects to be non-intersecting.

3.2.2 Simulating Process

The system automatically simulates the spatial layout based on UIC data and manual settings. Figure 5 shows a simulating process from random distribution (a) to the association (b) to aggregated state (c). In addition, the red lines in the diagram demonstrate the adjacent relationship between units that have adjacent requirements. A shorter length indicates higher value attractiveness and more optimized adjacency. The sum of spatial adjacency line distances can be seen as a parameter to measure and optimize the spatial adjacency of a solution [12]. As Fig. 5d shows, the curve shows how the sum of distances changes during the optimization process which can help designers to monitor the optimizing phase. The sum of the geometric distances among boxes with spatial proximity requirements tends to decrease over time and level off as the space reaches a more compact state. The system will give designers a completed calculated solution when the curve becomes stable.

3.3 Visualization Component

Turner proposed the Mean Shortest Path Length, which quantifies the accessibility of every location in the spatial system [13]. Based on his study, we define a colour range to represent different values of adjacency relationship to show the real-time average shortest path length visualisation. The lower values indicate closer adjacency and more

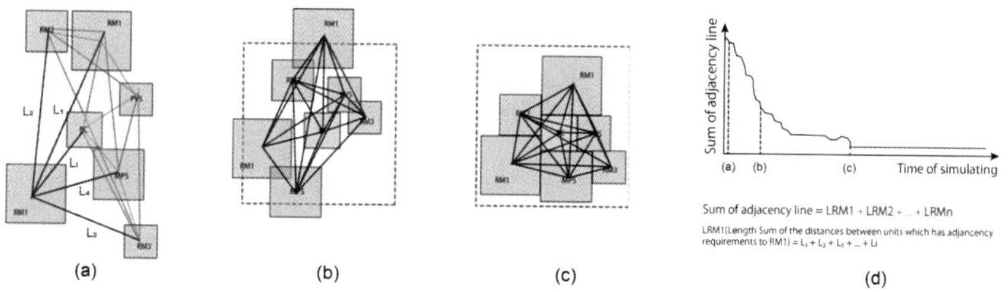

Fig. 5. Dynamic change of adjacency lines distance and optimization phase in simulating process

optimised coloured cool colours (Fig. 6), whereas higher values are hot colours. Users can consider and minimise this total distance to improve overall fitness. Designers can adjust and re-optimise the specific unit space in real-time by manual adjustments when getting a redpoint.

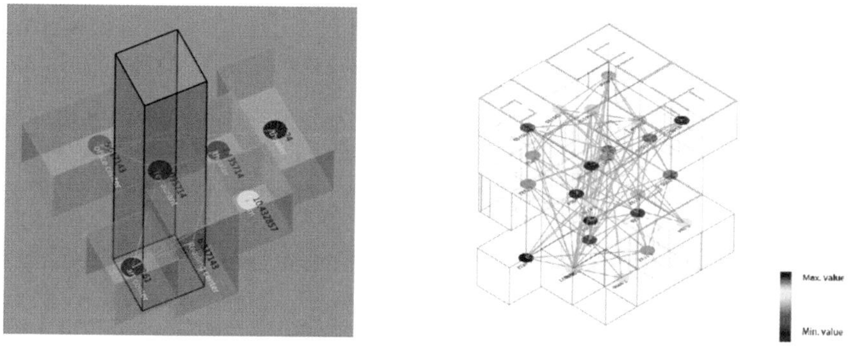

Fig. 6. Diagram of 3D spatial adjacency visualization

3.4 Designer Adjustment Component (Illustration of Interface)

During the simulation process, designers can manually move the different environment anchors to achieve different spatial layouts. Figure 7 shows an illustration of the designer adjustment interface. Users can change the scale, position, and shape of environment anchors' objects in the system. In addition, they can preview a real-time layout result of the unit's relocation under gravitational and repulsive forces after the adjustment. This workflow has been deepened into a tool called Autocat which is available on Food 4 Rhino.

4 Case Study

The Hong Kong social house project competition is selected as our case study. This is an architectural competition project completed by Archiford LTD in 2018 to satisfy complex area and spatial adjacency requirements by using a modular design approach.

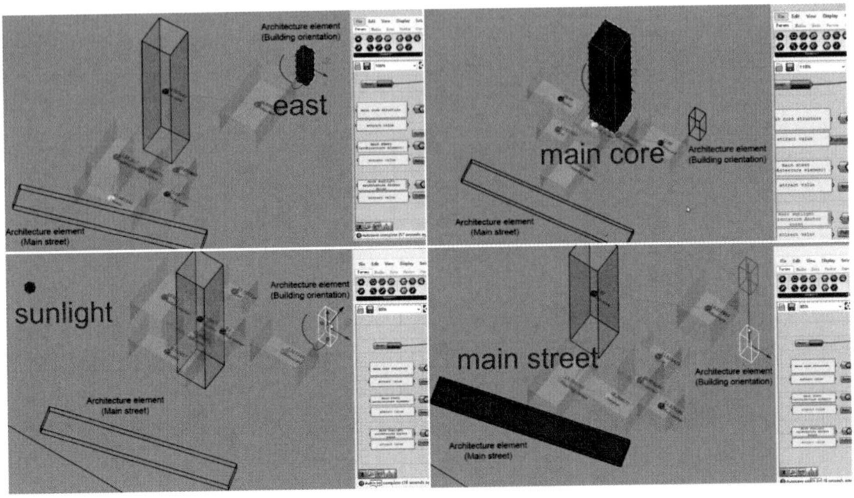

Fig. 7. Illustration of designer adjustment interface

We are trying to test our workflow based on the research and prototype design offered by Archiford LTD. In this case, we attempt to set different goals to verify whether our workflow can generate layouts that comply with the adjacency and set constraints. In addition, trying to verify if it is possible to guide the machine to generate the expected spatial layout by designers' manually adjusting the environment anchors.

Experimental rules: We expect each group to have a different layout and intention (details will discuss in the following setting section). We define different rules and expect these different rules to combine human decision-making and automated layout to decide the final layout results jointly.

Experimental Settings: Fig. 8a shows the base household type, and Fig. 8.b, c shows two sets of adjacency relations (unit and unit & unit and environment anchors).

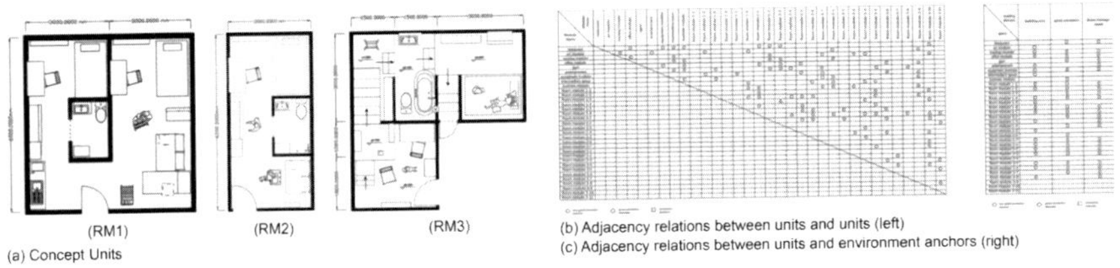

Fig. 8. Basic prototype and user requirement data of hongkong social housing project offered by Archiford LTD

Subject to the building units satisfying the adjacencies in Fig. 8, we propose five different prototype design tasks (PDT). These requirements are for the top-floor plan, which makes it easier to observe. Specific PDTs and environment anchors settings (EAS) are described below, EAS are shown in Fig. 9a.

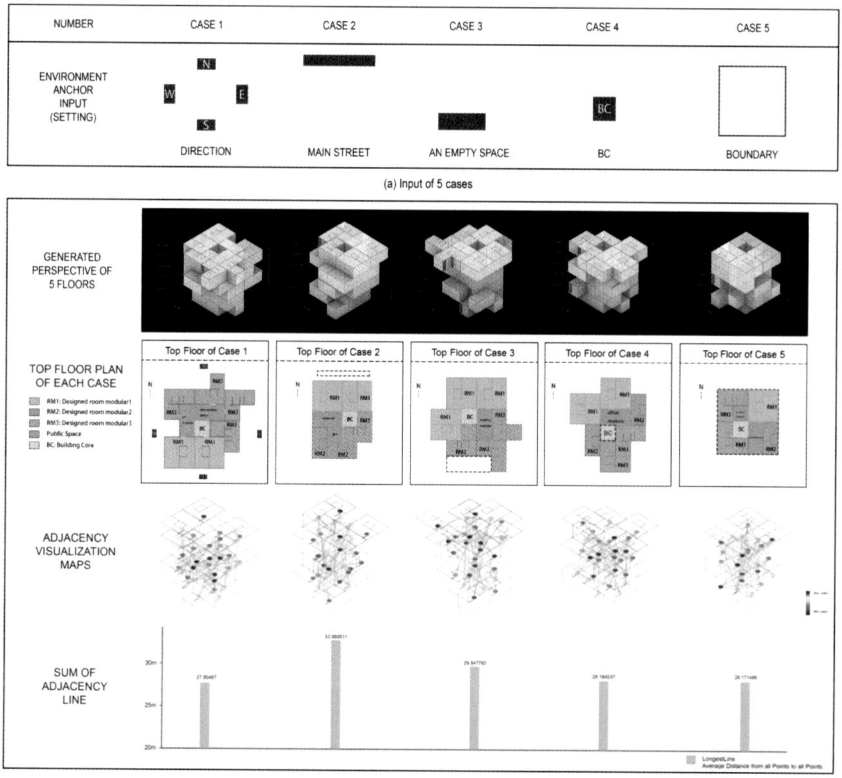

(a) Input of 5 cases

(b) Output of 5 cases

Fig. 9. EAS and the corresponding spatial layout results

Case 1: PDT: Try to make the RM3 unit lay on the north side of the plan. EAS: Design four direction anchors (North, South, West, and East) and increase the weights between RM3 and anchor 'North' (Fig. 9a).

Case 2: PDT: We intend the RM1 units to face or be further adjacent to the main street. EAS: Set the main street as the environment anchor. Adjust the position of the main street to check whether RM1 will change their positions accordingly (Fig. 9a).

Case 3: PDT: We intend to incorporate a terrace space in the southwestern part of the building. EAS: An 'empty' space. Check whether the terrace space will appear on the top floor of the building.

Case 4: PDT: Temporarily not adding other manual adjustments than the BC (Building core). EAS: BC. Use it to compare the effect of different environment anchor settings on the automated layout results.

Case 5: PDT: We expect it can be optimised to produce a plan-regular spatial layout. EAS: Boundary (Restrict the outline of the layout results). Check whether the layout of the top floor will be adapted to the boundaries we defined.

5 Result and Discussion

As shown in Fig. 10, the system achieves the rapid automated space layout, which the initial layout takes only 20 s to generate. From the result, we can find that most of the generated plans can satisfy the adjacency relationship, for example, the relationship between RM3 units and public space. Moreover, because of the manual modification, the accessibility to private units on each floor with building core and public spaces can also be satisfied.

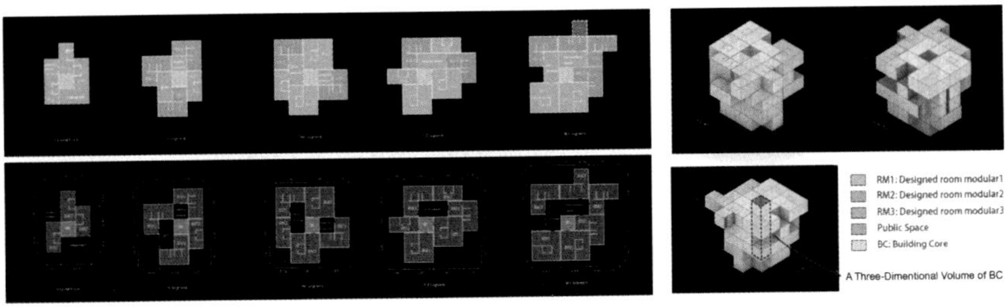

Fig. 10. Quick 3D spatial prototype layout result

Figure 9b demonstrates the results of PDT. Our workflow can basically meet the initial layout of different tasks depending on environment anchors set by designers. For example, the RM3 units in Case1 are finally laid on the north side. The RM1 units in Case 2 are adjacent to the main street and are changed along with the location of the main street. The final layout of Case 5 generates a square boundary. Furthermore, we can note that compared with Case 4, Case 3 generates an appropriate terrace space on the south side by the designers' manual adjustment and machine coordination. However, these detailed tasks are challenging to achieve with only automated algorithms.

Figure 10 shows the spatial layout of a solution with a five-storey building, which proves the potential of a 3D layout. We directly achieve the prototype layout, satisfying the adjacency relationship in a three-dimensional by establishing horizontal traffic space and vertical core.

Limitation. Our work still has some limitations. Our approach can only control environmental anchor points rather than each building module to drive building layout results. To control each unit, we need new algorithms for calculating adjacency and position information about the unit in real-time in the modelling software Rhino in conjunction with the physics simulation in Grasshopper. It ensures that the unit is always available for editing (area, shape, position), enabling architects to solve more complex architectural tasks. In addition, in unassembled buildings design, the shape of each unit is fixed in advance, and it is difficult to produce a variety of architectural forms, such as the plans in Fig. 10. For this problem, we can add more optimization objectives (number of building sides, side lengths, volume, etc.) The multi-objective optimization approach can be combined with current workflows to generate more diverse and adaptable building solutions.

6 Conclusion and Future Work

In conclusion, our research focuses on building a human–machine collaborative workflow to assist in the rapid generation of spatial prototypes to meet the designer's customisation needs. Our workflow enables fast layouts based on adjacencies, saving a great deal of time for designers in an early design stage. As for the Building prototype generation, our workflow shows the ability to solve different design tasks initially. It allows designers to manually adjust some nodes to determine the final layout with the machine to meet the different design tasks in real-time. The limitation of each unit's boundary calculation and control ability needs to be enhanced to solve more complex spatial layout tasks. It will provide a hybrid-intelligent possibility of human–machine collaboration for solving complex building design tasks in the future.

References

1. Anderson C, Bailey C, Heumann A, Davis D (2018) Augmented space planning: using procedural generation to automate desk layouts. Int J Archit Comput 16(2):164–177
2. Homayouni H (2000) A survey of computational approaches to space layout planning (1965–2000). Department of Architecture and Urban Planning University of Washington
3. Fricker P, Hovestadt L, Braach M, Dillenburger B, Dohmen P, Rüdenauer K, Lemmerzahl S, Lehnerer A (2007) Organised complexity
4. Hao H, Ting-Li J (2010) Floating bubbles: an agent-based system for layout planning. In: Proceedings of the 15th CAADRIA conference, pp 175–183
5. White R, Engelen G (2000) High-resolution integrated modelling of the spatial dynamics of urban and regional systems. Comput Environ Urban Syst 24(5):383–400
6. Dapogny C, Faure A, Michailidis G, Allaire G, Couvelas A, Estevez R (2017) Geometric constraints for shape and topology optimization in architectural design. Comput Mech 59(6):933–965
7. Rahbar M, Mahdavinejad M, Markazi AH, Bemanian M (2022) Architectural layout design through deep learning and agent-based modeling: A hybrid approach. J Build Eng 47:103822
8. Elsayed M, Tolba O, Elantably A (2016) Architectural space planning using parametric modeling-egyptian national housing project. In: ASCAAD conference proceedings, pp 45–54
9. Veloso P, Rhee J, Krishnamurti R (2019) Multi-agent space planning: a literature review (2008–2017). In: CAADFutures
10. Arvin SA, House DH (2002) Modeling architectural design objectives in physically based space planning. Autom Constr 11(2):213–225
11. Baraff D (1989) Analytical methods for dynamic simulation of non-penetrating rigid bodies. In: Proceedings of the 16th annual conference on computer graphics and interactive techniques, pp 223–232
12. Boon C, Griffin C, Papaefthimiou N, Ross J, Storey K (2015) Optimizing spatial adjacencies using evolutionary parametric tools: using grasshopper and galapagos to analyze, visualize, and improve complex architectural programming. Perkins + Will Res J 7(2):25–37
13. Turner A (2001) Depthmap: a program to perform visibility graph analysis. In: Proceedings of the 3rd international symposium on space syntax, vol 31, pp 31–12

Parametric Skin Design Method Based on Plane Crystallographic Group Operation Principle

Hao Zhang[1], Yuetao Wang[1](✉), Yuhan Tan[2], and Jilong Zhao[1]

[1] School of Architecture and Urban Planning, Shandong Jianzhu University, Jinan 250100, China
wyeto@163.com

[2] Jinan Foreign Language School, Jinan 250108, China

Abstract. Under the dual constraints of industrialization and digitalization, the building skin and structure are further integrated to form standardized units to meet the requirements of architectural performance, industrial prefabrication and "complexity" aesthetic characteristics. The complex and diverse forms of today's building skin hide profound mathematical logic relations and operation rules of form generation. Crystallographic group with regular symmetry and the operation principles reflected by it is one of the most important rules and methods of form and pattern processing in skin design. The study of the mural symbols in ancient Egypt, the murals in the Alhambra, the manuscripts of Escher and the window lattice in ancient Chinese architecture profoundly reflects the basic operation principle of crystal group in shaping the skin form of architecture. Abundant and diverse architectural skin forms can be formed through the operation of symmetry group on basic graphic units. On the basis of clarifying the basic principle of crystal group action, the operation matrix of crystallographic symmetry group can be transformed into parameterized operation steps through programming language for visual operation, and then the skin form with high complexity and leap dimension can be generated by geometric algorithm, and the design method of building skin generation based on crystallographic group is constructed. In the selection of operation form, combined with the calculation of building performance and structure, the construction skin can be used in practical engineering is generated. Based on crystallographic group operation, the unifications of building skin and the classification simplification of components can meet the requirements of modular and unifications design in the process of building industrialization, and meet the requirements of current building industrialization and digitization. It has great research significance and value in the aspects of design and construction efficiency and material economic cost.

Keywords: Parametric skin · Crystallographic group · Design method · Operation principle

P. F. Yuan et al. (eds.), *Hybrid Intelligence*, Computational Design and Robotic Fabrication,
https://doi.org/10.1007/978-981-19-8637-6_3

1 Introduction

At present, the architectural skin design tends to use parametric design tools to express its external artistry guided by the expression of complexity aesthetics under the operation of flat two-dimensional level. Under the development trend of building skin, the geometric pattern of skin structure design has developed from simple and orderly in the early stage to the nonlinear expression of composition and tends to pursue the nonlinear aesthetic characteristics of generating order by internal mathematical logic. In various geometric and mathematical models, this paper discusses the operation of translational symmetry based on crystal group in skin design.

The theory of translational symmetry was basically developed and logically perfected in crystallography. Symmetry is a very intuitive concept, which can be seen everywhere, such as plants, animals, human beings and natural minerals all have symmetry. Long before humans had the knowledge of group theory, many civilizations realized that two-dimensional crystals had only 17 symmetries, which was reflected in the field of architecture. Ancient Egyptian mural symbols, the murals in the Alhambra Palace, escher's manuscripts, and the window lattice in Chinese classical garden architecture were all great creations based on human cognition of the aesthetic sense of translational symmetry. The creation of murals and window mullions has some similarities with the architectural skin design in essence. In this paper, the parametric skin design method based on the principle of crystal group operation is studied. The 17 plane crystal groups of crystals are applied to the design of building skin to achieve the purpose of intensive cost efficiency of industrial production. At the same time, the nonlinear change supported by the logic of crystal group operation is pursued with the help of parametric design tools combined with algorithms. This paper breaks through the limitation of the design of the preformed skin under the crystal group operation logic and performs nonlinear operation on the preformed skin under the crystal group operation logic.

Based on crystal group design, building skin unit meets the requirements of modular and unitized design in the process of building industrialization and has greater research significance and value than the existing skin design research in terms of design and construction efficiency and material economic cost. Matrix can be obtained by the crystal group operation computer method for generating algorithm, through the generation algorithm based on crystal group operating principle and geometric algorithms of building skin design method can be in the future to create efficient economic and rich with nonlinear change of epidermis forms, with the research and application of high potential, can provide some enlightenment to the construction epidermis design of the future.

2 Algebraic Basis and Principle of Crystallographic Symmetry Group

Translational symmetry can be represented by the mathematical concept of group. Its research has been developed in the field of crystallography, and with it, it brings new design methods to architecture and art design in the process of digitalization. Plane space group is also known as "wallpaper group". Long before the group theory existed, the 17 symmetries of two-dimensional crystals had been widely applied in the field of architectural decoration, as reflected in the window lattice design of ancient Chinese buildings. With the development of industrialization and digitization of architecture, it is also suitable for building skin design. The translational symmetric operation matrix can provide operability for the parametric design of building skin through computer programming and form the parametric design process of building skin.

2.1 Crystallographic Plane Group and Symmetry Operation

Crystals in nature are composed of atoms, or groups of atoms, arranged regularly in three dimensions, and thus have a regular shape. The minimum repeating unit cells of the crystal are densely packed in three dimensional space, that is, the crystal has translational symmetry. Reaction in a graph, that is, the graph is composed of two or more parts, after a certain linear transformation, the whole graph remains unchanged after the transposition of each part.

Translation symmetry can be represented by the mathematical concept of group. Translational symmetry limits the crystal repetition element to only n = 1, 2, 3, 4, 6 rotation axes, namely the crystallography constraint theorem. The symmetry of the monocell limits the crystal to only 32 point groups. The combination of 32 point groups and translation operations in three-dimensional space determines that the crystal has only 230 space groups. In the two-dimensional case, n = 1, 2, 3, 4, and 6 rotation axes can be intuitively understood from the fact that only square, rectangle, regular triangle, and regular hexagon can be repeatedly filled with plane space, while 5-sided and N (>6) sided cannot be filled with plane space. Therefore, only 10 point groups can be obtained by adding mirror reflection. Only 17 kinds of two-dimensional space groups can be obtained by combining 10 kinds of point groups with translation operations in two-dimensional space. Operations that keep the whole figure unchanged are called symmetry operations, that is, operations in which the distance between any two points of the object remains the same before and after the operation. Symmetrical operation is divided into point symmetrical operation and non-point operation. Point symmetry operation has at least one point in the space does not move during operation, including identical operation, rotation, inversion, mirror reflection and rotation inversion (Table 1). Non - point operation includes spiral rotation and slip reflection. The matrix equation of point symmetric operation is expressed as:

$$\begin{bmatrix} \widetilde{x} \\ \widetilde{y} \\ \widetilde{z} \end{bmatrix} = \begin{bmatrix} w_{11} & w_{12} & w_{13} \\ w_{21} & w_{22} & w_{23} \\ w_{31} & w_{32} & w_{33} \end{bmatrix} \begin{bmatrix} x \\ y \\ z \end{bmatrix} \tag{1}$$

Abbreviated to:

$$\widetilde{x} = Wx \tag{2}$$

In the two-dimensional plane of crystallography, there are 17 kinds of plane space groups with 10 kinds of plane point groups and 5 kinds of plane lattice combinations. The plane point group is a combination of all symmetrical elements at one point, and the introduction of 10 plane point groups into the parametric design of building skin

Table 1. 5 Point symmetry operations

Type	Meaning	Graphic representation	matrix representation
Identity operation	No operation is performed		$E = \begin{bmatrix} 1\ 0\ 0 \\ 0\ 1\ 0 \\ 0\ 0\ 1 \end{bmatrix}$
Rotation	It's rotating 2 PI over n angles about some axis (N is the rotation axis, pure rotation)	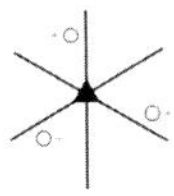	$C_{nz} = \begin{bmatrix} \cos\alpha & -\sin\alpha & 0 \\ \sin\alpha & \cos\alpha & 0 \\ 0 & 0 & 1 \end{bmatrix}$
Inversion	To change the right hand into the left by a central inversion Changed the right—hand orientation of the image	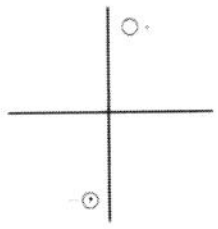	$i = \begin{bmatrix} -1 & 0 & 0 \\ 0 & -1 & 0 \\ 0 & 0 & -1 \end{bmatrix}$
Reflection	A reflection of a plane. Change the right—hand orientation of the graph	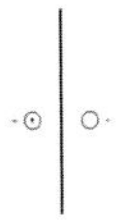	$\alpha = \begin{bmatrix} \cos 2\beta & \sin 2\beta & 0 \\ \sin 2\beta & -\cos 2\beta & 0 \\ 0 & 0 & 1 \end{bmatrix}$
Rotation and inversion	Compound operation The product of two operations, rotation and inversion	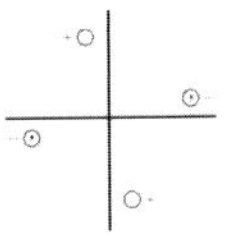	$S_n(z) = \begin{bmatrix} \cos\frac{2\pi}{n} & -\sin\frac{2\pi}{n} & 0 \\ \sin\frac{2\pi}{n} & \cos\frac{2\pi}{n} & 0 \\ 0 & 0 & -1 \end{bmatrix}$

can create a variety of building skin unit forms based on rotation and mirror reflection operation. Planar lattice represents the spatial lattice form of atoms arranged regularly in crystals, which is applied to the design of building skin and fits with the structural units of building skin, and is used to support and arrange and combine the skin filling units formed based on point group operation. The plane space group of plane point group and plane lattice is arranged and combined based on translational symmetry, which reflects that 17 kinds of permutations and combinations of skin unit types can be applied to form a complete building skin. The common characteristics of contemporary architectural skin design and two-dimensional space group research determine that if the minimum repeating unit of architectural skin is created through point group operation principle, and the space group operation principle corresponding to repeating unit is selected based on planar lattice, complete architectural skin prototype can be generated based on translational symmetry operation. The inclusion of late interference algorithms can create rich skin forms that adapt to both industrial and digital architectural design.

2.2 Crystallographic Plane Group and Architecture Skin Design

From crystal point group to lattice group and then to space group, it shows the characteristics which are consistent with the parametric design of architectural skin in the aspect of architectural industrialization. Simple to understand, the architect can select 5 kinds of lattice in the form of epidermal structure unit, and then design the corresponding lattice skin filling unit based on point group operating in the form of form, 10 kinds of planar point group internal repeating unit design combining with 5 kinds of planar lattice after some permutation and combination can be created based on 17 kinds of planar space group operating principle of the construction epidermis. The skin structural units based on planar lattice form have the possibility of translational symmetry operation on the building surface, which can form a complete parametric building skin with certain symmetry based on mathematical model. This series of operations can be realized by transforming the planar symmetric group operation matrix of crystallography into a complete parametric skin design process through computer programming language, and transforming the abstract mathematical model into a visual parametric skin design operation (Fig. 1).

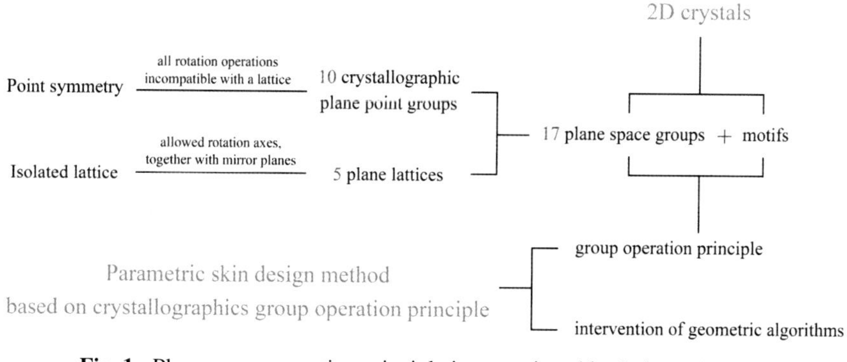

Fig. 1. Plane group operation principle intervention skin design schematic

3 Building Skin Form Based on Crystallographic Group Operation Principle

The parametric design method of building skin based on the operation principle of plane space group makes it possible to break the orderly order of architectural design under the restriction of the established geometric system of architectural design module system. As the key design research, translational symmetry of the unit design for operation, not only emphasizes the design unit in the repeat, also stressed that the skin cell and the overall relationship between the organization, is one of the most critical operation process contains profound mathematical logic and geometric relations, it also provides a precondition for the implementation of parameterized. The geometric transformation relationship of group operation is closely combined with building structure and building materials.

The architectural skin can be regarded as a two-dimensional plane formed by the arrangement and combination of elements, which has certain commonness and connection with the arrangement and combination of two-dimensional crystals, which indicates the feasibility of the translational symmetry operation principle of planar crystal group in the field of skin design. The architect can use 5 planar lattices as the operation base of the skin unit for parametric creation, and select the combination principle of symmetry operation such as rotation and reflection of 10 planar point groups to create a variety of skin unit pattern types for the building skin. The operation principle of translational symmetry of 17 kinds of plane space groups is taken as the parametric transformation and generation algorithm, and the unit patterns are subjected to the orderly repeated symmetrical operation in two-dimensional space. The architectural skin created is not only diversified in pattern, but also follows the internal mathematical logic of parametric design, in line with the development trend of building digitization and industrialization.

3.1 Plane Point Group and Building Skin Unit

After a series of translational symmetry operations, the plane point groups are arranged and combined on the lattice, and it can be concluded that there are altogether 17 space groups in a two-dimensional plane. From the perspective of operationalism, the architectural skin units generated based on the rotation of the symmetrical elements of the plane point group and the mirror reflecting the operation principle are characterized by rich and orderly patterns, limited types of units and convenience for industrial production and assembly of buildings. Then the complete skin pattern can be obtained by combining with the operation principle of plane space group of corresponding lattice translational symmetry on the plane.

There are 5 lattice types in two-dimensional crystal, which are presented as lattice shapes of Oblique, Rectangular, Square and Hexagonal, corresponding to 10 planar point group operations respectively (Table 2). In the parametric design of skin, if the skin filling pattern is the symmetrical element, then the lattice pattern is the structural unit for industrial prefabrication. If the structural elements are generated as symmetric elements, the lattice form becomes virtual body, and the structural elements suitable for various skin inlays can be generated. Lattice is the carrier of point group operation. Only plane point group under lattice constraint can form 17 plane space groups through translational symmetry operation. Therefore, whether the filling pattern of the skin or the structural rod of the skin is used as a symmetrical element for point group operation, the tightly laid complete building skin can be obtained by virtue of lattice constraints. The following table uses a symmetrical pattern as an example to enumerate architectural skin unit patterns that can be obtained by refining the symmetry operation principle of crystallographic plane point groups.

Table 2. 7 plane lattices and 10 plane point groups

Lattice type	Lattice symbol	Lattice diagram	Lattice parameters	Point groups	Point groups diagram (Case of skin unit)
Oblique	mp	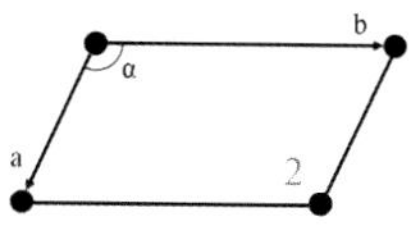	$a \neq b$ $\angle\alpha \neq 90°$	1 2	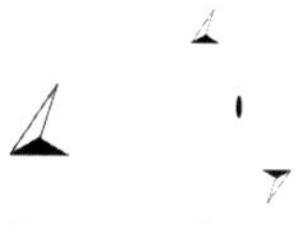 1 2
Rectangular primitive	op	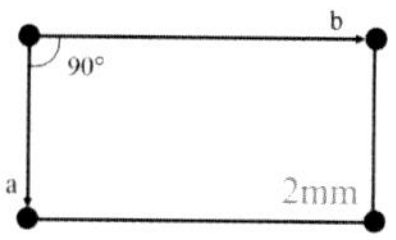	$a \neq b$ $\angle\alpha = 90°$	1 m 2 mm	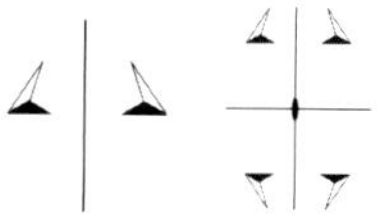 1m 2mm
Rectangular centred	oc	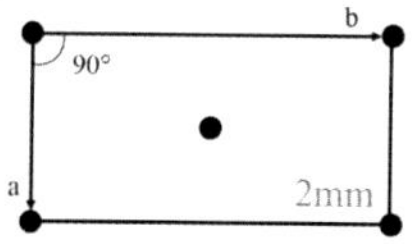	$a \neq b$ $\angle\alpha = 90°$	1 m 2 mm	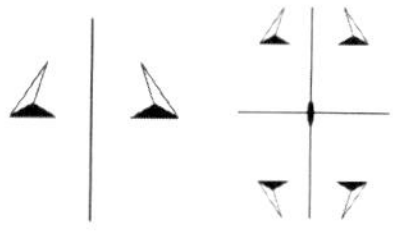 1m 2mm
Square	tp	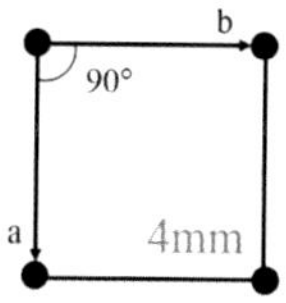	$a = b$ $\angle\alpha = 90°$	4 4 m	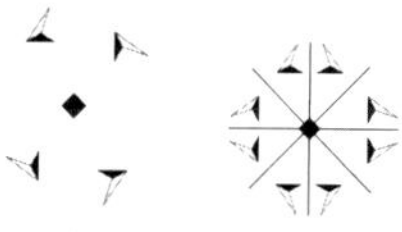 4 4mm

(*continued*)

Table 2. (*continued*)

Lattice type	Lattice symbol	Lattice diagram	Lattice parameters	Point groups	Point groups diagram (Case of skin unit)
Hexagonal	hp	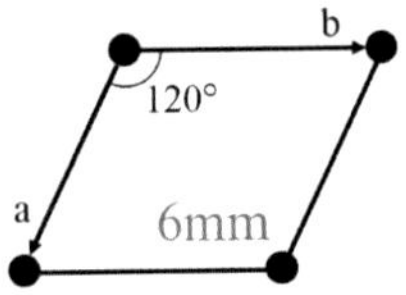	a = b ∠α = 120°	3 3 m 6 6 mm	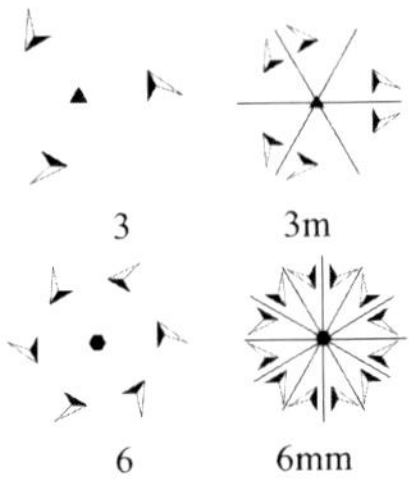

3.2 Plane Space Group and Building Skin Form

At present, patterns are increasingly becoming the prototype of nonlinear composition of building skin. The skin units generated by plane group principle can not only correspond to the skin Mosaic units, but also correspond to the structural units in the building skin to form the building skin with Mosaic characteristics. In this way, group principle becomes a mathematical logic and operation method to control and generate building skin. In transforming patterns into skins, in addition to the selection of repeating patterns themselves, the selection of lattice forms and the type of point group operation are also related to the physical properties of the building. Only when these elements are combined with the symmetrical pattern of crystal group operation, can the group operation of two-dimensional pattern be connected with the epidermal material, connection mode and function, bringing richer differentiation and design possibilities. Two-dimensional is not absolute two-dimensional. The combination of planar crystal group operation and planar concave and convex function can realize the jump dimension operation, which is more conducive to the physical performance of the building. The lattice types and rotation, translation and reflection operations involved in the 17 space group operations are shown in the table below (Table 3).

After comprehensive design of various design elements, firstly, lattice types for parametric operation are determined, and corresponding point group forms are selected. Then, repeated patterns are designed according to the principle of symmetric operation at different points to obtain the filling pattern units of the building skin. Finally, translational symmetry is carried out to obtain the laid out building skin. This skin design process based on the operation principle of plane space group can generate various forms of building skin prototype, providing objects for the next step of geometric algorithm intervention. The following figure illustrates the principle of skin pattern generation based on 17 kinds of planar space group operation by taking a symmetrical element as an example. Designers can create rich artistic effects of architectural appearance by designing symmetrical elements and combining architectural skin design elements (Fig. 2).

The interior tiles of The Stella Retail store in Manhattan are a typical example of using the p6 spatial group operation principle to generate a design with architectural

Table 3. P17 plane space groups

Symmetry group	IUC notation	Lattice type	Rotating degree	Reflection axis
1	p1	Oblique	With no rotation	No reflection
2	p2	Oblique	180° rotating	No reflection
3	pm	Rectangle	With no rotation	Parallel
4	pg	Rectangle	With no rotation	No reflection
5	cm	Rhombus	With no rotation	Parallel
6	pmm	Rectangle	180° rotating	90° apart
7	pmg	Rectangle	180° rotating	Parallel
8	pgg	Rectangle	180° rotating	No reflection
9	cmm	Rhombus	180° rotating	90° apart
10	p4	Square	90° rotating	No reflection
11	p4m	Square	90° rotating+	45° apart
12	p4g	Square	90° rotating*	90° apart
13	p3	Hexagonal	120° rotating	No reflection
14	p31m	Hexagonal	120° rotating*	60° apart
15	p3m1	Hexagonal	120° rotating+	30° apart
16	p6	Hexagonal	60° rotating	No reflection
17	p6m	Hexagonal	60° rotating	30° apart

+:All centers of rotation are located on the reflection axis
*: Not all centers of rotation are located on the reflection axis

requirements. Each Hexagonal tile is a repeating unit, which is the point group unit of the Hexagonal lattice generated by the Hexagonal rotation of petals as a symmetric element of point group. In accordance with the external artistic requirements, the thickness of each unit is concave and convex, and the symmetrical elements are convex, so that the raised point group pattern and the hexagon form an inclined curve in the honeycomb structure. The glass unit of Trutec Building in Seoul is also a Rectangle lattice stacked operation. The glass in the lattice unit is combined with the pattern elements to make different angles, refraction produces a kaleidoscope of light and shadow changes. The above cases show that the parametric design method combining skin design elements with planar crystal group operation principle can produce rich architectural skin effects (Fig. 3).

The above diagram illustrates the operation principle of skin pattern of 17 planar space groups through symmetry operation of single repeating element. Now that the skin filling pattern as action object can get rich form of epidermis, if the structure of the skin bar as a symmetric operation of symmetry elements for operation, through the computer bar gives the rods plane symmetry group symmetry matrix operation, set the boundary conditions, can make its evolution from a single cell iteration to spread the whole building skin. Many kinds of complete building skin forms can be obtained by

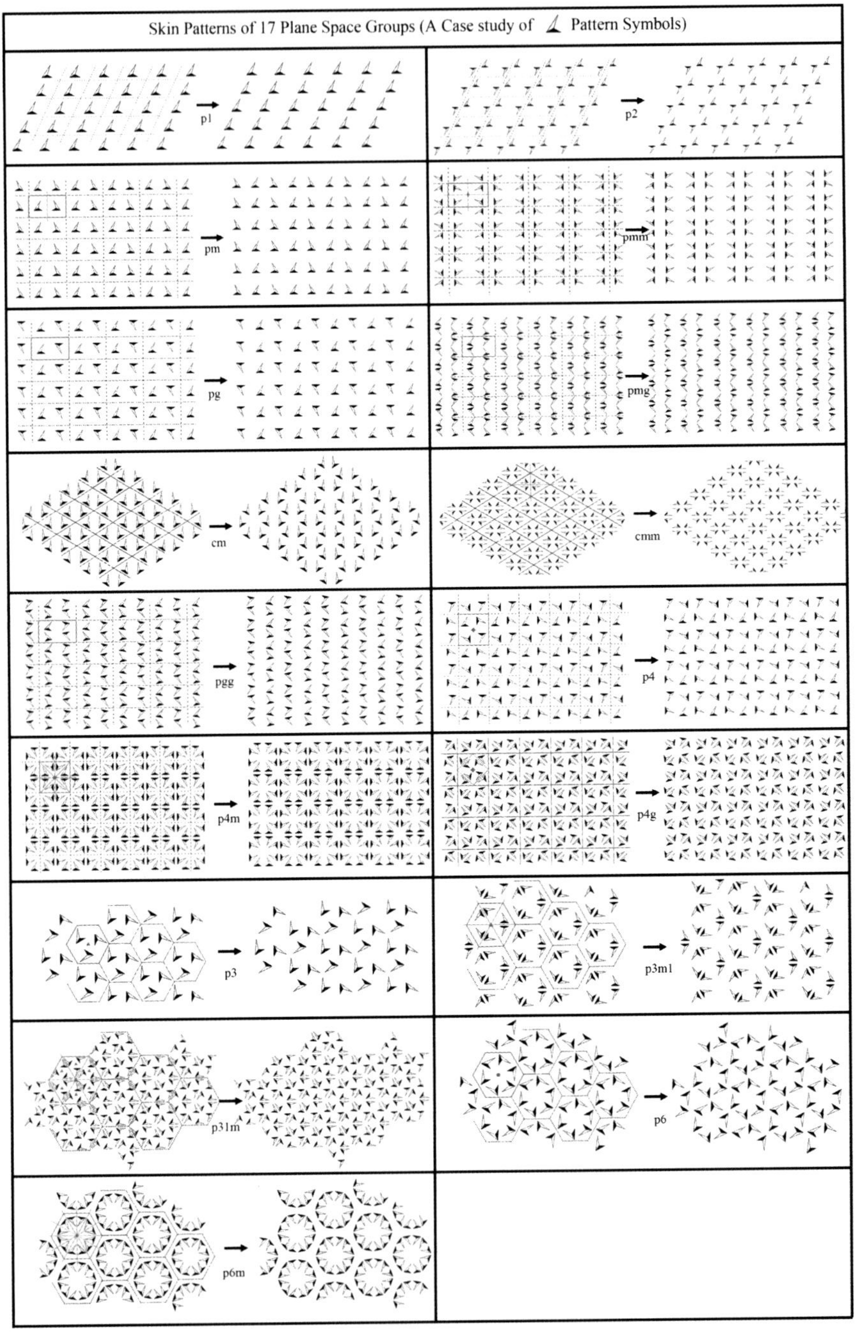

Fig. 2. Skin pattern generation principle for 17 plane space group operations

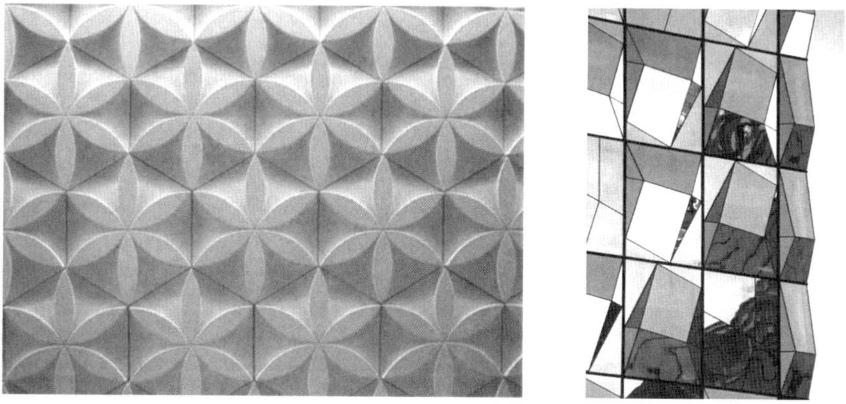

Fig. 3. The tile of the Stella retail store in Manhattan and the glass facade of the Turtec Tower in Seoul

selecting different forms of unit bars as symmetrical elements for translational symmetry operation. P2 space group operating principle, for example, select the corresponding parallelogram lattice, and then place the symmetry element as bar in the crystal lattice, based on planar point group operating principle of the rotation, mirror image operation generates epidermis local structure unit, finally, based on the operation principle of space group p2 translation and translation operation such as sliding reflect an entire construction epidermis keel structure, Finally, the epidermal filler was inserted with other constraints (Fig. 4). The generated epidermis keel rod has fewer types and rich Mosaic forms, which is convenient for industrial production and assembly of building epidermis.

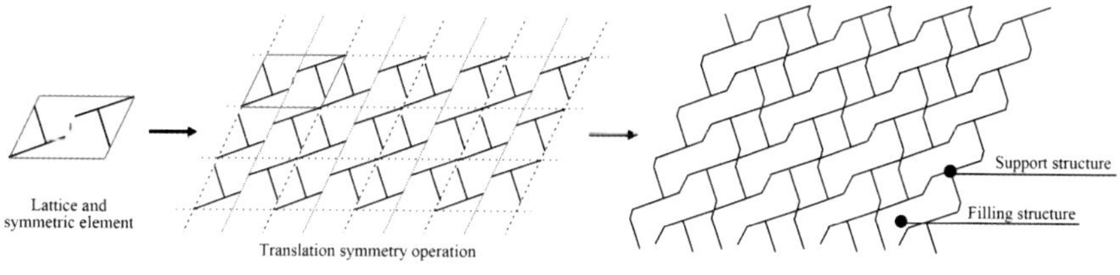

Fig. 4. Skin generation logic for translational symmetry operation of structural bars as symmetrical elements

The skin patterns generated by the p2 operation principle of plane space show the law of periodic Mosaic, and the operation principle of crystal group explains the mathematical basis of previous skin Mosaic research. Compared with the periodic Mosaic of regular polygons of building skin, the operation of building skin based on the principle of translational symmetry of planar crystal group can produce richer Mosaic combination forms.

4 Intervention of Interference Algorithm

Based on the above research on planar crystal group operation, the building skin prototype generated based on crystal group operation principle is both mathematical logic and external artistic, and is an ideal algorithm input parameter. Therefore, the geometric algorithm closely related to the principle of crystal group operation can be introduced into the architectural skin design method based on the principle of crystal group operation to pursue more diversified architectural skin pattern forms. The advantage of geometric algorithm intervention lies in that it can break the limitation of human brain and get uncertain or even infinite output results through limited input steps. The intervention of set algorithm can break the limitations and order of the building skin formed by the operation principle of crystal group and find a better solution from the essence of the problem. Algorithm intervention in the pursuit of complex forms of skin does not interfere with the logic of building skin generation. Complex geometric forms of skin are derived from the strategy of plane group generation. Similarly, complex geometric algorithms can produce intuitive skin pattern forms that are both rhythmic and artistic. Taking the interference algorithm as an example, the skin parameterization design method is discussed in combination with the skin generation algorithm based on the principle of plane group operation.

In the field of architectural design, interference algorithms can be used to intervene the generation of architectural form, space and skin. Intervention will interfere with the algorithm to the prototype design of generated on the basis of the principle of crystal group operating skin, break through the crystal group operating logic into epidermis both the limitations of design itself and stability, and tend to pursue to follow the inherent nonlinear aesthetic characteristics on the basis of mathematical logic to generate order, the pursuit of crystal group of logic operation supported by nonlinear changes, The established skin nonlinear operation under crystal group operation logic is carried out. By inputting a set of parametric variables that produce regular changes under the control of specified points, lines or patterns, the interference algorithm can regulate the skin elements at the macro level and output the skin variation forms under the control of several specified factors. Based on planar crystal group of the operating principle of choosing different forms of the lattice grid as a unit, under different interference sources intervention, control skin partial or whole follow certain regularity of the change, break of design on the basis of the principle of planar crystal group of the stability of the construction epidermis and regularity, output both external artistic form and inherent logic of the skin.

Point interference algorithm is to use one or several points to control a group of parameter variables, and with the interference points as the center, to produce interference deformation in the form of noise diffusion in the lattice skin unit for the preformed skin based on planar crystal group operation. The lattice skin element deformation operation is divided into two steps. Based on the interference parameters, the edge of the element is changed symmetrically to the center of the element, and then the deformed element is enlarged or reduced as a whole to obtain the final overall skin deformation effect. The deformation degree is centered on the interference point and gradually diffuses and changes gradually. Through the computer platform, the algorithm matrix can be realized on the building skin. Line interference algorithm uses one or several lines to control multiple groups of parameter variables, which can also be understood as a series of points on the line under the control of the set. The skin change under the control of the line interference algorithm is the interference effect of the line centered outward diffusion gradual change of the skin element (Fig. 5). Image interference algorithm can extract the gray scale of the image as a parameter variable to determine the degree of interference to the element deformation. The gray scale data makes the morphological changes of the building skin more random and natural, and can present the artistic state of three-dimensional fluctuation on the two-dimensional building skin.

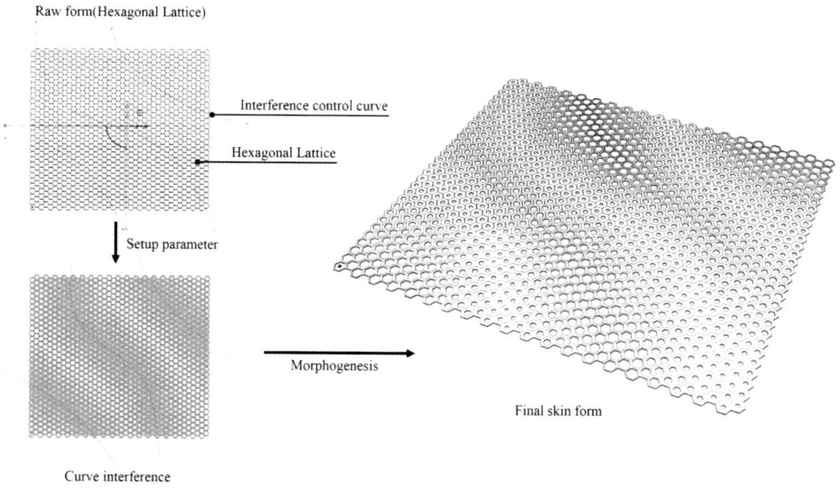

Fig. 5. Schematic diagram of curve interference algorithm for skin design

5 Conclusion

Based on the symmetry operation principle of planar crystal group, a new parametric design method of building skin based on mathematical basis is discussed by means of method explanation and examples. The generated building skin has both internal mathematical logic and external artistic quality. Selection based on planar crystal group of 5 kinds of planar lattice, 10 kinds of planar point group and 17 kinds of planar space group operation principle, this paper explains its operation principle in building skin design contains the inherent logic and diversity, demonstrated combined with various design elements applied in the field, the feasibility of building skin design. Through the disassembly of planar crystal group symmetry operation process, find the way to combine with the problems and elements in the design of architectural skin, transform the crystallographic symmetry group operation matrix into parameterized operation steps through programming language for visual operation, and generate skin prototype. Then the skin prototype is operated by geometric algorithm to generate the skin form with high complexity and leap dimension, and a design method of building skin generation based on planar crystal group is constructed.

The parametric design method of building skin based on the operation principle of planar crystal group has great research potential and space. It has the following characteristics: it fits with the process of building industrialization and digitalization; Generate rich skin effects with mathematical logic; Feasibility of auxiliary design; Feasibility of generation process; Complexity of generating results; Constructability of generative modes. In the stage of skin design, rich skin prototype forms can be generated by using the translational symmetry of planar crystal group, and further interference operation of parameters and variables can be carried out by intervening geometric algorithm. Based on the operation of mathematical logic, the external artistic expression of skin can be enriched, and rich architectural skin forms can be generated. At the same time, the building physical environment and skin structure can be optimized by combining performance simulation. In the skin construction phase, the symmetrical operation of the abstract planar crystal group can be used to deepen the skin entity, so that the construction of the building skin is compatible with the industrial prefabrication. In the future, the application of the operation principle of two-dimensional or even three-dimensional crystal group in architectural skin and space needs more research and attempts, which is also the focus of future exploration.

References

1. Cao ZX (2019) One of geometric series crystal: crystal point group and space group. Physics 2:4
2. Kizilörenli E, Maden F (2021) Tessellation in architecture from past to present. In:IOP Conference Series: Materials Science and Engineering, vol 1203(3). IOP Publishing, p 032062
3. Zaera-Polo A (2009) Patterns, fabrics, prototypes, tessellations. Archit Des 79(6):18–27
4. Powell RC (2010) Symmetry, group theory, and the physical properties of crystals. Springer, New York
5. Xu YJ (2019) Tessellation facade: a parametric design method of architectural facade based on polygon periodic tessellation patterns. Huazhong Architecture 37(2):5
6. Jiang JC, Liu SX (2018) Performance-based generation of openings in parametric surface design. New Architect 4:5
7. Wang YJ, Liu SX (2017) Tessellation in building surface. New Architect 2:5
8. Wang H, Cao K (2009) Applications of tessellation in contemporary architectural surface design. J Zhejiang Univ (Eng Sci)
9. Wang H, Wang RR (2015) Architectural surface design based on sectioning convex uniform honeycombs. J Zhejiang Univ (Eng Sci) 2015(7):6
10. Chang W (2018) Application of tessellation in architectural geometry design. In: E3S web of conferences, vol 38, p 03015. EDP Sciences

A Slime Mold System Driven by Skeletonization Errors

Yufan Xie[1], Jingsen Lian[2], and Yufang Zhou[2](✉)

[1] University of Southern California, 850 Bloom Walk, Los Angeles, CA 90089, USA
[2] Central Academy of Fine Arts, Huajiadi South Street 8, Beijing 100102, China
zhouyufang@cafa.edu.cn

Abstract. This paper proposed a new way to generate slime mold patterns using a typical voronoi-based skeletonization method. As a recursive system, it redraws and expands the resulting trails of skeletonization and feeds them back as an image source for skeletonization. Through iterations, it utilizes the difference before and after skeletonization to generate slime-mold-like patterns. During the whole process, we tested different growth types with different parameter settings and environmental conditions. Since most researches on skeletonization focus on minimizing errors, on the opposite side this method utilizes errors of skeletonisation (e.g. subtracted skeletons at "branch" areas of the bitmap are different from the original brush trails or the best result we expect) as the basis of the generative process. The redraw process makes it possible to reconnect skeletons via intersected brushes, continuously changing the topology of the network. Unlike the traditional slime mold algorithm which operates on every single agent, our method is driven by image-based solutions. On the output side, this system provides a condensed vector result, which is more applicable for design purposes.

Keywords: Slime mold · Physarum · Skeletonization · Generative · Error

1 Background

Skeletonization algorithms are generally used in pattern recognition and image subtraction. There are many precedents on pixel-based and vector based models, ranging from 2 to 3D space. By far, there is no known exploration using the skeletonization process as a generative system. Meanwhile, to realize the slime mold system, a general method is using an agent based model to simulate the process. (The physarum model by Jones [1] is one of the most widely used frameworks.) A collection of points are defined, with detection ranges and movements, reacting to each other to form a connected dynamic network like an ant farm. These two algorithms—one for subtraction and another for generation—seem to be non-related in most conditions. Our research discovered a new simple, but effective way to integrate both through errors.

P. F. Yuan et al. (eds.), *Hybrid Intelligence*, Computational Design and Robotic Fabrication,
https://doi.org/10.1007/978-981-19-8637-6_4

This exploration was an accidental result I discovered when learning skeletonization for image data processing. After doing research on skeletonization, I found most skeletonization methods are not “perfect”. Under some conditions, they even partially distort features of the original figure, which are considered as problems and errors to be solved in most research. In an opposite way, this research is not an optimization, nor efficient-oriented exploration, but to discover a new possibility inspired by errors. It simply asks—can we use error as a key factor to drive a system, or design? Can subtraction algorithms be used in a generative way?

2 The System

Errors of skeletonization can be easily observed and paths are not perfect, although most features of the figure are preserved. Our research started with analyzing the performance of 2D skeletonization. We implemented a voronoi-based skeletonization method [2]. It subtracts points from the gray-scale silhouette of the figure to generate a voronoi diagram, from which the voronoi edges are selected based on their intersection relationship to the silhouette. Additionally we also remove unnecessary paths by selecting segment lengths. We compared a figure with its skeletonization result (Fig. 1). Deviations can be found in these specific areas:

1. Sharp corners, even secondary corners are taken into account, which results in extra skeleton paths. Though from the result we generally consider them as unwanted.
2. "Branch” areas of a figure are not as sharp as the path we expect or the original path used for “drawing the figure”.
3. Once two figures touch each other, two figures are recognized as one connected figure and skeletonized as continuous paths—although as humans we conditionally consider it as “not connected”.
4. Due to the error of pixel distribution of the figure, extra short skeletons are created around the “tip” of a round shape.

By utilizing the difference in areas above, we can develop a recursive generative system by simply connecting the start and the end of the process, with a redraw operation. The base framework of the skeleton-based slime mold system (Fig. 2) consists of four steps for each iteration:

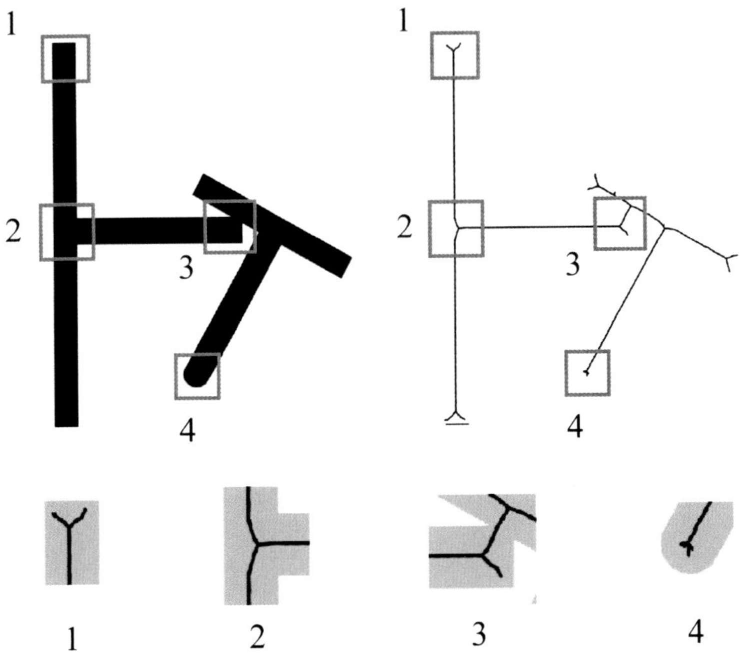

Fig. 1. Details of skeleton errors in different areas. By overlapping the two, we can observe the difference between the original figure and the best path we expect/perceive.

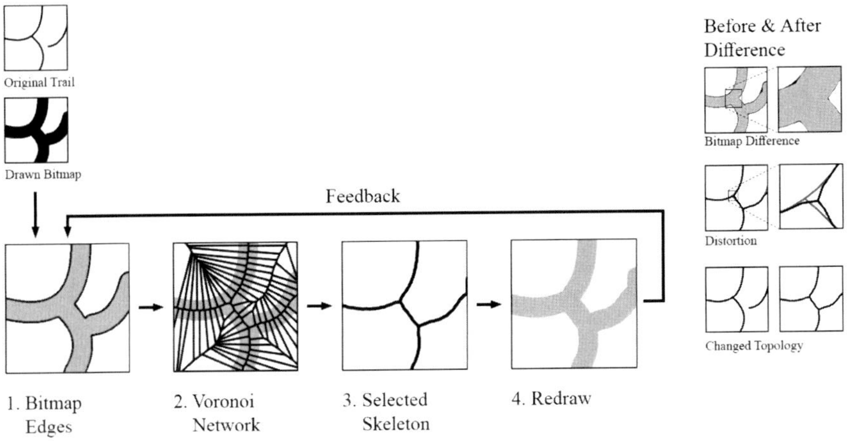

Fig. 2. Pattern generation by accumulated errors.

Input: image, length *L*, redraw diameter *D*, background color, foreground color.

```
Output:  redrawn image
1. edges = new mid-value edges of the grayscale mesh of image
2. centers = new points by subdividing edges with length L
   voronoi units = new voronoi network based on centers
3. for each segment of each voronoi units
      if any of the following are false, delete the segment:
         a. both ends of the segment are in the foreground color area of image
         b. the length of the segment is smaller than redraw diameter D
4. fill redrawn image with background color
   draw segment in diameter D as foreground color to redrawn image
   image = redrawn image
repeat 1.
```

This system is developed on the Grasshopper platform [3]. C# and Rhinocommon API [4] are used for edge operations and voronoi generation. In the base framework we use a round brush to redraw the skeleton. In our experiments (100*100 size and 240*240 resolution mesh, with a 720*720 bitmap, $D = 5$, $L = 5$), maximum speed could reach 15 fps. The efficiency is largely related to the amount of branches in iterations, which directly affects the amount of redraw and skeletonization.

Based on the previous analysis of skeletonization errors, as well as details of the pattern generation, we found four behaviors that determine the pattern generation process: bending, merging, reconnecting and branching (Fig. 3).

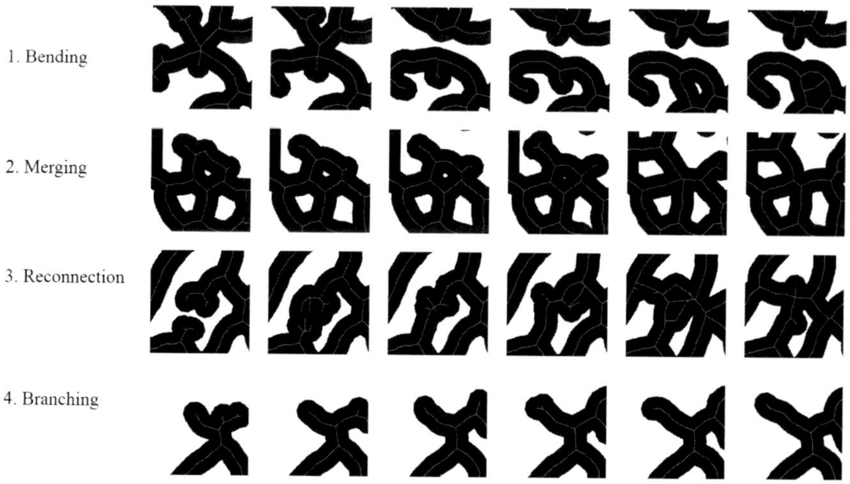

Fig. 3. Detail of pattern formation.

1. We found that skeletons of branched paths are slightly bent to approach evenly divided branch nodes. For example, if a supplemented branch node has 3 paths as 180°, 90° and 90° distribution—the distribution of this node in a skeletonized result is slightly closer to regular 60°, 60° and 60°. This is also a key factor affecting an enclosured unit to "bend" inwards, which slightly scales down the unit by iterations.
2. Merging occurs in closed units of the skeleton. Such "circles" are more likely to shrink than other parts, when angles of branches on the "circle" bend the shape inwards to the center. They vanish when the unit size is smaller than the redraw brush diameter—in which the unit is filled by redrawn paths.
3. Reconnection occurs at the tips, when ends of two paths accidentally touch each other (even just a few pixels). Such processes extend paths at tips. This behavior is triggered by errors in pixels and density of voronoi units, which affects the resolution of the skeleton. Once reconnections produce enough connection to form closed units,
4. Branching generally occurs at the convex side of paths, especially when the redrawn figure is wide. New short branches are created and extended. Larger the curvature a path has, the more likely it creates branches. The width of the figure, or redraw brush diameter also increases branching possibilities.

Four types of pattern formations are related, each of them are causes of others.

Parameters define different generative processes. When we increase the ratio of redraw diameter D to segment length L, the system shows more splitting and branching. In our analysis, when D increases, the accuracy of the network reduces, as details of branches and corners are replaced, and even merged by a thick brush. When L increases, the resolution of subtracted edges/voronoi units reduces, and more extra branches are kept in skeletonization. This feature can be seen in skeletonization, when the original figure is too wide—especially when the width of the figure is uneven, skeletons are less accurate in wide areas than narrow ones. From the pattern formation process (Fig. 4), we found that if no extra information or control is applied to the system, by iterations, the pattern will lose features of the original figure and result in a homogenized distribution. After the pattern fills the canvas, the density is balanced at an approximate ratio, despite skeletons still continuing to grow and split. After a certain amount of iterations, the pattern is stabilized and generation stops. Under such conditions, only two parameters—the redraw diameter and segment length can effectively change the pattern. Simply controlling these two parameters are limited, which doesn't meet our goals in design, only affecting the density and connectivity of the pattern, not responding to the original figure, or any additional context. In the next step of our research, we attempt to make the system more reactive to additional brush information, transform information, pixel feeds and vector feeds. Especially because the system is based on image processing, we have different options to manipulate the process by changing the drawing methods.

2.1 Changing Brush Shape

In our earliest analysis on the errors of skeletonization, small protrusions of the figure are a key factor resulting in extra unwanted paths. The sharp corners of brushes are key

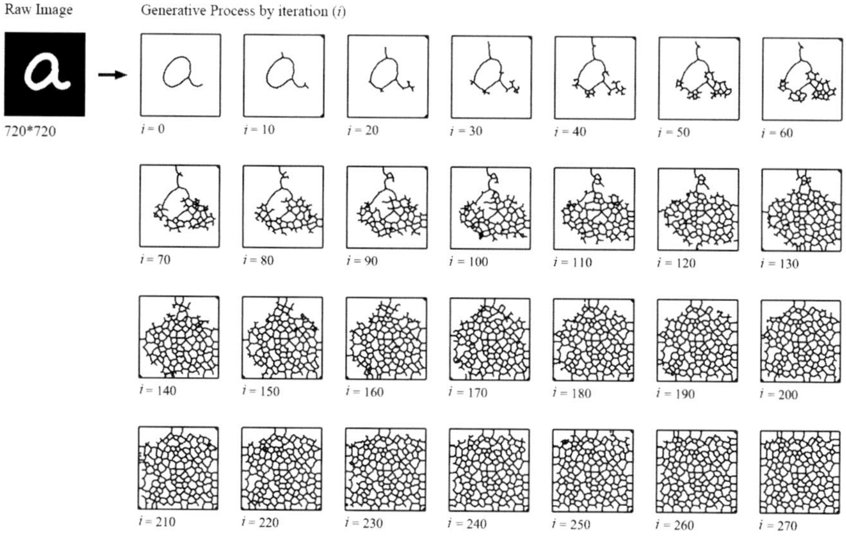

Fig. 4. Slime mold pattern with larger D/L ratio—pattern quickly expands from the original figure "a" and fills the canvas.

factors guiding branches to grow. As the brush we used in our base framework is round, in later experiments we attempted to use other brush shapes to give the redrawn figure more sharp corners. For instance, the square brush biasedly guides skeletons to grow along corners, which turns out to be more branches (Fig. 5). As we reduce the brush density and define the direction of square brush as orthodoxical, the direction of growth is more guided along four uniformed corners.

Fig. 5. Square brush redrawing the path as a saw-shaped figure, further guiding the skeleton to grow in other directions.

2.2 Additional Transforms

When we apply transforms to skeletons at each iteration, the variations of patterns are different in aspects below:

1. When scaling is applied, patterns are stretched or merged based on brush size and trail connectivity with their neighbors, affecting branching possibilities. While scaling is non uniform (e.g. scale along x axis), the pattern is more stretched in one direction than others, which results in a series of near-parallel paths.

2. When rotation is applied, patterns are cut off by the boundary of the image, since no information is given out of the image. Result of rotation usually turns out to be a circle-shaped pattern with a diameter equal to the short side of the image. Paths within the circular culling area are not affected by rotation.
3. When movement is applied, the pattern will continuously move from one side and disappear on the other side. While new paths are generated on the side it moves away from, the connectivity of the main pattern is stable. If there is any skeleton drawn on the edge of the opposite side of the vector, a stretched pattern will be continuously generated. Basically, differences made on network topology by movement are less obvious than scaling and rotating, since the relative distance between paths does not change.

Difference made by scaling information is the most effective one among others (Fig. 6). In our analysis, the density of the pattern is maintained, if redraw diameter D and segment length L are not changed. As scaling constantly changes the density of pattern/the distance between a segment of skeleton and its neighbors, paths merge or grow to maintain the "balanced" density. The further a segment is to the scaling center, the bigger absolute movement is caused by scaling, the more possible it is going to split. Vise versa, the closer it is, the more possible it is going to merge. Unlike multi-agent systems, the pattern information contained in canvas is largely limited by the resolution of the image. This feature is obvious when transform is applied.

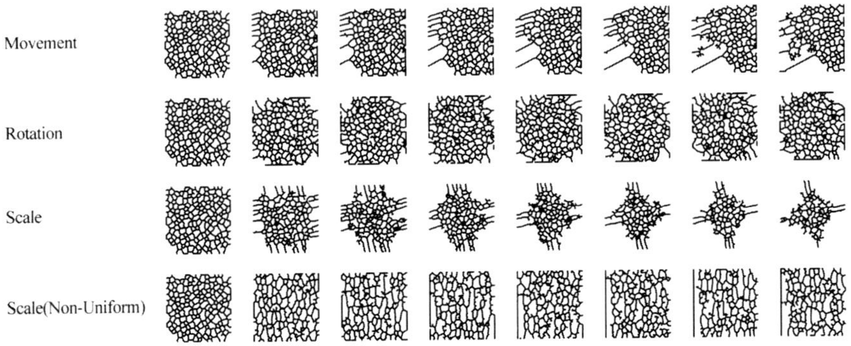

Fig. 6. Different progression of the same network, by transforming the skeleton in each iteration.

2.3 Additional Contexts

We attempted to use external bitmaps and vector trails to make this system more adaptive for design contexts. Before feeding the redrawn skeleton into the next iteration, overlaying a specific area with background or foreground information before the next iteration will largely change the way the system responds. In our experiments, we found the way the system responds to figure-ground context is more dynamic and adaptive than directly culling areas from a general non-context pattern.

For instance (Fig. 7), under the settings of "skeletonizing white figure from a black background", continuous feeding a context figure image of black building blocks can define non-design areas which omit all generated paths. Meanwhile, moving the skeleton along a direction will turn the pattern into a shape adapted to the image context. On the "frontside"—sides facing against the movement direction, skeletons moving towards the context figure are mostly omitted, but merged along the edge of the context figure. On the "backsides"—sides facing along the movement direction, blank spaces are created like the backside of obstacles in a flushing fluid field, because no information is provided in the non-design area. Based on the ratio between D and L, the patterns on the "backsides" grow to backfill culled areas. The second variation of the experiment is, under the same skeletonization settings, redrawing the skeleton with additional vector curves (e.g. a circle or rectangle) to provide a continuous growing source from curves. In this case, whatever the pattern is, even when a huge amount of scaling is applied (e.g. scaling down around the center), paths are not strictly defined by given curves or moved away, they are still re-generated approximately around fed curves. The results of both experiments turned out to be a solution similar to road generation in an urban/landscape context.

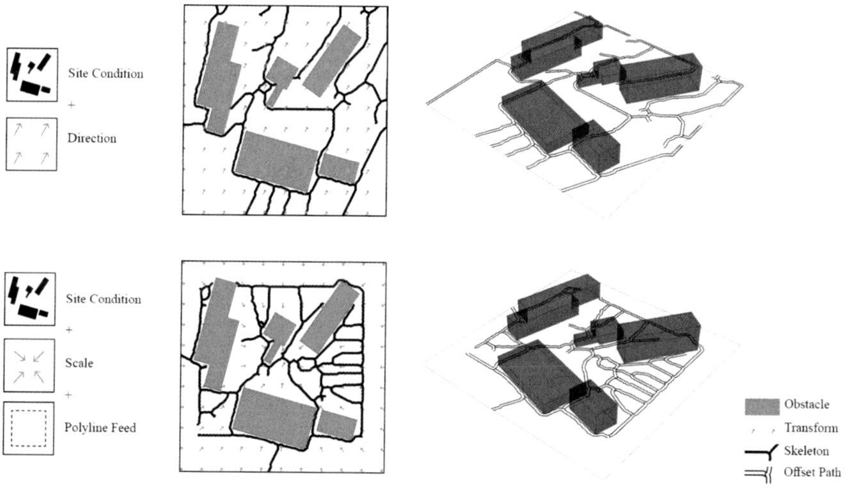

Fig. 7. Two experiments on implementing the system to a site, with obstacles and transform settings.

In our research, we also attempt to realize different pattern density by controlling redraw diameter with a gradient color map (Fig. 8). The gray scale information is remapped in range, to control redraw diameter. In this case, the patterns are more operable to meet design needs.

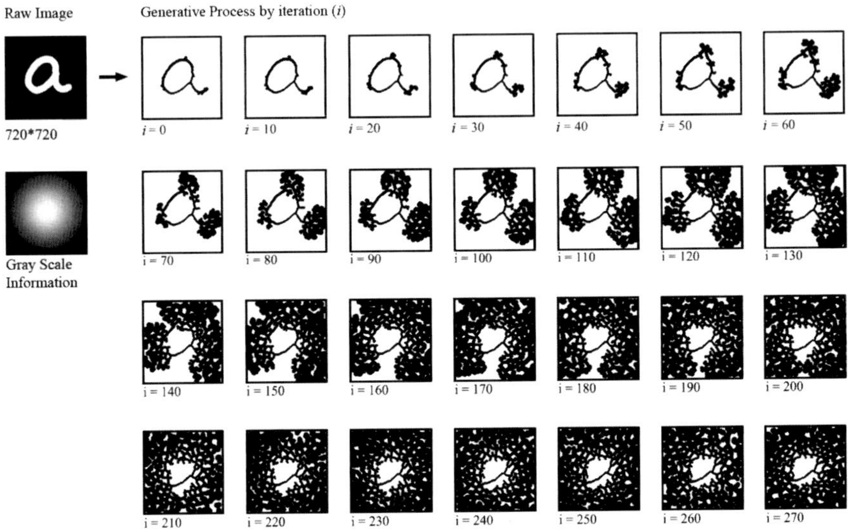

Fig. 8. Grayscale to pattern density. Networks in areas of thicker redraw diameter are more likely to grow.

3 Result Analysis

After manipulating the process, we analyze the final pattern for design purposes. Generally, to subtract paths from a traditional multi-agent slime mold system, additional operations are required to extract the medial axis from discrete point clusters. Since the patterns in our research are skeletonized, outputs are clean vector lines rather than a series of points. The paths are usable directly after removing duplicate paths—since the system uses voronoi-based skeletonization, the direct branch results are two overlapped paths in opposite directions from neighboring voronoi units. By analyzing the topology, segment length and connectivity of the network (Fig. 9), design operations can be more specific—such as generating a varied rectangle pattern along the direction of each unit.

Even though we can conduct similar analysis on a typical agent-based physarum system by voxelizing/redrawing/expanding each agent into a solid figure, and skeletonize the result, it is still not a process driven by attributes of skeleton (topology, curvature, density etc.). Influenced by the image-based method, this system shows features below:

1. Results are the medial axis, ignoring the uneven thickness/width of the foreground figure. But patterns are more likely to change, when the figure is wider, which is less controllable.

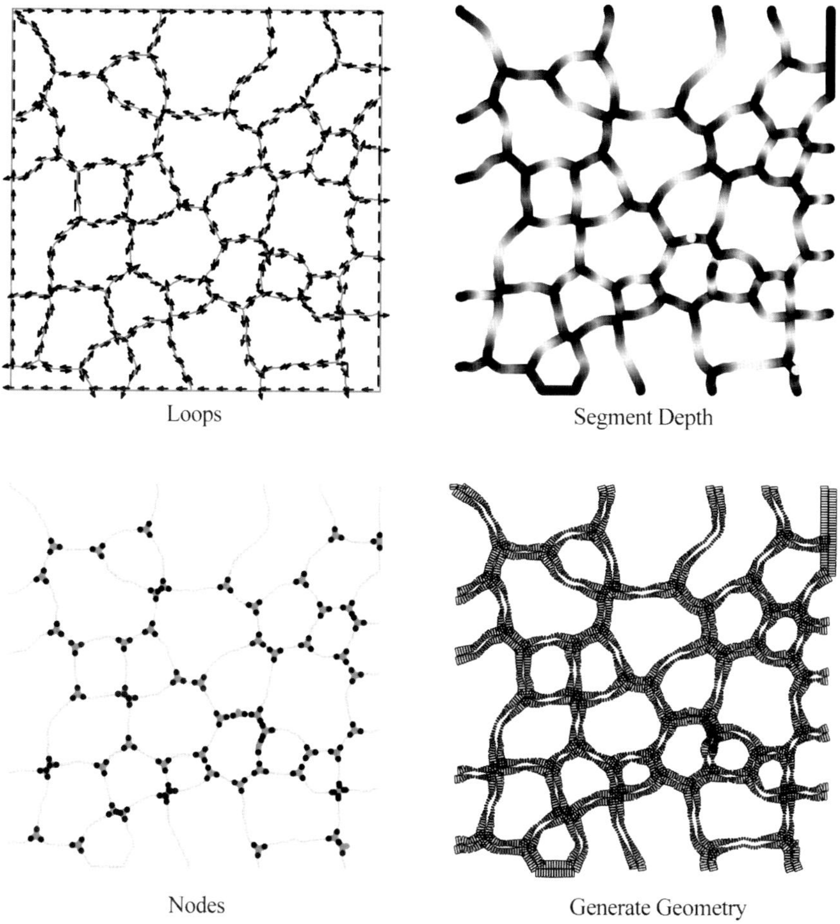

Fig. 9. Analysis of a generated network. New geometries can be generated along the path, and based on their relationship to network nodes. We can notice that three-connection nodes are most common in generated patterns (bottom left), which are also approximately averagely divided.

2. Attributes of redraw brushes (size,shapes) define the behavior of the whole system.
3. The results are connected vector trails, instead of disconnected points/agents. Directions and lengths of skeleton segments can be quickly sorted.
4. The system is reactive to context, allowing different controls over parameters.

Meanwhile, influenced by this image-based solution, we also found challenges in the aspects below:

1. Difficult to track the growth history of a network segment, like tracking an agent in a standard physarum system.

2. The accuracy of the network generation is limited, which is affected by image-related parameters (e.g. the system cannot provide precision smaller than redraw diameter, since neighboring paths are merged by the redraw process).

4 Conclusions

Our research turned out to be a similar pattern to a traditional physarum/slime mold pattern, but in a totally different process. Skeleton is defined not just as a result, but the core of the generative process.

This method is not aimed to optimize, nor as efficient as existing slime mold or skeletonization algorithms. It attempts to creatively reveal an undiscovered possibility by errors in efficiency-driven models. The result turned out to be interesting. However, there are some limitations—the practical value of this system is still unclear, since in the design field, most physarum/slime mold research are experimental or conceptual projects. It is still undeniable that this system proposed a creative way of utilizing skeletonization and a new alternate workflow to generate slime mold/physarum pattern, in a different data format.

By far we have not tested this system with pixel based skeletonization methods. It is still unclear if the generative mechanism of the system works with the same efficiency and behavior, as they run with different precision and speed. The next step of this project is examining the practical use, such as path optimization and connectivity analysis. We are also planning to test its compatibility with real-time images from devices such as webcams, to integrate the image-based process with image-based interaction—we expect the exploration to be "imprecise", as more noises and errors from reality and hardwares are accounted for. Meanwhile we will continue exploring the 3-dimensional version, which is expected to be volumetrically thickening 3D skeletonization and feeding back to the loop.

References

1. Jones J (2010) Characteristics of pattern formation and evolution in approximations of physarum transport networks. Artif Life 16(2):127–153
2. Mayya N, Rajan V (1994) Voronoi diagrams of polygons: a framework for shape representation. J Math Imag Vis - JMIV 6:638–643. https://doi.org/10.1109/CVPR.1994.323787
3. Grasshopper3d, Grasshopper. https://www.grasshopper3d.com/
4. Rhino3d, RhinoCommon API. https://developer.rhino3d.com/api/RhinoCommon/html/R_Project_RhinoCommon.htm

Research on the Spatial Layout Design of University Educational Buildings Based on Rule Screening and Multi-agent System

Yixuan Zeng[1], Qiaoming Deng[2], and Yubo Liu[2,3](✉)

[1] College of Architecture and Urban Planning, Tongji University, 1239 Siping Rd, Shanghai, China

[2] School of Architecture, South China University of Technology, Guangzhou, China
liuyubo@scut.edu.cn

[3] State Key Laboratory of Subtropical Building Science, School of Architecture, South China University of Technology, Guangzhou, China

Abstract. Unlike traditional empiricism-based building design, the data-oriented quantitative analysis method is more rigorous and intuitive, taking into account a variety of factors such as site conditions, functional requirements and design specifications, and combining computer technology to propose a more rational and efficient design strategy. This study takes the design logic process and algorithm rule screening as the entry point to explore the design method of using multi-agent body algorithm planning to realize the spatial layout of university education buildings. Based on multi-agent algorithms, spatially rich and morphologically complex architectural solutions can be quickly generated, and new designs with generality and universality can be produced by changing the initial shape and syntax rules. The authors attempt to design a program based on a multi-agent body system, where architects only need to set initial parameters to quickly construct a variety of initial volume solutions, offering a wide range of possibilities for initial design.

Keywords: Multi-agent · Self-organization · Bottom-up · Generative design · Educational buildings

1 Introduction

The design of a university is a large-scale urban design for small education, which is realized for such a combination of buildings, a multi-body structure or a multi-faceted design, as well as a multi-body design and multiple decent design perspectives. Teachers need to spend more time participating in the whole process of design.

Computer-aided design has been used in the field of architectural design maturely, such as auxiliary drawing, modeling, statistics and so on. These auxiliary tools can not only help designers better improve the model, but also combine with other related disciplines to complete the full-cycle design of construction projects, so that various disciplines can better coordinate in actual projects. With computational design, the designer

P. F. Yuan et al. (eds.), *Hybrid Intelligence*, Computational Design and Robotic Fabrication,
https://doi.org/10.1007/978-981-19-8637-6_5

only determines the general goals of the design, and automatically solves it by setting a series of generative rules and generative logic. Multi-agent is a bottom-up approach to model building. In the multi-agent method, the design will gradually spread out with the interaction effect between the elements, and can ensure that each result is the evolution of the previous result, and can ensure that the design process is bottom-up and evidence-based. Based on the self-generating theory of multi-agent systems, the method of generating educational buildings in colleges and universities is explored by means of rule screening and algorithms. Designers only need to adjust input conditions or parameters to obtain various design results in a short time. This paper attempts to explore the use of computational design thinking to solve campus architectural design problems, and to tap the potential of computational generative design in architectural design.

2 Discussion on the Generation Method Based on Multi-agent

The concept of the multi-agent model was extended and evolved from the study of cellular automata. In 1940, John von Neumann and Stanislaw Ulam proposed a device for self-reconstruction and regeneration [1]; in 1970, John Conway simplified their vision and designed a "survival game" [2]. The color of the squares is used to indicate the "survival" or "death" of the cells. After determining the number of grids, the proportion of initially generated cells, and the rules of cell change, each unit cell will adjust its state according to the state of its neighbors. The system eventually reaches equilibrium.

Multi-agent system is a computing system composed of institutions that consists of multiple agents interacting in an environment. Its concept was first proposed by Minsky at MIT in the 1980s [3]. A multi-agent is a dynamic system in which all internal intelligences interact and interact with the environment. The multi-agent system focuses on the interaction, cooperation and conflict between multiple intelligences, emphasizing the group nature of intelligence rather than individual intelligence. The application research of multi-agent system in architecture mainly focuses on urban morphogenesis, human flow simulation and functional topological relationship generation. The ETHZ-CAAD laboratory has completed a practical project in the station square in Groningen, the Netherlands, with a multi-agent agent acting as a pillar being generated by a column grid arrangement of a planar non-orthogonal system [4]. The team of Li Biao developed the "CUBE101" simulation agent's life and death game to generate the floor plan and spatial form of the collective housing [4]. Guo Zifeng from Southeast University studied multi-agent systems constrained by functional topological relations, explored functional topological multi-agent systems in three-dimensional space, and realized the layout of three-dimensional functional space [5].

Architecture is a very complex subject, involving many influencing factors, and there is no clear judgment and evaluation standard for its results. Based on the multi-agent system, under the premise of the same number of input rules, as long as the evaluation criteria and goals are set, the best results can be obtained under the input conditions. At present, there are few researches on the generation design of educational building layout based on multi-agent, and research can be carried out on issues such as building spatial layout, functional topological relationship, and building group generation. At the same time, the bottom-up generative design idea of multi-agent system is an effective way to solve the layout and functional space layout of educational buildings. Time-consuming issues under some specific conditions. Although the multi-agent system generates building plans, it is still necessary to split the design steps, perform program calculations in batches, and perform manual intervention on the results in time to achieve results that meet the set expectations. But it is undeniable that the potential of multi-agent systems in architectural design has not been fully tapped.

3 The Rule Deduction of Space Combination

3.1 Modular Unit Design

The author studies and analyzes the main space of common teaching activities in colleges and universities, such as ordinary classrooms, amphitheatre classrooms, book reading rooms and laboratories. These regular teaching spaces provide students with a place to study and are also important places for students to carry out public activities and daily interactions. Analyze the basic dimensions of these common functional spaces, take the 8-m column span as the basic modulus, and take the 3 × 3 modular standard unit as the basis, and organize the main teaching space into three modular units of 1 × 1, 2 × 1, 2 × 2, and 2 × 3.

3.2 Unit Plane Translation and Growth Rules

In the 3 × 3 modulus unit, the author divides the path, and uses this path division as the main plane generation in the future. The path division method can be expressed by a program algorithm. We determine the position of the path by the variables n and m, and denote the side length of the module as quantitative mod, then the vector of the X-axis is X(1,0,0), and the vector of the Y-axis is Y(0,1,0), the position of the initial point is determined by the above parameters (Fig. 1).

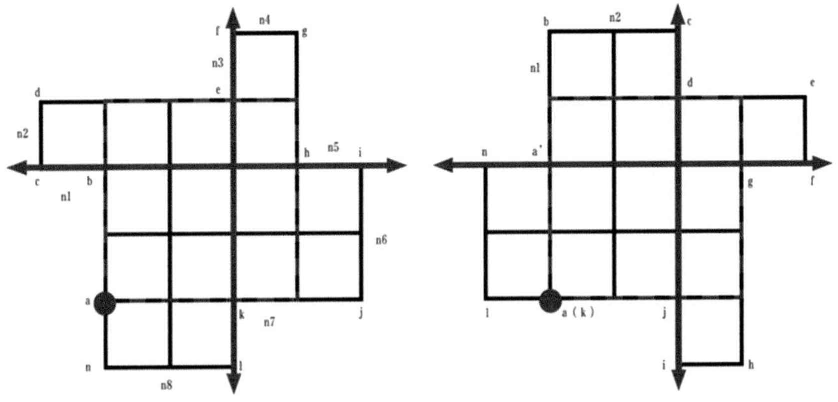

Fig. 1. 3 × 3 modulus unit growth rule (drawn by the author)

A simple modular unit can be derived from a variety of plane generation schemes according to three growth rules. The first rule is the number of changes. Depending on the way the path is divided, the core module can grow into different branches. The second rule is the direction change. The growing branches are generated in a clockwise or counterclockwise direction, and different growth directions determine the next generation law. The third rule is step size. According to the different random seeds, the step size modulus of the length and width of each branch is determined. This rule is the main factor that affects plane richness. The way the module grows rules can be expressed in a program. The following four growth directions are used as examples to illustrate, and the same is true for other directions (Fig. 2). Select an initial control point a, deduce the step length from this point to the X-axis/Y-axis, generate the next point, and repeat in turn, until the enclosing and generate a single plane, and thus obtain the path, path intersection and port, and port direction.

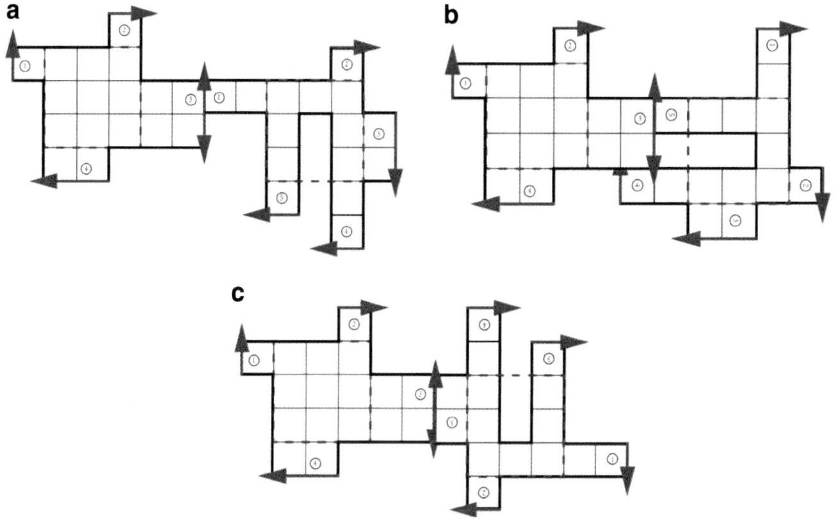

Fig. 2. (**a**) Directly connected when the ports are parallel. (**b**) Directly connected when the ports are vertical (**c**)

3.3 Unit Connection Rules and Translation

Rotation rule: After the monomers are generated by the program algorithm, according to the relationship between the ports of each monomer and the directions of the ports, the monomers are spliced in pairs (Fig. 2). When the angle between the ports of the monomer A and the monomer B is 180°, the rule of direct connection is satisfied. When the angle between the ports of the monomer A and the monomer B is 90° or 270°, the monomer B is clockwise or Select 90° counterclockwise to connect. When the angle between the ports of monomer A and the monomer is 180°, the monomer B rotates 180° to connect. Through these three rules, the principle that the planes of the monomers are connected can be satisfied.

Connection compensation rule: When the four monomers are connected to each other and exist in the form of a closed loop, due to the randomness of shape generation, it cannot be ensured that the shape finally exists in an enclosed form. Therefore, a connection compensation rule is added (Fig. 3). The stabilized form will retrieve the surrounding blocks of the interface. When the surrounding blocks overlap or border, the interface will automatically close, and when the surrounding blocks do not have the above situation, it will remain open.

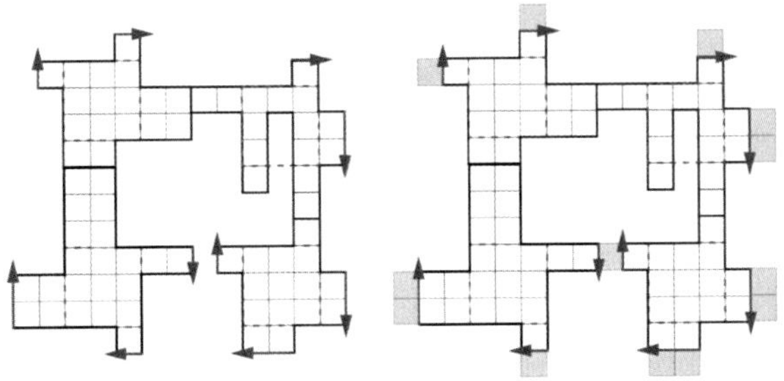

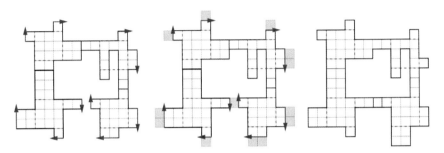

Fig. 3. Monomer connection compensation rules: detect the blocks around the interface, if they overlap or border, the blocks are connected

Overlapping and merging rules: When four monomers are connected to each other and exist in the form of a closed loop, due to the randomness of shape generation and the final positional relationship of each shape, there may be overlapping between monomers (Fig. 4). Therefore, overlapping merge rules were added. The stabilized shape will search for the overlapping area between the monomers. If the overlapping unit modules are less than or equal to 2 units, they will be automatically combined into a complete shape, but if this requirement is not met and there are too many overlapping units, this scheme will The generation is invalid, and a new scheme is recalculated.

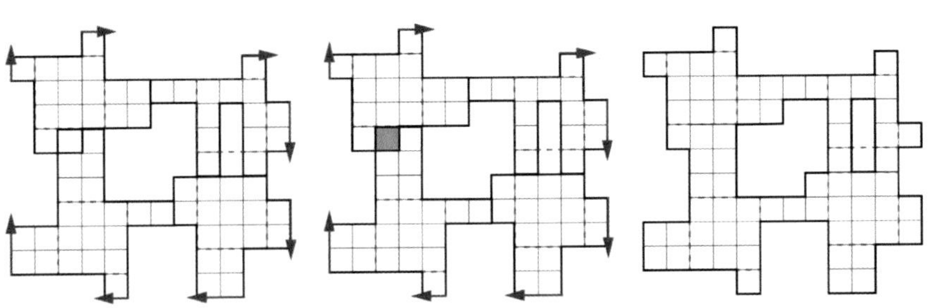

Fig. 4. Monomer overlapping merge rules: detect overlapping areas, if less than 2 units, generate a flat plan

3.4 Core Tube and Layout

Auxiliary modules (toilet and traffic space) are placed in each 3 × 3 core, the location is determined by the location of the intersection of the paths. Auxiliary modules are added at the connection of the monomers, and at the same time, the fire protection distance requirements are met. The distance between the fire stairs in the strip corridor should not be greater than 40 m (5 module units), and the space at the end of the fire stairs should not be greater than 22 m (3 module unit). After the path direction and the position of the core tube are determined, the basic functional space is divided, and the large space modules are placed on the plane first, and then the small space modules are placed, and the aisles are kept connected to each other (Fig. 5).

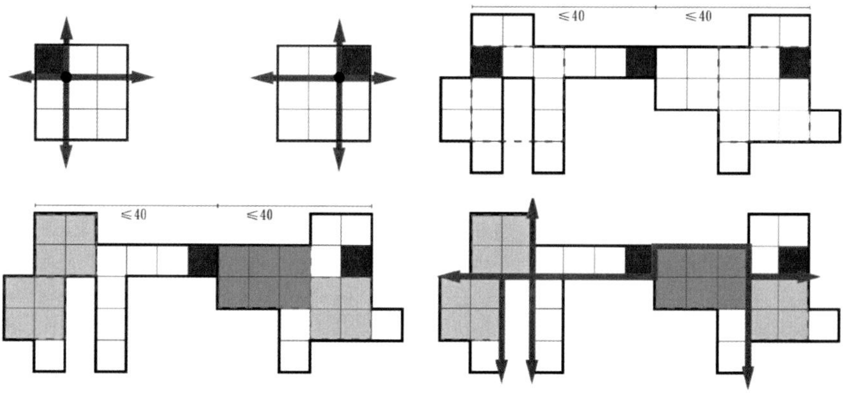

Fig. 5. Core tube and layout rules

4 Exploration of Space Composition and Scheme Comparison

4.1 Program Design Ideas

Based on a campus site in the south, this research combines the design process with multi-agent algorithm technology, translates the traditional design links into computer programs, and realizes the generation of university building space layout based on multi-agent system. The main work links of this experiment are as follows: First, the site element processing, the site is subjected to vectorized grid processing and digital output, the site is assigned grid cells and the complete site grid is screened out. Second, set the initial parameters, through the basic parameter control terminal, set the initial point position, random seed number, and arrangement rules. Third, agent translation and operation rule setting, set different agent arrangement forms and data attributes, obtain the plane layout, path, and core position of each layer, and form the final layer of each layer according to the compensation and merging rules. Architectural Design. Fourth, generate a comprehensive score for the scheme, and select the appropriate scheme based on the objective basis of plot ratio, boundary richness, block richness, and comprehensive index. Fifth, based on the rationality of the spatial functional layout, optimize the block relationship and functional layout for the optimal solution.

4.2 Evaluation Index Factors

Based on the grasshopper platform, the above rules are simplified into four basic battery packs: initial battery module, interface direction battery, initial connector battery, and process connector battery. Only need to set the initial plane generation position, adjust the parameters as needed, the positional relationship between the monomers and the connection between the monomers will change, and a variety of changeable layout forms will be generated. Taking the side length of 216 × 216 m as the simulation site, the intelligent unit is placed in the simulation site for calculation. After many tests, four types of generative layouts have been concluded, namely, the whole is centered and not out of bounds; the whole is scattered and not out of bounds; the center is arranged and out of bounds, and the whole is scattered and out of bounds (Fig. 6). These four types of randomly generated layout schemes are difficult to judge by human judgment. The richness and standardization are the goals pursued, and they are refined into floor area ratio, boundary richness, volume richness, and transboundary coefficient., the overlap coefficient of these five basic indicators.

A (Floor area ratio) is the ratio of the plane area of the generated building to the plane area of the site, and this ratio determines the accommodation density relationship of the site. B (Boundary richness) is the ratio of the sum of the perimeters of the generated plane to the sum of the perimeters of the 3 × 3 core space plane. This ratio determines whether the scheme is stretched or intensive, scattered or centered. C (block richness) is the ratio of the overhead or setback area of each floor of the generated plane to the area of the floor. This ratio determines the spatial experience of the scheme. When a scheme has more overhead or setback space, a certain To a certain extent, it can be explained that the scheme has better space tour and building roaming experience. D (out-of-bounds coefficient) is to judge whether the generated plane exceeds the building red line. This value determines whether the generated plan is reasonable. When the building does not exceed the red line, the value is 1, otherwise, the value is 0. E (overlap coefficient) is to judge the overlap relationship between randomly generated planes. Due to the randomness of the building plane generated by the algorithm, the location of the building units at each step is uncertain, so there is a situation where the building units overlap each other. This value determines whether the generated scheme exists or not. Set a comprehensive mean Score, the formula is

$$Score = (a + b + c)^{*} d^{*} e \quad (1)$$

The larger the value of Score, the better the performance of each index of the scheme, and the better it can meet the design requirements of richness.

4.3 Comparison and Selection of a Campus Design Scheme

After extracting the vector boundary of a university site in the south, a grid with a modulus of 8 is built into the site, and a basic grid frame is divided according to the road network

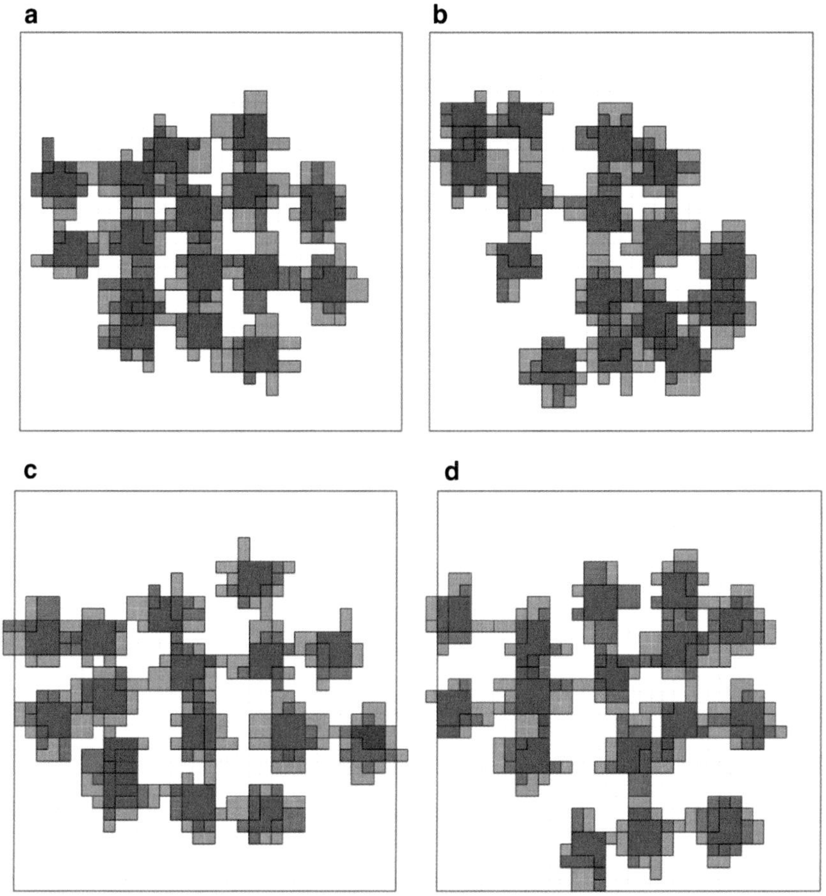

Fig. 6. Four layout types: **a** the whole is centered and not out of bounds; **b** the whole is scattered and not out of bounds; **c** the center is arranged and out of bounds; **d** the whole is scattered and out of bounds

relationship, and a point in the lower left corner is selected as the initial point for shape generation (Fig. 7). The architect only needs to set the initial parameters in the program, and then the design scheme can be generated, which provides more possibilities for the initial architectural design. Among them, the quantitative parameters are: site boundary A, grid modulus B, control grid C, and the variable parameters are: control point U, random parameter V, generation method W. These factors determine the final mass generation.

The author divides the generation schemes into three categories: single arrangement, staggered arrangement, and forward and reverse arrangement. Monolithic layout, that is, each unit is generated by a control point, each building volume is like an independent small house, and the position between each unit is determined by the initial point. The multi-storey building layout has been spatially designed to The first floor is the benchmark, and 0 ~ 2 step units are gradually retreated upward to ensure that the building

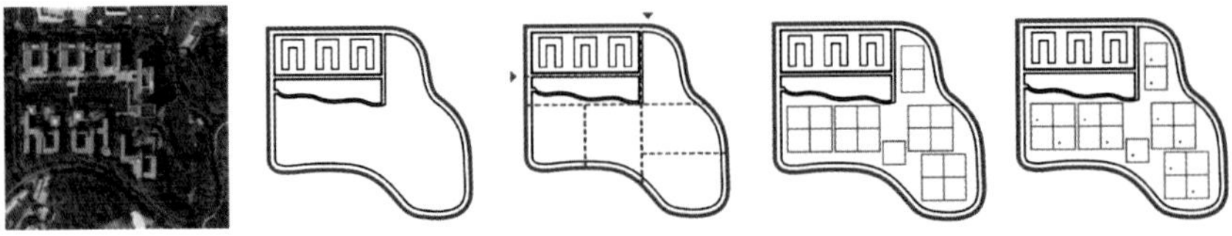

Fig. 7. Site meshing and setting initial points

space on each floor has more dislocations, and the building space on each floor can get better lighting. However, the connection between the monomers is not strong, the enclosures between the monomers are arranged in a "well" shape, and the enclosure relationship is relatively monotonous (Fig. 8).

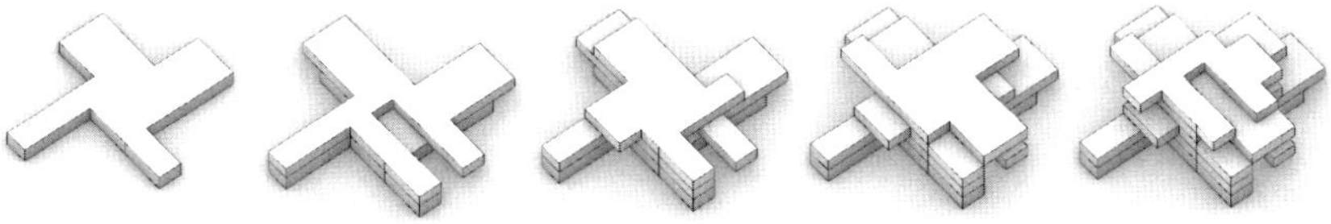

Fig. 8. Single arrangement: layer by layer to get a single building

Staggered arrangement, that is, the building volumes on each floor, like building blocks, are staggered in the order of "horizontal-vertical-horizontal-vertical". A control point can determine the arrangement of 2 unit blocks. The spatial layout of the multi-storey building is staggered horizontally and vertically, and 0~2 step units are retreated up by layer to ensure the dislocation relationship and lighting of the building space on each floor. The generated architectural plan has a relatively close connection in the horizontal and vertical layout, showing rich features like "weaving" (Fig. 9).

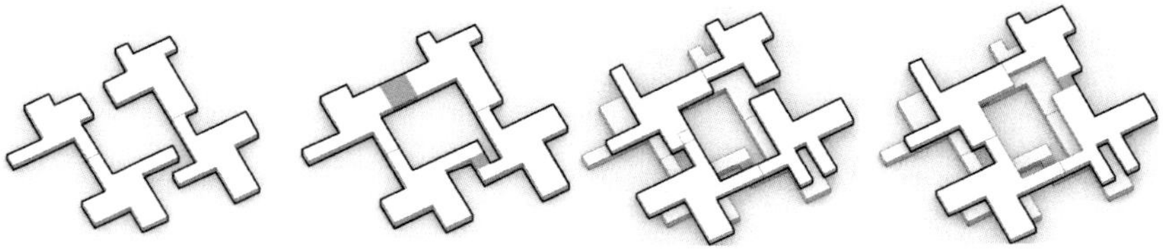

Fig. 9. Staggered arrangement: criss-cross arrangement of blocks

The forward and reverse arrangement, that is, the building volume of each floor, like a greedy snake, is enclosed in a clockwise direction, and then in a counterclockwise direction, and so on. One control point can determine the arrangement of 4 unit blocks, and only 5 control points are needed to complete the layout of the entire site. Based on the basic rule of "step back and stagger each other", each floor is set back 0 ~ 2 step units, ensuring the horizontal and vertical staggered arrangement of the multi-storey building space. The connection between the four monomers is very close. As long as the initial building block changes, the enclosing relationship of the entire building will also change. The generated shape not only has the characteristics of criss-cross texture, but also reflects the "maze". layout features (Fig. 10).

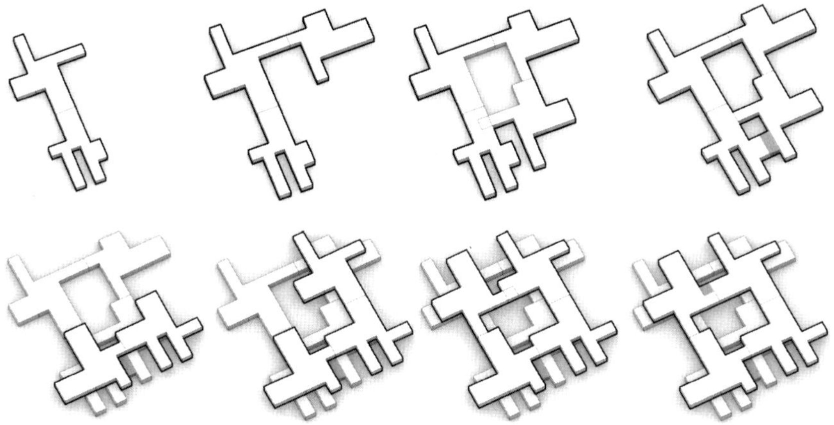

Fig. 10. Forward and reverse arrangement: arrange blocks in a clockwise-counterclockwise direction

Based on the annealing algorithm of galapagos, the author obtains the optimal solution of weighted value through multiple iterations, counts various numerical indicators in the experimental process, and obtains the optimal solution of weighted value for three generation methods. Then, the numerical data of the optimal solutions of the three schemes are sorted and summarized, and a horizontal comparative analysis is carried out (Fig. 11). In order to make a more comprehensive evaluation, the author organizes the obtained data and summarizes it into five indicators: plot ratio, boundary richness, block richness, open space value, open value, and comprehensive value. It can be seen from the data results that the forward and reverse scheme is higher than the other two schemes in terms of boundary degree and block richness, indicating that the generated scheme has better spatial effect. At the same time, in terms of volume ratio, it is the same as the staggered type, and is much larger than the single type. However, its vacant land value is small and its openness is high, which means that the site utilization rate is high, the volume is richer, and the overall performance is better. In the analysis of physical environment performance, the lighting performance is also worthy of appreciation. Therefore, the optimal solution of the forward and reverse can be used as the final detailed design scheme (Table 1).

5 Strengths and Weaknesses

Based on the self-organization characteristics and "bottom-up" logic characteristics of multi-agent systems, this research quantifies constraints such as various normative regulations and people's subjective orientations, and translates buildings into systems that can operate within the system, It is an agent that restricts and interacts with each other,

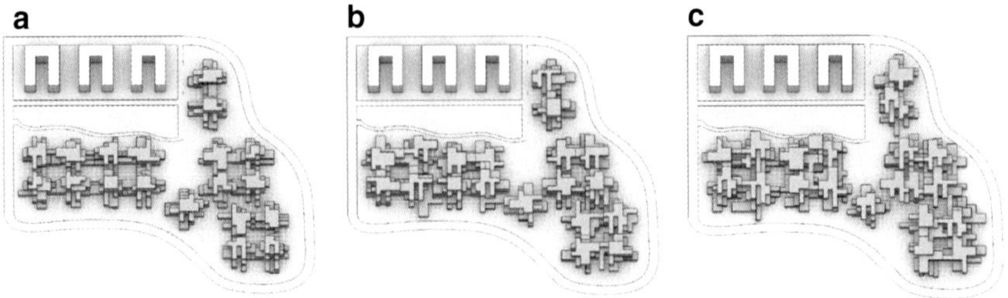

Fig. 11. Comparative analysis of three methods, **a** Single arrangement; **b** Staggered arrangement; **c** Forward and reverse arrangement

Table 1. Indicators for the three programmes

	Single	Staggered	Forward and reverse
Volume rate	1.24	1.15	1.13
Boundary	2.22	2.33	2.24
Bulkiness	4.51	3.79	3.31
Openness	3.44	3.06	3.12
Comprehensive	4.48	4.14	3.61

and can interact with the environment, so as to complete the spatial layout of college campuses. At the same time, this is a universal architectural design generation system. The design does not emphasize the location and surrounding environment of the site, but pays more attention to the richness of its building volume.

Compared with traditional architectural design, the multi-agent system used can obtain endless solutions, avoiding the continuous attempts of manpower for a large number of solutions, which can greatly improve the generation efficiency, expand the possibility of forms, and make it easier to find and screen. Better solution. It provides a large number of options and possibilities for architects in the early stage of sketch design, which is conducive to enriching the architect's design material library and stimulating the designer's creative potential. At the same time, the generated building group is rich in volume and has a strong sense of spatial hierarchy. While doing daily scientific research and study, students can also stroll through the building space and relax. These rich and interesting building volumes will become places for students to communicate and communicate.

To summarize the process and results of this research, there are still many imperfections. At the programming level, this research realizes the multi-agent system experiment and volume generation based on Grasshopper and Python programming languages, but the functional plane deepening is redrawn in Revi, and cross-platform operation will cause the problem of model information conversion, and The program cannot achieve the effect of real-time linkage on multiple platforms. At the same time, because the program code behind the model is not concise enough, it takes a long time to generate the optimal solution. At the level of interaction design, this program code does not have a concise interactive interface and clear operation instructions, and cannot be directly oriented to the underlying design workers without programming foundation. Furthermore, the scoring data cannot be converted into standard charts for display in real time, and it is impossible to intuitively analyze and interpret the generated design scheme rationally. At the architectural design level, the building volume automatically generated by the program is too staggered, resulting in misalignment of the upper and lower column grids, dark rooms, and poor building lighting. This reduces the richness of algorithmically generated architectural solutions and goes against the original pursuit of algorithmically generated design.

6 Future Outlook

The generation system based on multi-agent and rule screening does not yet have the ability to self-optimize, and the future optimization space will consider adding core methods such as machine learning to make it have the potential for self-improvement. Multi-agent model systems are widely used in many fields such as architecture, planning, landscape, and transportation. Although these research applications are still mainly in the academic field, they have not been able to be applied and promoted in the industrial field. However, as more and more designers with programming foundation and thinking join the digital design team, the results of multi-agent systems in the field of digital generative design will be more fruitful. These results can also be packaged into practical tools with commercial value and delivered to the hands of every architectural design practitioner. It is hoped that in the near future, computational generative design will become an essential "weapon" for thousands of architects. Architects can design a variety of rich architectural solutions through a simple interface and the use of complex and efficient programs behind it. So that the creativity of architects can truly be liberated from complicated and meaningless drawings, and the spark of thinking of architects can truly shine in the field of design.

References

1. Biao L (2012) Building generative design—Research on computer generation method of building design based on complex system (in Chinese). Southeast University Press, Nanjing
2. Eckel W (2015) Creating an artificial world with a new kind of cellular automata. arXiv preprint arXiv:1507.05789
3. Doulgerakis A (2007) Genetic programming + unfolding embryology in automated layout planning. UCL, London
4. Scheurer F (2005) A simulation toolbox for self-organisation in architectural design. In: Sariyildiz S, Tuncer B (eds) Innovation in architecture, engineering and computing (AEC), vol 2. Delft University of Technology. Faculty of Architecture, Rotterdam, pp 533–543
5. Biao L, Jingping Q (2009) Research on the generation method of "cellular automata" architectural design—Taking the generation tool of "Cube101" as an example (in Chinese). New Architect 2009(03):103–108
6. Zifeng G (2017) Research on building generation method limited by functional topological relationship (in Chinese). Southeast University

Common Ground—Online Platforms for Bottom-Up Collaborative Decision Making in Design Education

Nicolas Stephan(✉), Marine Lemarié(✉), and Kristina Schinegger

i.sd Structure and Design, Faculty of Architecture, University of Innsbruck, Technikerstrasse 21C, 6020 Innsbruck, Austria
{nicolas.stephan,marine.lemarie}@uibk.ac.at

Abstract. Co-creation and real-time collaboration have always been an integral potential of digital design methodologies and have been accelerated by the rapid digitalization of teaching due to current societal developments. This paper discusses the prototype of a real-time multiplayer building platform as a video game developed for a first-year design studio impacted by pandemic-related teaching restrictions. The aim was to develop a methodology that enables first-year students to meet peers, build models collaboratively, and teach implicit design knowledge such as aesthetics and formal analysis while allowing individual creativity within the populous class. Through a combination of a step-by-step iterative design system and a real-time decentralized multi-player platform, students can work collaboratively on common digital designs. The design method is based upon building units and individualized strategies of aggregation and differentiation that are built up into larger structures. Special focus is paid to how new online platforms created for architecture education can migrate the advantages of physical intuitive design methods to a digital setting and eventually fill the gap of lacking implicit knowledge pedagogies.

Keywords: Co-creation in design · Collaborative design · Real-time platform · Crowdsourcing · Game engine · Mass-customization

1 Introduction

With a 170-students-large design class of first-year students who do not know their peers amid the 2020 Covid-19 lockdown, we, as instructors, wondered how we will offer a studio culture experience during a 5-week workshop while communicating both explicit and implicit/tacit knowledge fundamental to the architecture discipline [14].

1.1 Explicit and Implicit Skills

To teach how to develop "good" design is to impart two forms of skills: explicit and implicit/tacit [16] skills. Explicit design intentions are easy to articulate and summarize in written form or drawings, as opposed to implicit design criteria which cannot be

P. F. Yuan et al. (eds.), *Hybrid Intelligence*, Computational Design and Robotic Fabrication,
https://doi.org/10.1007/978-981-19-8637-6_6

objectified or explained precisely, such as aesthetic preferences. In the discussed workflow, explicit knowledge encompasses disciplinary concepts such as typology, scale, and building blocks. The aim of the studio framework was that students understand that the design process of a building involves multiple steps, with various levels of detail, from small scale to large scale and vice versa.

The teaching of implicit/tacit knowledge is usually done via a design studio system where students design as a team and regularly meet the instructors to evaluate the results in a conversation. The process of learning how to design involves complex skills and decision-making (e.g., articulating design intentions through words and visuals, synthesizing often conflicting design intents and translating concepts into spatial elements and effects, abstracting specific ideas into abstract design diagrams and techniques) and is, to a large part, supported by a tacit acquisition like observing peers and tutors during designing. Even the surrounding studio filled with models, images, and artifacts play a major role [1]. According to Nigel Cross, design knowledge resides in people, processes, and products of design [3]. Consequently, it is necessary to learn in situations where all three of them are around or can be observed. A common understanding of design knowledge is that it is "person- and situation-orientated" [11], especially in the Anglo-American understanding [10] (which has been adopted also in most Austrian Architecture faculties), it is encapsulated and expressed in a "studio culture."

This well-established teaching method of learning by observing and reenacting was not fitting to the new teaching challenges: a very large class of first-year students who have never entered an architecture school, during online teaching. Despite these problems, we wanted to ensure the students' acquisition of implicit knowledge and a high level of individuality within their projects. We wanted to train their ability to self-evaluate design decisions and recognize the moment when "design" happens [4].

1.2 Teamwork in a Virtual Teaching Environment

The progress of remote and open-source collaboration in design and the consequences for production, decision-making, authorship, and concepts like authenticity or originality is not a new discourse [7, 17], but it has been accelerated by the Covid-19-related shift to remote working. Despite recent advancements, more intuitive hands-on processes such as collaborative model-making in environments that enable the transfer of tacit knowledge are hard to implement. We propose that online platforms specifically created for architecture education can transfer many advantages of in-presence design studios to a digital setting.

1.3 Project Intention

To address the challenges of virtual teaching in Covid-19 lockdown, we decided to develop our collaborative building platform in the form of a video game (Fig. 4.) and a corresponding design methodology (Fig. 1). The game had to be decentralized, accessible, easy to use, and motivating, enabling students to work collaboratively and facilitate social contact with peers. It was intended to be embedded into a consistent iterative design system that still allowed students creative exploration.

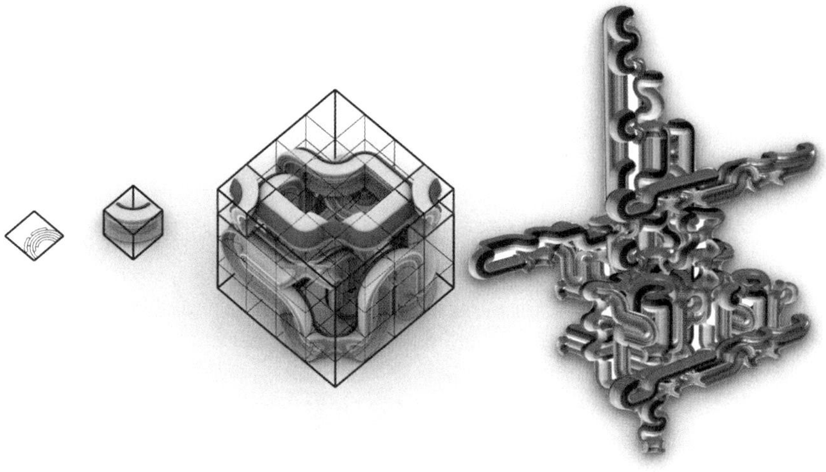

Fig. 1. A four-step process: 1. Curve → 2. Voxel → 3. Module → 4. Aggregation

2 State of the Art

Collaborative online games have been successfully adapted in science education and research: A prime example is "Foldit", an online puzzle video game, which made laypersons find solutions for protein folding through a trial-and-error crowdsourcing approach [2].

In architecture, real-time building platforms are developed mostly as experimental design tools: Urs Hirschberg and his team at ETH Zurich developed a web-based system allowing users to work simultaneously on a common design project while being separated across three continents [7]. At TU Graz, Alexander Grasser and Alexandra Parger introduced concepts of configuration into the real-time building to inform the collaborative design process with three-dimensional reinterpretations of structuralist geometric operations [5]. Valerie Messini, Damjan Minovski, Dominik Strzelec, and Dominic Schwab developed a multi-user painting platform, where players use personalized brushes to "paint" collaboratively on a spatial "canvas" and tested it with students at the University of Innsbruck [12].

These platforms are great advances in collaborative design, but they lack certain qualities needed for an application in architecture education: 1. A mechanism within the application itself or its context to evaluate design decisions. 2. Integration in a larger pedagogical methodology. 3. The possibility to implement unique user-made objects. They consequently do not address the users' engagement with aesthetics, missing out on the potential to let the users develop their approach towards design and form analysis.

We would like to fill this gap by focusing on the potential of real-time building platforms to teach implicit/tacit knowledge in an architecture design studio. This research process is based on three guiding principles: *collaborative bottom-up decision making*, a *grid-based design strategy*, and *gaming in architectural education.*

3 Concept

To elaborate on the three guiding principles and identify possibilities of implementing them in the game and design methodology, we set up a theoretical framework, which is briefly highlighted in this chapter.

3.1 Collaborative Bottom-Up Decision Making

Collaboration in Architecture: In their book "Open-Source Architecture" Carlo Ratti and Matthew Claudel advocate for the future of architecture discipline to become more collaborative, inclusive, and network-driven, based on the current online-technological advancements, such as open-source culture, crowdsource, and open-access principles [17]. The notion of open-source architecture is a novel way of designing, as opposed to the "starchitect" culture. Here, the architect serves as a curator and educator to many other people, designers, or non-experts. This inclusive approach to spatial design enables collaborative use of design tools by both professionals and laypersons: the term "citizen-centered design" emerges. The citizen-centered design movement is a key aspect of collaborative design at the intersection of design and public policy.

Decentralized Real-Time Collaboration: Urs Hirschberg acknowledges that collaboration across continents is flourishing with the accelerated development of internet-based open-source software and his research enables shared platforms for collaborative projects, opening up new avenues for collective authorship [8].

Findings for the authors' tool and methodology: The core intentions for *Common Ground* were based on Carlo Ratti's concept of open-source architecture. Urs Hischberger's decentralized multi-author approach proved to be a suitable way of implementing these intentions in a practical tool. Online collaboration and a free, accessible, and customizable platform were postulated as necessary and very valuable in this situation. Further, we saw the potential to open the platform to non-experts, in this case, first-year students, to expand the "citizen-centered design movement" within the setting of video games.

3.2 Grid-Based Design Strategy

Grid-Based Geometric Transformations: An early example of a grid-based design system is the educational practice of Jean-Nicolas-Louis Durand, who made his students draw on squared paper. Durand started by letting them design a catalog of stylized building elements, like porches, vestibules, halls, staircases, or courtyards, to develop a range of geometric vocabulary. The grid allowed precise positioning of standardized elements to be composed [15, p. 44].

Encapsulation of Design Decisions: When using tools and, especially, computers for design education, there is a potential of focusing on specific design decisions within the discussion with the students, by embedding pre-made decisions within the software. When operating the software, users automatically build on these decisions and use the design knowledge already embedded within the system. Andrew Witt calls this "encapsulated knowledge." Users do not need the design knowledge encapsulated in the tool, but only need to know how to access it [20].

Findings for the authors' tool and methodology: Like Durand, the authors' design methodology uses a grid to situate and measure compositions. To guide students through the complex process of designing, setting up a structure removes weight and leads the students faster towards creative work. This structure is especially relevant in large classes to guide the thought process of many students. Encapsulating knowledge and decisions within the tool helps strengthen this structure.

3.3 Gaming in Architectural Education

Component-Based Games in Architecture Education: Although component-based games have become popular within the last years in architecture, they have been used in education for centuries [13, p. 83]. An early architectural example is the "Riesenspielzeug" ("Giant Toy"). It was invented by German architect Gustav Lilienthal in the 1880s after seeing prefabricated houses in Australia. The game consisted of standardized wood elements and metal wedges, designed to accurately represent the elements and systems of technical building construction. Their large size allowed for constructions up to 1 m high and offered various connection possibilities. The system was intended for children and architects alike to create miniatures of real buildings [13, p. 99].

The Logic of Continuity: In the 1970s graphic designer Ken Garland developed "Connect," a game of 140 cards with red, blue, and black lines and curves of different colors. Players had to place the cards on the floor in a way that the curves formed a continuous loop while winding them around furniture or other obstacles. "Connect" is a great example of how a design language and aesthetics can be embedded in a game while allowing autonomous decision-making; it was adapted for design education [6].

Video Games as Educational Tools: Today, there are countless architecture/city planning-themed video games, but most are not suitable for architectural education. City-building games usually depend on a strict set of rules and relationships, forcing the player to build a "realistic" city. To develop and teach original architectural design, it is more suitable to leave the outcome unpredictable and let the players decide if the outcome is successful. As architect and educator Damjan Jovanovic puts it: "Play does not have to be always goal-oriented, and although most games do have a goal (the 'win' state), more and more the inherent specificity of experience leads to the player being content with merely 'existing' within a game. Immersion does not depend on and is more likely even disturbed by direct calls for action towards reaching a goal." [9, p. 33].

Findings for the Authors' Tool and Methodology: "Gaming" as a didactically premeditated methodology, which allows to channel design options and increases the students' understanding of their design decisions and consequences [19]. The component-based system of the "Riesenspielzeug" is a great precedent for the building logic because it allows for three-dimensional aggregations in many configurations. "Connect"s logic of continuity became one of our main principles when developing the design logic for the studio. We wanted to implement these two aspects into the framework of an open non-goal-oriented video game.

4 Practical Implementation of the Concept

4.1 Design Methodology

Based on these three key features, we developed a design methodology, including a step-by-step design and evaluation process, wherein groups of students collaboratively designed their building components and used those to build large aggregation in an online video game setting. To break down the complexity of large aggregations, the design method is made of steps that build upon the last step, gradually increasing the level of detail. Breaking the task down into smaller steps helps slowly move towards an implicit knowledge of designing complexity. The game helps communicate this form of knowledge playfully while simultaneously providing the fun part of playing the video game with peers. In the next paragraphs, this process will be explained in detail (Fig. 1).

Step 1 "Curve": The process starts with a simple 2-dimensional curve drawn in Rhinoceros 3D, fitting a 3-by-3 m bounding area. Students were encouraged to develop different versions of curves for further testing and evaluation in Step 2 (Fig. 2).

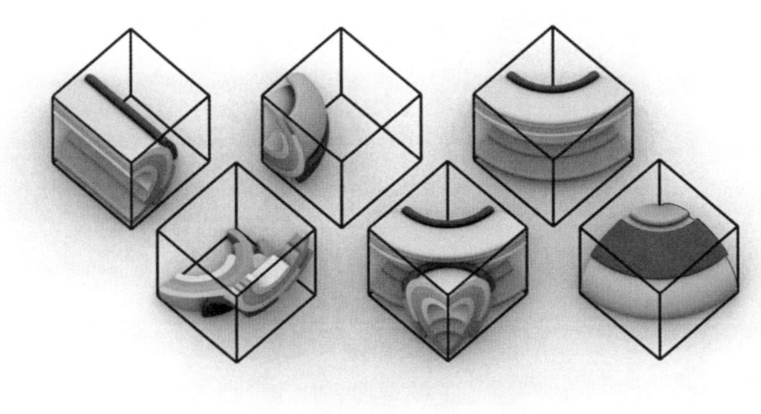

Fig. 2. A minimum set of different voxel types are needed to fill the modules: straight, left angle, right angle, multiple connections, up, down

Step 2 "Voxel": Students three-dimensionalized their previously developed curves with various manipulations, such as linear and non-linear extrusions, rotations, and boolean operations to fit a $3m^3$ bounding box. They had to make six different voxel typologies that can be rotated to ensure the continuous connection between the voxels for the next step: a straight connection, left corner, right corner, joint right and left corner, down corner, up corner (Fig. 3).

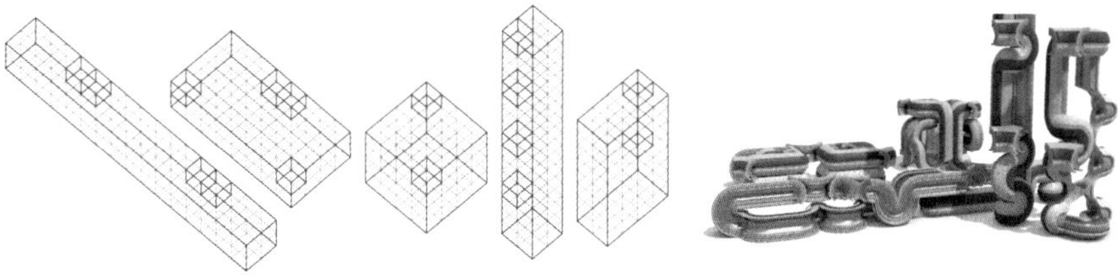

Fig. 3. The five-module types with connection blocks and the same modules are made of continuous voxels, leaving open ends for module-module connection

Fig. 4. In-game collaborative aggregation process

Step 3 "Module": To achieve a diversity of possible decisions for the final aggregation, we defined five larger "module" bounding volumes in different dimensions based on a 3 m x 3 m 3D grid. Each module contains 64 voxels. The modules have several connection blocks needed for future continuous assembly in the final step. Students now used their previously designed $3m^3$ voxels to fill the space between the connection points inside the module in a continuous procedure, leaving loose ends only at the connection blocks. Not every empty position has to be filled with a voxel object. Meanwhile, it is challenging to select the correct "corner voxel" to create continuity within the tight limitations of the bounding volume (Fig. 4).

Step 4 "In-game Aggregation": The modules from Step 3 are uploaded to the game engine and form a 5-piece collection of building elements within the game. Students then log in as a group and choose from the modules to assemble them collaboratively and evaluate their results in the context of a city environment. Connection blocks on the modules guarantee continuity between modules as well as a range of aggregation possibilities due to their alternating position within the module (Figs. 5 and 6).

Co-Evaluation of Design Outcomes: In between each step, students are encouraged to follow an evaluation cycle that starts with the design, then evaluates its eligibility for the next step. Design choices like low density vs. high density or simplicity vs. complexity were left to the students. If it succeeds, the design continues to the next step. If not, it will be redesigned, re-evaluated, and so on until it is declared eligible for the next step.

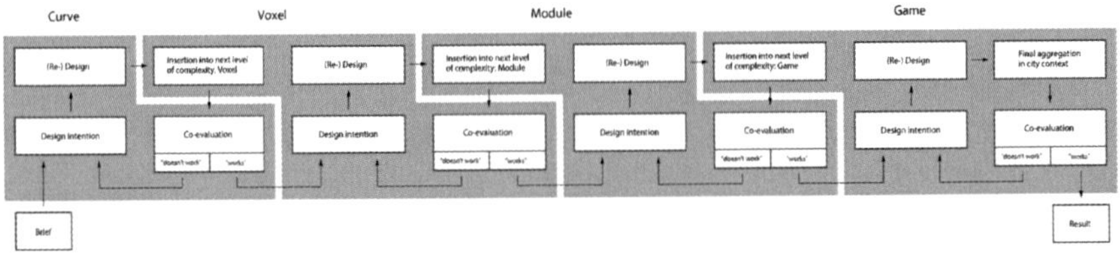

Fig. 5. Diagram of the iterative design method

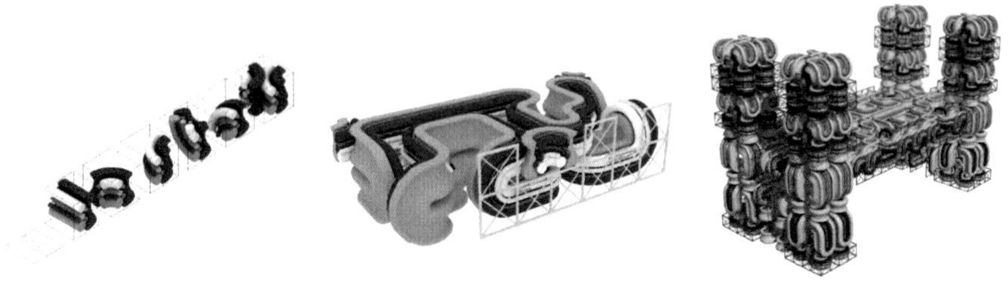

Fig. 6. Evaluation renderings of the same project in various stages. (*Credit* Miriam Meyer, Matti Schlenther, Noel Melmer, Maximilian Rieder, Julia Muschler)

Hence students get to slowly integrate how to design best, which feeds their implicit knowledge. Designing is a quality that cannot be taught as straightforward as a history class which is why such methods of self-evaluation and re-adaptation are necessary.

4.2 Game

The name "Common Ground" suggests that users share the same site, basic building elements, and game rules. Important aspects are the multi-player feature, the in-game communication through the chat, and the realistic representation of the 3D city environment through shadows and materials. The players are given a free-roaming camera from an aerial viewpoint to view the whole scene.

Preparation & Export: Users need to design assets in a 3D modeling software of their choice, within the predefined bounding boxes of the modules, their size, and the location of the connection blocks. (Fig. 3) Next, they export the modules and add them to the game engine.

Download & Installation: The game can be made publicly available and downloaded from a conventional server or file-sharing site such as OneDrive, Wetransfer. After unzipping the file, the application is operative.

Login & Group Selection: Upon starting the application, the users enter a log-in screen. They can choose a nickname and the number of their group, which is linked to their five previously designed modules. They then proceed to a lobby room where they can create new groups of collaborators or join already existing groups. After pressing "start," they enter the virtual building environment.

Navigation & General User Interface (GUI): Inside the building environment, players can use their mouse to navigate around the virtual construction site, pan left and right, and zoom in and out. The GUI of the game includes a set of buttons, which can be used to select the desired modules for the construction or deletion of modules. Selecting a part from the catalog activates the building mode and moving the mouse cursor points to the desired building location. Further options include "Quit Game", "Save Scene", "Load Scene", "Toggle Site", "Chat", and "Cancel Build". Players can see a list of their collaborators and a chat window for communication.

Design & Communication: The real-time aspect is key for users to react rapidly to their teammate's new placements. When building, new blocks quickly appear in different places. Some users concentrate on one area, others try to build higher, wider, more compact, horizontal slabs; every player has their unique playing and building behavior. In addition, users can communicate verbally as well over the chat function (Fig. 7).

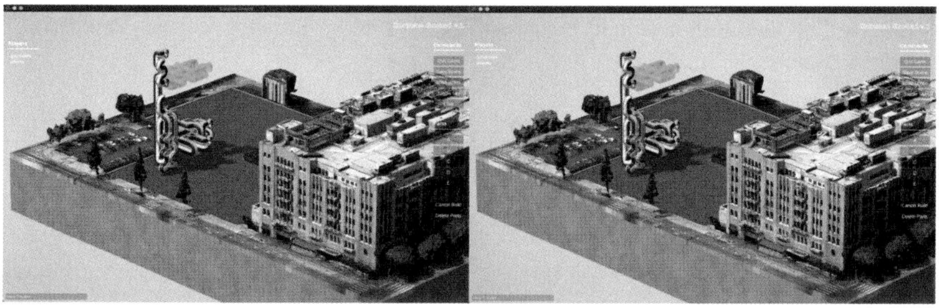

Fig. 7. Possible positions are rendered green, impossible ones are rendered red

Progress Saving & Logout: Users can save their progress using the corresponding button in the GUI. They can then log out of the game and continue working on their saved progress in a later session.

4.3 Implementation of Guiding Principles

The game is set up to allow collaborative bottom-up decision making, a grid-based iterative design system in education, and gaming in architectural education.

Bottom-Up Decision Making: The game space hosts up to 100 players simultaneously, who get real-time updates when new parts are added or deleted by other users. This makes communication essential and encourages collaborative decisions about the development of the structure.

Grid-based Design Strategy: To condense the learning experience, specific design decisions and geometric operations, which would usually require a high understanding of 3D software, were embedded into the game. This enabled the students to focus on the essential tasks, without having to deal with too much complexity.

Gaming in Architectural Education: The simplified GUI and playful art style of the game encourage students to experiment without the fear of being wrong. Choices on aesthetics and grade of abstraction were made by the students, allowing them to approach the design problem more pragmatically or experimentally. Even though these decisions were discussed intensively, the game did not judge them (Fig. 8).

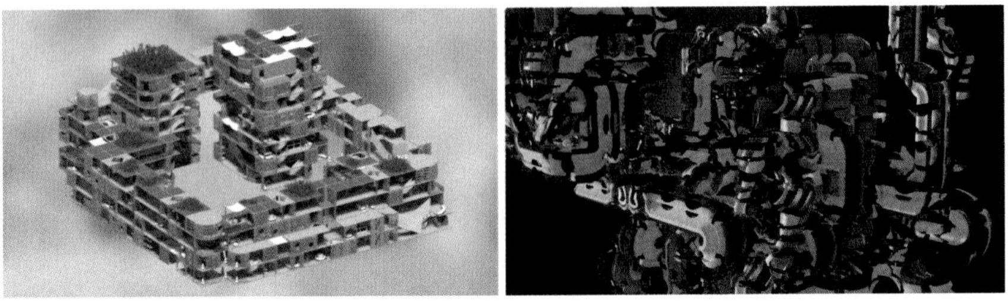

Fig. 8. The workflow allows for the expression of subjective design preferences and different degrees of abstractness. (*Credit* Antonia Hornauer, Hannah Rainer, Irinia Radeva, Katharina Rauch, Christian Rehnisch and Leander Gasteiger, Mara Ruperti, Vincent Reichardt, Bert Landsmann, Samuel Schmid)

5 Survey

Survey Setup: To evaluate the effectiveness and learning experience of our pedagogical method and tool, a survey was sent to the 170 participating students. The questions were directed towards the performance of the method and tools used in the context of distance learning, collaboration, imparting of explicit and implicit design knowledge, and the general group working experience. The survey was structured as a written interview including questions such as: "Are you satisfied with the way distance learning was handled?", "How is it a restriction or an improvement against physical class?", "Do you have any feedback on the course material and structure? What was the most successful part?".

Survey Results: 70% agreed that in the context of distance learning, the teaching method was comparable to the advantages of physical group work. Examples of answers are "I have been feeling extremely lonely and isolated this semester and this course was the one exception." or "In our group, we could work together at any time and place. That was efficient." Regarding the online pedagogy, one reply was: "Really great that you can get a project going in this short time only via Zoom, Miro, and Vimeo". The final video game was successful as the questionnaire showed 80% satisfaction with the game as a design-pedagogical tool. Students replied: "The most successful parts were the game and in the beginning the design of each voxel.", "I like the idea of starting simple and increasing the project to a large building." or "Most successful part—to see the whole project "grow" and create different aggregations in the game".

The feedback indicates that students enjoyed seeing the project grow by starting with simple voxel designs, increasing the scale, and finally creating in-game aggregations. They were also excited about having their own designs as actual building elements in a video game that they played together with 5–10 players. Chatting in-game and video calling while playing nurtured peer contact (Fig. 9).

Fig. 9. Two examples of teams of five students aggregating the elements together into architectural figures in real-time in the game Common Ground. (*Credit* Andreas Lederbauer, Ariana Gosalci, Chiara Koch, Lena Jenn, Sophie Gruner and Dominikus Schlögl, Theresa Riedmann, Sarah Rieder, Kilian Rietzler, Poledt Cedillo Peralta)

A suggestion for improvement was a clearer outlook on the final product of the studio to help the students improve their objects' fitness progressively.

6 Conclusion

The three key features of this project are *collaborative bottom-up decision-making*, a *grid-based design strategy*, and *gaming in architectural education*. Addressing the combined factors of Covid-19 lockdown, a class of 170 spatially separated first-year students, while teaching online, the project successfully fulfilled its pedagogical aim: to develop a method and tool for collaborative design decision-making for a large class, teaching implicit skills and effectively establishing a virtual studio culture.

The students developed a series of designs implemented in the online multi-player real-time video game "Common Ground". The game is accessible as a multi-user tool in the design context. Thanks to the user-friendly nature of its interface, it is ideal for distance education, facilitates student–teacher and student–student communication, allows students to simultaneously design complex aggregated structures, and encourages them to communicate through the in-game chat function. This step-by-step design cycle method improves the students' implicit skills while the boundness to the 3-dimensional grid structure, the voxels, and the modules, helps open a framework for creativity and freedom. In the end, the students produced 36 unique sets of 5 modules made of continuously attached voxels and even more aggregations in the game.

The project is a continuation of the use of game engines in the architecture field. In this case, the game was the last step at the end of the semester while the rest of the design process focused on creating the assets for the game with conventional 3D modeling software. In the future, it would be interesting to weave 3D assets earlier in the game as a constant back and forth manner to test how they perform in the final environment. The learner could evaluate its fitness through the game itself.

Finally, the step-by-step process in combination with the game initially developed for educational purposes could be used for "citizen-centered design with "real world" communities. It allows people with no background in architecture education to evaluate their own design decisions in collaboration with their peers.

Acknowledgements. The authors would like to thank Julian Edelman (UIBK) and Dipl.-Ing Nina Hütter (UIBK) for their continuous support throughout the semester for providing skills, time, and effort to assist students and make this class possible. The research-led design course was conducted at i.sd Structure and Design, University of Innsbruck and contributes to its research agenda on advanced computational tools for the early design phase and computational immediacy (SFB project F 77 funded by the Austrian Research Fund FWF).

References

1. Avermate T (2020) Communities of tacit knowledge architecture and its ways of knowing [Internet] 2020 [cited 2022 March 12]. https://tacit-knowledge-architecture.com/research-programme/
2. Cooper S, Treuille A, Barbero J, Leaver-Fay A, Tuite K, Khatib F, Snyder A, Beenen M, Salesin D, Baker D, Popovi Z, Foldit Players (2010) The challenge of designing scientific discovery games. In: FDG '10: proceedings of the fifth international conference on the foundations of digital games, vol 2010, pp 40–47. https://doi.org/10.1145/1822348.1822354
3. Cross N (1999) Design research: a disciplined conversation. Design Issues 15(2):5–10. https://doi.org/10.2307/1511837
4. Cross N (2001) Designerly ways of knowing: design discipline versus design science. Design Issues 17(3):49–55. http://www.jstor.org/stable/1511801
5. Grasser A, Parger A (2021) Reappraising configuration and its potential for collaborative objects. In: Stojaković V, Tepavčević B (eds) Towards a new, configurable architecture, Proceedings of the 39th conference on education and research in computer aided architectural design in Europe (eCAADe), vol 1, Novi Sad, Serbia, September 8–10, vol 39(1), pp 181–188
6. Green C (1979) Playing design games. JAE 33(1):22–26. https://doi.org/10.2307/1424460
7. Hirschberg UL, Schmitt G, Kurmann D, Kolarevic B, Johnson B, Donath D (1999) The 24 hour design cycle. In: CAADRIA '99, Shanghai, pp 181–190
8. Hirschberg UL (2020) Collaboration. In: Atlas of digital architecture: terminology concepts methods tools examples phenomena. Birkhäuser, Basel, pp 629–641
9. Jovanovic D (2016) Fictions: a speculative account of design mediums. In: Sheil B, Migayrou F, Pearson L, Allen L (eds). Drawing futures: speculations in contemporary drawing for art and architecture. Riverside Architectural Press, pp 28–33
10. Kuhn S (2012) Learning from the architecture studio: implications for project-based pedagogy. Int J Eng Educ 17(4, 5):349–352. https://www.ijee.ie/articles/Vol17-4and5/Ijee1214.pdf
11. Mareis C (2012) The epistemology of the unspoken: on the concept of tacit knowledge in contemporary design research. Design Issues 28(2):61–71. https://www.jstor.org/stable/41427826
12. Messini V, Minovski D, Strzelec D, Schwab D (2022) Safari (performed at the University of Innsbruck, Studio3). [Internet] 2020 [cited 2022 March 3]. https://wiedenski.org/safari/
13. Muñoz J (2017) Through a technique of building. Icon, vol 23 (2017), pp 83–112. https://www.jstor.org/stable/26454977
14. Oxman R (2008) Digital architecture as a challenge for design pedagogy: theory, knowledge, models and medium. In: Design studies [Internet], vol 29(2). Elsevier BV, pp 99–120. https://doi.org/10.1016/j.destud.2007.12.003
15. Picon A (2000) From "Poetry of Art" to method: the theory of Jean-Nicolas-Louis Durand. In: Durand J-N-L. Précis of the lectures on architecture. The Getty Research Institute; Los Angeles, pp 1–68

16. Polanyi M (2009) The tacit dimension. University of Chicago Press, Chicago
17. Ratti C, Claudel M (2015) Open source architecture. Thames Hudson, London
18. Sanchez J (2015) Temporal and spatial combinatorics in games for design. In: Computational ecologies, proceedings of the 35th annual conference of the association for computer aided design in architecture (ACADIA), Cincinnati, Ohio, October 19–25, 2015. pp 512–523
19. Collaborative SH, Processes D (1979) JAE 33(1):18–22. https://doi.org/10.2307/1424459
20. Witt A (2010) A machine epistemology in architecture. Encapsulated knowledge and the instrumentation of design. Candide. J Architect Knowl 03:37–88

Heritage Information Modeling: The Case of Chellah's Gate

Sharif Anouar[1](✉), Adam Anouar[2], and Ayoub Lharchi[3](✉)

[1] International University of Rabat (UIR), Salé, Morocco
sharif.anouar@uir.ac.ma
[2] Vacken, Rabat, Morocco
[3] Centre of Information Technology and Architecture (CITA), Copenhagen, Denmark
alha@kglakademi.dk

Abstract. This paper aims to propose an integrated workflow for the digitization of the built cultural heritage. To this end, we leverage the power of computational tools and the relevancy of Building Information Modeling (BIM) process to overcome the limitations and challenges faced by Scan-to-BIM. We describe the automatic generation of an as-built BIM model of a heritage building in a three-step procedure. Firstly, we outline the data acquisition method of the point cloud. Secondly, we describe the automatic processing and segmentation of the point cloud according to architectural elements using Machine Learning. Then, we tested and compared various meshing algorithms and utilized a combination depending on the desired level of details. Lastly, the resulting geometry is converted into a BIM object that will be subsequently semantically labeled. We used a UNESCO world heritage in Morocco—Chellah, as a case study to test the robustness of our protocol.

Keywords: Built cultural heritage · Digitization · Point cloud · Machine learning · Building information modeling

1 Introduction

Built heritage carries the legacy of humanity. It encapsulates the memory of the people, their accomplishment, and their aspirations. However, architectural heritage worldwide faces severe threats from anthropogenic hazards and natural disasters, making its preservation, documentation, and dissemination one of the challenges of our time [10]. The semantic digitization of the architectural heritage has provided an effective solution to this issue [14], while using remote sensing technologies in conjunction with BIM authoring tools (Scan-to-BIM) has become standard procedure [13].

In recent years, Building Information Modeling (BIM) has become widely accepted within the Architecture, Engineering, and Construction (AEC) industry [5]. When implemented correctly, it facilitates integrated workflows covering design, construction, and management processes [7]. Despite this integration in design practices, the use of BIM for remodeling and analysis of historic buildings remains an unexplored area [4]. The

P. F. Yuan et al. (eds.), *Hybrid Intelligence*, Computational Design and Robotic Fabrication,
https://doi.org/10.1007/978-981-19-8637-6_7

complex nature of such buildings and the logistical constraints that usually arise when addressing structural and preservation considerations. Therefore, existing BIM workflows for digitization hit their limitations. This is particularly the case in the generation of reusable models that feature different Levels of Details (LOD) and a high number of irregular geometries and patterns [3], which is typically the case in old construction. Although Scan-to-BIM workflow is well established, it still requires additional complex manual modeling, which is often time-consuming and prone to errors [17]. Besides, the BIM platform offers neither the possibility to process massive point clouds data nor advanced modeling tools for the high-fidelity reconstruction of the architectural artifact, which in return leads to the usage of several expensive specialized software.

To tackle these shortcomings, this paper will aim to describe an agile, zero data loss methodology to digitize the built heritage. Its novelty lies in the use of Machine Learning and a reduced set of software deployed under a BIM platform. The end goal is to achieve an as-built BIM model, robust and flexible enough to be used for further analysis and visualization scenarios.

2 Background

Heritage Building information modeling (HBIM) is a multidisciplinary and evolving system. It consists of a reverse engineering process based on creating parametric objects derived from remote survey data of existing constructions [11]. These objects represent architectural elements and hold comprehensive information, including geometrical, material, texture, and historical and structural data. As an application of BIM to heritage buildings, HBIM opens promising prospects for managing the ever-increasing complexity of the conservation practice [8]. Moreover, the understanding of undocumented heritage, its analysis, maintenance, restoration planning, and simulation, communication, among others, become then achievable.

Remote sensing technologies such as 3D scanning and photogrammetry are the foundation of HBIM. The procedure of transferring the acquired spatial data and its transformation into BIM models is commonly called Scan-to-BIM. This procedure is threefold: It starts with the survey and mass data capture. Then comes the processing phase, which encompasses point clouds cleaning, segmentation, and modeling. Finally, the last step involves the conversion of the generated geometry into BIM objects that will be subsequently semantically labeled [16].

While studies related to the survey side of Scan-to-BIM show established techniques and frameworks, research about the two following phases is growing, with a multitude of approaches being investigated. This interest is explained by problems inherent to the acquired point cloud, software used, and the BIM procedure.

The difficulties in the cleaning and segmentation of point cloud datasets have been addressed by employing a deep learning framework using either DGCNN (Dynamic Graph Convolutional Neural Network) for point-based approaches [12], voxel-based (Zhou and Tuzel 2018), or multiview-based methods [2]. Others worked on the automatic detection and segmentation of building components from the point cloud using the Hough transform algorithm [6]. In contrast, others investigated semantic-based methodology [9].

Recent studies have investigated the possibilities of (semi-) automating the geometric reconstruction of as-built models of built heritage. Some authors explored the parametric generation of NURBS-based 3d models from mesh [18], others focused on the automatic conversion of Terrestrial Laser Scanning (TLS) and Structure from Motion (SfM) point cloud data into textured BIM objects [1]. Despite the intense focus on automatic cleaning of noisy point clouds and the recognition and segmentation of architectural elements, there is, to the best of our knowledge, a lack of studies leveraging the power of Machine Learning (ML) in HBIM-based workflows.

3 Methodology

Building upon the standard procedure of the Scan-to-BIM process, we propose an integrated workflow for Heritage BIM (Fig. 1) enabled by ML.

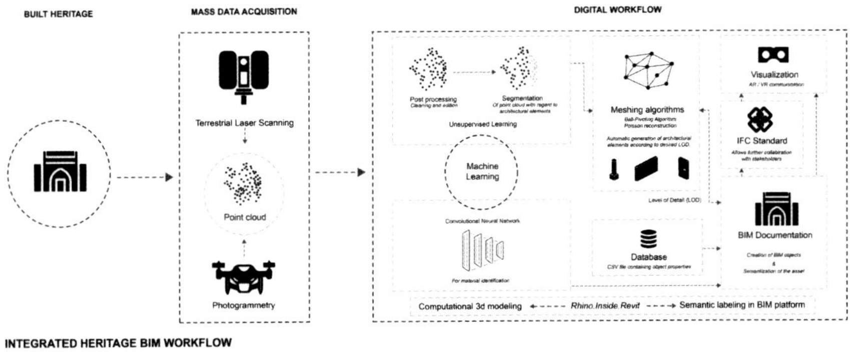

Fig. 1. Overview of the proposed workflow.

3.1 Case Study: The Main Gate of Chellah

The necropolis of Chellah (Fig. 2) is an iconic cultural site of the capital city of the Kingdom of Morocco. The oldest building in Rabat, was successively a Roman city, a Berber city, sometimes a capital, and finally a Marinid necropolis. Its rich heritage can be discovered through the multiple vestiges of the civilizations that succeeded each other over almost three millennia. Classified as a historical monument since 1913, it integrates in 2012 with a set of other remarkable buildings of the capital, the UNESCO world heritage list. The gate of Rabat's most venerable building seemed to us to be the most legitimate to inaugurate the phase of digitization of the architectural heritage, which is undoubtedly in Morocco, only at its beginnings.

3.2 Data Acquisition

The data acquisition was achieved using two different methods depending on the intervention scale. In the field of mass data acquisition for larger-scale artifacts, Terrestrial

Fig. 2. Chellah's gate.

Laser Scanners (TLS) appear to be the preferred choice. Despite being an expensive piece of equipment necessitating a high degree of expertise, TLS can output highly accurate spatial data. The information requirements for the project determined the required scan data quality. A LOD of 500 has been chosen to facilitate the automatic reconstruction of the architectural artifact in the following stages of the workflow.

The survey was conducted with a FARO Focus M 70. After setting the resolution parameters, the scanning speed, and precision, multiple scans of Bab Chellah have been performed from different perspectives. This strategy produces overlapping point clouds data yet avoids the problems of occluded areas. Unfortunately, the cumulative point clouds that were obtained were not usable. Too much noise was picked and due to the current restoration works in Chellah, we could not perform additional scans to improve the quality of the point cloud. The resulting point clouds were not pre-processed using commercial software like FARO Scene,[1] Leica Cyclone[2] or RecapPro[3] but directly transferred to Rhinoceros 3D. On a smaller scale, we used a Microsoft Kinect Azure (MKA) to create 1-2 m scan section of elements of interest that were further processed and segmented (Fig. 3).

1 https://www.faro.com/fr-FR/Products/Software/SCENE-Software.

2 https://leica-geosystems.com/products/laser-scanners/software/leica-cyclone.

3 https://www.autodesk.com/products/recap/overview.

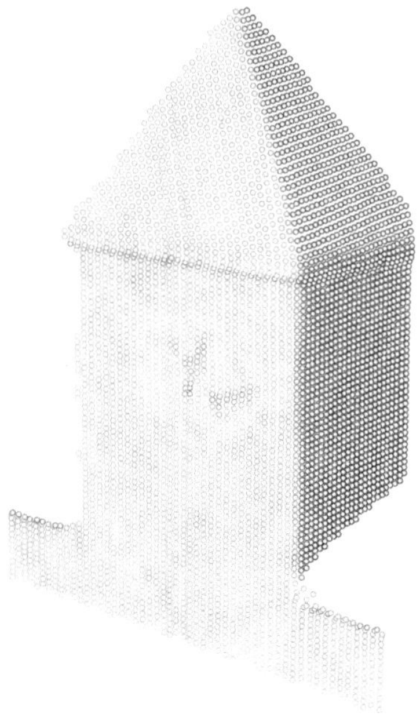

Fig. 3. Small scale point cloud obtained using a MKA and a laptop.

3.3 Processing and Segmentation Using RANSAC

There are many algorithmic methods for grouping point cloud data. In the case of Chellah, two parameters must be considered. The nature of the data to be grouped and the size of the project.

The cloud point here is an unsorted set of unlabeled data; therefore, the clustering method is preferred to the classification method, which is not feasible. The size of the Chellah model to reconstruct is relatively small, and it can be subdivided into a not very high number of primitive geometries, in our case, 2D planes. For this purpose, several methods are available to us: First, we discard kernel methods like k means, etc., they have the disadvantage of assuming somewhat rounded cluster shapes, and above all, they require giving the number of clusters as input. Geometric diversity, even in the smallest and simplest projects, prohibits us from doing this, which would significantly reduce the automation of the cloud point segmentation process. The unsupervised algorithm to group unlabeled data will also have to do without the number of clusters as input. In addition, the algorithm must not be density-based because the clusters to be produced will be adjacent (the faces being adjacent to each other). Therefore, there will be no gaps separating the different groups. The clustering algorithm must therefore be able to detect the primitive shapes. Thus, the final choice was between the clustering Region Growing approach and the Model Fitting method. We preferred the latter and especially the Random Sample Consensus (RANSAC) method because it works well with noisy data, unlike the former, which requires a preliminary cleaning step. Concretely, we imported the Point cloud in Grasshopper[4] platform and used the RANSAC component of its free plug-in Cockroach [15] for the detection and clustering of the plane distribution of the point cloud.

3.4 Mesh Generation Procedure

To generate a usable 3D mesh from the collected set of point clouds, we used three different algorithms:

- **Ball-Pivoting Algorithm (BPA):** It is based on the use of a "Virtual Ball" that is rolled across the point cloud with a variable diameter. Ideally, the diameter should be larger than the average distance between the points. With continuous rolling and pivoting, triangles are generated and finally combined to constitute the mesh.
- **Poisson Reconstruction (PR):** This method tries to envelop the point cloud with a "cloth" that keeps shrinking depending on the depth parameter.
- **MeshFromPoints (MFP)** feature included in the modeling software (Rhinoceros 3D).

For the first two algorithms, we used the open-source Python library "Open3D" [19]. Since our modeling software does not support natively using CPython libraries, we developed an intermediate external script, that was capable of loading the XYZ file format that contains the points coordinates and the faces normals. Once loaded, the point cloud was meshed, and then reimported within Rhinoceros 3D. It is critical to note that extensive testing was necessary to optimize the script and find the correct parameters for each algorithm to obtain a valid mesh with minimal artifacts (Fig. 4).

Fig. 4. Comparison of the obtained mesh with the 3 meshing methods. From Left to Right: BPA, MFP, PR.

While the MFP might seem to be the best option, tests showed that the PR method is the most effective as it produces satisfactory results, much faster, and is the least affected by noise in the input data. Although PR is generating an irregular mesh topology, it was

[4] https://www.grasshopper3d.com/.

trivial to simplify it and regenerate different topologies (Fig. 5), which is a necessary step for further BIM element generation.

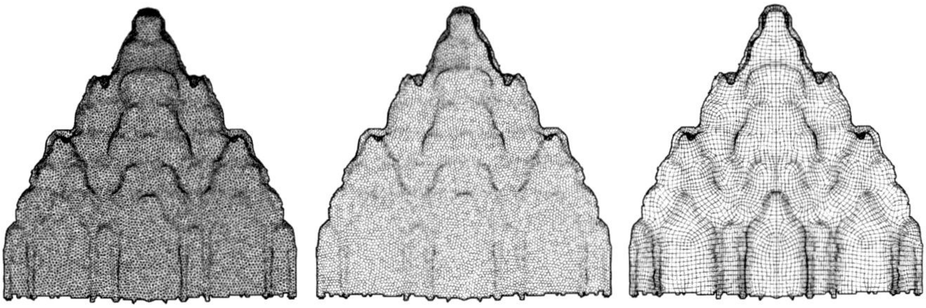

Fig. 5. Processed results. From left to right: Triangulated, Dual Ngon, Quadrated meshes.

3.5 BIM Documentation

All the meshes produced already form a faithful representation of the model. However, to transfer it to Revit and develop an informed model, it is necessary to create a family. To do this, the mesh must be converted into a NURBS surface since the families on Revit only accept closed poly-surfaces. To this end, the first code in Grasshopper deconstructs each mesh into its many points and boundaries and then reconstructs it into a single surface, or at least into a minimum of surfaces. The process is then applied in a loop to all the meshes representing all the faces of the model. Among the different mesh typologies created and tested, the unstructured triangulated mesh is preferred because triangles surfaces are planar, and Revit has a greater tolerance to planar surfaces, their insertions in families, and their conversions into walls and other native Revit elements. Moreover, Revit is sometimes quite reticent and unpredictable to accept surfaces created with Rhinoceros 3D and Grasshopper, even when they are qualitative and well defined. This matter is explained by the difference the two software have in handling tolerances; where Rhinoceros 3D can go to tolerances of 0.001 mm, Revit can only accept 1-mm tolerance.

Each project has its own physical and material characteristics that distinguish it, both in size and complexity of form, requiring a particular and adapted approach. Therefore, the following approach is not intended to be the universal method to be adopted but the best option available to us in the context of this particular project. Starting from the characteristics of the heritage artifact that is the Chellah Gate, it appeared to us that the best approach to follow would be to algorithmically create a mass via the "conceptual mass" family template using the Rhino.Inside plugin, through which the surfaces, thickened on occasion so that they become solids with a thickness, will be transferred to Revit. Inside Revit finally, a code on Dynamo; the visual programming language of Revit, will take care of generating via the "wall-by-face" tool, the walls from the faces of the conceptual mass generating the model of Bab Chellah with Revit native elements. The door itself was modeled using. Each element is finally generated according to its nature (the wall modeled with the wall tool, the door created with the door family etc.).

Regarding the LOD (levels of detail), dealing mainly with the issue of the graphic level of detail (representations of muqarnas, low relief frescoes on walls), it was decided that they will be generated via in-place family, categorized as a wall, they will be produced according to the same process as for the walls, namely, by creating a conceptual mass, and its automatic export from Grasshopper to Revit via a code using the plugin Rhino.Inside.Revit, and thereby they will be placed within the in-place family as a component.

Separating the walls from their ornamentation allows a practical and straightforward management of the LOD at the time of the export and exchange of information by the IFC, according to what is defined by the "information requirements" (Figs. 6 and 7).

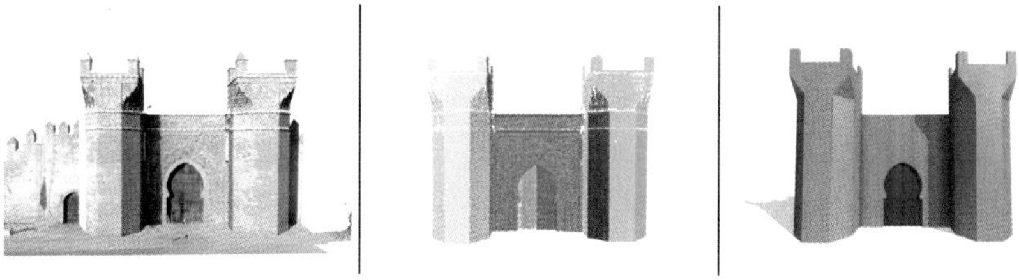

Fig. 6. Three stages of Chellah's gate reconstruction.

From left to right: Raw point cloud, clustered point cloud, BIM model LOD 100.

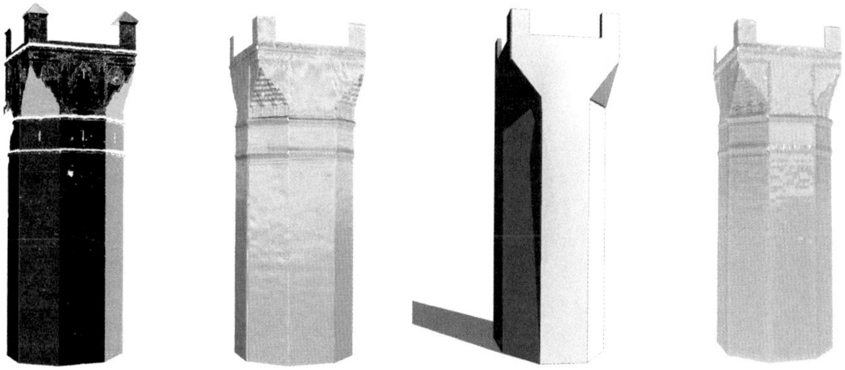

Fig. 7. Chellah tower. From left to right: Segmented point cloud, mesh reconstruction, BIM model LOD 100, BIM model LOD 400.

Finally, given the quality of the point cloud and the complexity of the architectural elements, it was decided to reconstruct them from the existing survey as here in the case of the muqarnas (Fig. 8).

The level of detail of the created mesh is directly correlated to the resolution of the point cloud. The survey should represent the existing building; the maximum LOD corresponds to the LOD 500 ''as Built". To propose different levels of detail (and therefore necessarily lower, LOD 400, 300, 200, etc.), we can play on the resolution of the generated surfaces by reducing it at the time of the conversion of the mesh into the surface. This produces simplified surfaces given with washed-out ornamentation, allowing a saving of computational resources when a high level of detail is not necessary.

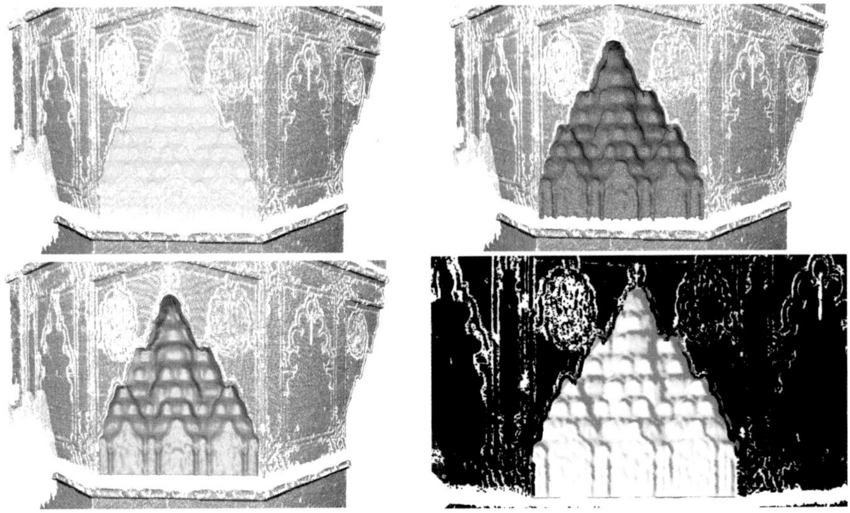

Fig. 8. Muqarnas detail reconstruction (from top left to bottom right): Point cloud segmentation, mesh generation, NURBS-based surface conversion, Revit in-place family (wall).

Afterwards, a text-based CSV file (Comma-Separated Values), containing the mechanical properties of the architectural elements is linked to the BIM model to automatically label its different parts.

4 Results

The outcome of this work is an attempt to automate the process of Scan-to-BIM while reducing both the number of software involved and the time spent on cleaning, segmenting, modeling and labeling based on an initial point cloud. Furthermore, the workflow proposed based on Rhinoceros 3D-Revit maintained robust interoperability, allowing the designer to have complete control of the overall process and giving the possibility of using Machine Learning to speed up the process while reducing human involvement. Finally, the compatibility and flexibility of the tandem Rhino-Revit linked by Rhino.Inside through Grasshopper platform enables flexible strategies for transferring more or less detailed geometry depending on the LOD required and needed for the IFC export.

5 Conclusion and Discussion

HBIM is gaining momentum within the AEC industry. Although studies focusing on the management of point cloud data or the automatic reconstruction of architectural heritage exists, there is still a lack of an exhaustive and comprehensive framework for HBIM. This paper tries to narrow this gap by proposing an integrated workflow relying mainly on close interoperability between a BIM platform and a 3D modeler enhanced by ML methods. This combination showed to be a powerful tool for the culturally built heritage.

Future work could expand on the reconstruction of other building types. It might as well integrate different phases into the workflow, such as fabrication. Besides, 3D laser scanning being expensive, we expect to experiment with photogrammetry using DSLR cameras and smartphones mixed with free, open-source software to propose an Open-HBIM workflow that can be vendor agnostic.

References

1. Andriasyan M, Moyano J, Nieto-Julián JE, Antón D (2020) From point cloud data to building information modelling: an automatic parametric workflow for heritage. Remote Sens 12:1094. https://doi.org/10.3390/rs12071094
2. Boulch A, Saux BL, Audebert N (2017) Unstructured point cloud semantic labeling using deep segmentation networks. In: Eurographics workshop on 3D object retrieval, p 8. https://doi.org/10.2312/3DOR.20171047
3. Brusaporci S, Maiezza P, Tata A (2018) A framework for architectural heritage Hbim Semantization and development. In: The international archives of the photogrammetry, remote sensing and spatial information sciences. Presented at the ISPRS TC II Mid-term symposium towards photogrammetry 2020 (vol XLII-2) - 4;7 June 2018, Riva del Garda, Italy, Copernicus GmbH, pp 179–184. https://doi.org/10.5194/isprs-archives-XLII-2-179-2018
4. Capone M, Lanzara E (2019) Scan-to-Bim Vs 3d ideal model Hbim: parametric tools to study domes geometry. In: The international archives of the photogrammetry, remote sensing and spatial information sciences. Presented at the 8th international workshop 3D-ARCH 3D virtual reconstruction and visualization of complex architectures - 6;8 February 2019, Bergamo, Italy, Copernicus GmbH, pp 219–226. https://doi.org/10.5194/isprs-archives-XLII-2-W9-219-2019
5. Chen, Chao & Tang, Llewellyn. (2019). Development of BIM-Based Innovative Workflow for Architecture, Engineering and Construction Projects in China. International Journal of Engineering and Technology. 11. 119-126. 10.7763/IJET.2019.V11.1133.
6. Díaz-Vilariño L, Conde B, Lagüela S, Lorenzo H (2015) Automatic detection and segmentation of columns in as-built buildings from point clouds. Remote Sens 7:15651–15667. https://doi.org/10.3390/rs71115651
7. Eastman, C., Eastman, C. M., Teicholz, P. et al. (2011). BIM Handbook: A Guide to Building Information Modeling for Owner, Manager, Designer, Engineers and Contractors (p. 243). Hoboken, NJ: John Wiley & Sons.
8. Gigliarelli E, Calcerano F, Cessari L (2017) Heritage Bim, numerical simulation and decision support systems: an integrated approach for historical buildings retrofit. Energy Proc 133:135–144. https://doi.org/10.1016/j.egypro.2017.09.379
9. Hmida HB, Cruz C, Boochs F, Nicolle C (2013) From 3D point clouds to semantic objects an ontology-based detection approach. arXiv:1301.4783 [cs]
10. Holman N, Ahlfeldt G (2015) No escape? The coordination problem in heritage preservation. Environ Plan A 47:172–187. https://doi.org/10.1068/a130229p
11. Murphy M, McGovern E, Pavia S (2009) Historic building information modelling (HBIM). Struct Surv 27:311–327. https://doi.org/10.1108/02630800910985108

12. Pierdicca R, Paolanti M, Matrone F, Martini M, Morbidoni C, Malinverni E, Frontoni E, Lingua A (2020) Point cloud semantic segmentation using a deep learning framework for cultural heritage. Remote Sens 12:1005. https://doi.org/10.3390/rs12061005
13. Rocha M, Fernández F (2020) A scan-to-BIM methodology applied to heritage buildings. Heritage 3:47–67. https://doi.org/10.3390/heritage3010004
14. Salvador García E, García-Valldecabres J, Viñals Blasco MJ (2018) The use of Hbim models as a tool for dissemination and public use management of historical architecture: a review. Int J SDP 13:96–107. https://doi.org/10.2495/SDP-V13-N1-96-107
15. Vestartas P, Settimi A (2020) Cockroach: a plug-in for point cloud post-processing and meshing in Rhino environment. EPFL ENAC ICC IBOIS.
16. Volk R, Stengel J, Schultmann F (2014) Building information modeling (BIM) for existing buildings — literature review and future needs. Autom Constr 38:109–127. https://doi.org/10.1016/j.autcon.2013.10.023
17. Wang Q, Guo J, Kim M-K (2019) An application oriented scan-to-BIM framework. Remote Sens 11:365. https://doi.org/10.3390/rs11030365
18. Zheliazkova M, Naboni R, Paoletti I (2015) A parametric-assisted method for 3d generation of as-built Bim models for the built heritage. Presented at the STREMAH 2015, A Coruña, Spain, pp 693–704. https://doi.org/10.2495/STR150581
19. Zhou Q-Y, Park J, Koltun V (2018) Open3D: a modern library for 3D data processing. arXiv: 1801.09847 [cs]

Cellular Automata, Memory and Intelligence

Alberto Fernández González[1,2(✉)]

[1] UCL The Bartlett School of Architecture, UCH Faculty of Architecture and Urbanism, 22 Gordon Street, London, UK
alberto.fernandez.11@ucl.ac.uk
[2] 84 Portugal, Santiago, Chile

Abstract. Understanding memory as the faculty by which a system stores and remembers information from the past to a new purpose with shapes that are emerging as "collective designs"(a repository of built information), this research works with the demonstration in how CA can generate a trace of its existence as memory based on the activation and deactivation of the discrete system in which grows, like a footprint in the affected area of intervention, improving a "stigmergic operation" in the field, conditioning the following steps in the collaborative growing of this basal structure. Based on sets of digital experiments, a set of CA using Langton Ants generates different solutions based on the activation and deactivation of rules according to information coming from patterns, creating spatial solutions that deal with built memory three-dimensional emergent structures.

Keywords: Cellular automata · Memory · Intelligence · Digital design · Generative design · Digital theory

1 Introduction

In nature, we can find some examples of bottom-up design application in contrast to a top bottom classic design approach, creating shapes that are complex but efficient at the same time, dealing with problems collaboratively with optimal use of available resources and a maximum performance related with shapes that are emerging as "collective designs" being the fruit of this collective interaction (Fig. 1). Examples of this design way from nature are, for instance, social insects like Ants that are building intricate underground structures that are built by generations of coordinated builders through the whole structure, which acts as collective memory (as a catalyst of development, brings the opportunity to any system natural or artificial to retain information from past events to possible new steps). This structure (as a repository of built information) continuously optimizes resources by the feedback action from the memory storage to a new action or movement, working as a predictive manual of probabilities instead of a rigid process of rules (deterministic).

This process is the first step through which any organism can learn, as a set of continuous improvements over existing reality. The previous aspect (as a predictive model) can develop a growth strategy as a target to follow from the starting point or state 0. The learning process can never end if the environment is dynamic and in constant change,

P. F. Yuan et al. (eds.), *Hybrid Intelligence*, Computational Design and Robotic Fabrication,
https://doi.org/10.1007/978-981-19-8637-6_8

generating a continuous review in the whole generative process. If the environment does not have changes due to external or internal actions (by the action of exogenic elements like whether or endogenic like our system immerse in this context), the system can not evolve, going into a dead state due to lack of stimulus, entering in a close loop (repeating the previous step deleting the last one).

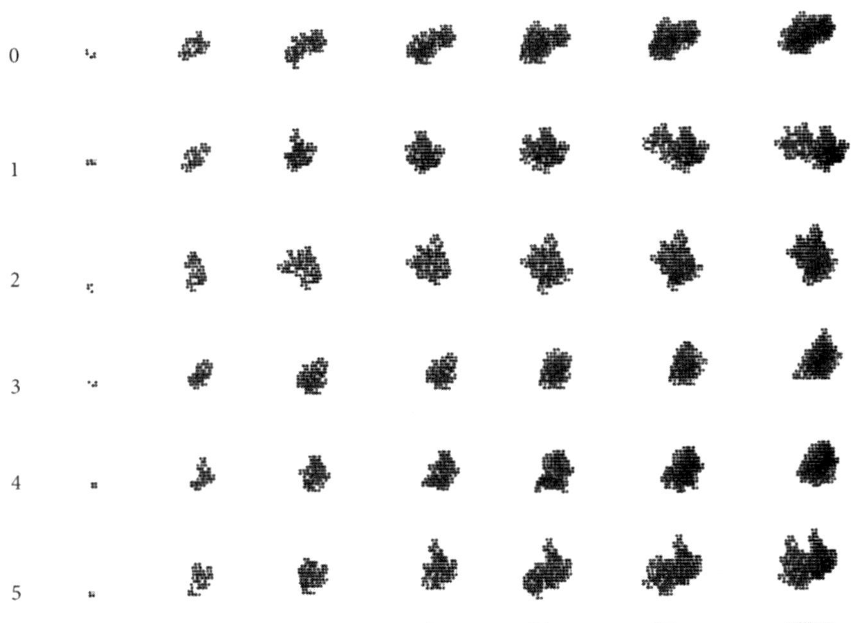

Fig. 1. 3D movements and five 4D generative variations using 6 directions (x, y, z, −x, −y, −z) based on a time line, recording the previous steps as visible generative memory. A discrete set of 3D rules working in a 3D space can build a 4D movement map. Image from the author

2 Memory and CA

Understanding memory [1] as the faculty by which a system stores and remembers information from the past to a new purpose 1. In the architecture field, this concept can be translated as a design that can store information and process it, with the possibility to evolve from an initial state 0 to a new stage: reading, learning and applying from its context of development. In that sense, memory can be used as a design tool, learning how to create a specific scenario of evolution compatible with the project nature. In this context, it is necessary to store the information of each evolutionary step for newer generations of growing and, following this methodology; architecture can iteratively evolve into a complex system being itself its own memory.

In that context, we can find cellular automata as one of the complex systems that can create shapes from 1D, 2D and 3D by the application of these bottom-up design principles, in where at the end, our role as designers is out of the final object, instead of that we can design the small relationships between every element of the system, discovering our final results as part of a try and error debugging process of our designs.

CA as a generative physical simulation system works with time, in which past, present and future are agents of redesign, working in the simulated nature as associates, where the feedback becomes from the future to the present as information processed, developing structures and patterns over and over again as a no-end cycle of design, being at the end, the synthesis of memory in the system. Each layer of information (from the base to the top) need to be recognized easier by the new generations in the system; as it is easy to recognize layers of growing in a tree, slicing its trunk, reading each step of development from the begging to the end, as well as in other example, crystals can create amazingly complex and intricate forms, which a clear starting point and a clear end, when the system can not grow more due to the lack of conditions to develop more steps, following the original expansion rules.

CA works like an enormous computing architecture [2] that can scale to high-resolution patterns, based on finite-state machines (also named finite-state automaton, it is a computation model that simulates a sequential logic), arranged in a discrete network, allowing local interactions with its neighbors [3]. These geometric orders (coming from crystal growing simulations) are the base in which the atoms are arranged in nano-metric scale by bounding phenomena (atomic stickiness), from pure orders in which the inner lattice between atoms is perfect (crystalline) to small crystals arranged in different directions (polycrystalline) to solids without arrangement (amorphous). Any of these kinds can coexist, building intricate shapes, increasing resistance to stress if more complex is the alignment between atoms. Understanding the previous examples, it is possible to design new rules, creating patterns into specialized data fields, using criteria of morphology, physiology, anatomy, behavior, origin, and distribution as integrated rules in the design and its relations with contextual changes.

The lattice and array rules can simulate these complex and hi-resolution geometries, extrapolated from this natural world to the design approach, understanding similarities in the representation language (graphic code) with other rules from the complex world of Escher [4] and the geometrical theories of William Huff [5]. In both references, the authors read the complexity as a set of rules, retaining the initial form information, from simple to high-density patterns to where any shape can go.

The first set of rules from these geometrical transitions are one of the initial keys in the growth as an evolutive design from an existent geometry (in this case, an origin A) to a possible destiny (the evolution to a B shape). According to Hofstadter [6], the transformation rules in a 2D deformation device are: *lengthening or shortening a line; rotating a line; introducing a "hinge" somewhere inside a line segment so that it can "flex"; introducing a "bump" or "pimple" or "tooth" (a small intrusion or extrusion having a simple shape) in the middle of a line or at a vertex; shifting, rotating, expanding, or contracting a group of lines that form a natural subunit; and variation of these themes... "a line" or "a vertex" is actually a reference to a line or vertex inside a unit cell, and therefore, when one such line or vertex is altered, all the corresponding lines or vertices that play the same role in the copies of that cell undergo the same change.*

We need to understand that points, lines, and containers (a unit cell as original) are the minimal geometrical units for a transformation in the parquet systems rules. The original shape is the masterpiece in all the evolutive logic because it contains the starting set of movement. It is possible to go from one origin to a different destination or vice versa. Using this method, it is feasible to achieve a high level of intricacy in the geometrical progression, understanding the possible movements (rotations and translations), as generative rules with a certain number of variations and steps of evolution, as an incremental factor of complexity (Fig. 2).

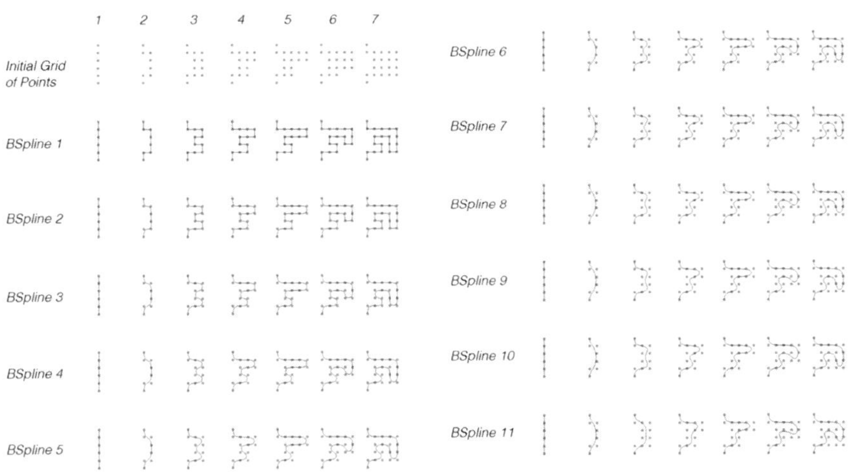

Fig. 2. Discrete evolutive rules in a 2D grid applied to a parquet system using a set of 7 evolutions and 11 BSpline types, Based on Kraig S. Kaplan works. Image from the author

Based on the previous concept, it is possible to understand how powerful this tool is to achieve memory capacity in architecture, using the topological approach as the container of pre-existent shapes, colors and textures, being feasible to adapt these geometries to a new evolutive scenario. Following these logics, any pre-existent shape can be used as initial material (such as an in-material building block) only if the next steps in the evolutive result are coherent with the first topological container, generating a dynamic structural progression as the prime structure for the whole system.

Nature always tries to find the most efficient shape available for each specific problem; in this case, shapes, as a result of memory accumulated in the system, are answer redefined in each moment, being perfected over time. Each growing pattern identified from the natural kingdom has a strong relationship with the geometrical world, matching many times with formulas that explain the procedure which made possible the existence of this shape in the natural world: "How these natural patterns are constructed may tell us something about how the far more complicated forms of animals and plants are created by a progressive division and subdivision of space, orchestrated by nothing more than simple physical forces".

Linking this geometrical approach with the natural abstraction of minerals (one of the original ideas from John Von Neumann), it's possible to cross these two "different universes" in the same proposal. As the first point, we need to understand each tessellated geometry from the mineral arrays as a tridimensional "topological container" which is designing the initial set of rules of the artificial growth in the project. Secondly, from the site, each container can reload specific information to be evolved inside the growing, as a memory heritage for the following steps in the system. Points and lines are unable to store information without coordinated work as growing sequences. With this collective skill, they can reach the ability to store memory as a graphical map with clear coordinates, a first step to translate these generative shapes to a real project.

These shapes are linked with a specific function in the environment, evolving from circles and spheres (as the most efficient form) to triangles, squares, hexagons and many others by necessity, only looking at structural competence according to specific problems (like a honeycomb based on hexagonal patterns). For the previous reason, it is possible to argue that, from the same shape, it is feasible to obtain different results changing only a few factors, in many cases linked with probabilities and combinatory.

3 Self-intelligence

CA is linked with [7] artificial intelligence (AI), in particular with the simulation origins of artificial life, is one of the most representatives self-organizing artificial systems, together with neural networks and genetic algorithms, working as a structural analogy of intelligence, coming from large systems developed with simple elements having only local interactions, (a seed interacts with its own neighbors, which are usually just the cells closer to the seed as an activator), like crystal growing elements from a saturated media, working as a self-organized system [8], highly coherent in the whole generative structure, showing patterns and behaviors in a decentralized non-controlled system (Fig. 3).

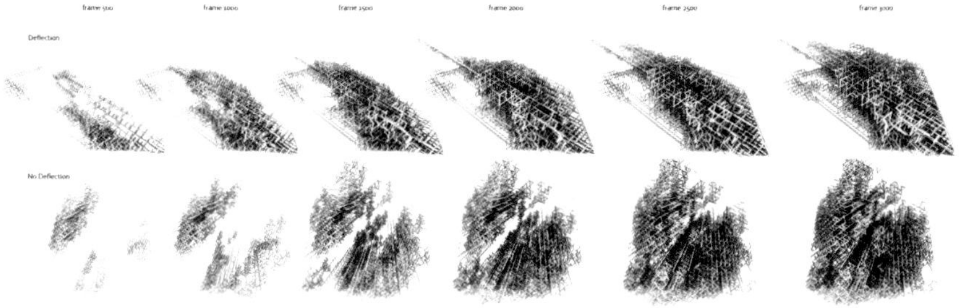

Fig. 3. Discrete CA rules in a 3D grid applied to a Basic Langton's Ant, A seed as constant activator between 0 and 1 and from 1 to 0 as Asynchronous CA (where only one cell is active at each time step, and the transition rule specifies the fate of the activation) Image from the author

The architectural body has changed from a static matter to a dynamic element, being possible to imagine a system that can learn and read complexity from the 4D space in which they are "growing", assuming movement and evolution as inherent parts in the design project. This concept of "live organism" has been defined as "hyper-body" [9], in which the building is a live system with capabilities of evolution, change, adaptation and interaction in where any form must have a certain degree of memory retention, recognizing its origin and proposing new spaces based on this concept of dynamic space deformation that can be understood as contextual project adaptation.

By convention, self-intelligence [10] is the ability to acquire and apply knowledge and skills by itself, belonging to this the artificial intelligence meaning, in where is possible to identify two different kinds: the vertical intelligence (from the top to the bottom of the system) and the collective intelligence (well known as bottom-up). The first one is highly structured and hierarchical by concept, which means that some control systems can take decisions over the whole components, affecting the total performance in a non-democratic action (as a god finger controlling everything) or a panoptic system all at the same time). On the other hand, the collective intelligence works as a self-generated organization, composite by agents as minimum organizational structure, by which the information shared between each member as local or collective trace can produce a highly diverse and complex result, only using simple rules shared with one or more system members. In this second system, the result could be read as a graphic or constructed historical registry of events, which at the end is the collective memory, built layer by layer in the life sequence. The sequence can be read by old or new agents, being reloaded for the following growing generations. The adaptability mechanism is founded in the system flexibility (collective intelligence) instead of the rigid organizational structures (vertical organization rules). In the end, adaptability is the catalyst of evolution for live organisms, from a 0 state to the next step in a generative way over the evolution of all its components which are interaction by simple local relationships, emerging at the end complexity behaviorally and spatially.

In that sense, this research works with the demonstration in how CA can generate a trace of its existence as memory [11] based on the activation and deactivation of the discrete system [12] in which grows, like a footprint in the affected area of intervention, improving a "stigmergic operation" [13, 14] in the field, conditioning the following steps in the collaborative growing of this basal structure (Fig. 4).

In synthetic systems (non-natural), memory has been related to how artificial structures are connecting current and new behaviors with old experiences, dividing the memory into two main groups: Short Term STM (containing the working memory) and Long Term LTM (collecting declarative and procedural memories). These memories are implicit or explicit, such as Ants, Bees, Termites, in which many times the STM records in the LTM during the time. Understanding that the working memory has an individual effect in each subject belonging to a colony, the constant repetition of behaviors can record a collective LTM, generating a spontaneous Procedural Memory, which is, by definition: non-conscious accessible memories like habits behaviors and stimulus–response conditioning. This kind of memory has a wide application in computer-simulated systems, firstly like Ruled-based systems, where explicit computational constructs encode a process, operation, or temporal experience. Secondly, the online learning of artificial neural networks is based on experience in neuroscience-inspired architectures. Thirdly,

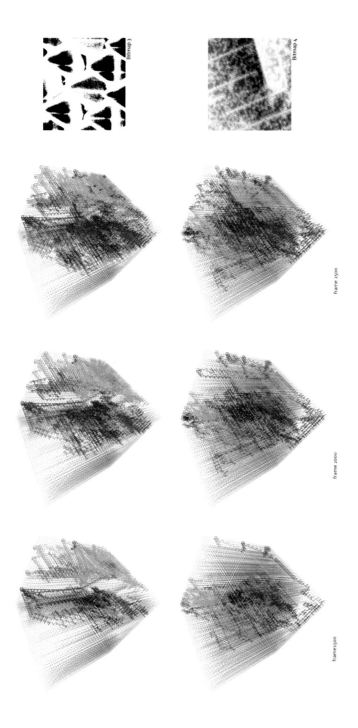

Fig. 4. Discrete CA rules in a 3D grid as a stigmergic operation working on an environment created by grayscale images. The red structures follow the structures from the different previous generations as "memory", optimizing the steps from the bottom-up growing direction. Image from the author

developmental robotics models, particularly systems for acquiring sensorimotor coordination, are relevant given the explicit necessity of experience. In the case of these experiments, the generated structure works as an LTM for each Asynchronous CA seed, reading the previous steps as "paths" to follow. The CA seed is programmed as a "map reader", expanding the traditional range of neighborhood interaction; usually, one additional cell surrounds each seed according to Von Neumann and Moore models (Fig. 5).

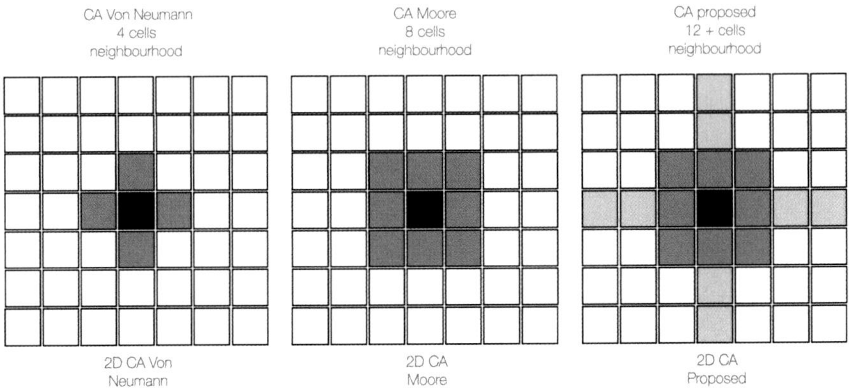

Fig. 5. Von Neumann, Moore and Proposed CA neighborhood interaction model from a CA seed

Based on sets of digital experiments, a set of CA based on Langton Ants [15] as an asynchronous model, together with the expanded neighborhood criteria, the CA system can create a final approach in which the activation and deactivation of rules according to information coming from patterns, can effectively build spatial solutions that deal with built memory three-dimensional emergent structures (Fig. 6) as a coordinate system in which is possible to create certain levels of interaction between "seeds", adding to the system some conditions of Separation, Alignment and Cohesion in a primitive way [16]. In this sense, the assignation of intelligence is quite near to the system capacity to apply the knowledge based on previous experiences in ongoing or further scenarios, no matter if the structure has a large number of different experiences accumulated along with a range of time, the important are principles running and how they are threading relations between two or more agents in the whole organization, as the rules working in a swarm: move in the same direction as your neighbors, remain close to your neighbors, avoid collisions with your neighbors.

On these experiments (as artifice life digital simulations), memory and self-intelligence can be appreciated in similar ways as in the termite mound plaster negatives of Dr. Turner [17] as a highly complex structure, generated by the action of termites (as collective intelligence) working with a simple implicit set of rules linked with specific features of their environment. Stigmergy is the indirect communication taking place among individuals in social insect societies [18], has the labor of the main generator of collective intelligence, as a repetitive work along centuries, generating pheromone traces, which allows the communication by trace, producing specificity in the colony group members, and building the history layer by layer, generation by generation. In

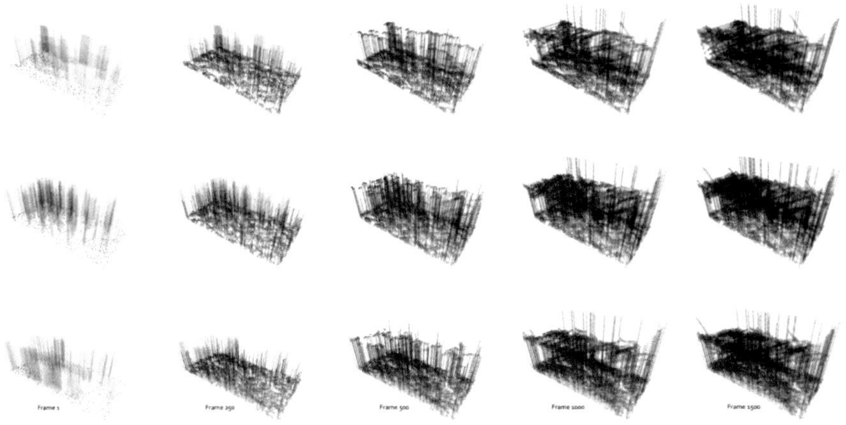

Fig. 6. Proposed CA neighborhood interaction model from a CA seed adding separation, alignment and cohesion on a 3d environment

that sense, the CA generated lattice structure works as a communication media between seeds and collective memory and crowd Intelligence simultaneously.

4 Conclusions: CA as Bottom-Up Architecture with Memory

Memory has more than only a unique dimension from which the architectural design can grow towards a new dimension, less deterministic and more responsive to new possible scenarios. As a design tool for projects, this concept can bring coherence to the broken link between design and the built environment in the fine grain of the neighborhood, learning how the geometrical structure can learn from the urban data field as useful generative information for the system, in a coherent way.

Learning from natural systems, it is possible to understand how the shapes are arranged in the environment, being adaptive and highly scalable due to efficient subcomponents as fundamental building blocks of growing (primary forms). Applying geometrical rules from these previous steps has been possible to develop small structures to read, learn and create strong support for the design project. Using points and lines as a geometrical base is suitable to merge the worlds of nature and science with architecture and design, making it necessary to review our design protocols, changing the classic disciplinary conception to a new model of design, with fewer preconceptions and more perceptions (capturing and analyzing all the available dimensions of information from our project site).

Every new step in the growing process can increase the result definition, adding frames to project development. This is the key in the system simulation: more time = more quality = more accuracy. On the other hand, it is impossible to accelerate the final solution only by increasing the number of growing seeds in the space because more elements can exponentially increase the number of interactions in the site as an evolutive mainframe, distorting the final solution.

CA can generate a trace of its existence based on the activation and deactivation of the discrete system that grows, like a footprint in the affected area of intervention, improving a "stigmergic operation" in the field, conditioning the following steps in the collaborative growing of this basal structure. This cooperative way of space construction was and is one of the essential elements in this research; without this, the ever-increasing by the collective action of seeds could be unfeasible, without historical registry of expansion or contractions in the whole system, and at the end with less or no coordination between cells as a choreography of termites building a termite mound. As pseudo-natural forms, this way of design has memory in itself, storing its historical evolution, feeding the following steps on this system, conditioning the use of rules and paths inside the project development, focused on design data field conditions instead of only one unique result.

Maps of movement and evolution are the hidden information of our eyes, which can be switched from negative to positive as a structural result (points as articulations and lines as connections), being a final link between virtuality (the hidden) and reality (the visible). From the built environment reality, points, lines and shapes are how the city infrastructure dialogues with itself and the environment in a broad sense. A field condition ruled more than only translated by these simple geometrical elements. Finally, we can understand that memory and Intelligence are working together in a distributed way, instead just only as a concentrated model with one input and a set of possible outputs, like models based on machine learning that are working similarly to natural neural systems and the way biologic neurons convey information by integrating incoming inputs. CA can be referred to as self-Intelligence organization, to the overall coherence of the generative structure, displaying patterns and behaviors in a decentralized, non-controlled system, in which complex systems are constructed using essential pieces with just local interactions as a structural analogue, building Intelligence in the process.

References

1. Definition for memory-Oxford Dictionaries. http://oxforddictionaries.com/definition/memory
2. Wolfram S (2002) A new kind of science. Wolfram Media, Champaign, IL
3. Hoekstra AG, Jiri K, Peter S (eds) (2010) Simulating complex systems by cellular automata. Springer Nature, Cham
4. Escher MC (2016) The graphic work: 1898–1972. Taschen Editors
5. Hoeydonck W (2019) William Huff's parquet deformations: a Viennese experiment. In: Conference: symmetry: art and science. 11th congress and exhibition of SIS special theme: tradition and innovation in symmetry, Katachi Kanazawa, Japan
6. Hofstadter D (1985) Metamagical Themas questing for the essence of mind and pattern. Basic Books; Book Club (BCE/BOMC)
7. Definition for Artificial intelligence-The Turing archive for the history of computing. http://www.alanturing.net/turing_archive/pages/reference%20articles/what_is_AI/What%20is%20AI09.html
8. Bonabeau E, Dorigo M, Guy T (1999) Swarm intelligence: from natural to artificial systems. Oxford University Press, New York, N.Y.
9. Oosterhuis K (2003) Hyperbodies: towards an e-motive architecture/Kas Oosterhuis. Birkhäuser, Basel; Boston; Berlin

10. Wang P (2007) The logic of intelligence. In: Goertzel B, Pennachin C (eds) Artificial general intelligence. Cognitive technologies. Springer, Berlin, Heidelberg. https://doi.org/10.1007/978-3-540-68677-4_2
11. Von Neumann J, Burks AW (1966) Theory of self-reproducing automata. University of Illinois Press, Urbana
12. Frazer J (1995) An evolutionary architecture. Architectural Association, London, England
13. Wood R, Paul B, Tony B (2012) A review of long-term memory in natural and synthetic systems. Adapt Behav 20(2):81–103
14. Meyboom A, Reeves D (2013) Stigmergic space, ACADIA 13: adaptive architecture [Proceedings of the 33rd annual conference of the association for computer aided design in architecture], Cambridge, pp 200–206
15. Langton CG (1996) Artificial life, the philosophy of artificial life. Oxford University Press, New York and Oxford, pp 39–94
16. Reynolds C (1987) Flocks, herds and schools: a distributed behavioral model. In: SIGGRAPH '87: proceedings of the 14th annual conference on computer graphics and interactive techniques. Association for Computing Machinery, pp 25–34
17. Turner S (2011) Termites as models of swarm cognition. Swarm Intell 5:19–43. https://doi.org/10.1007/s11721-010-0049-1
18. Theraulaz G, Bonabeau E (1999) A brief history of stigmergy. Artif Life 5:97–116. https://doi.org/10.1162/106454699568700

InterspeciesForms

Natalie Alima(✉)

Victoria St., Carlton, Building 100, Melbouorne, VIC 3000, Australia
alimanatalie@gmail.com

Abstract. InterspeciesForms hybridizes mycelia's agency of growth with architectural Design intention in the generation of novel, crossbred designed outcomes. In order to establish a direct dialogue between architectural and mycelia agencies, robotic feedback systems are implemented to extract data from the physical and feed it into the digital realm. Initiating this cyclic feedback system, mycelia growth is scanned in order to computationally visualize its entangled network and agency. Based on the logic of stigmergy, computational agents trace around the organisim's patterns of growth, forming entangled and complex networks. Through this unification of biological growth and computational agencies, non-indexical crossbred outcomes begin to emerge. Bringing this hybridized computational form back into the physical realm, form is 3D printed with a customized mixture of mycelium and agricultural waste. Once the geometry has been extruded, the robot, patiently waits for the mycelia to grow and react to the living extrusions. The architect then responds with a counter move by scanning this new growth and continuing the cyclic feedback system between nature-machine and architect. This procedure demonstrates form emerging in real time according to the co-creational design process and dialogue between architectural and mycelia agencies.

Keywords: Keywords are separated by half-angle origin

1 Introduction

Mycelium are threadlike fibrous root systems made up of hyphae, that form the vegetative part of a fungus [7]. Known as the hackers of the wood wide web [10] mycelia forms complex symbiotic relationships with other species that inhabit our earth, as Michael Lim states "Fungi redefine resourceful-ness, collaboration, resilience and symbiosis" [7, p. 14]. When wondering around the forest to connect with other species or searching for food, fungi form elaborate structures by spreading their hyphal tips. Darwin illustrated that root tips act like a brain as they link perception and action, and determine the trajectory of growth. Sheldrake [9] links this behaviour to fungal hyphae, as data is streamed through the organism's tips, determining the speed and direction of growth. The Latin root of the word intelligence means "to choose between" [9, p. 73], suggesting a form of intelligence expressed through the mycelia's 'decision gates'. Due the organisms ability to solve problems, communicate, make decisions, learn, and remember, fungi are indeed an intelligent species with a unique aesthetic that must not be ignored. In drawing on these concepts, I refer to the organism's ability to search for, tangle and digest its

P. F. Yuan et al. (eds.), *Hybrid Intelligence*, Computational Design and Robotic Fabrication,
https://doi.org/10.1007/978-981-19-8637-6_9

surroundings as 'mycelia agency of growth'. It is this specific behavioural characteristic that is the focus of this research, with which I as the architect set out to co-create and hybridise with.

2 Mycelium and Architecture

Over the years, a vast amount of interest and research has developed, testing fungi's compressive strength, acoustic absorption, fire safety properties and application to architecture. Due to the organisms ability to up-cycle materials and biodegrade them, fungal mycelium is often considered a sustainable alternative to synthetic materials. Mycelium-derived materials have several key advantages over traditional synthetic materials, including their: low cost, density, and energy consumption; their ability to biodegrade; and their overall low environmental impact and carbon footprint [3]. David Benjamin for example has successfully converted mycelium into known architectural applications such as bricks. Displayed in the courtyard of MoMA's PS1 space in New York Hy-Fi Tower (2014), Benjamin demonstrates the fabrication of organic, biodegradable bricks made of farm waste and a culture of fungus that was grown to fit a brick-shaped mold. In a similar approach, Phillipe Block's MycoTree installation (2017) is made of load-bearing mycelium components that replaces know architectural applications such as columns. Comprised of mycelium-infused waste that is ordinarily weak in tension, Block's team hacked into the mycelia's structural capabilities through form [4]. Advancing this fabrication method further, Pulp Faction by Ana Goidea, David Andreen and Dimitrios Floudas, demonstrates the robotic extrusion of mycelium infused with agricultural waste. This innovative research utilizes local organic waste in order to replace the existing petroleum based plastics, that are commonly used for 3D printing. To convert the living organism into a material that is suitable for the architectural field, the following process occurred with all three precedent projects listed above: The mycelium medium was initially grown on agricultural waste which provide nutrients for the organism to flourish. In the case of Phillipe Block's MycoTree and David Benjamin's Hy-Fi Tower, the mixture was then infused into predetermined digitally fabricated moulds in order for the organism to adopt the generated form. Following the growth period, the composite mixture is removed from the mould and either hot-pressed or oven dried which dehydrates the material and neutralizes the fungus [6]. By applying extreme heat to the fungus, this process converts the once living organism into a pre-formulated static building material by ensuring that it does not grow past the required shape [5]. In the case of 'Pulp Faction' mycelia is grown on agricultural waste and robotically extruded, illuminating the need for a digitally fabricated mould. In a similar process to MycoTree and Hy-Fi Tower, once the living organism has been 3D printed, extreme heat is applied in order to ensure that the organism is no longer able to grow, adapt, or respond past its required geometric shape.

2.1 The Agency of Mycelium

InterspeciesForms however posits that by making nature inert or converting it into static materials, 'appropriate' for the building industry, a crucial, biological- driven design process is being ignored. By converting mycelium into static non responsive materials, I posit that fungi's agency and true contribution to architecture has yet to be explored. Rather than ignoring natures agency, this research seeks to examine the potential contributions of the fungus's aesthetics to architectural de-sign. From this perspective, the mycelia's properties of growth are essential for the creation of hybridized, novel forms. InterspeciesForms therefore explores the possibilities of fungi that transcend its application of a sustainable material. This research explores novel ways to partner with the fungus in the cocreation of hybridized crossbred forms. The purpose of this partnership is to not only expand the imagination of the designer, but to generate novel forms which otherwise would not have been generated if designing individually. Michael Lim explains this notion that organism's should not be understood in isolation when stating "Fungi teach us that we are all interdependent. When we finally surrender our separateness, we realize that we are not outside of nature, but with it" [7, p. 15]. Fungi has been scientifically proven to partner and form symbiotic relation-ships with other species on our planet such its symbiosis with algae that generates lichen. This research questions, can architects and mycelium hybridize in the same way? This developed dialogue and feedback systems between the architect and mycelia agency is described in what follows.

3 Technical Workflow

Methodologically, this project involved the following stages: (i) preparing and applying the mycelium for growth on petri dishes; (ii) scanning the form; (iii) 3D printing the derived form using mycelium extraction, clay and agricultural waste; and (iv) incubating the forms and (vi) continuing the cyclic feedback systems of scanning mycelia data and extruding so that form is generated in real time. These processes involved the introduction of a scanning technology, developed computational algorithms, as well as developing a new mycelium based mixture for 3D printing the forms. In order to work with the organism intricate patterns of growth on a micro-scale, mycelium was initially grown on a series of petri dishes containing agar. Demonstrated in Fig. 1, this medium of growth was selected due to its capability to support a micro-scale growth pattern known as rhizomorph. To cultivate rhizomorph growth, mycelium was grown and sliced into 3 mm $\times$ 3 mm pieces. In a sterile environment, one slice of spawn was transplanted into a petri dish containing an agar medium. During the mycelium's growth, the petri dishes where stored in a dark, temperature controlled, and humid room to encourage cultivation.

In order to convert these intricate patterns of growth from the physical to the digital realm, a series of web cameras and microscopes where installed to track and computationally map the mycelium's dense fibrous network at the micro scale. Through a developed algorithm based on the logic of edge detection, computational agents traced over mycelia's patterns of growth, converting the once static image into a complex labyrinths of polylines. Using an additional process of color detection, the algorithm eliminated

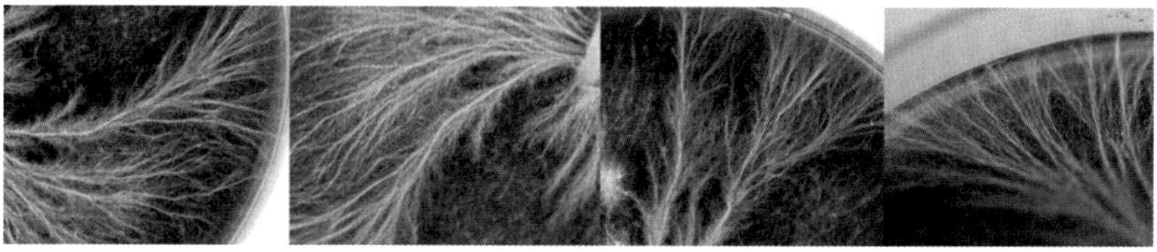

Fig. 1. Mycelium growing inside a series of Petri dishes. Demonstrating qualities of rhizo-morph entangled patterns of growth, mycelia spreads its hyphae tips in order to explore its surroundings.

unnecessary background interference, by filtering out forms that did not represent the mycelium's distinct white flourishing color. This procedure enabled the mycelium's intricate physical data to be accurately represented in the digital realm. As a result, a series of delicate computational drawings were generated, representing the organisms agency and autonomy of uninhibited growth. These computational drawings are exhibited in Fig. 2. Here computational form accurately captures characteristics of mycelia's rhizomorph growth such as: branching, fusing, entanglement, bifurcation and webbing, all which are visible at the micro-scale. Particularly noticeable are the interweaving hyphae tips as they bifurcate, separate and form new connections, resulting in root clusters of entanglement, attracting itself- to itself. In order to hybridise architectural design intention with the organism's agency, an additional algorithm based on the logic stigmergic principals was implemented in order to intertwine the volatile nature of mycelium with restrained computational algorithms. The aim of this process was to form a shared inter-species space to which, each creator may contribute, according to their unique affordances. To do so, InterspeciesForms utilizes stigmergy as a methodology for co-creation in self-organizing systems. Through self-organizing algorithms that are attracted to predetermined paths, stigmergy is utilized in novel ways that enable architectural design intent to follow the paths set by the fungus. In analogy to the ant trail, here too the mycelium's 'pheromones' lead the architect along their trails and the architect responds to these trails through a series of developed rules and restraints. These encoded rules and restraints may not only seed architectural design aesthetic but add a sense of organized complexity to mycelia's patterns of growth. Architectural intention is therefore imbedded by orchestrating the local interactions and micro decisions of computational agents. To do so, the developed algorithm was programed to include the following behavioural protocols: cohesion, separation, flow along curve, seek trail and evaporation of trail. The following pseudo-code describes this application and computational behaviours.

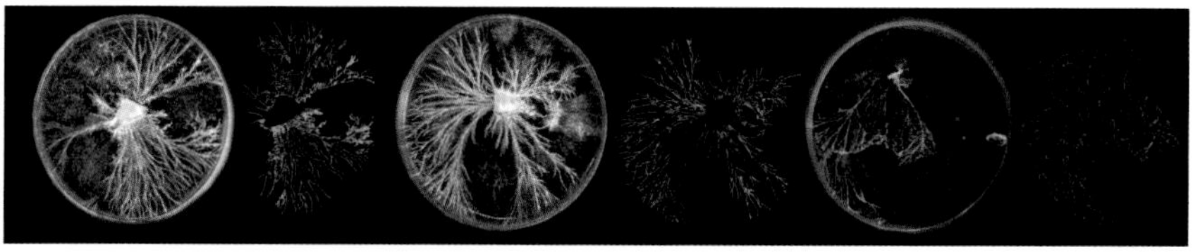

Fig. 2. Petri dishes containing mycelia growth (left) and the generated computational scan (right). Computational drawings assisted by Hanying Zhao and Christine O'neill.

Cohesion	*If* the computational agents move too close together, separate. *If* the computational agents move too far away, then seek the nearest mycelia pathway.
Separation	*If* the distance between the mycelium curve and the computational agent is <200 separate. *If* the distance between the mycelium curve and the computational agent is >200 move closer.
Flow Along curve	*If* the computational agent range is < 50, flow along mycelium curve. *If* the computational agent range is >100, do not flow along the mycelium curve.
Seek Trail	*If* value is >0.1 Seek mycelium trail. *If* value is < 0.1 Do nothing.
Evaporation	*If* the agents have been tracing around the mycelia's curves for 100 iterations. Evaporate.

Responding to the organisms agency, computational agents follow along the designated mycelia trails and simultaneously generate intricate entangled fibrous networks between each hyphae tip. This additional process adds a sense of designed complexity to the mycelium's web, by generating entangled connections between the hyphae tip. The sequential steps of growing the organism within the petri dish, the computational scan and applying the stigmergic algorithm which is seeded with architectural design intention is demonstrated in Fig. 3.

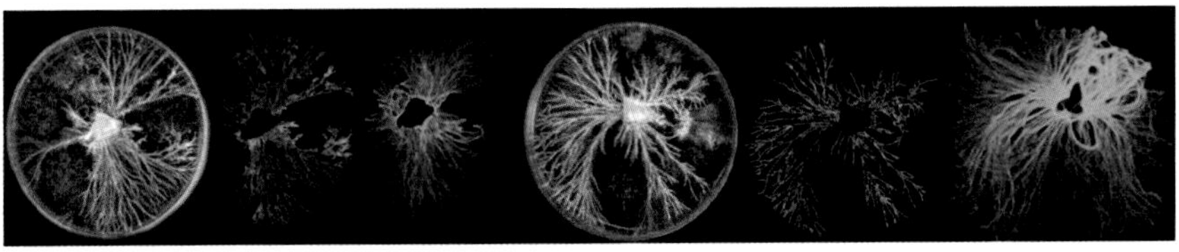

Fig. 3. From left to right, Petri dishes containing mycelia growth, computational scan and mycelia drawings and hybridized stigmergic outcome.

For the purposes of form finding, the goal of this process was to draw out the hybrid emergent characteristics between the natural and artificial realms. The results of this hybridization between mycelia and architectural agencies represent novel outcomes, which are non-indexical back to either mycelia's scan or computational agency. Whilst mycelia's original polylines have therefore been morphed, mutated and manipulated into an unrecognizable result, the organisms original features of fibrosity, delicacy and complexity still remain. Demonstrated in Fig. 4, a delicate balance of agency is achieved as characteristic of mycelia growth and stigmergic processes are still present, but have morphed into something new.

In order to bring these hybridized output forms back into the physical environment and continue this cyclic feedback system between the natural and the artificial, InterspeciesForms where robotically extruded with the mycelium medium it-self. However,

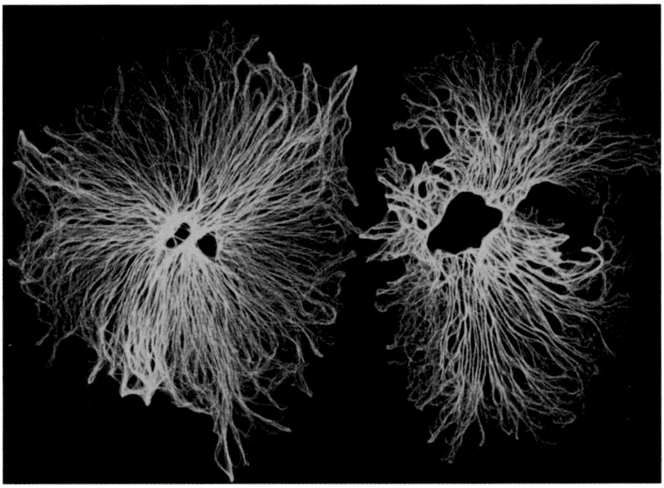

Fig. 4. Hybridized outcomes of the entanglement between mycelia's and architectural agencies.

in comparison to existing projects that 3D print mycelium, mycelia was kept alive in order to enable its patterns of growth to thrive and contribute to the create of form and robotic movement. Demonstrated in Fig. 5, a feedback system and direct dialogue is developed between biological, architectural and robotic agencies as mycelia growth becomes impute into computational form and robotic movement. This technical process is described in what follows:

A customized mixture of mycelium, clay and agricultural waste was created in order to robotically fabricate the biological medium and encourage growth. This mixture consisted of agricultural waste, which provided nutrients of the organism to thrive and clay which acted as a natural binder for the living fibbers to adhere to. Each hybridized form was 3D printed at the scale of 300 mm × 400 mm. The size of the forms could not identically match the original size of the petri dish de-rived forms (200 mm diameter), due to limitations set by the nozzle size and the need to prevent clogging. The following steps were taken to test the points mentioned in the creation of the form:

Mycelium was firstly inoculated and grown on a mixture of wood chips and paper pulp. Over a seven day period this mixture was incubated with controlled temperature (24–30 C) and humidity (90%), under a greenhouse tent. This chamber was kept sterile to prevent bacteria growth, while enabling sufficient natural light to pass through. Ph.D. RMIT diss.

Once the mycelium flourished in growth, over a seven-day pe-riod, it was introduced to the clay medium. Earthenware clay was utilized due to its porous aerated structure, which enabled the mycelia to seep through and even-tually degrade the substrate. Both the clay and mycelium bio composite mixture were fused together, generating a living paste to robotically extrude.

Utilizing a customized clay extruder on a Universal Robot (UR), the mycelium mixture was 3D printed according to the following protocol: (i) Nozzle Height: 23.1 mm

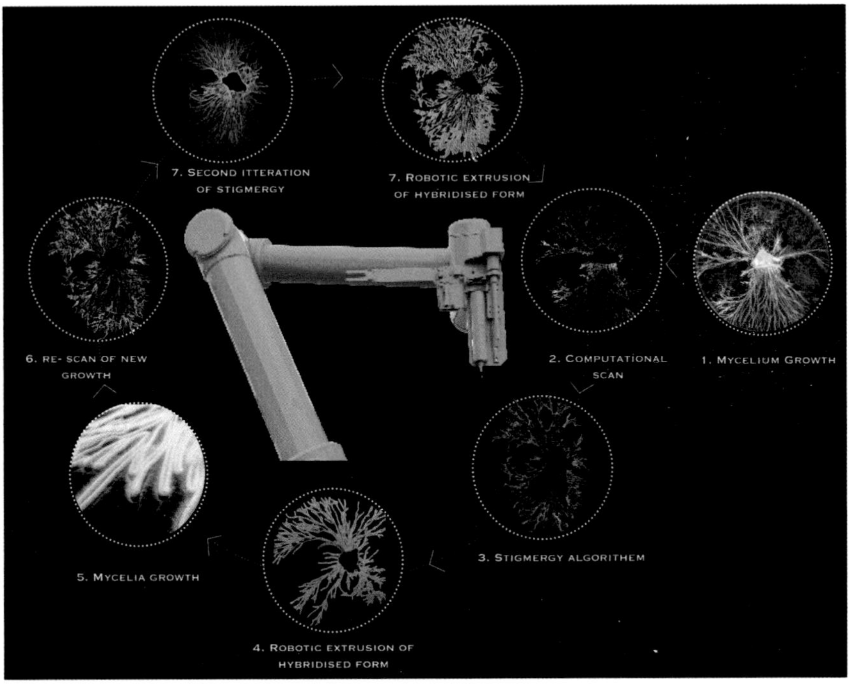

Fig. 5. The developed cyclic feedback system between mycelia growth, robotic intervention and computational form. Initiating this feedback system is the growth of mycelia with-in a series of petri dishes and agar cultures. The organism's physical patterns of growth are scanned and become inputs for computational form. Stigmergic algorithms are then applied to the mycelia's polylines, generating hybridized non indexical results. These outcomes are then 3D printed with the biological medium itself. Over time, the robot waits for the organism to grow and responds to fungi growth by initiating the feedback system of scanning and extruding once again.

from Surface; (ii) Nozzle Width: 5 mm; (iii) Print Speed: 10 mm/s; (iv) Flow Rate: 58%; (v) Layer Height: 1.5 mm; and (vi) Air Pressure: 50 PSI.

Images of the clay-mycelium medium being robotically extruded and growing are shown in Fig. 6.

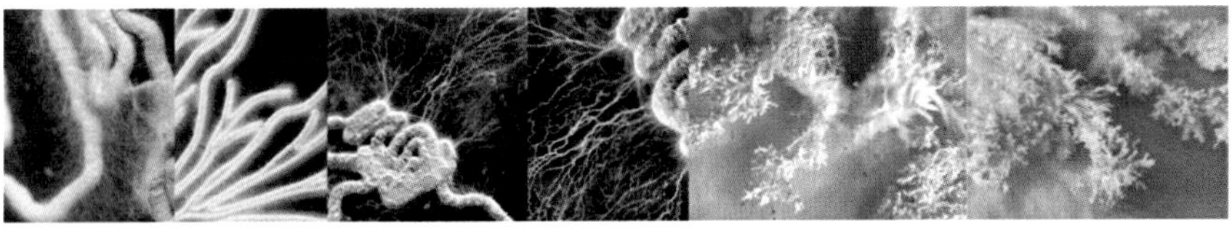

Fig. 6. Clay infused with mycelia robotic extrusions. Overtime, the mycelium began to grow from the organic mixture and wondered around the surface area in search for additional nutrients.

4 Results

Once the organism was 3D printed, the organism began to grow from the living extrusions by extending its hyphae tips away from its designated path that the robot extruded. This behavior not only exposed the organism agency but asserted its autonomy, that the living cannot be completely controlled. During this period, three types of growth behaviour were observed. The first was noted by change of texture and color. Whilst initially the designed form had a smooth grey exterior—resembling the appearance of clay, over time the extrusions turned hairy, fluffy and furry, resembling the texture of hyphae, imbedded in the extrusion. Figure 6 present this formation.

A second noted change was that the form began to fruit. Long cylindrical mushrooms began to blossom and tower over the tapestry. Finally, it was observed that over the course of seven days, the mycelium increasingly biodegraded the clay mixture substrate. When doing so, it began wondering around its surrounding in search for additional nutrients to absorb. This was fascinating to observe, as the fungus was no longer constrained by the extruded form from which it originated, but rather began to affirm its own agency of growth. A distinct set of generative patterns of growth and characteristics such as branching and fusing began to emerge. This volatile behavior resulted in vein like formations, not set by the architect. Here the fungus clearly asserted its autonomy over the design. Once the organism began to divert and grow beyond its set boundaries, an additional scan was conducted in order to repeat the cyclitic feedback prosses of growing-scanning and extruding (Fig. 5). Resulting from this feedback systems, a catalogue of forms shown in Fig. 7 were generated exposing emergent qualities as form mutates and evolves over time.

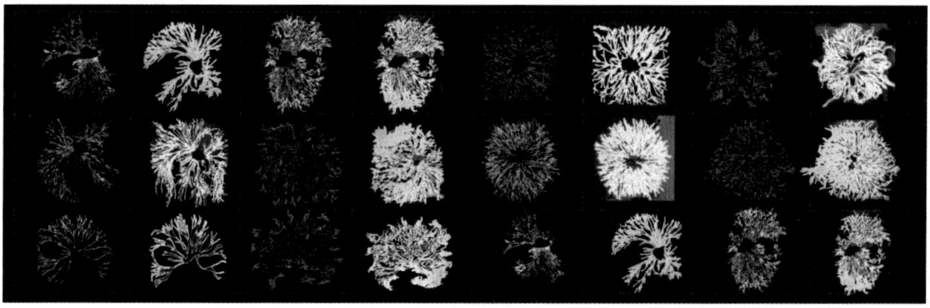

Fig. 7. Computational Matrix showcasing mycelia's original growth and computational scan conducted. Once mycelia has grown beyond its set boundaries, an additional scan is conducted in order to capture its new agency of growth.

5 Conclusion

Through this developed feedback system the formation of architecture is directly driven by mycelium behavior, rather than an a priori parametric model or generative algorithm. Methodologically, the contributions of InterspeciesForms include the scanning mycelia growth in order to computationally visualize its patterns of growth, utilizing mycelia growth as impute to developed algorithms based on stigmergic behaviors, generating hybridized outcomes from the co-creation of architectural and mycelia agencies, and developing behavioral based feedback systems. The findings of this research conclude that by applying stigmergy into biodesign, new opportunities for co-creation may arise. By embracing non-human aesthetics within architectural design, multi-species and biocentric forms may arise. Furthermore, it appears that the allocation of high levels of design autonomy to the fungus, has led to the creation of a new set of forms that embrace the strange and highly volatile nature of the mycelium. No two from outputs in this process were the same, where each fabricated design presented a contrasting set of biological patterns of growth. Reflecting upon the limitations of this research, the hybridized results remained on the two-dimensional scale and did not eventuate into three-dimensional forms. Due to a lack of access to high resolution vision systems that could capture the organism growth in the 3D realm, the scanning of mycelia's was limited to the two-dimensional scale. Thus the stigmergic response mirrored this scale and generated two dimensional hybridized outcomes. In order to advance this established feedback systems further, higher resolution vison systems are required to capture the mycelia's growth within the petri dish on the three-dimensional scale. Rather than responding in the two-dimensional realm, the architect may generate 3D forms which emerge from the interaction of architectural aesthetic and the behavior of mycelia growth. This research demonstrates and offers new methodologies of co-creating with non-human organisms that may give rise to new non-indexical formations for architectural design purposes.

References

1. ALIMA, N. (2022). Interspecies formations [RMIT University]. https://researchrepository.rmit.edu.au/esploro/outputs/doctoral/Interspecies-formations/9922152713401341#file-0
2. Grassé PP (1959) The automatic regulations of collective behavior of social insect and stigmergy. J Psychol Norm Pathol (Paris) 57:1–10
3. Haneef M, Ceseracciu L, Canale C, Bayer IS, Heredia-Guerrero JA, Athanassiou A (2017) Advanced materials from fungal mycelium: fabrication and tuning of physical properties. Sci Rep 7(1)
4. Heisel F, Lee J, Schlesier K, Rippmann M, Saeidi N, Javadian A, Nugroho AR, Mele TV, Block P, Hebel DE (2017) Design, cultivation and application of load-bearing mycelium components: the MycoTree at the 2017 Seoul biennale of architecture and urbanism. Int J Sustain Energy Dev 6(1): 296–303. https://block.arch.ethz.ch/brg/files/HEISEL_2017_WCST_design-loadbearing-mycelium-structure_1546891598.pdf. Accessed 12 Nov 2020
5. Holt GA, Mcintyre G, Flagg D, Bayer E, Wanjura JD, Pelletier MG (2012) Fungal mycelium and cotton plant materials in the manufacture of biodegradable molded packaging material: evaluation study of select blends of cotton byproducts. J Biobased Mater Bioenergy 6(4):431–439
6. Jones M, Mautner A, Luenco S, Bismarck A, John S (2020) Engineered mycelium composite construction materials from fungal biorefineries: a critical review. Mater Des 187:108397
7. Lim M, Shu Y (2022) The future is fungi: how fungi can feed us, heal us, free us and save our world. Port Melbourne, Vic, Thames & Hudson Australia
8. Navlakha S, Bar-Joseph Z (2011) Algorithms in nature: the convergence of systems biology and computational thinking. Mol Syst Biol 7(1):546

9. Sheldrake M (2021) Entangled life: how fungi make our worlds, change our minds & shape our futures. Random House, S.L.
10. Simard SW, Perry DA, Jones MD, Myrold DD, Durall DM, Molina R (1997) Net transfer of carbon between ectomycorrhizal tree species in the field. Nature 388(6642):579–582
11. Snooks, R (2014) Behavioural formation: multi-agent algorithmic design strategies. PhD RMIT diss

Simulation Algorithm Based on Weathered Rock Morphology and Optimization Algorithm for Design Applications

Wei Ye[1], Xiayu Zhao[1], and Weiguo Xu[2(✉)]

[1] Tsinghua Shenzhen International Graduate School, Shenzhen 518055, China
[2] School of Architecture, Tsinghua University, Beijing 100084, China
xwg@mail.tsinghua.edu.cn

Abstract. The rich organic pore spaces of weathered rocks bring inspiration to architectural design. Based on the existing research on the natural formation mechanism of weathered rocks, this paper proposes two algorithms that achieve natural formation mechanism simulation and morphology simulation. Firstly, this study deeply explores the intrinsic characteristics of weathered rocks; secondly, the basic framework of iterative cyclic calculation by multiple weathering forces is built to make the calculation results of 3D point cloud close to the real morphology of weathered rocks; subsequently, this study innovatively introduces a 2D stacked layer algorithm for optimization while maintaining the morphological characteristics; finally, the architecture design application of the optimization algorithm is verified. Compared with the 3D point cloud simulation algorithm, the 2D layered algorithm can greatly reduce the computational time complexity and control the generated space's utilization.

Keywords: Weathered rock · Natural form simulation algorithm · Algorithm-based design

1 Introduction

With the application of new technologies in the field of architectural design, computer algorithms can create a large number of architectural forms with similar features in a short time, and change the detailed shape according to the actual needs under the control of some parameters. Computational design methods assist designers to generate buildings that mimic natural forms. There are generally two ways to develop the natural form simulation algorithm: one is to summarize the intuitive characteristics of natural form and directly simulate it; the other is to study the laws of how it is formed and use the mathematical formula to develop algorithms [13]. The former focuses on the external shape, while the latter focuses on internal reasoning.

The natural form selected in this study, tafoni, is composed of countless non-Euclidean surfaces at the microscopic level, which would be a huge workload for human modelling alone and difficult to restore its natural morphological characteristics through manual control. The introduction of the algorithm can be interpreted as a "natural hand"

P. F. Yuan et al. (eds.), *Hybrid Intelligence*, Computational Design and Robotic Fabrication,
https://doi.org/10.1007/978-981-19-8637-6_10

to restore the organic form and maintain the characteristic elements of tafoni from the beginning to the end.

Different methods have been used in the previous study on morphological simulation algorithms of natural shape. Xu Weiguo et al. took the example of mutualistic leaf order morphology and described the whole process from the analysis of biological mechanism to the simulation algorithm and architectural design application [14]. Z. Feng et al. introduced a generative design method that integrates two techniques of computational fluid dynamics (CFD) simulation and bidirectional progressive structural optimization (BESO) based on the steady-state effect of Taihu stone [4].

This paper first explores the intrinsic characteristics of weathered rocks and reclassifies them based on architectural geometric elements, then proposes two algorithms that achieve natural formation mechanism simulation and morphology simulation. The algorithm can simulate the natural form and has great application value on architecture design.

2 Study of the Natural Morphology

The study of this paper is focused on mining the morphological characteristics of tafoni and applying them in architectural design. The point, line, and plane are the basic geometric elements to express space and entity. Therefore, this study classifies the morphology of tafoni into three types: point base, line base, plane base. The natural pictures and form description diagrams are shown in Fig. 1.

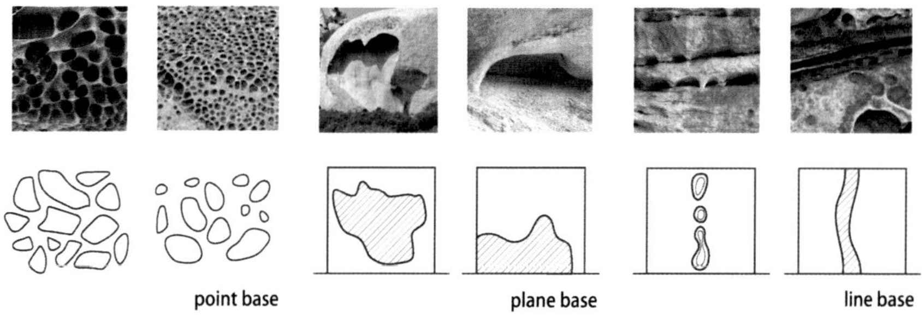

Fig. 1. Natural forms of weathered rocks and Forms description diagrams.

2.1 Literature Review

Weathered rocks were studied and classified as early as the nineteenth century, and the first such honeycomb cavities larger than 0.5 m in diameter were called tafoni [12]. Paradise defined tafoni as "lace-like, honeycomb, bowl, or pan-shaped cavities occurring in a variety of rock types that show a commonly unique assemblage and morphology" [11]. Previous studies have proposed classifications such as alveoli, stone lace, honeycombing, caverns, pitting, and other terms based on the different characteristics of weathered pore morphology in terms of size, morphology, and distribution [6]. The classification of this study is from the perspective of design.

2.2 Morphological Classification

Point base tafoni has irregularly concave holes of different sizes and organic shapes, with the inconsistent depth of the concave surface.

Line base tafoni has irregular holes of different sizes and are fused together, and different holes meet in the form of grooves, the whole is of different lengths, uneven thickness, bifurcation, irregularly curved strip-like grooves, and the depth of the grooves is not uniform.

Plane base tafoni has hook-like sides, concave in the middle and lower parts, rounded and smooth inner walls, and protruding like a cap tongue at the top.

3 Simulation Algorithm of Weathering Process

Earlier studies on tafoni indicate that tafoni develop in different types of bedrock, most commonly in granitic rocks [12]. Granitic weathering cavities are essentially the product of differential weathering and are closely related to the granular structure, homogeneity, weak permeability, and ease of disintegration of granites [9]. Burridge et al. propose a model for the formation of craters produced by corrosive gases that can be analogous to the weathering process: all solid sites are given an intensity value. If a corrosive particle occupies a site near the surface and its next randomly chosen step will bring it to the surface, the particle has the probability to decay [2].

Collectively, the main factors that control and influence tafoni development are microclimatic changes, salt weathering, and valley wind erosion [7]. Valley wind erosion is a physical weathering and salt weathering is a chemical weathering. In this study, these two weathering processes were simulated separately using algorithms.

3.1 Methodology

Based on the fact that the bedrock of tafoni is generally granular in structure [12] and homogeneous [9], this study chose to use Orthogonal 3D point cloud to simulate the weathering process of tafoni.

3.2 Physical Weathering

Physical weathering is accomplished by valley wind erosion [7]. Physical weathering is the same process in three types: point, line, and surface. An iterative mathematical model of physical weathering is established. The study classifies the forces applied to the 3D point cloud into three forces: inter-particle force, wind force, and gravity. The combined force on each point is calculated. Under the action of external force (wind force), the points with the smallest combined force in the point cloud are eliminated after each iteration. The vector value of the wind changes randomly in each iteration. The iteration will continue until all points disappear. This process simulates the actual process of physical weathering. The particle force analysis is shown in Fig. 2.

The plausibility of the physical weathering forces discussed in this paragraph has been verified and implemented in the ESO and BESO algorithms, and the benefits of the optimized structure are retained by removing ineffective or inefficient materials [15].

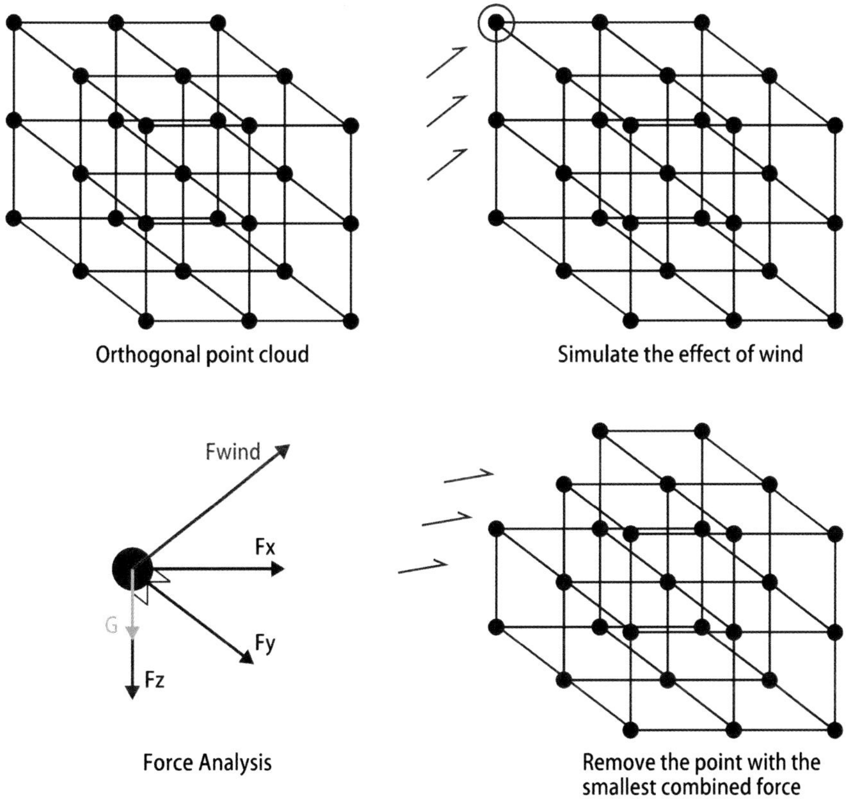

Fig. 2. Force analysis model

3.3 Chemical Weathering

Brandmeier's study showed that chemical weathering rates were found to be related to differences in lithology, microfracture, alteration, and cementation [1]. In which alteration and cementation are chemical weathering, so our simulations are focused on these two processes. A mathematical model of chemical weathering is built basically based on the two steps of chemical weathering: the decay of internal particles due to humidity changes and enhancement of external particles due to gelation [8]. The two steps are corresponding to alteration and cementation.

For point base tafoni, first input the original mesh, generate the orthogonal point cloud inside the original mesh and assign a value x to each point, when x is less than or equal to zero, remove this point. Secondly, randomly reduce some points from the point cloud. Thirdly use the two evolutionary rules according to the two steps of chemical weathering to calculate:

(i) *the values of the neighboring points A in the vanishing particles decay*
(ii) *the values of the surface layer(exposed) points S increase.*

For point A,

$$x_a = \mathrm{x} - \mathrm{N} \tag{1}$$

For point S,

$$x_s = \mathrm{x} + \alpha\mathrm{N}\ (0 < \alpha < 1) \tag{2}$$

For a point X in point cloud,

if x < 0, remove point X from point cloud

(N is the decay integer constant, α is the enhancement ratio constant.)

For plane base tafoni, the basic process of chemical weathering is similar to point base tafoni. Due to the large surface area of each cavity of the faceted weathering cavity, the temperature and humidity at different locations where its surface is located also vary [10], and the different temperatures and humidity have a greater influence on the chemical weathering. Most chemical weathering occurs on the more obscure surfaces of the rock where the drying rate is lower. The upper part of the tafoni is not exposed to direct sunlight and has a much lower drying rate than the lower part, so the weathering is more pronounced. Therefore, compared with the point-like evolution rule (i) and rule (ii), rule (iii) is added:

(iii) *decay degree of upper points* a_1 > *lateral points* a_2 > *lower points* a_3

For point a_1 **point** a_2 **point** a_3,

$$x_{a1} = \mathrm{x} - \mathrm{N} \tag{3}$$

$$x_{a2} = \mathrm{x} - \beta\mathrm{N} \tag{4}$$

$$x_{a3} = \mathrm{x} - \gamma\mathrm{N}(0 < \gamma < \beta < 1) \tag{5}$$

For point S,

$$x_s = \mathrm{x} + \alpha\mathrm{N}(0 < \alpha < 1) \tag{6}$$

For a point X in point cloud,

if X < 0, remove point X from point cloud

(N is the decay integer constant, β and γ is the decay ratio constant, α is the enhancement ratio constant.)

For line base, the chemical weathering process is different due to stress [5]. The process of linear type weathering is affected by stress, and no decay occurs for particles with higher stress and decay occurs for particles with lower stress. Therefore, the linear chemical weathering process needs to extend a straight line on the surface of the point set to randomly delete points, which are less stressful points. After the evolution rule (i), add rule (iv):

(iv) *When there is no connection point on both sides of a point, set the point and all the points on its z-axis (stress points) as non-attenuating points.*

Set the set of stress points as δ

For point A,

$$x_a = \mathrm{x} - \mathrm{N} \tag{7}$$

For a point X in point cloud,

if X < 0, remove point X from point cloud

if $x \in \delta$, do not remove point X from point cloud

The mathematical model of the above algorithm is shown graphically in Fig. 3. The constants used in calculation can be adjusted according to the specific weathered rock characteristics.

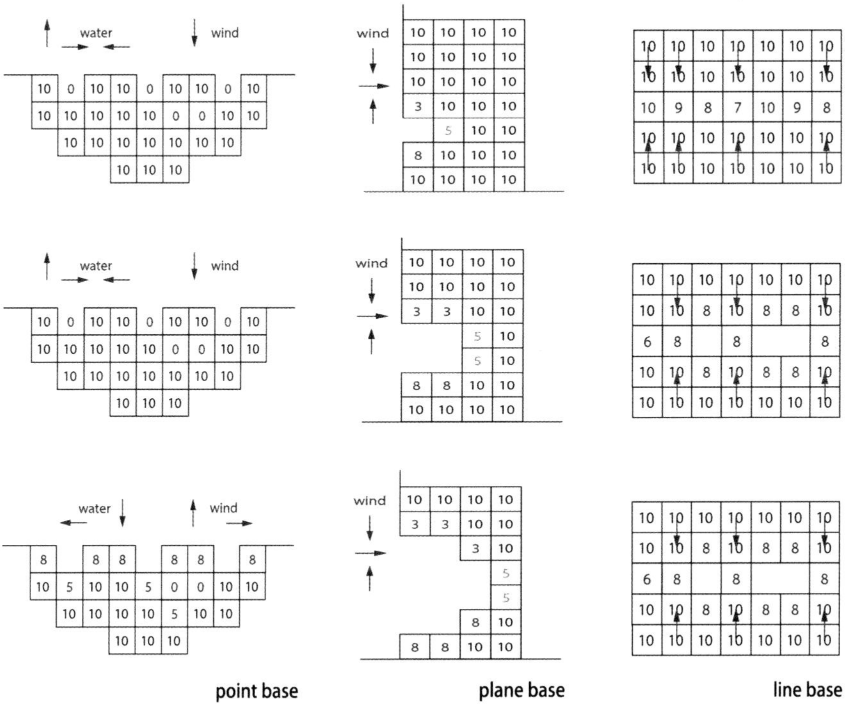

Fig. 3. Chemical weathering

3.4 Combination of Physical Weathering and Chemical Weathering

In order to better simulate the natural weathering process of tafoni, this study needs to develop the summative algorithm combining physical and chemical weathering. After studying physical weathering and chemical weathering processes separately, it can be seen that shaping the different forms of weathered rocks (point base tafoni, line base tafoni, plane base tafoni) mainly relies on chemical weathering. In the simulation of physical weathering, the angle of the external wind is constantly changing due to the valley wind erosion process over a long period of time. So, the role of physical weathering is more to soften the cavities. In order to weaken the effects of physical weathering in the combination, the physical weathering process is simplified as a rock surface softening process. *Isosurface* is a smooth surface through all points between voxels, and the algorithm is easy to apply, so *isosurface* algorithm is used to simulate physical weathering effect in this study. The process combining chemical weathering and physical weathering is to first calculate the chemical weathering for the point cloud-first, and then apply *isosurface* to simulate the physical weathering. The tafoni simulation algorithm flowchart is shown in Fig. 4. The results of the simulation for the three types are shown in Fig. 5.

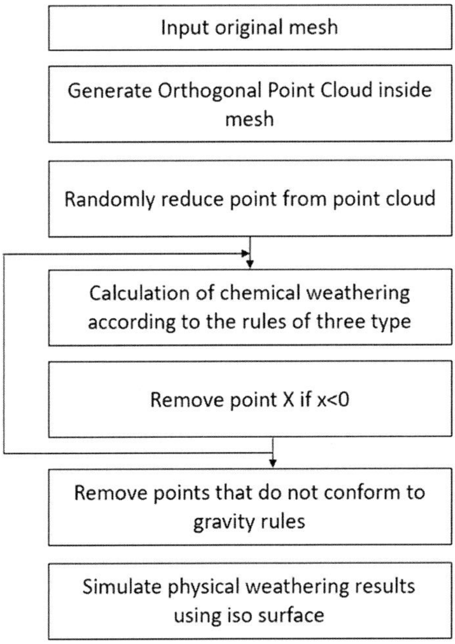

Fig. 4. Algorithm

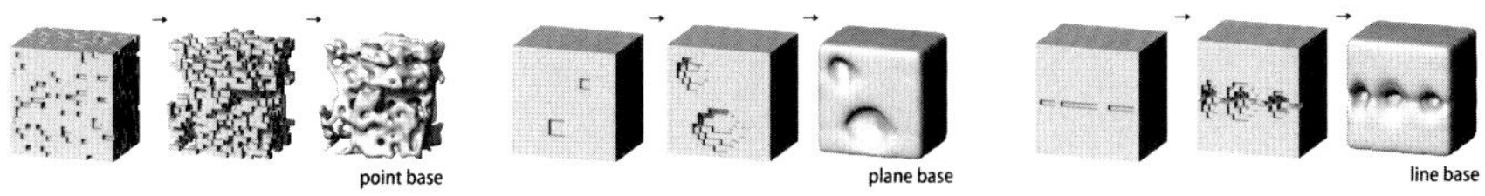

Fig. 5. Chemical weathering mathematical model

4 Optimization Algorithm

The tafoni simulation algorithm developed above is calculated iteratively to simulates the evolution of tafoni. However, the algorithm has high computational time complexity. For the input mesh volume, when the point cloud density ρ is low, this simulation algorithm can be easily calculated, but when a high-precision simulation is required, the point cloud density increase and the computation time will greatly increase. The previous algorithm is not feasible to be applied to practical design. For the purpose of applying the algorithm to architectural design, the algorithm needs to be simpler and more operable.

This study optimizes the algorithm by reducing 3D computations to 2D computations and then layering them to form point sets.

4.1 2D Stacked Layer Algorithm

The 2D stacked layer algorithm steps are as follows. The algorithm process is shown in Fig. 6 with diagrams.

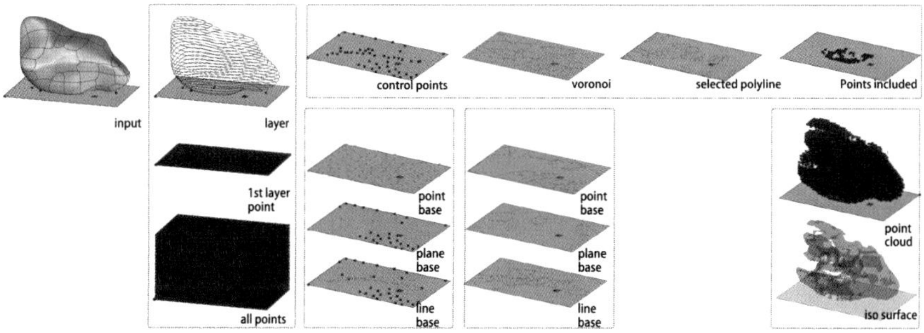

Fig. 6. Optimized algorithm process

1. Input the initial volume mesh of the rock to be weathered;
2. input the number of layers x, the number of layers can be converted into point cloud density parameters (because the point cloud for XYZ direction equally spaced orthogonal point set, so the initial volume of the number of layers to convert the overall point density in the point cloud);
3. Generate the initial point cloud;
4. Generate 2D Voronoi control points by point type, line type, and surface type features, these control points are randomly generated by random number n control;
5. Generate the control Voronoi polygons and extract the polygons representing point type, line type, and surface type respectively;
6. the three corresponding polygons are divided, rounded, offset, and other operations;
7. Merge into the final set of closed curves;
8. Calculate whether the point cloud and the curve set are contained or not, and remove the points contained by the curve set;

9. Loop x times to get the points kept in each layer, these points form the final point set. In the calculation, it is found that the more the number of layers, the more laminar the generated form is. In order to eliminate the laminarization, a formula is experimentally derived that, let when the number of layers is x, for a random number n, every (x − 7)/7 cycles, n = n + 1.

$$n = n + \left[\frac{\mathrm{x} - 7}{7}\right] \tag{8}$$

10. Using Isosurface to convert the points to mesh, the effective range of iso is calculated by comparing and setting the number of layers to x as follows.

$$\text{effective range} = \frac{3000}{\mathrm{x} - 53} \tag{9}$$

As can see in Fig. 7, When the value of n is larger, which has larger density ρ of points in the volume, the more detailed carvings the generated shape has. The smaller the point density is, the rougher the generated shape is. Also, with different seed, multiple different simulated rock shape can also be generated from the same initial volume.

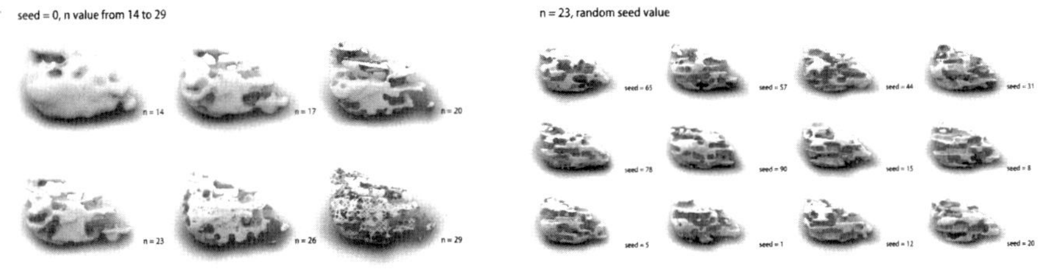

Fig. 7. Different input of n, seed and the generated results

4.1.1 Computational Time Complexity

Next, the computational time complexity is estimated for the described 3D algorithm as well as for the optimization algorithm of the 2D stack. For the same input mesh, whose volume is V, set the point cloud density to ρ, the number of points is calculated by multiplying the volume by the density as N.

For the algorithm that directly simulates the weathering process, each point needs to be calculated with N points, a total of N points is calculated. It takes N^2 steps to calculate through all points, and N steps to assign the value. The time complexity can be estimated as N^3, ignoring constant V, the time complexity O_1 is as follows.

$$O_1(\rho) = \rho^3 \tag{10}$$

For the 2D stacked layer algorithm, the number of points per layer is $\sqrt[3]{N^2}$, and the loop operation is $\sqrt[3]{N}$ steps from the bottom loop to the end of the topmost layer. The time complexity can be estimated as N, ignoring constant V, the time complexity O_2 is as follows.

$$O_2(\rho) = \rho \tag{11}$$

Comparing time complexity O_1 and O_2, the 2D stacked layer algorithm can greatly reduce the time complexity of the algorithm since $\rho > 1$.

4.1.2 Practical Application Value

This optimization further clarifies the design logic and improves the practical design application value of the generated forms. Since there are different spatial attributes and spatial characteristics required for specific functions in architectural design, the layer-by-layer calculation and then superposition calculation method can make the computer-generated morphology more consistent with the design language, which can better serve the architectural design and actual construction, and then form a space truly used for human living, dwelling and sharing, and the human feeling in the space is more comfortable.

4.2 Algorithm Design Applications

Apply the 2D Stacked Layer Algorithm in to architecture design. First generate pore spaces in a rock, then the pore space is further designed as an office building. The input mesh, generated result, the architecture elevation, the indoor and outdoor renderings are shown in Fig. 8.

5 Prospects

The algorithm-controlled morphological simulation will allow the otherwise uncontrollable natural forms to grow as the designer or the people want, and also facilitate the storage of information needed for spatial orientation. The rich organic pore form of tafoni, the multi-directional penetration of horizontal and vertical space, the flowing non-linear curved form, the reasonable force structure, and consumable rate all reflect the application value and potential of this form, which will bring infinite possibilities for future urban planning and architectural creation.

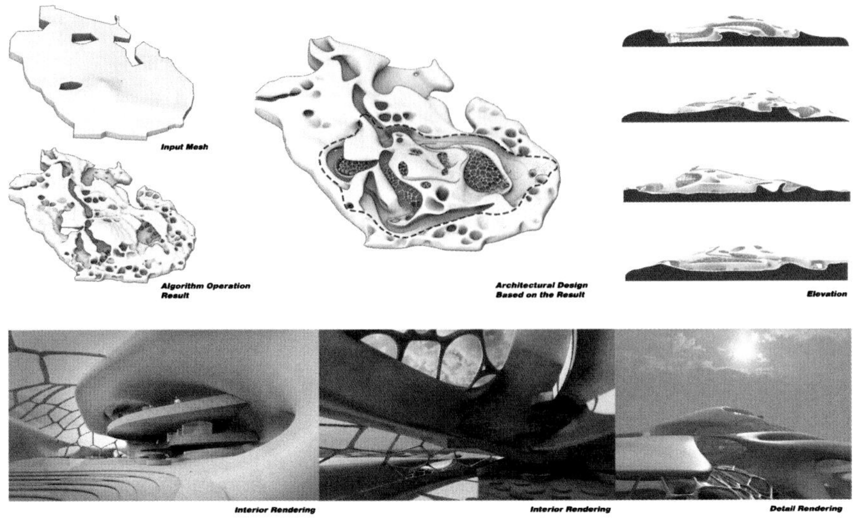

Fig. 8. The input mesh, generated result, the architecture elevation and renderings

References

1. Brandmeier M, Kuhlemann J, Krumrei I, Kappler A, Kubik PW (2011) New challenges for tafoni research. A new approach to understand processes and weathering rates. Earth Surface Process Landf 36(6):839–52
2. Burridge J, Inkpen R (2015) Formation and arrangement of pits by a corrosive gas. Phys Rev E Stat Nonlin Soft Matter Phys 91(2):022403
3. Collins GR (1963) Antonio Gaudi: structure and form. Perspecta. 63–90
4. Feng Z, Gu P, Zheng M, Yan X, Bao DW (eds) (2022) Environmental data-driven performance-based topological optimisation for morphology evolution of artificial Taihu stone. Springer Singapore, Singapore
5. Filippi M, Bruthans J, Řihošek J, Slavík M, Adamovič J, Mašín D (2018) Arcades: products of stress-controlled and discontinuity-related weathering. Earth Sci Rev 180:159–184
6. Groom KM, Allen CD, Mol L, Paradise TR, Hall K (2015) Defining tafoni. Progress Phys Geogr Earth Environ 39(6):775–793
7. Huang R, Wang W (2017) Microclimatic, chemical, and mineralogical evidence for tafoni weathering processes on the Miaowan Island, South China. J Asian Earth Sci 134:281–292
8. Huinink HP, Pel L, Kopinga K (2004) Simulating the growth of tafoni. Earth Surf Proc Land 29(10):1225–1233
9. Li D, Cui Z, Li H, Nan L (2003) Mechanism of granite weathering cave formation and environmental significance in northern China. J Nanjing Univ (Natural Science Edition) 01:120–128
10. Matsukura Y, Tanaka Y (2000) Effect of rock hardness and moisture content on tafoni weathering in the granite of Mount Doeg-Sung, Korea. Geogr Ann Ser B 82(1):59–67
11. Paradise TR (2015) Tafoni and other rock basins. reference module in earth systems and environmental sciences
12. Penck A (1894) Morphologie der erdoberfläche: J. Engelhorn
13. Rossi M (2006) Natural Architecture and constructed forms: structure and surfaces from idea to drawing. Nexus Netw J 8(1):112–122

14. Xu W, Li N (2016) Algorithms and illustrations Digital illustration of biomorphology. Time Archit (05):34–39
15. Xie Y, Huang X, Zuo Z, Tang J, Rong S-H (2011) Recent developments in progressive structure optimization (ESO) and bidirectional progressive structure optimization (BESO) methods. Adv Mech 41(04):462–471

Simulation and Optimization

Research on Real-Time Interactive Spatial Element Optimization Method Based on EEG Signal—Taking Indoor Space Color and Window Opening Size as the Optimization Object

Zihuan Zhang(✉), Zao Li(✉), and Zhe Guo(✉)

Hefei University of Technology, Hefei, China
540530969@qq.com, lizao72@hotmail.com, guogal@hotmail.com

Abstract. In recent years, the research on digital design and perceptual evaluation has gradually become a hot topic in the field of digital design. Based on digital space optimization theory and perceptual evaluation tools, this study attempts to establish an optimization method to optimize built space elements in real-time using human psychological indicators. This method takes the specific indicators of the Meditation value and Attention value in the human EEG signal analyzed by the TGAM module as the optimization objective, the architectural space color and the window size as the optimization object, and the multi-objective genetic algorithm as the optimization tool. To realize this optimization method, this research combines virtual reality scene and parametric linkage model to establish tool platform and workflow. Taking the optimization of typical residential space as an example by recruiting 50 volunteers to participate in the experiment, this study concludes that this method is effective and feasible through experiment and quantitative analysis of experimental results and lays the foundation for more EEG indicators and more complex spatial element optimization research in the future.

Keywords: EEG · Spatial optimization · Real-time interaction · Multi-objective genetic algorithm · TGAM module

1 Introduction

The sudden spread of COVID-19 in 2019 challenged human physiology and had a great impact on human psychology. The long-term isolation has caused extensive socio-economic losses during the epidemic, and the loss of income and livelihood is causing social and psychological distress. On the one hand, people with mental illness are more likely to be infected with neocoronavirus [1], on the other hand, mass isolation measures and mental health factors such as anxiety, depression, and stress are the causes of infection with the neocoronavirus [2].

P. F. Yuan et al. (eds.), *Hybrid Intelligence*, Computational Design and Robotic Fabrication,
https://doi.org/10.1007/978-981-19-8637-6_11

Besides, the design of the space environment needs to meet human physiological and psychological needs, especially for children, the elderly, and pregnant women in human vulnerable groups. The satisfaction of various elements in architecture and space environment to their mental health is particularly important. Architectural space is closely related to people's psychological space. The color of the enclosure, the light in the environment, the outline of objects, and even the style of buildings will affect people's psychological feelings [3, 4].

Previous studies on the correlation between EEG signals and psychological quantities such as human emotion and psychological stress found that it is theoretically feasible to judge people's emotional characteristics [5] and quantify people's degree of relaxation and stress through EEG signals [6, 7]. To sum up, the mental state represented by specific EEG characteristics can significantly positively impact human psychology and physiology [8–10]. Enhancing people's meditation training and enhancing the meditation value corresponding to the TGAM EEG module can reduce people's stress and anxiety to a certain extent [11, 12].

In the field of perceptual engineering, EEG monitoring methods have great research potential in the fields of architecture [13, 14], landscape [15, 16], urban design [17], and art [18]. These studies also fully illustrate the feasibility of applying the EEG method to building evaluation and space design. Through the above research, it can be concluded that human beings will produce different brain wave states under different spatial atmospheres or elements. EEG signals in different states can correspond to different psychological states [19].

However, compared with the field of building physics [20, 21], building structure [22], and other areas [23, 24], the relationship between human physiological data and spatial elements established in the above research does not seem to be accurate to the quantitative level, and the overall optimization suggestions prefer qualitative suggestions.

1.1 Research Aim

This study aims to create an effective architectural design optimization method, which is a quantitative optimization method of architectural space elements based on the physiological data of human emotion and psychological elements. Figure 1 shows the goal vision of this approach.

This method is used to form a closed-loop optimization participated by people. People do not need to make active judgments in the optimization process. During optimization, the experimentee observes the virtual reality scene, and the EEG signal is monitored in real-time by the TGAM EEG module. The optimization algorithm changes the parametric linkage model in real-time via referring to the EEG value, to change the virtual reality scene observed by the experimentee. In this study, the specific indexes of meditation value and attention value in human EEG signal are analyzed based on the head ring of TGAM module as the optimization goal, the color of building space and window

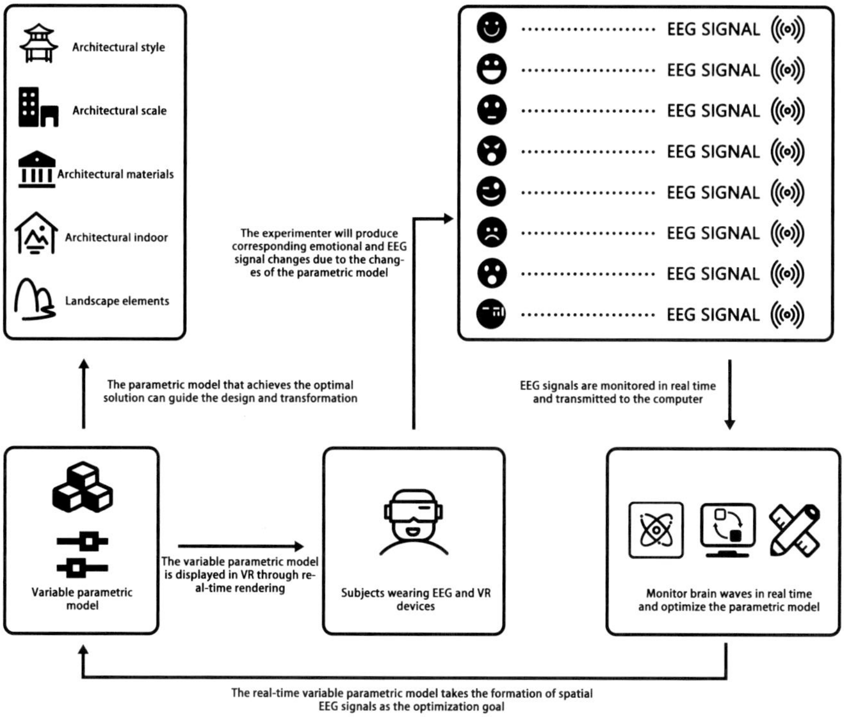

Fig. 1. EEG-based spatial optimization method

opening size as the optimization object, multi-objective genetic algorithm as the optimization tool, combined with virtual reality scene and parametric linkage model, the tool platform and workflow to realize this method is established. The realization of the research purpose of this study consists of the following three points:

- Develop a tool platform for real-time interactive spatial element optimization based on EEG signals. The tool platform is composed of hardware system and software system.
- Establish the workflow of real-time interactive spatial element optimization based on EEG signals.
- Carry out multiple groups of optimization experiments of real-time interactive spatial element optimization methods based on EEG signals, complete the evaluation of the optimization methods and propose improvement goals through quantitative analysis of experimental data and optimization results.

2 Method

In the experiment, this research involves electrical signal communication, real-time rendering of virtual reality scene, optimization and linkage of parametric model, and the workflow is relatively complex. The required work platform is divided into hardware platform and software platform.

On the hardware platform, this study assembled a single electrode ear clip brain wave head ring through TGAM EEG module, which can monitor human brain waves in real-time α Wave β Wave γ Wave, etc., and calculate people's meditation value and attention value through the black box (meditation here represents people's sense of calm and pleasure, which is a relaxing EEG feature, and attention here represents concentration value, which is an EEG feature generated when people pay attention and tension in the brain).

In this study, the TGAM EEG module will continuously send human EEG signals through Bluetooth in real-time during the experiment. The EEG signals are preprocessed through the Arduino development board, and the processed meditation value and attention value are sent to the computer serial port. The computer is used to run the optimization program and transmit the changing parametric model to oculus rift s virtual reality glasses through real-time rendering.

On the software platform, the preprocessing of the uploading program of the Arduino development board is completed by writing C language in Arduino ide. Its purpose is to capture and process the original EEG data sent via the TGAM EEG module, convert hexadecimal into binary language and input it into computer serial port. Based on Grasshopper, this research uses the data read from the serial port as the reference of the optimization algorithm in real-time, establishes the standard bedroom unit with variable color and window hole size, links it to the real-time rendering software based on twinmotion platform through the program, and transmits the virtual reality scene to oculus rift s virtual reality glasses through twinmotion. In this study, wallacei multi-objective genetic algorithm based on the Grasshopper platform is used as the main optimization algorithm to optimize the color and window opening size of typical bedroom units with real-time reference to the optimization objectives. Figure 2 shows the workflow of the optimization tool platform.

2.1 Workflow

This section includes the establishment of the initialization virtual scene, the establishment of a variable library, and the design of the basic experimental process. The design of the basic flow of the experiment includes the methods of establishing the key steps of this research, such as the real-time reading and processing of EEG data on the Grasshopper platform, the real-time rendering linkage of parametric model and virtual reality scene, and the adaptation of black-box optimization algorithm on the Grasshopper platform.

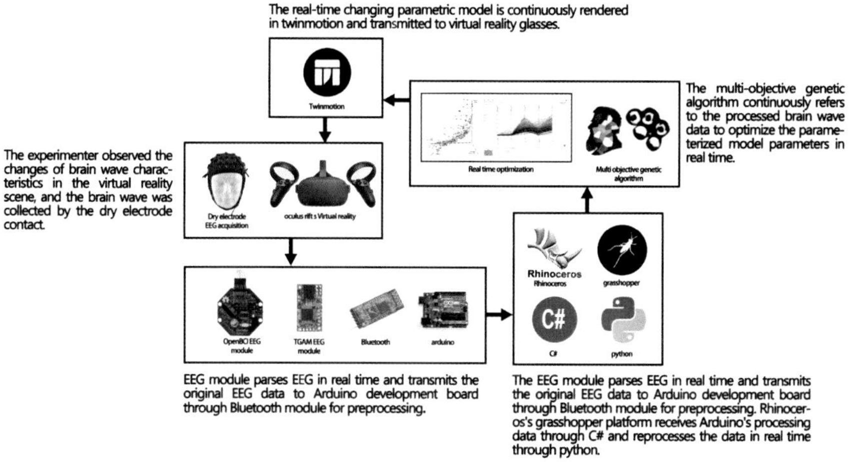

Fig. 2. EEG-based spatial optimization tool platform workflow

2.1.1 Establishment of Initial Virtual Reality Scene

The real-time interactive optimization experiment has high requirements for the linkage between software platforms. In this study, a bedroom model with a bay of 4 m and a depth of 5 m, and a height of 3 m is established on the Grasshopper platform. The window is located on the south side, and the initial window size is to scale the south wall shape on its plane to 0.5 times of the original shape. The initial indoor color of this bedroom model is white (all three RGB values are 255). The interior model does not contain other objects, and the color and material characteristics of the floor, wall, and ceiling are the same.

2.1.2 Establishment of Variable Database and Optimization Objectives

As mentioned above, the initialization scene is the optimization subject of this study. The typical bedroom model of this scene will be observed through indoor observation in this study. This subject contains five optimization variables, namely, the variable parameters controlled by the program in the optimization process.

In previous studies, it has been found that light and color will have a great impact on people's psychological feelings in the scene [4, 5]. Therefore, in this study, the window hole size and indoor color that control the amount of light will be optimized, in which the variables controlling the window hole size are X-axis zoom and Y-axis zoom, and X-axis zoom controls the width of the window hole, The Y-axis scaling amount controls the height of the window opening. The color variables in the control room are R, G, and B. Attention represents the degree of tension and concentration, the smaller the better in the calculation, while the value of meditation represents people's degree of relaxation and pleasure, and the larger the better. However, in the experiment of this study, because the characteristic of a genetic algorithm is to find the minimum value of the optimization objective, it is calculated with its negative value in the calculation process.

2.1.3 Arduino Board Program Uploading Method

The EEG signal based on the TGAM EEG module in this study needs to be preprocessed in the Arduino development board to make the Arduino board can read EEG data in real-time and capture the attention value and meditation value in the TGAM module, and then transmit them to Grasshopper platform through the serial port.

Before that, it needs to upload the program to the Arduino board. The EEG mode of receiving the TGAM EEG module on the Arduino board is Bluetooth reception. The pin connection method during program uploading is Arduino 5 V—VCC; Arduino GND—GND; Arduino Pin10—TXD; Arduino Pin11—RXD. After uploading, connect TXD and RXD to pin0 and Pin1 respectively (Fig. 3).

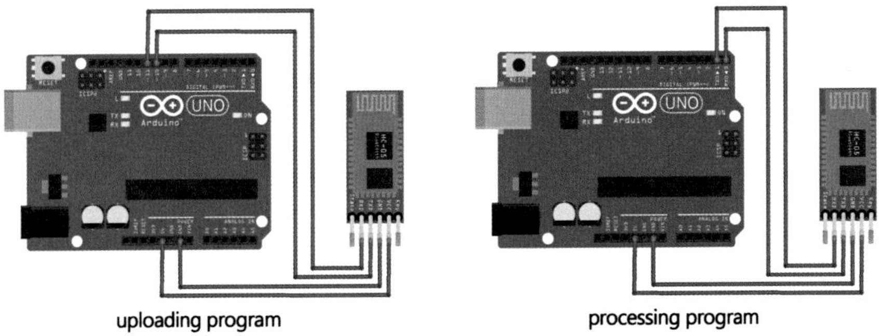

Fig. 3. Arduino board program uploading method

2.2 Adaptation of Optimization Algorithm on Grasshopper Platform

The TGAM EEG module collects the EEG signals of the subjects through the electrodes of the prefrontal lobe and sends the subjects' current meditation value and attention value every second through the module calculation. Both values range from 0 to 100. This study hopes to obtain the space that can make people have higher meditation value and lower attention value through optimization calculation to reduce people's sense of tension and anxiety in the environment and increase the sense of relaxation.

During the experiment, to ensure that the experimenter can produce more reasonable EEG feedback corresponding to the scene during observation, each scene is stopped in front of the experimenter for 5 s, the EEG values of the first and last seconds are discarded, and the average value of EEG in the middle section is taken as the reference value of genetic algorithm (objective). The specific formula is as follows [4, 13, 15].

$$Average\_Meditation = \frac{\int_{t_0}^{t_1} Meditation \cdot dt}{t_1 - t_0} \tag{1}$$

Based on the above, a group of EEG processed data can be generated every five seconds during the experiment as the optimization reference of the optimization algorithm. In this research, human observation participation is generated, so the optimization algorithm needs to wait in the process of optimization calculation. Here, the problem is solved by the "time. Sleep (5)" algorithm. Within the time when other calculations stop for 5 s, the meditation value and attention value will accumulate 5 s of data, and at the end of 5 s, input it as a list to calculate the average value (1). The time of the calculation process is proved to be within 0.5 s after many tests, which is lower than the time of the previous second discarded in the average calculation, and will not be superimposed with time, so it will not affect the accuracy of the optimization algorithm. Figure 4 shows the experimental operation interface.

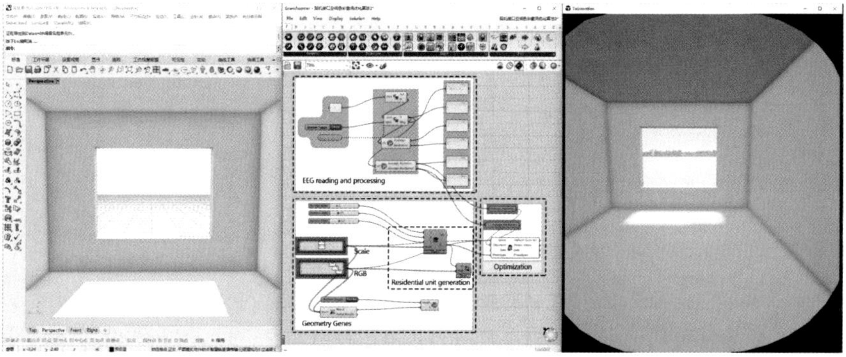

Fig. 4. Experimental operation interface

3 Experiment

To test the applicability of the optimization method in this study, 50 volunteers of different ages were recruited for the optimization experiment. Of the 50 volunteers, 52% were male and 48% female, with 4% children, 76% youth, 16% middle-aged, and 4% elderly. This section will discuss the experimental environment preparation, experimental process, experimental results, and analysis.

As shown in Fig. 5, the experimental environment is a special EEG and Eye movement laboratory. The laboratory provides the subject with an environment with less external interference and allows the researcher to observe the experimental process. Before the experiment, the subjects wear EEG equipment first and then wear VR glasses. After the EEG runs stably, the experiment can be started. The genetic algorithm in the experiment is initially set to iterate (Generation Count) 20 times, with 10 biomass (Generation Size) per generation. The experimental time is estimated by the software to be about 18 min and 31 s.

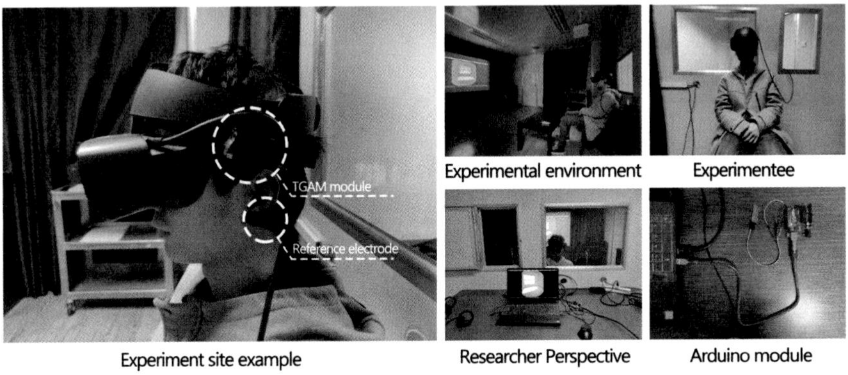

Fig. 5. Experiment site example

3.1 Experimental Results and Quantitative Analysis

Figures 6 and 7 respectively show the parallel coordinate plot and optimization target distribution diagram of 10,000 optimization target data in the optimization experiment of 50 volunteers.

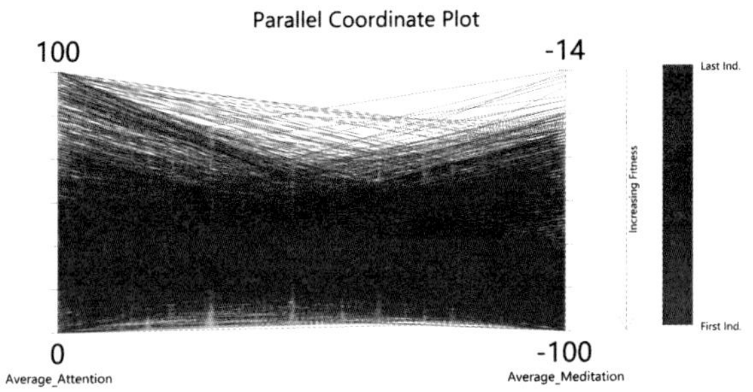

Fig. 6. Parallel coordinate plot

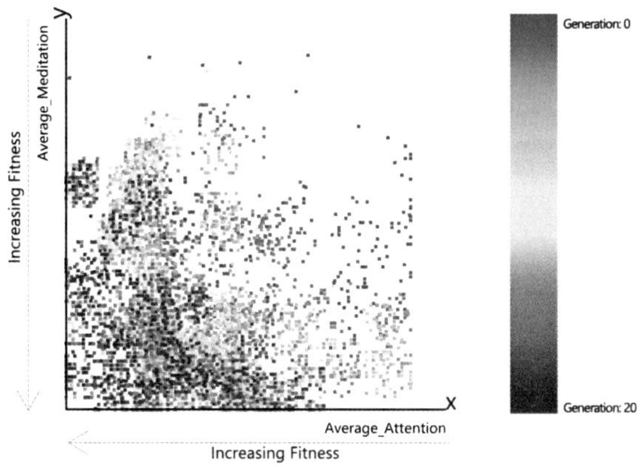

Fig. 7. Objective space

In the parallel coordinate graph (PCP), each line represents the optimization target value of a single optimization individual, the red line represents the results at the bottom of the ranking, and the blue line represents the results at the top of the ranking. It can be seen from Fig. 4 that the optimization effect of the genetic algorithm is obvious. Through optimization, even one of the experimentees has produced better results with attention as low as 2.5 and meditation as high as 98.5. By indexing the result, it can be got that the index is (Gen: 12 | ind: 3) (Gen: 19 | ind: 3), that is, the 4th individual of the 13th generation and the 4th individual of the 20th generation (the index in the computer starts from 0 by default).

The Objective Space (OS) remaps the optimization target value of the analog output and specifies a different axis for each target. Average_Attention and average_ meditation are displayed on the X and Y axes respectively. In the optimization experiment, the non-dominant Pareto optimal polyline can be calculated from the optimization target distribution map, and the better optimization results of a single experimenter can be obtained. Taking one of the experimentees as an example, it can be got from the results that there are nine non-dominant Pareto optimal solutions of the 20th generation results, and one of them coincides with the optimal solution obtained from the above PCP analysis. To sum up, nine excellent optimization results of the experimentee can be obtained through PCP and OS analysis methods. Table 1 shows the example of optimization results from one of the experimenters.

From the whole optimization process, referring to the generation average value trend chart (MV) of the optimization target, as shown in Fig. 8, it can be got that with the optimization iteration, Average_Attention value decreases obviously in the optimization process, while the Average_Meditation calculated by negative value shows a significant decrease in optimization, that is, the positive value of this value increases significantly. The larger red dot in Fig. 8 represents the average value of the optimization target generation of the results of the 20th generation. It can be seen that the optimization of window opening size and indoor color based on EEG signal in this study is effective.

Table 1. Example of PCP and OS optimization filter results

X_Scale:0.3	X_Scale:0.5	X_Scale:0.4	X_Scale:0.3	X_Scale:0.3
Y_Scale:0.7	Y_Scale:0.7	Y_Scale:0.4	Y_Scale:0.7	Y_Scale:0.7
R:78.0	R:185.0	R:241.0	R:68.0	R:185.0
G:193.0	G:195.0	G:201.0	G:193.0	G:195.0
B:226.0	B:168.0	B:226.0	B:226.0	B:226.0
Average_Attention:50.75	Average_Attention:28.5	Average_Attention:168.75	Average_Attention:39.0	Average_Attention:4.2
Average_Meditation:50.75	Average_Meditation:82.75	Average_Meditation:56.6	Average_Meditation:89.0	Average_Meditation:45.6
Gen:19 | Ind:0	Gen:19 | Ind:1	Gen:19 | Ind:2	Gen:19 | Ind:3	Gen:19 | Ind:4

X_Scale:0.3	X_Scale:0.3	X_Scale:0.3	X_Scale:0.3
Y_Scale:0.7	Y_Scale:0.7	Y_Scale:0.7	Y_Scale:0.7
R:68.0	R:62.0	R:185.0	R:68.0
G:193.0	G:193.0	G:195.0	G:178.0
B:167.0	B:167.0	B:226.0	B:231.0
Average_Attention:43.5	Average_Attention:31.4	Average_Attention:14.5	Average_Attention:35.8
Average_Meditation:93.25	Average_Meditation:88.2	Average_Meditation:79.75	Average_Meditation:88.6
Gen:19 | Ind:5	Gen:19 | Ind:6	Gen:19 | Ind:7	Gen:19 | Ind:8

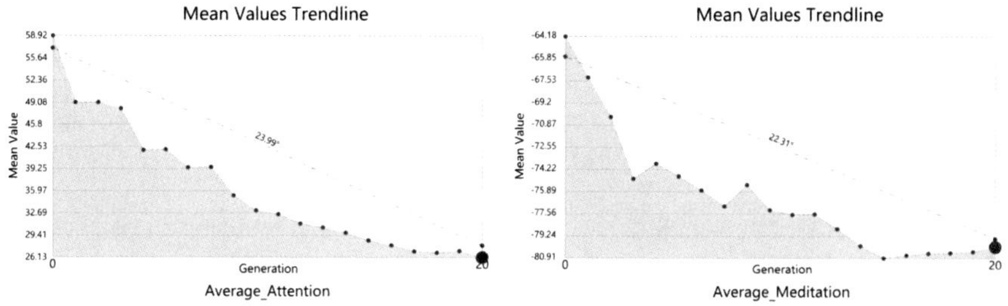

Fig. 8. Mean value trendline of Average_Attention and Average_Meditation value

4 Conclusion and Discussion

In this paper, it can be found that it is feasible to take human-specific EEG signals as the optimization goal of optimizing building space elements. In the experiment of taking the size of the window opening and the color of the indoor space as an example, the optimization results are selected according to the Pareto optimal method. The optimization effect of the genetic algorithm with the optimization objectives of meditation value and attention value is obvious. This study will further analyze the optimization results in this experiment in further research, and try to analyze the characteristics and scope of the optimization results by clustering.

The closed-loop optimization model established in this study from EEG equipment to grasshopper platform, and then to the real-time rendering engine to VR equipment will be used to further optimize a wider range of scenes involving building volume, architectural style, and landscape elements, etc. The applicability scenario of the optimization algorithm in further research will be expanded and the development of more possibilities of man–machine coupling design in artificial intelligence technology will be promoted. Based on this research, complex multi-objective and multi-factor real-time interactive optimization will become possible.

References

1. Alshammari MA, Alshammari TK (2021) COVID-19: a new challenge for mental health and policymaking recommendations. J Infect Publ Health (9393)
2. Yj A, Ts A, Pz B, Ja C (2021) Mass quarantine and mental health during COVID-19: a meta-analysis
3. Huang WX, Xu WG (2009) Interior color preference investigation using interactive genetic algorithm. J Asian Archit Build Eng 8(2):439–445
4. Li Z, Sun X, Zhao S, Zuo H (2021) Integrating eye-movement analysis and the semantic differential method to analyze the visual effect of a traditional commercial block in Hefei, China. Front Archit Res 10(1)
5. Msa A, Szaj B, Ku C (2021) Analyzing recognition of EEG based human attention and emotion using machine learning
6. Katmah R, Al-Shargie F, Tariq U, Babiloni F, Al-Mughairbi F, Al-Nashash H (2021) A review on mental stress assessment methods using EEG signals. Sensors 21(15)
7. Devi D, Sophia S, Janani AA, Karpagam M (2020) Brain wave-based cognitive state prediction for monitoring health care conditions

8. Sampaio C, Magnavita G, Ladeia AM (2021) Effect of Healing Meditation on stress and eating behavior in overweight and obese women: a randomized clinical trial. Complement Ther Clin Pract 45:101468
9. Delorme A, Grandchamp R, Curot J, Barrangan-Jason G, Valton L (2020) Effect of meditation on intracerebral EEG in a patient with temporal lobe epilepsy: a case report. EXPLORE J Sci Heal 17(3)
10. Fell J, Axmacher N, Haupt S (2010) From alpha to gamma: electrophysiological correlates of meditation-related states of consciousness. Med Hypotheses 75(2):218–224
11. Amha B, Hb A, Cp A, Skab C (2021) Sustained effects of mantra meditation compared to music listening on neurocognitive outcomes of breast cancer survivors: a brief report of a randomized control trial. J Psychosom Res
12. Kanchibhotla D, Sharma P, Subramanian S (2021) Improvement in gastrointestinal quality of life index (GIQLI) following meditation: an open-trial pilot study in India. J Ayurveda Integr Med 12(1)
13. Zuo HW (2019) A visual quantitative study on "Second Contour" of the historic blocks: taking the Tunxi ancient street. Anhui Ex Modern Urban Res 01:88–93
14. Cheng CZ, Li H (2018) A comparative study of Chinese and foreign architectural styles based on EEG technology. Huazhong Archit 36(4):4
15. Li Z, Munemoto J (2010) Comparative study on waterscaped and non-waterscaped spaces using electroencephalogram analysis audio-visual experiment on outer spaces of Chinese residential quarters basing on EEG measurement. J Archit Plan (Transactions of AIJ) 75(647):67–74
16. Liu BY, Fan R (2013) Research on visual attraction elements and mechanism of landscape space. Chin Landsc Archit 29(5):6
17. Li K, Gong C (2015) Application of EEG technology in the analysis of landscape differences of tourist trails. Central South For Inven Plan 34(2):6
18. Macruz A, Bueno E, Palma GG, Vega J, Palmieri RA, Wu TC (eds) (2022) Measuring human perception of biophilically-driven design with facial micro-expressions Analysis and EEG Biosensor. Springer Singapore, Singapore
19. Kandel E, Schwartz J, Jessell T, Siegelbaum S, Hudspeth AJ (2013) Principles of neural science, 5th edn. Principles of Neural Science
20. Lakhdari K, Sriti L, Painter B (2021) Parametric optimization of daylight, thermal and energy performance of middle school classrooms, case of hot and dry regions. Build Environ 204(2):108173
21. Suga K, Kato S, Hiyama K (2010) Structural analysis of Pareto-optimal solution sets for multi-objective optimization: an application to outer window design problems using Multiple Objective Genetic Algorithms. Build Environ 45(5):1144–1152
22. Li L (2012) The optimization of architectural shape based on Genetic Algorithm. Front Archit Res 1(004):392–399
23. Aljalal M, Ibrahim S, Djemal R, Ko W (2020) Comprehensive review on brain-controlled mobile robots and robotic arms based on electroencephalography signals. Intell Serv Robot 13(3)
24. Shen X, Wang X, Lu S, Li Z, Wu Y (2021) Research on the real-time control system of lower-limb gait movement based on motor imagery and central pattern generator. Biomed Signal Process Control 102803

A Sunlight Duration Time Driven Multi-objective Optimization Method for the Layout of High-Rise Residential Quarters Based on NSGA2 Algorithm

Ze Zhang and Zhengwang Wu(✉)

School of Architecture, Huaqiao University, Xiamen, Fujian, China
wuzhengwang@126.com

Abstract. Extending sunlight duration time by optimizing the layout of the high-rise residential quarters during the early design stage is one of the most effective approaches to reducing carbon emissions. This paper proposes a multi-objective optimization method for high-rise residential quarter layout based on the NSGA2 algorithm. The method is aimed to maximize the first floor's sunlight duration time and its uniformity both. A simulated plot in Xiamen is taken as an example for multi-objective optimization. After the optimization, the layouts are analyzed and the better one is selected. The results show that the proposed method can achieve higher overall sunlight duration and its uniformity rate and maximize floor area ratio in the early design phase. However, the proposed method has its drawbacks. This method requires the pre-design of the building plan. The algorithm generates a lot of invalid solutions during the optimization. The optimization time increases dramatically with the quantity increase of input parameters. According to the above, there is still room for improvement in the proposed method.

Keywords: Residential quarter · Sunlight duration time · Multi-objective optimization · WallaceiX

1 Introduction

The urbanization process has accelerated since the Reform and Opening, resulting in cities full of skyscrapers with a typical example of high-rise residential quarters. In the design of high-rise residential quarters, sunlight is a crucial factor that determines whether the project can continue. According to statistics, the nationwide building life-cycle carbon emissions in 2018 were 4.93 billion tons of CO_2, and that of the operation phase is 2.11 billion tons, accounting for 42.8% of building lifecycle carbon emissions [1]. Maximizing sunlight duration during the daytime by adjusting the layout in the early design stage to provide users with a good daylighting environment can effectively reduce the energy consumption and carbon emission of daytime lighting.

P. F. Yuan et al. (eds.), *Hybrid Intelligence*, Computational Design and Robotic Fabrication,
https://doi.org/10.1007/978-981-19-8637-6_12

As shown in Fig. 1, the traditional workflow relies on designers' experience, but they can hardly obtain a plan that is optimal in multiple dimensions in a short time. Most of the designs just meet compulsory codes, while other criteria are neglected. Generative design research has become a trend in recent years to solve these problems. In this way, designers can improve the efficiency in the early design stage, and optimize building performance to reduce carbon emissions.

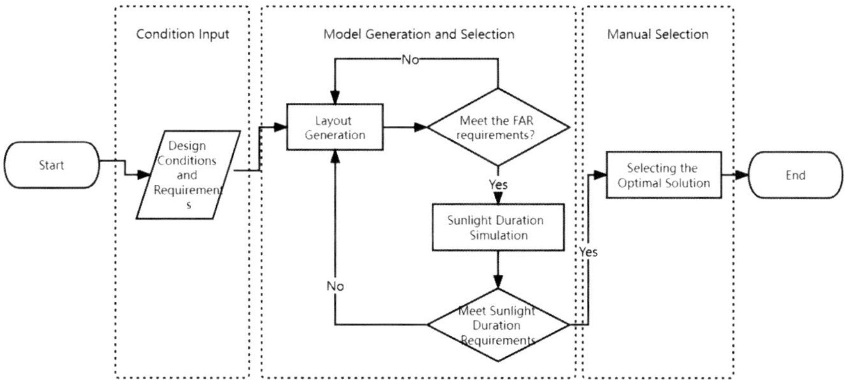

Fig. 1. Traditional design workflow

There is much research about building group layout optimization, and the methods can be classified into three categories: agent-based model (ABM), genetic/annealing algorithm and artificial intelligence (AI). ABM: JI Guohua et al. explored the optimization of residential quarters based on Netlogo [2]. LI Biao et al. used the highFAR to optimize the layout of a residential quarter and calculated the shadow area [3, 4]. YUAN Feng proposed an ABM based on Rhino, GH, and python to generate wind environment performance optimal building group layout [5]. Genetic/Annealing algorithm: GAO Fei optimized the layout driven by sunlight duration based on Geco and Galapagos [6]. LIU Ke et al. used genetic algorithm engine with Galapagos to perform energy efficiency optimization of residential quarter layout [7]. AI: Xcool Technology has made great strides in residential quarter master plan design relying on AI [8]. DENG Qiaoming et al. implemented the generative design of campus layout through generative adversarial networks (GAN) [9].

The methods above are gradually moving to application. ABM and AI are difficult to shift to the industry in a short time due to their high coding demand and genetic/annealing algorithms are preferred due to their out-of-the-box feature. A lot of the research focuses on single-objective optimization, while in real design scenarios, the optimization objectives of design problems are often multiple conflicting indicators. Therefore, based on the Grasshopper platform, this study proposes a multi-objective optimization method for the layout of high-rise residential quarters based on sunlight duration time using the NSGA2 algorithm plugin WallaceiX [10] and building simulation plugin Ladybug Tools [11], then conducts an analysis of the optimization results with inbuilt tools to test the optimization capability. With the help of such methods, designers can efficiently complete the design of the residential quarter layout in the preliminary design phase and can add other building performance simulation tools for energy, ventilation, and microclimate simulation on this basis to further optimize the overall building layout, optimize building performance, and reduce carbon emissions.

2 Research Methodology

The problem can be translated into an optimization problem: within a given site, the layout of the residential quarter is generated by controlling the location, orientation, and the number of floors, to maximize the total sunlight duration, its uniformity, and FAR.

As shown in Fig. 2, the optimization process consists of five parts. The basic calculation and parametrize section calculates the plot area, building projection area, FAR, and building density and optimize the maximum number of each building type with Galapagos, as a reference for the optimization. In generating model section, the program generates residential models defined by building location, orientation, and the number of floors. In constraint calculation section, the building spacing is represented by no-entry zones according to the relevant codes of Xiamen City. In the sunlight duration calculation section, the sunlight duration of the test surface is analyzed by Ladybug Tools to generate corresponding values. Then determine whether the duration meets the needs. In the genetic algorithm & merit selection section, WallaceiX is used to start and stop optimization and control parameters, together with merit selection and breeding.

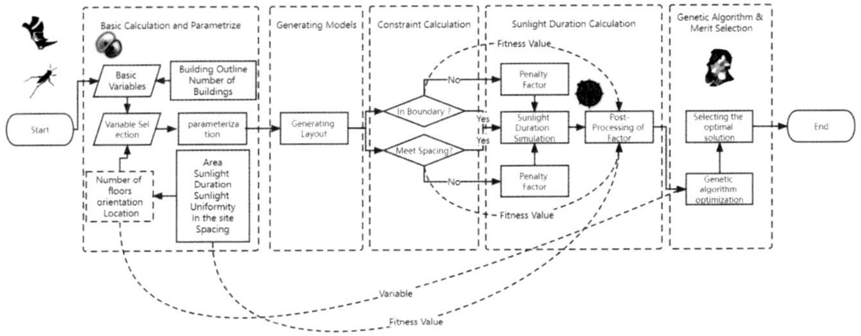

Fig. 2. Workflow of the proposed multi-objective optimization

2.1 Basic Calculation

The experiment simulates an L-shaped irregular plot in Xiamen. The site area is about 33,671.6 m^2. Outside the site in the northeast is an existing residential area, with three 6-story residential buildings, and the sunlight duration is up to standard. According to the regulations, the property line setback according to the upper limit of 18 m in the north–south direction and 15 m in the east–west direction (Fig. 3).

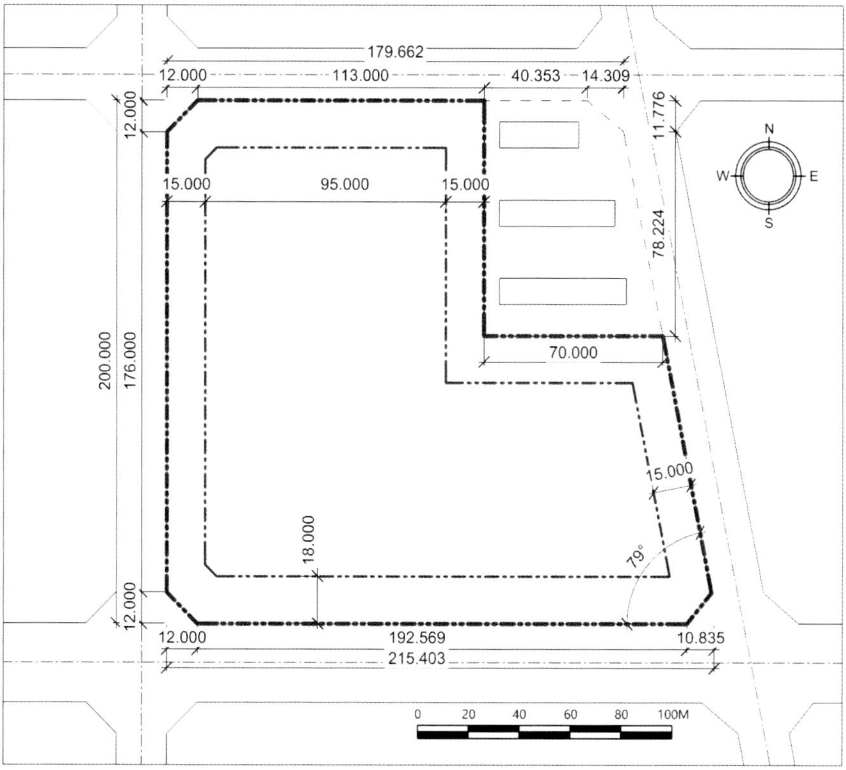

Fig. 3. Plan of the plot

Residents have higher requirements for living quality, reflected in the demand for housing width resources, therefore, Large-width, slab-type buildings are more popular. Two similar types of buildings are arranged within the setback Line: "long-slab type" 53 m*15 m, and "short-slab type" 26 m*15 m, both simplified into rectangles. According to the "Technical Regulations for Urban Planning and Management of Fujian Province", high-rise residential quarters with a land area of more than 30,000 m^2, FAR shall under 2.9, and a building density of 20%, so the maximum building area is 97,647.7 m^2, and the maximum building projection area is 6734.3 m^2. Using Galapagos to optimize the number of two types of buildings mentioned. Getting a result of 5 short-slab buildings and 6 long-slab buildings. To reduce the building density, it is changed to 3 short-slab buildings and 4 long-slab buildings. In this step, the result generated by the algorithm is only for the designer's reference.

2.2 Parameters of Model

In this optimization, the variables are designed according to the problem: building location, building orientation, and the number of floors. The center of building projection is the reference point to the building position. There are two ways to generate the reference point position: coordinate method and point grid method. As shown in Fig. 4, for complex shaped site, method 2 has no points outside the site, saving computing power. Generating a total of 67,011 points in the site at an interval of 0.5 m. Rotation axis point is also the center of the projection, the default angle is 0°. Floor number of building is expressed as an integer.

In order to improve the efficiency of the program, the continuous variables are discretized to reduce the search space. The step of position is 0.5 m. The rotation angle of the building is −45° to 45°, with a step of 1°. The number of floors varies from 8 to 27 floors, with the same step of 1, and the story height is 2.9 m.

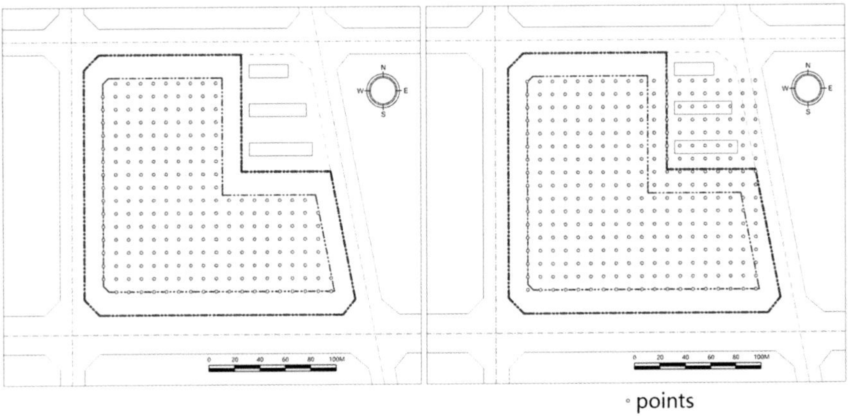

Fig. 4. Two ways generating points

2.3 Constraint Design

The generated plans have building overlap and mass intersection problems, so it is necessary to construct a constraint mechanism that generates penalty coefficients to clarify the optimization direction of the program and improve the computing efficiency. There are two constraints in this optimization: boundary line constraint and spacing constraint.

1. Setback line constraint. The generated buildings need to be in the setback lines. Set the area of the setback line as S. The area that represents merged surfaces of the setback line and building projections is set as S_1. If $S_1 - S = 0$, which means all the buildings are in the setback lines, the program proceeds with the actual value. If $S_1 - S > 0$, then there is at least a part of buildings out of the setback line, output the penalty coefficient to all the fitness values.
2. The spacing constraint is based on the Codes of Xiamen City Urban Planning Management. According to the height of the building on the south side of each building, or, its own height (if there is no building on the south side), offsetting 14–24 m forward and backward, and offsetting the hill wall 6.5–8 m. (as Fig. 5) The area of a single no-entry zone is set as S_i. The sum of them is set as $\sum_{i=1}^{7} S_i$. The area of

merged no-entry zones is set as S. If $\sum_{i=1}^{7} S_i - S = 0$, then there is no overlapping of no-entry zones and the layout is valid, proceeding with the actual value. If $\sum_{i=1}^{7} S_i - S > 0$, it means that there is overlapping of no-entry zones, output the penalty coefficient to each fitness value.

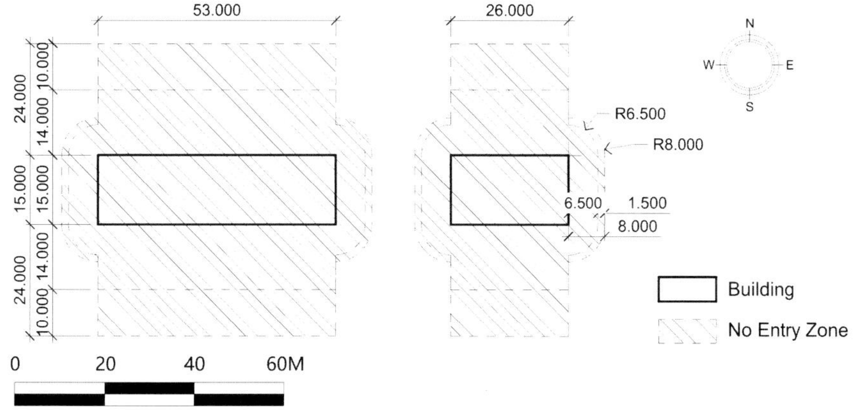

Fig. 5. No entry zone map

Only when layouts satisfy both the setback line and spacing constraints can the algorithm output actual values. If not, the fitness values are multiplied by penalty coefficients to make them higher, thus prompting the engine to optimize to the smallest value.

2.4 Sunlight Duration Calculation

According to the requirements of the Standard for Urban Residential Area Planning Design. GB 50,180–2018, the sunlight duration of the first-floor windows is at least 2 h on January 20th [12]. A test surface was generated at 1.35 m height of the south elevation and divided into multiple sampling points. There are three existing residential buildings in the northeast corner, so there will be shading influence between proposed buildings and existing buildings. The sunlight duration of each sampling point on the test surfaces on that day can be measured. Whether each measurement point meets the duration requirement, the overall sunlight duration, the uniformity of sunlight duration, and the total sunlight duration of existing buildings can also be measured. The effective sunlight time on January 20th is 8:00–16:00, a total of 8 h.

2.5 Fitness Value Design and Multi-objective Optimization

In the optimization, 7 objectives are set, which are: whether each sampling point meets the duration requirement, total sunlight duration, sun-light duration uniformity, total floor area, sunlight loss of existing buildings, whether any building exceeds the setback line, and whether any buildings' no-entry zones overlapped. The objectives are converted into fitness values that suit the engine feature of the WallaceiX. Then WallaceiX will optimize the fitness to the minimum.

"Whether each measurement point meets the duration requirement" is translated into the number of measuring points with sunlight hours less than 2 h. "Sunlight uniformity" is transformed into the value of 1 minus the ratio of the minimum value to the mean value. "Total sunlight duration" is transformed into the reciprocal of the sum of sunlight duration at sampling points of planned buildings. "Total floor area" is transformed into the difference between the maximum floor area and the floor area of planned buildings. "Sun-light duration loss of existing buildings" is transferred to the number of measuring points with sunlight hours less than 2 h on existing buildings. Whether there is a building crossing the setback line and whether there is a building not meet the building spacing requirements have been described in the previous section.

The location, rotation angle, and the number of floors of each building are input into the WallaceiX as variables and the above 7 fitness values to the optimization objectives. Then set up the genetic algorithm config: the optimization last 500 generations, with 50 results per generation, and the rest of the settings follow the default. When optimization starts, the WallaceiX automatically changes parameters in the gene pools and forms the building layout, and will output the fitness values, recording the relevant results and data in the memory for retrieval in the subsequent analysis. The optimization will stop automatically after reaching the number of iterations. Afterward, select the optimal layout with the help of GH definition and built-in tools.

3 Result and the Optimal Layout

The optimization shows convergence overall. The optimization runs for about 5 h, generating a total of 25,000 phenotypes. As shown in Fig. 6, the overall change curve of each fitness value shows a fluctuating decreasing trend, and the standard deviation value also tends to be a low value. For most of the fitness values from the beginning to 500 generations of the optimization process, its standard deviation decreases rapidly and enters a stable state. The standard deviation curve of most fitness values shows a narrow peak and the peak of the blue curve is shifted to the left compared with the peak of the red curve, indicating better mean performance. The overall trend of FO5 (out of the building plot red line) is different from that of the other fitness values, which oscillates in the process of optimization. In summary, after 500 generations of iterations, the optimization shows a trend of convergence and stability, and better layouts can be obtained by selection.

Using the Pareto frontier function to find the set of layouts where the optimal layout lies, there exists a certain number of Pareto optimal layouts in such multi-objective optimization problems. The layout on the Pareto front is selected as the better layout. Since the weights of all the fitness values in the optimization are the same, it is necessary to perform a secondary selection of these layouts according to the priority of each value in the designer's mind. A total of 50 layouts are output using this method. They are all 500th generation, as shown in Fig. 7.

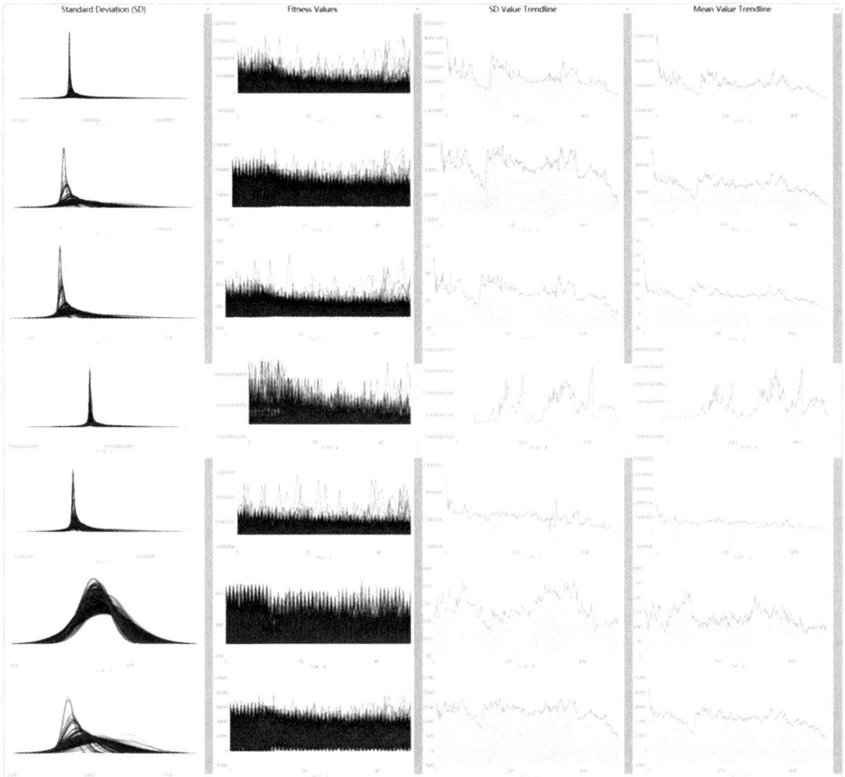

Fig. 6. Built in optimization analysis chart

Corresponding fitness values are extracted from the selected results for comparison, and further optimization is carried out. The layouts that meet the two constraints at the same time with optimal sunlight duration are selected. These layouts are analysed with the diamond diagram analysis tool. The diamond diagram reflects the fitness rank of the layout. The closer to the center point, the lower the fitness rank is, and the better the layout is. Through the comprehensive comparison of the 7 objectives, the optimal layout is selected from the layout set. As shown in Fig. 8, the layout numbered Gen499 Ind29 has the best comprehensive fitness values, and the number of points for insufficient sunlight duration is 0; The sunlight duration uniformity is 0.33; the sum of the sunlight duration is 846 h; the building area is 100,815 m^2 (the maximum is 97,648 m^2, so it needs to be adjusted); 5 test points of existing buildings cannot meet the sunlight duration requirement. The area out of the setback line is 166.8 m^2, and the overlapping area of the no-entry area is 318 m^2. In conclusion, this layout is the optimal layout for this optimization, but it still needs to be further optimized. Adjusting the No. 4 building plan to the short slab type and rotating No. 3 and No. 7 buildings to 0°. The adjusted layout basically meets all the constraints and maximizes the floor area and total sunlight

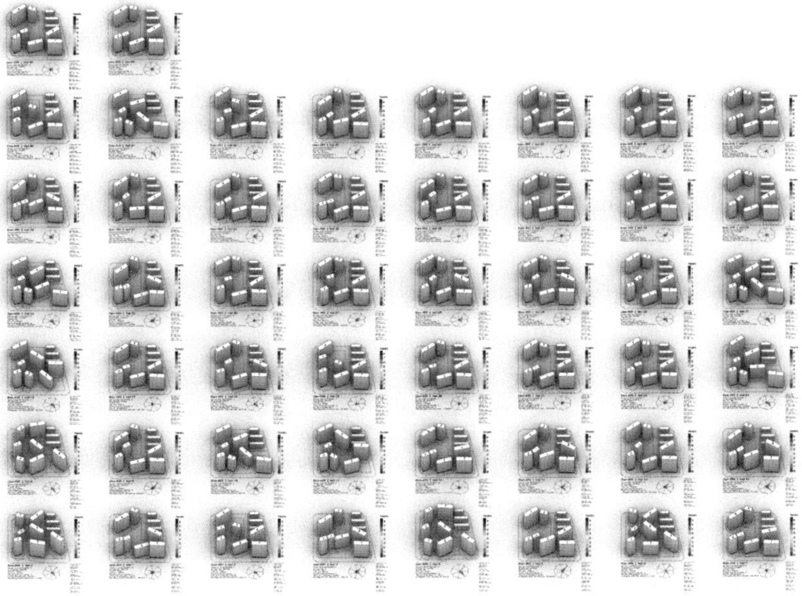

Fig. 7. Output better layouts

duration on the first floor. After adjustment, the objectives are more reasonable, and in this layout, buildings have different rotation angles and floors. Such a layout is difficult to derive from traditional means.

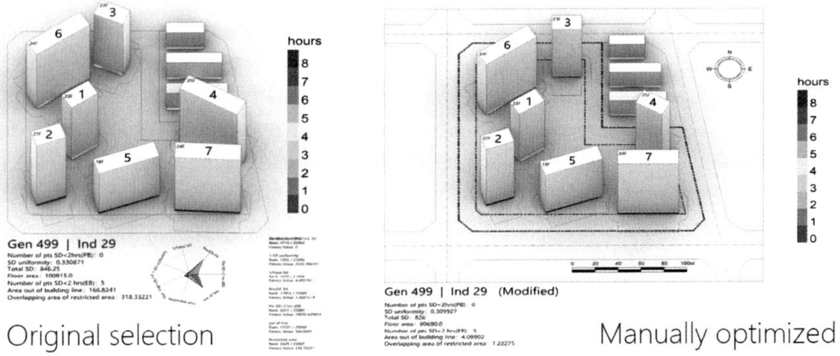

Fig. 8. Optimal layout diagram

4 Conclusion

The optimization results show that this method can significantly optimize all aspects of the sunlight duration and other indicators, thus obtaining better effects during the preliminary design of the residential quarter, which is more efficient than the traditional approach.

4.1 Advantages of Multi-objective Optimization

In traditional design workflow, architects can only consider the minimum requirements of sunlight duration while considering the building layout, ignoring other subtle objectives such as the uniformity of sunlight duration, etc. The experiment demonstrates the overall improvement of multiple indicators. The algorithm can break through the limits of human brainpower and experience to find optimal solutions. This type of design process is only limited by the constraints set, without the limitations of human cognition, and can even find a global optimum layout that humans can't.

The optimization shows strong potential for sunlight duration-driven objective optimization in residential quarter layout design. In this paper, three variables are selected and optimized roughly due to time and computing power limitations, and also achieve better results for the sunlight duration. In the future, designers can add more variables (CFD, noise, view, microclimate) for the automatic generation and optimization of the residential quarter according to the design requirements, and assist them with building energy simulation, which can achieve a more significant energy-saving effect.

4.2 Shortcomings of This Algorithm at this Stage

In this experiment, there are about 4.9*1064 possible solutions in the search space. Although this method is designer-friendly, simple, and effective, it is still time-consuming. When the number of objectives is greater than 3, the selection pressure of the NSGA2 algorithm in this plug-in decreases, thus the optimization outcome becomes unsatisfying. This problem is solved in the NSGA3 algorithm, so the NSGA3 algorithm should be used in the subsequent multi-objective optimization.

Due to the combinations of multiple genetic pools of different values, there will be a large number of invalid layouts. The more parameters applied, the more buildings generated, and the greater the probability of such cases, resulting in the computing power waste on such layouts. When the designer is not clear about the design problem, it could be even worse. Some records show that when the number of buildings increases, the time consumed rises from 1.5 to 60 h, and there is no guarantee of an optimal solution [13]. In this experiment, there are about $4.9*10^{64}$ possible solutions in the search space. Although this method is designer-friendly, simple, and effective, it is still time-consuming. When the number of objectives is greater than 3, the selection pressure of the NSGA2 algorithm in this plug-in decreases, thus the optimization outcome becomes unsatisfying. This problem is solved in the NSGA3 algorithm, so the NSGA3 algorithm should be used in the subsequent multi-objective optimization.

This experiment provides a prototype of the residential layout design workflow for practitioners. The energy conservation and carbon emission reduction of new housing needs to be paid attention to, and it is very important to optimize the layout of the residential quarter in the early stage of design. Maximizing the use of solar energy by design is still the first choice for energy saving. If such a workflow can be generalized, a huge amount of carbon emissions can be saved.

References

1. China Building Energy Consumption Annual Report 2020 (2021) Build Energy Effic 49(2):1–6. https://doi.org/10.3969/j.issn.2096-9422.2021.02.001.chi
2. Liu H, Ji G (2009) Automatic planning of residential quarter under insolation condition based on multi-agent simulation. Archit J (z1):12–6. https://doi.org/10.3969/j.issn.0529-1399.2009.z1.004.chi
3. Biao L, Jingping Q (2011) Exploration of generative method based on MAS for architectural design: highFAR. New Archit (3):99–103. https://doi.org/10.3969/j.issn.1000-3959.2011.03.022.chi
4. Biao L (2008) A generative tool base on multi-agent system 335 a generative tool base on multi-agent system subtitle: algorithm of "HighFAR" and its computer programming
5. Jiawei Y, Chenyu H, Yuan PF (2020) Research on the generation method of building group driven by environmental performance based on multi-agent system. In: Quan, Guogaodengxuexiaojianzhuxuezhuanyezhidaoweiyuanhui, 'editor'. 2020 National conference on architecture's technologies in education research, 1 Jan 2020, pp 152–7
6. Fei G (2014) High-rise residential automatic layout based on sunshine effect. Nanjing University
7. Ke L, Xu X, Wei W (2021) Research on automatic optimization method of residential district layout based on energy efficiency. Ind Constr 51(8):1–10, 27. https://doi.org/10.13204/j.gyjzG20111404.chi
8. Wanyu H (2019) From competition, coexistence to win-win—relationship between intelligent design tools and human designers. Landsc Archit Front 7(02):76–83
9. Qiaoming D, Wenqiang L, Yubo L, Lingyu L (2021) Exploration of generative design of campus general layout based on generative adversarial network: taking primary school campuses as example. World Archit 09:115–119
10. Makki M, Showkatbakhsh M, Tabony A, Weinstock M (2019) Evolutionary algorithms for generating urban morphology: variations and multiple objectives. Int J Archit Comput 17(1):5–35. https://doi.org/10.1177/1478077118777236
11. Pak M, Smith A, Gill G (2013) Ladybug: a parametric environmental plugin for grasshopper to help designers create an environmentally-conscious design
12. Standard for urban residential area planning design. GB 50180-2018.2018. Chinese
13. Hui-xing L, Ran Z, Guo-hui F, Kai-liang H, Chi-hong C (2015) Comparison of air conditioning load between a designed nearly zero-energy building and a common building in cold region. Build Energy Effic 43(6):10–2. https://doi.org/10.3969/j.issn.1673-7237.2015.06.002.chi

Mapping Plant Microclimates on Building Envelope Using Environmental Analysis Tools

Ana Zimbarg(✉)

Florida International University, 11200 SW 8th Street, Miami, FL 33199, USA
aczimbarg@gmail.com

Abstract. Can we build our cities not only for humans but also for all living systems? How can we consider other species occupants of the built environment? Planning cities as an element of the natural domain can reshape our relationship with nature and help redefine sustainability in architecture. Although current design strategies of reducing energy use does not rectify past/continuing im-balances in the natural environment. Landscape architect John Tillman Lyle expanded the regenerative design concept based on a range of ecological concepts. The environment's complexity, and the urge to use resources smartly, encouraged him to think about architecture and the environment as a whole system. John Lyle's regenerative design strategies scaffold a conceptual framework of treating the building as part of the landscape. Environmental tools such as Ladybug can map out the different conditions surrounding the building's envelope. This information can assist in selecting and populating a building façade with suitable plant species. The framework presents the building as a feature in the landscape, creating microclimatic conditions for various plant habitats. This conceptual workflow has the potential to become a tool to include regenerative principles in the urban context.

Keywords: Regenerative architecture · Bio digital architecture · Sustainable architecture · Urban ecology · Environmental analysis

1 Introduction

Can we use architecture to create new ecological relationships to improve nature[1] in urban areas? What if buildings supported native plants and animals? Human development modifies the surface and landscape soil, destroying small ecosystems or flooding entire valleys. Such interventions accelerate the loss of vegetation, impacting the ecology of the non-human inhabitants of these areas. Planning cities as an element of the natural domain can reshape our relationship with nature and help redefine sustainability in architecture.

[1] With linguistic evolution in popular language, scientists and philosophers have remained cautious with this word. In this paper 'nature' is a term referring to scientific concepts such as "biodiversity", "evolution", "ecosystem", "landscape", "wildness", "population", "community"; the whole of material reality, considered as independent of human activity and history [19].

P. F. Yuan et al. (eds.), *Hybrid Intelligence*, Computational Design and Robotic Fabrication,
https://doi.org/10.1007/978-981-19-8637-6_13

Instead of separating developed zones from nature, it is worth considering our cities as another layer of the landscape, an "open system" of living things that constantly interact with the physical environment [1].

The climate effects of cities result from the urban form of the landscape (extent, materials, building dimensions and density) and urban function (energy, water and material use through the urban metabolism). Increasing vegetative cover and incorporating natural landscape features into cities are the best means of managing climate effects in developed areas [1]. This situation allows us to reconsider current contemporary planning and design methods for sustainable urban growth and recognise the need to design and plan with nature. The current scenario requires a behaviour change by acknowledging our responsibility and vulnerability, to changes in the natural environment [2].

John T Lyle (1934–1998), a professor of landscape architecture at Cal Poly Pomona from 1968 to 1998, developed an approach to design that he called regenerative. His strategy was based on improving resource efficiency and considering solutions that would have a circular life cycle. His design system inspired a conceptual framework to design the built environment for humans and also to consider nature as occupants of the urban space.

Planning cities as an element of the natural domain is a way of rethinking our relationship with nature and redefining sustainability in architecture. Instead of separating developed zones from nature, it is worth considering our cities as another layer of nature. Analogous to a tree, the microclimate on the top of a building will differ from the bottom, requiring different architectural solutions. Environmental tools such as Ladybug can map out these different microclimates throughout a building's exterior. The environmental analysis set to evaluate the whole building's climatic conditions can result in the information necessary to populate the building envelope with adequate plant species. Architectural elements can be inserted into the design to protect and benefit the building's green walls and roof. This conceptual work envisions activating ecological relationships within the built environment, aiming to improve the urban ecosystems instead of focusing on humans and building interiors.

2 Regenerative Design

During the 1970s, the awareness of the impact of human civilisation on the environment changed the way people think about the resources. Altering the way, we think would help us recognise that the natural systems would have to be part of the design process [2]. Regenerative design theories emerged from early sustainable development, which attempted to integrate environmental responsibility, social equity, and economic viability [3]. Many regenerative design outlines appeared since the 1970ties, such as permaculture, developed by David Holmgren, Robert Rodale's regenerative organic agriculture or ecological design, where Sim Van Der Ryn describes restoration and regenerative strategies.

Sustainable design, to John Lyle, endeavours to minimise harm and have a neutral impact on the environment; however, he pointed out that it did not address the harm that human growth had caused to the environment [2]. He explained that landscape throughout time had been conceived in a superficial manner, where the architects were only interested in what was seen on the surface, only concerned visually. Lyle called this approach "shallow forms". All parts within the ecosystem should be considered for a system to be sustainable [6].

John T Lyle expanded the concept by incorporating ecological theories into design [4]. The findings of the environment's complexity, and the urge to use resources more smartly, encouraged him to think about architecture and the environment as a whole system. Lyle's theory focuses on designing landscapes to support ongoing supplies of energy and materials for habitat, daily living, and economic activity, replacing the linear material flow system with a cyclical use of resources [5].

2.1 Regenerative Design Strategies

Considering that natural and social processes make the design more com-plicated, requiring a solid framework. Lyle developed a set of hierarchies in his book Regenerative design for sustainable development to address such complexity. Everything that exists in the natural and built environment has an order.

Lyle's regenerative design, this organisation is called "structural order", which describes the composition between living and non-living. In considering the structure of an ecosystem, all its elements are recognised: rocks, soil, plants, and animal species [7]. Each species inhabits a particular niche in natural eco-systems and maintains ongoing interactions with other species.

The more complex the regeneration goal, the greater are the challenges for de-sign, requiring a different pattern of thought. John Lyle listed some of his de-sign strategies in his book as a "tentative effort to summarise the experience" he had [7]. Nature is highly evolved in water conversion, distribution, filtration assimilation, and storage processes if undisturbed. As a result, using natural processes as a regeneration strategy requires an investigation of on-site resources and processes.

There are current projects that applies strategies introduced by Lyle such as Illura Apartments (Fig. 1) model how architects can contribute to ecology. The Australian building has a green façade that was seeded with grassland species that have been extinct in the area since colonisation, resulting in a resilient green façade that restored the native ecosystem. Projects like this illustrate that incorporating the plants at the early design stages can result in restorative design.

Lyle's second mode of order is the "functional order", related to the flow of energy and materials that distribute the necessities of life to all the species belonging to an ecosystem. These flows determine the dynamic of ecologies and often explain the change it undergoes. Every landscape receives energy from the sun, which is absorbed or reflected by the planet's surface, warming the atmosphere and water, and powering the water cycle. The energy is fixed into living matter by photosynthesis, making its way through the food web and supplying other creatures with energy [7].

Industrial systems use resources independent of the overall flow. Fossil fuel, for instance, makes minimal use of solar energy to maintain heat balance within a building, virtually independent of the larger regional and global heat balance [7]. Lyle proposes that understanding the flows of a locality can promote the development of the site location. In controlling the flow of stormwater, the form of the landscape can hold the water as in a bowl while it penetrates the underground storage, for example.

Fig. 1. Illura apartment buildings **by Elenberg Fraser** in Melbourne, Australia. on the right, the detail of the green facade. Photo: **Peter Clarke**

The last order proposed by John Lyle is the "Locational Patterns". John Lyle suggests that each location has its requirements, and that the development of each area should follow the local parameters. When life appeared on Earth, its evolution was guided by several climate and geological compositions. The species that thrive in a desert are very different from those that thrive in a rainforest. Industrialisation changed the urbanscape. Buildings and cities were an integral part of their location, whereas now, a high-rise office building in South America is not too different from a high rise in North America. Both [7].

Green infrastructure intends to control urban sprawl and protect nature [8] Initiatives to convert urban environments into gardens to increase the ecological base, such as "Green Scaffolding (GS) (Fig. 2), a concept for a modular system that wraps around the façade of an existing building, provide multiple ecosystem services and environmental amenities [4]. Such structures can be established in different sizes, types, and scales. Depending on the design, the structures can provide thermal comfort, help with urban acoustics, or serve as shelter for local species from invasive species [5]. Although these initiatives are a positive manifestation of addressing ecology and climate issues in urban areas, they do not consider nature as a whole. They are just adding patches of green in the city, accentuating the clear separation between built-up areas and nature. No restoration or significant improvement to the natural environment is achieved with Green Scaffolding.

3 Environmental Analysis, Microclimate, and Vegetation

John Lyle's regenerative design inspired a framework to include the vegetation as an 'occupant' of the built environment. Following Lyle's design strategy of letting "Nature does the work" improve the natural flow of energy [7], the proposed workflow will perform an environmental analysis on a building geometry to map out the existing outdoor conditions of a building. Plants thrive where the conditions are correct. It requires the right amount of sunlight, filtered sun or shade. They also depend on the correct

Fig. 2. Eco Boulevard was designed and built in Madrid, Spain, by Ecosistema Urbano. Photo by Emilio P. Doiztua

humidity levels [9]. The evaluation will determine the plant species suitable for each surface area and assist in designing the building to protect the green wall.

3.1 Environmental Analysis

The majority of microclimate evaluation software is not integrated into mainstream architectural software. However, open-source platforms such as Grasshopper and Dynamo, along with their plugins, have made the integration of environmental data more accessible. A better understanding of climate data and human comfort indices impacts designing resilient and energy-efficient buildings [10]. Plugins such as Grasshopper's (Rhino 3D) environmental analysis plugins are used within architectural design environments to offer a socio-ecological assessment during the design decision-making [11]. This tool allows the designer to know the environmental conditions that characterise a given urban environment to mitigate the adverse effects and exploit the positive ones to ensure optimal comfort conditions [12].

Ladybug tools are one of the standard environmental tool analyses, and it has been selected as they can cope with complex and straightforward geometry [13]. Ladybug imports standard EnergyPlus Weather files (.EPW) into Grasshopper and Dynamo. It provides a variety of 2D and 3D interactive climate graphics that help the decision-making process during the early stages of design. The plugin supports evaluating initial design options through solar radiation studies, view analyses, and sunlight-hours modelling. Honeybee provides detailed daylighting and thermodynamic modelling that tends to be most relevant during the mid and later design stages. Specifically, it creates, runs, and visualises the results of daylight simulations using Radiance, energy models using EnergyPlus/Open Studio, and heat flow through construction details using Berkeley Lab Therm/Window. It accomplishes this by linking these simulation engines to CAD and visual scripting interfaces such as Grasshopper/Rhino and Dynamo/Revit plugins. Butterfly is a Grasshopper/Dynamo plugin and object-oriented python library that creates and runs computational fluid dynamics (CFD) simulations using OpenFOAM.

At present, OpenFOAM is the most rigorously validated open-source CFD engine. It can run several advanced simulations and turbulence models (from simple RAS to intensive LES). Butterfly is built to quickly export geometry to OpenFOAM and run several standard airflow simulations applicable to building design. This includes outdoor simulations for urban wind patterns and indoor buoyancy simulations for thermal comfort and ventilation effectiveness [10].

3.2 Microclimate

Climate is a critically significant factor for an organism to thrive, and it involves many components such as temperature, rainfall, relative humidity, and winds. The microclimate on the top of a building will differ from the microclimate on the bottom. To include the natural order in building design is important to understand that the building itself should be treated as a feature in the landscape that creates microclimatic conditions for various plant habitats.[2] Environmental tools such as Ladybug can be a way of mapping these different microclimates throughout the exterior of a building, providing the information required to determine the multiple plant habitats.

The microclimate describes the conditions of sun, shadow, temperature, and humidity in each precise location within a macroclimate. Plants and animals experience the microclimate of the exact location where they live, for example, in full sun and wind at the top of a tree or in the shade, stillness, and humidity at its base. It is different from the meteorologist's climate (macroclimate). The physical processes and factors creating a microclimate: solar radiation, air humidity and temperature, and wind velocity vary widely depending on the physical conditions of the surroundings and affect the local distribution of plants and animals and their communities and the local survival of populations [14].

Changes in temperature and humidity are most significant near the ground, absorbing a high proportion of the earth's share of the sun's energy. When the sun shines on the ground, the surface temperature rises, and a temperature gradient is set up. Since the relative humidity of the air is related to the saturation vapour pressure, which is directly related to the temperature, the gradients of temperature and near the surface typically mean that gradients of relative humidity also exist. Dennis Unwin, from the Department of Zoology at Cambridge University, explains that to paint a picture of the climate requires knowing how the temperature the humidity parameters vary. Since the driving force behind the whole system is the sun, it is also essential to measure solar radiation and its variation with time [15].

The climate parameters measured in this study are sun hours, wind, humidity, and radiation. The average temperature of the habitat is not as central as its extremes, which

[2] The habitat is the condition in which an organism completes its life cycle. Habitats have two types of components: the abiotic, which are the nonliving components, such as climate, soil, latitude, altitude, and disturbances such as fires, flood, and avalanches; and biotic are the factors that are living such as the plant itself, other plant species, animals, fungi, protists, and prokaryotes [3].

are the lowest winter and highest summer temperatures [16]. Latitude contributes to many factors to the abiotic environment, such as hours of sun exposure throughout the day. It also influences how the light strikes the plant. Light strikes the Earth obliquely when the sun is low, and less energy is received per square meter. High altitude regions have higher winds and more intense ultraviolet light [3].

The proposed workflow involves presenting to the grasshopper plugin, Ladybug, a context geometry, and a subject building. The subject location presented in this paper is a random site in Melbourne, Australia (Fig. 3). There were no urban planning considerations as the purpose is exclusively to illustrate the framework.

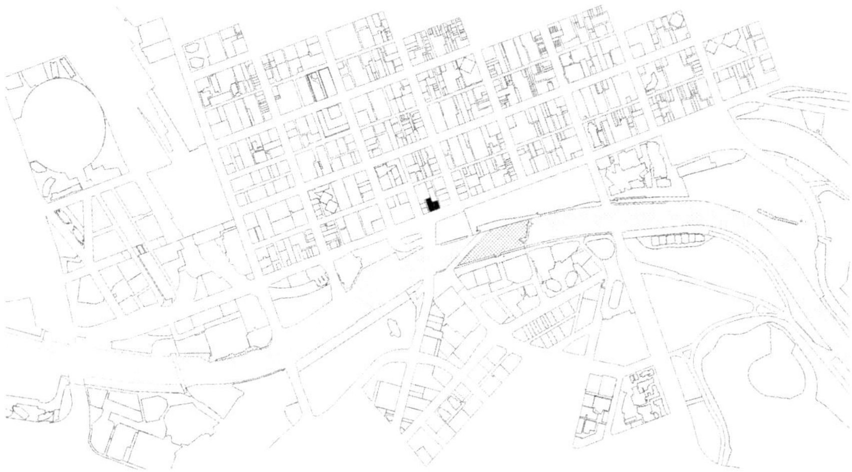

Fig. 3. A hypothetical site in Melbourne, Australia marked in black (image by author)

The environmental tool is set to analyse the whole building's humidity, radiation, and sun exposure. The analysis period relevant to determining a microclimate is the hot and cold, extreme temperatures. Consequently, the evaluation period will be 24 h during midsummer (Fig. 4) and mid-winter (Fig. 5). Each investigation will build a set of data: the sun hours, the radiation throughout 24 h and the relative humidity (Fig. 6) during the analysed period.

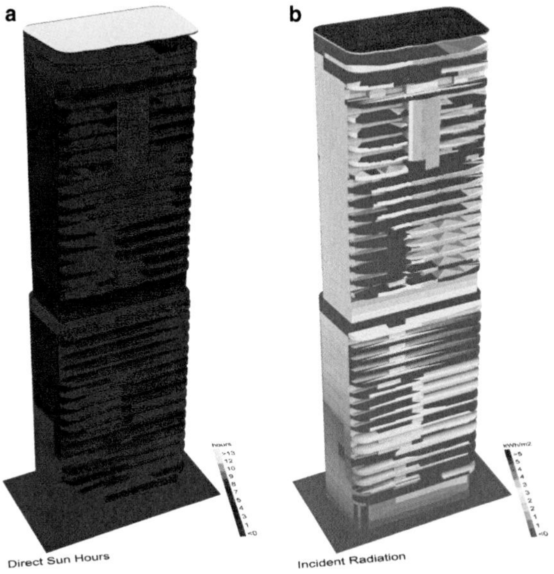

Fig. 4. Analysis of direct sun hours (**a**) and radiation (**b**) in mid-summer. (3D image by author)

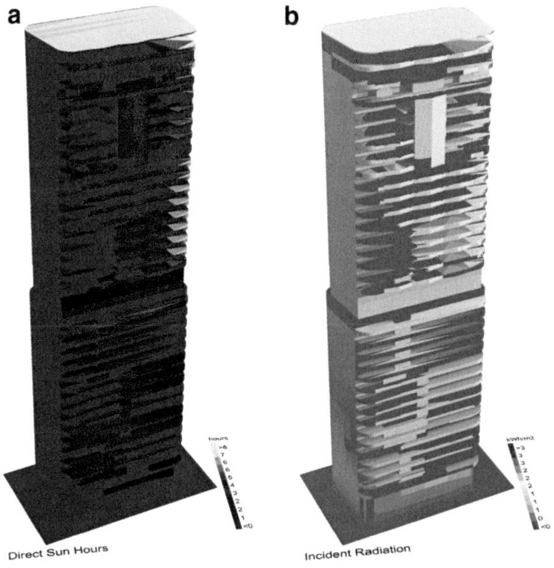

Fig. 5. Analysis of direct sun hours (**a**) and radiation (**b**) in mid-winter. (3D image by author)

The surface temperature calculation requires a material. For the example pre-scented in this paper, a generic material was selected to simplify the calculations. A python script

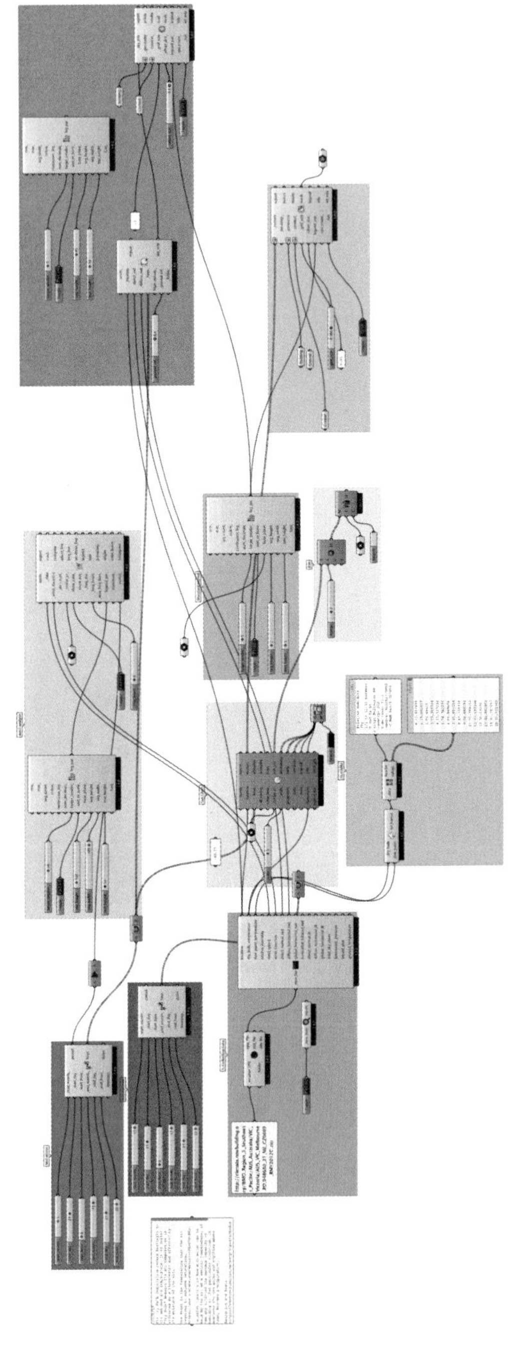

Fig. 6. Grasshopper/Rhino 3d definition showing environmental analysis. Relative humidity data is highlighted in green (image by author)

for Grasshopper compared the analysis result with the parameters that define microclimates (Table 1). The outcome is a coloured geometry, where each colour represents specific climate conditions (Fig. 7).

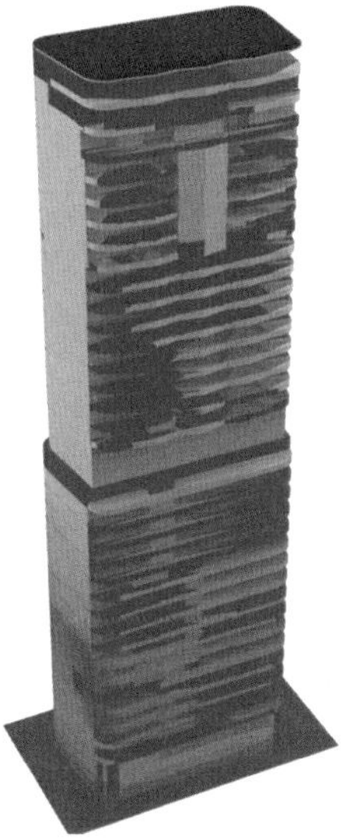

Fig. 7. Grasshopper/Rhino 3d definition showing environmental analysis. Relative humidity Each colour represents a different microclimate. The colours result from the cross-reference between the environmental analysis and the microclimate parameters. (3D image by author)

Table 1 The table shows the parameters that define the different microclimates. The information was extracted from the Australian Bureau of Meteorology [17]

Climate	Station number	Monthly mean daily global solar exposure (kwh/m^2)		Min temperature celsius of year (2012–2022)	Max temperature celsius of year (2012–2022)
Climate A	83085	1.39	7.83	−10.4 to −7.6	21.5–28.2
Climate B	96033	3.67	4.08	−8.1 to −14.2	26.6–32.3
Climate C	75032	4.89	5.31	−4.6 to −2.2	39.5–47.2
Climate D	15135	6.00	6.44	6.2–8.5	41.9–45.6
Climate E	14932	6.11	6.50	4–8.5	40.6–43.1
Climate F	28004	5.78	6.17	5–10.3	35.7–41.7
Climate G	14142	5.47	6.03	12–15.9	35.8–36.8

This information will help select and populate the building façade with suitable plant species, as the analyses provide details of the building envelope's particular conditions (Fig. 8). The wind analysis will assist in the development of the façade elements. These components will direct the airflow where plants benefit from wind and protect the vegetation that requires protection from the weather (Figs. 9, 10 and 11).

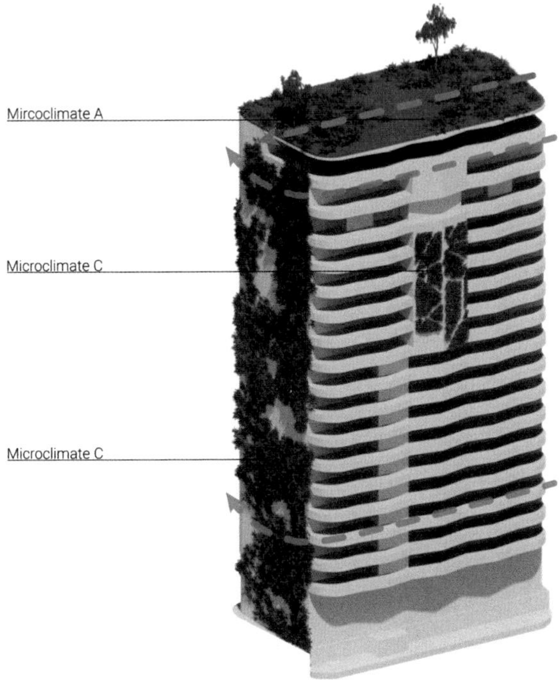

Fig. 8. Each microclimate will define a different plant species that will populate the building envelope. The dotted lines above show the wind direction. (Model by author, rendered by Taylor Ristevski)

4 Conclusion

The climate effects of cities result from the urban form of the landscape and urban function. Increasing vegetative cover and incorporating natural landscape features into the urban design are the best means of managing urban climate effects at all scales [1].

This situation serves as an opportunity for us to reassess current planning and design methods for sustainable urban development and recognise the importance of ecological systems as an integral part of our planet. Consequently, the current scenario requires a behaviour change by acknowledging our responsibility, vulnerability, and reconnecting with nature [18]. Although sustainability reduces environmental harm, it does not rectify past/continuing imbalances in the natural environment. There is a clear separation

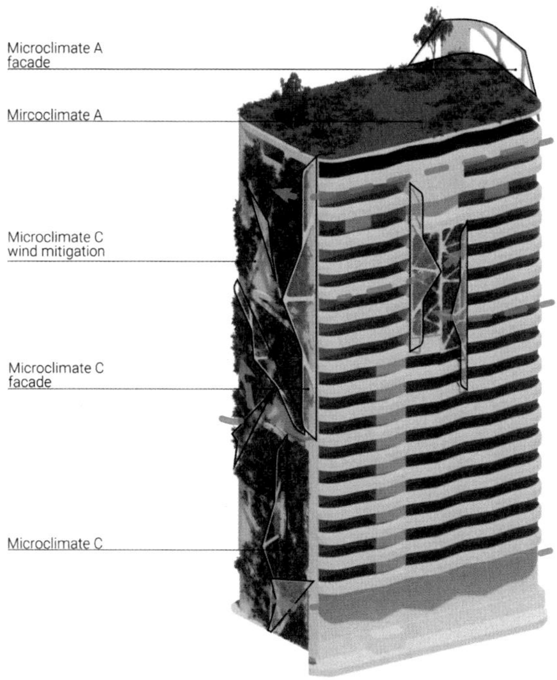

Fig. 9. Façade design to protect vegetation. (Model by author, rendered by Taylor Ristevski)

between developed areas, and nature and green infrastructure are initiatives exclusively human-centred.

Environmental tools such as Ladybug can be a way of mapping out these different microclimates throughout a building's exterior. The environmental analysis is set to evaluate the whole building's climatic conditions, resulting in a coloured geometry, where each colour represents a distinctive microclimate. This information can help select and populate the building façade with suitable plant species.

This assessment can be used to select the suitable plant species for the building façade, and it can also assist in decision-making concerning the overall configuration of the building. The wind evaluation is performed after the plant selection, guiding the design of the façade. The façade elements will perform the task of protecting or channelling the wind towards the plants. Advancing this environmental tool is valuable to facilitate the incorporation of regenerative solutions in the built environment and to provide opportunities to strengthen existing ecosystems. The difference between adding a traditional green wall to a building and this framework is that the design is done for the vegetation to thrive instead of populating buildings with plants to suit the indoor conditions only.

Fig. 10. Detail of façade elements. (Model by author, rendered by Taylor Ristevski)

Fig. 11. Building in context. (Model by author, rendered by Taylor Ristevski)

One of the challenges to developing this framework is the environmental factors that need to be measured for more accurate results, such as the air temperature near the building surface, which are difficult to predict accurately. That happens because the current tools are focused mainly on indoor comfort. This method also addresses the philosophical issue of our anthropocentric behaviour as it includes nature not only as part of the design but also as a "user" of the architectural space.

References

1. Oke TR, Mills G, Christen A, Voogt JA (2017) Urban climates. https://doi.org/10.1017/9781139016476
2. Dunlap RE, Mertig AG (2013) The evolution of the U.S. environmental movement from 1970 to 1990: an overview. Am Environ US Environ Mov 1970–1990:1–5
3. Roös PB (2021) Regenerative design, ecology as teacher BT-re-generative-adaptive design for sustainable development: a pattern language approach. In: Roös PB (ed). Springer International Publishing, Cham, pp 103–112
4. Motloch J (1995) Regenerative design for sustainable development: John Tillman Lyle. Landsc Urban Plan 32:198–201
5. Zari MP (2018) Regenerative urban design and ecosystem biomimicry. Regen Urban Des Ecosyst Biomim. https://doi.org/10.4324/9781315114330
6. Blanco E, Zari MP, Raskin K, Clergeau P (2021) Urban ecosystem-level biomimicry and regenerative design: linking ecosystem functioning and urban built environments. Sustainability (Switzerland) 13:1–12
7. Lyle JT (1994) Regenerative design for sustainable development. Wiley, New York
8. Arcidiacono A, Ronchi S (2020) Cities and nature ecosystem services and green infrastructure perspectives from spatial planning in Italy
9. Jones HG (2013) Plants and microclimate: a quantitative approach to environmental plant physiology, 3rd ed. https://doi.org/10.1017/CBO9780511845727
10. Schaller J, Gnädinger J, Reith L, Freller S, Mattos C (2017) GeoDesign: concept for integration of BIM and GIS in landscape planning. J Digit Landsc Archit 2017:102
11. Liu X, Wang X, Wright G, Cheng JCP, Li X, Liu R (2017) A state-of-the-art review on the integration of building information modeling (BIM) and geographic information system (GIS). ISPRS Int J Geo Inf. https://doi.org/10.3390/ijgi6020053
12. Gherri B, Maretto M, Guzhda A, Motti M, Zannetti G (2018) Early-stage environmental modeling: tools and strategies for climate based design
13. Naboni E, Meloni M, Coccolo S, Kaempf J, Scartezzini J-L (2017) An overview of simulation tools for predicting the mean radiant temperature in an outdoor space. Energy Procedia 122:1111–1116
14. Stoutjesdijk P, Barkman JJ (2014) Microclimate, vegetation & fauna, 1st ed. Koninklijke Nederlandse Natuurhistorische Vereniging, Stichting Uitgeverij
15. Unwin DM (1978) Simple techniques for microclimate measurement. J Biol Educ 12:179–189
16. Birkeland J (2020) Design for nature exemplified. Net-positive design and sustainable urban development, pp 136–152
17. Weather Maps. http://www.bom.gov.au/australia/charts/?ref=ftr. Accessed 15 May 2022
18. Zari MP (2018) The importance of urban biodiversity–an ecosystem services approach. Biodivers Int J 2:357–360
19. Ducarme F, Couvet D (2020) What does "nature" mean? Palgrave Commun 6:14

Embedding Design Intent into Performance-Based Architectural Design—Case Study of Applying Soft Constraints to Design Optimization

DongLai Yang, Likai Wang(✉), and Ji Guohua

Nanjing University, Hankou Road 22, Nanjing, Jiangsu, China
wang.likai@nju.edu.cn

Abstract. The lack of consideration of subjective design intents hinders the application of performance-based design optimization to architectural design because building performance is not the only aspect that designers need to solve. In response, this study proposes a method integrating subjective design intents into performance-based design optimization using soft constraints. To demonstrate the method, a case study is presented, where the design optimization continuously provides feedback to the designer and helps them reformulate and redefine the design problem. The case study shows how the application of design optimization and soft constraints is able to assist designers in identifying implicit and hidden design problems and stimulate design exploration at the early design stage.

Keywords: Performance-based design · Design intent · Soft constraints · Optimization · Co-evolution

1 Introduction

Performance-based design optimization, which integrates parametric models, building performance simulations, and evolutionary optimization, has been widely considered an effective design tool for sustainable architectural design. Many studies have demonstrated its role in addressing complex performance challenges in building design. However, other factors in architectural design, such as functionality and aesthetics, are often omitted in research. This tendency is also reinforced by the notion that design intents are difficult to quantify [5]. The claim greatly affects the application of performance-based design optimization to real-world architectural design tasks. As a result, the design optimization is often conducted after the design scheme is determined, thereby, separating it from the conceptual development process.

In architectural design, designers have to integrate various factors into the design synthesis process, including functionality and aesthetics that are judged subjectively and building performance factors that are evaluated objectively. In order to incorporate subjective intents into the optimization, several approaches have been explored, such as using interactive genitive algorithms (IGA) [2, 3] or aesthetic-related constraints [9].

P. F. Yuan et al. (eds.), *Hybrid Intelligence*, Computational Design and Robotic Fabrication,
https://doi.org/10.1007/978-981-19-8637-6_14

The use of these approaches enables architects to intervene in the optimization process and allow architects' personal preferences to be integrated into the process of design optimization, but they are not without problems. First, using these methods can still result in a huge number of unfeasible designs generated if using under-constrained or naive generative models. Second, interactive approaches, such as IGAs, require architects to spend considerable time and energy to select or score the generated designs, which can disrupt architects' design processes. Last but not least, the feedback loop between architects' design development and performance-based design optimization is also absent from most existing studies.

1.1 Paper Overview

Considering the limitation of the previous studies, this study proposes an approach to integrating subjective design intents into performance-based design optimization using soft constraints. With the use of soft constraints, specific design intents, such as view, building forms, and site constraints, can be formulated into the fitness evaluation by using penalty or award functions [1]. As a result, designers are enabled to navigate the optimization search and make the optimization produce more desieable designs that can both satisfy the performance objective and design intent.

To demonstrate the efficacy of the proposed approach, a case study is presented in the paper, where the design is started only considering performance factors. Then, through reflecting on the optimization result, we iteratively insert factors related to functionality into the fitness evaluation and make the optimization result achieve an acceptable compromise between the performance improvement and design intent. The case study shows that the proposed design approach can strengthen the feedback loop between designers and computers, making the designer more engaged in the design development process informed and inspired by performance-based design optimization. This design process can be viewed as a "meta-optimization" process where the objective is not merely focused on performance improvement but also to achieve a "co-evolution" between designers and computers to attain a well-rounded design.

2 Method

Early-stage architectural design is widely accepted as an iterative design exploration process. Therefore, it is also critical for the computational design tools or design approaches to support the iterative and *human-in-the-loop* design process. In light of this, the proposed design optimization is envisioned as continuously providing feedback to the designer from the outset of the design process rather than offering specific and determined solutions (Fig. 1). In other words, it encourages architects to reflect on the optimization result and iteratively reformulate the design objective with the use of computational design optimization.

The workflow is built on the Rhino-Grasshopper platform. EvoMass and other building performance simulation tools, such as Ladybug and ClimateStudio, are used in the

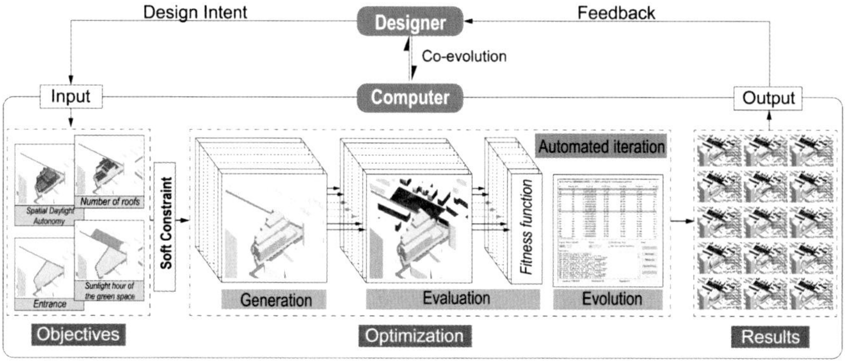

Fig. 1. Proposed optimization-based design workflow

design optimization workflow. The combination of these tools has already been applied to performance-oriented design optimization. However, previous applications fell short of the pursuit of satisfying subjective design intents. Therefore, to make architects' design intents to be included in the optimization process, we introduce the application of soft constraints, which can effectively embed design intent into the optimization.

2.1 Design Generation and Optimization

In the proposed design optimization workflow. EvoMass serves as the building design generator and the optimization solver [8]. When using EvoMass for building design optimization, the designer first customizes the generative component in EvoMass to adapt the generated building massing design to the building site. There are two generative components in EvoMass built on the additive and subtractive form generation principles, and both components can generate diverse building massing designs, which can facilitate the optimization process to identify site- and task-specific solutions for various design projects.

Second, the generated building massing design is assessed by different design evaluation functions. For performance-based design optimization, simulation tools, such as Ladybug and ClimateStudio, are often used to measure the performance of the generated design. The design evaluation function will guide the optimization search direction. Therefore, in addition to performance factors, other design factors can be also included in the design evaluation function and, thereby, steer the search direction.

Third, the performance and the evolution of the design are converted into a fitness score and sent back to the optimization algorithm. When using EvoMass, the embedded evolutionary algorithm—SSIEA (Stead-State Island Evolutionary Algorithm), is used as the optimization solver to evolve the design population and identify the high-fitness solutions [8].

EvoMass can produce optimization results with diverse solutions that best satisfy the optimization objective. Furthermore, designers can re-evaluate the optimization result and modify the optimization objective. Previous applications of EvoMass mostly focused on building performance, while the capability of generating diverse building massing forms makes EvoMass an ideal form-finding tool for combining performance-based design and architectural design. Hence, this study further explores the potential of EvoMass in architectural design and investigates how the design optimization workflow can be intertwined with architects' design loop for conceptual development.

2.2 Soft Constraint

In evolutionary computing, constraint handling, including direct and indirect constraints, plays a critical role in solving optimization problems [1]. For direct constraints, the constraint is embedded into the design generation stage instead of in the design evaluation stage, using methods such as repair functions. This approach can effectively prevent invalid and chaotic solutions from being generated, while it often reduces the variability of the design generation, and it is possible to exclude promising solutions from the design search space [7].

For indirect constraints, the constraint is embedded into the design evaluation stage, using methods such as penalty or award functions. Regarding design applications, we further divide indirect constraints into hard and soft constraints. For hard constraints, designs that cannot meet the constraint will be directly eliminated and "killed" from the design population, which can rapidly narrow down the search scope and speed up the convergence of the optimization process. However, when using hard constraints, the population diversity will drop rapidly, and promising designs that even slightly violate the constraint will also be removed from the pool of recombination. It is because, for evolutionary optimization, the design optimization process heavily relies on the re-combination, namely crossover, of the genotype from different designs (typically two designs). Thus, to fully explore the design space, the evolutionary process needs to maintain an adequate population diversity that allows for the recombination of the design with heterogeneous genotypes.

In comparison, the application of soft constraints has the advantage of maintaining the population diversity during the optimization process. When using soft constraints, the fitness of the design that violates the constraint will be proportionally decreased to reduce its chance of surviving in the subsequent evolutionary process. Since the design remains in the design population, it can still be recombined with other designs. More importantly, if this design contains key parameters (genomes) that are essential parts of the genotype of high-fitness designs, the recombination with other designs may produce the offspring design that does not violate the constraint while having an advantageous fitness.

The application of soft constraints provides a feasible way for the designer to navigate the optimization process by converting the design intent into the design optimization process. Thus, in the following case study, we demonstrate how different design intents can be integrated into the optimization process using soft constraints and how the design optimization result shows the response to the design intent.

3 Case Study

In this study, we present a case-study design consisting of three stages and assume that the designer begins with the use of design optimization only focusing on performance factors and then iteratively integrates the design intent into the optimization process, including responding to the surrounding environment, functionality, and aesthetics. This case study describes an office building design in Nanjing, China. The building is located in the city center and is imposed a 50-m height limit (Fig. 2). There are several residential buildings on its west and north sides, urban green space on its northwest side, a main road on the southern side, and several high-rise office buildings on its east and west sides.

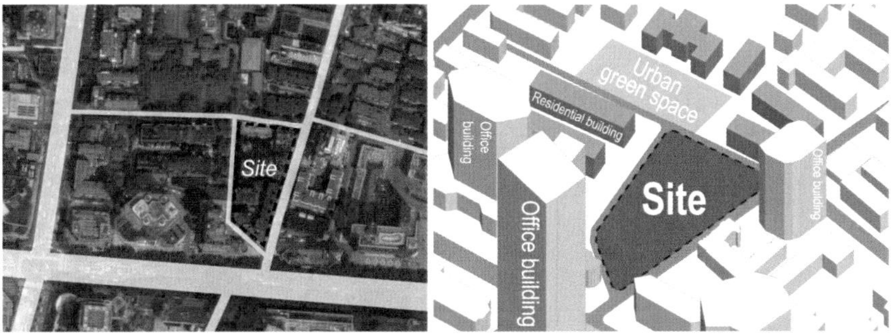

Fig. 2. Site overview

Within such a complex urban environment, only considering the building performance is insufficient. The irregular site boundary and the high-density urban environment pose a great challenge to the designer. In addition, as widely accepted as a "wicked" problem, architectural design often faces many hidden constraints that are not explicit to be identified at the outset of the design. As Schön [4] stated, conceptual design is a "*moving-seeing-moving*" process, where designers often discover implicit problems or constraints when manipulating the design object. As such, architects can leverage design optimization as an approach to uncovering hidden design problems and reformulate the design objective by superimposing the information gathered from design optimization.

In terms of building performance, daylight factor, spatial daylight autonomy, and discomfort glare have been commonly used in design evaluation. In this case study, the surrounding high-rise buildings cast a large shadow that can affect the daylighting quality of the target building. Thus, the spatial daylight autonomy (sDA) is first taken as the evaluation metric, simulated by ClimateStudio in Rhino-Grasshopper. Additionally, two soft constraints are applied to control the gross floor area (GFA) and density of the design.

According to the above-mentioned objectives, the initial fitness function of the first stage is shown in Fig. 3, where *p_area* and *p_den* represent the penalty function for GFA and density. For GFA, it calculates the difference between the actual GFA of each generated design and the target GFA (40,000 m^2) and proportionally decreases the fitness value according to the GFA difference. For the density, it punishes the design with a density outside the range of 0.6 to 0.8 and also proportionally decreases the fitness value based on the difference between the actual density and the target density.

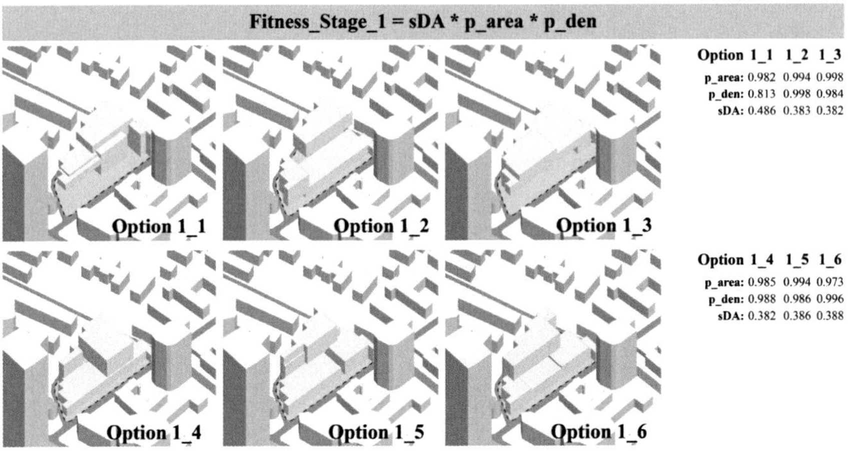

Fig. 3. The fitness evaluation function (top) and the optimization result in the first stage (bottom)

The above fitness function shows that when using soft constraints, there can be multiple objectives that need to be optimized. To handle multiple optimization objectives, the conventional approach based on Pareto optimization becomes inefficient as the goal of seeking as many trade-off (non-dominated) designs as possible can hinder the optimization progress. In addition, using Pareto optimization often results in too many design options, making it difficult to analyze and extract design information. In this regard, when using soft constraints, a more advisable approach is to use weight-sum and -product approaches. In this case study, we adopt a weight-product approach to integrate different optimization objectives. In comparison to weight-sum approaches, weight-product approaches do not require normalization as the change in each value can equally affect the overall fitness.

Figure 3 (bottom) shows the optimal design options found by the optimization process. It is not difficult to notice that these design options tend to keep away from the high-rise buildings in the south to escape from the shadow cast by these buildings. However, as the building massing typically gathers to the north to enhance its daylighting accessibility, this tendency significantly undermines the sunlight and daylight accessibility for other buildings and the surrounding public space

In the second stage, to decrease the adverse effects on the surrounding environment, the sunlight hour of surrounding residential buildings and the urban green space are included in the optimization. Sunlight hours are simulated using Ladybug in Rhino-Grasshopper, and the simulation only calculates the sunlight hour during winters. Therefore, by integrating the factors considered in the first stage, the fitness function of the second stage is shown in Fig. 4, where *sunlight_rsd* and *sunligh_green* respectively indicate the sunlight hour of the residential buildings and the green space. As a result, with the inclusion of the two new soft constraints, the fitness of the design is decreased according to the amount of sunlight blocked by the generated building massing form.

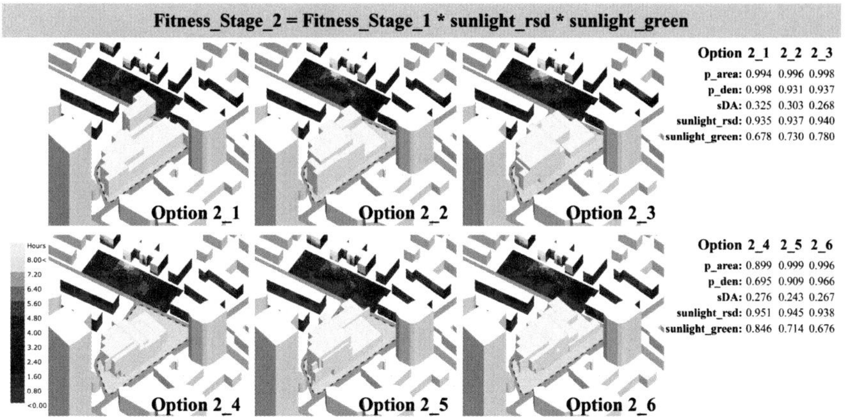

Fig. 4. The fitness evaluation function (top) and the optimization result in the second stage (bottom)

As shown in Fig. 4 (bottom), to reduce sunlight obstruction, the optimal design options typically have the massing volume retreating from the north, which can reduce the shadow cast on the residential building and the green space. However, to meet the GFA requirement, most of the massing volume is stacked and accumulated in the south, resulting in the building being too close to the surrounding office buildings. From an urban design perspective, this tends to make the outdoor space in the south of the target building over-crowded, which may also reduce urban ventilation.

In the third stage, four additional design intents are included. First, we place the entrance on the southern side of the target building to reduce the crowdedness in this area. Hence, we define a volume for the entrance space and punish the design based on how much the entrance space is "invaded" by the building's massing volume. Second, to enhance people's well-being, the percentage of the unobstructed view is also included in the optimization. Third, to allow for more outdoor and semi-outdoor spaces for the people working in this building, we award the design with more roof surfaces. The roof surface can be used for resting and viewing. Finally, we calculate the standard deviation of the roof surface area as a measure of the difference between all roof surfaces. To prevent the design from creating oversized roofs, we award the design with a smaller standard deviation value.

As a result, by integrating the factors included in the first two stages, the fitness function of the third stage is shown in Fig. 5, where *p_entrance* is the penalty function for the entrance space, *num_roof* is the number of roof surfaces, and the *view* is the average value of the unobstructed view on the surface of the building massing, *p_roof_area_diff* indicates the penalty function that punishes the design with a large difference of roof surface areas.

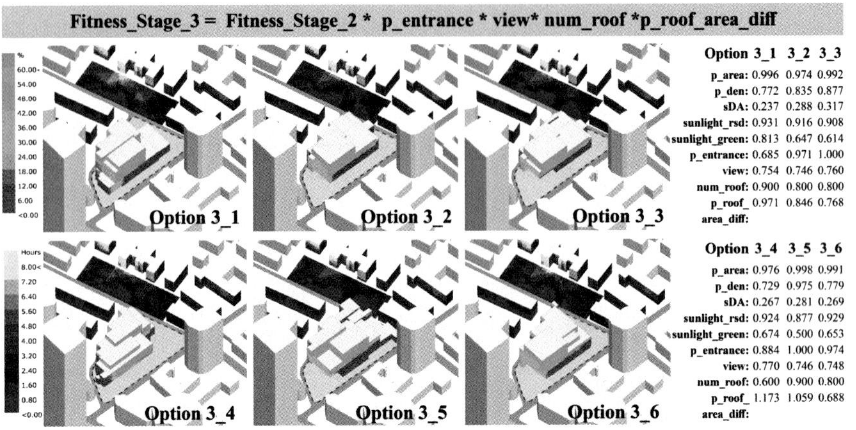

Fig. 5. The fitness evaluation function (top) and the optimization result in the third stage (bottom)

As shown in Fig. 5 (bottom), the optimal design options tend to place the major building massing volume in the middle of the site, which is the result of the compromise under multiple constraints. First, the building massing volume needs to make room for the entrance space in the south, and some design options feature an overhanging block above the entrance. Second, the building massing volume still maintains a distance from the residential building and green space in the north to mitigate the sunlight obstruction. Third, the building massing volume also stays away from the high-rise buildings in the south to enhance the unobstructed view. Fourth, the building massing volume tends to have more roof surfaces at different heights to increase the rooftop area. Finally, the area distribution of the roof surface becomes more evenly distributed, making the building more balanced visually. However, with the inclusion of the new soft constraint, the optimization has to make more compromises on other design aspects, which is evident by the drop in other optimization metrics.

4 Discussion and Conclusion

The presented case study shows that the use of the computational design help designers identify implicit design constraints and problems, and thereafter, further stimulates an iterative design exploration process. Regarding the proposed workflow, the use of soft constraints facilitates the designer to embed their design intents into the optimization, while the weight-product approach allows the new design intents can be integrated with the existing ones. Hence, the pre-existing design implication can be preserved but also dialed down with new constraints included.

Figure 6 summarizes the numerical change of the optimization result regarding different optimization metrics during the three stages. We select the best 15 designs from each design stage and calculate the average value for each metric. Left of the dotted line is the optimization objective included in the corresponding stage. At the same time, we also evaluate the optimal design options found in the earlier stage against the optimization metric included in the later stage to provide a holistic view of the change in optimization metrics. To compare the change in the value, we use the metric of the last stage as a benchmark (0.50) and recalculate the average value of the metric at earlier stages.

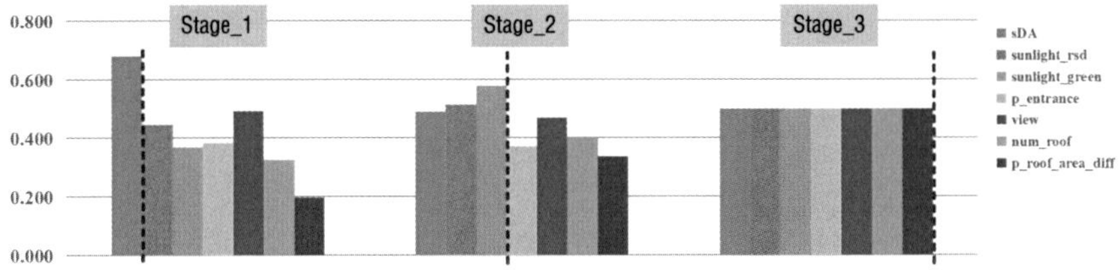

Fig. 6. The change in the value of each metric in the three stages

It is noticeable that the inclusion of each new metric in the optimization typically lowers the metric already in the optimization. This highlights that the design options produced in the final stage are the trade-off that achieves an acceptable compromise among various design aspects. Nevertheless, it should also be stressed that the design options produced in the final stage are more architecturally appealing and rational compared with those produced in earlier stages.

To conclude, this study is aimed to further extend the application of performance-based design optimization in architectural design and demonstrates how subjective design intents can be incorporated into the optimization process by using soft constraints. As demonstrated in the presented case study, the proposed design optimization workflow incorporating the application of soft constraints enables a more integrated human–computer design process. The optimization can effectively stimulate design exploration and assist designers in defining the design problem. Thus, by strengthening the feedback loop between designers and computers, a co-evolutionary design process emerges, where the application of performance-based design optimization provides designers with a "medium of reflection" in the early-stage design ideation and conceptual development.

Acknowledgements. This study is funded by China Postdoctoral Science Foundation (2021M701664) and National Natural Science Foundation of China (52178017).

References

1. Eiben AE, Smith JE (2003) Introduction to evolutionary computing, evolutionary computation.Springer Berlin Heidelberg (Natural Computing Series), Berlin, Heidelberg
2. Huang W, Xu W, Wang T (2011) Structural form generation using interactive genetic algorithm. In: Circuit bending, breaking and mending-proceedings of the 16th international conference on computer-aided architectural design research in Asia, CAADRIA 2011, pp 751–760
3. Marsault X (2013) A multiobjective and interactive genetic algorithm to optimize the building form in early design stages. In: Proceedings of building simulation, pp 809–816

4. Schön DA (1992) Designing as reflective conversation with the materials of a deisgn situation. Knowl-Based Syst 5(1):3–14
5. Shi X et al (2016) A review on building energy efficient design optimization from the perspective of architects. Renew Sustain Energy Rev 65:872–884. https://doi.org/10.1016/j.rser.2016.07.050
6. Wang L et al (2020) Algorithmic generation of architectural massing models for building design optimisation: parametric modelling using subtractive and additive form generation principles. In RE: anthropocene, design in the age of humans-proceedings of the 25th international conference on computer-aided architectural design research in Asia, CAADRIA 2020, pp 385–394
7. Wang L, Janssen P, Ji G (2018) Utility of evolutionary design in architectural form finding: an investigation into constraint handling strategies. In Gero JS (ed) Design computing and cognition '18. Springer, Cham, pp 177–194. https://doi.org/10.1007/978-3-030-05363-5_10
8. Wang L, Janssen P, Ji G (2020) SSIEA: a hybrid evolutionary algorithm for supporting conceptual architectural design. Artif Intell Eng Des Anal Manuf 34(4):458–476. https://doi.org/10.1017/S0890060420000281
9. Yousif S, Clayton M, Yan W (2018) Towards integrating aesthetic variables in architectural design optimization. In 106th ACSA annual meeting proceedings, the ethical imperative. ACSA Press, pp 430–436. https://doi.org/10.35483/ACSA.AM.106.68

Effect of Morphological Indicators on the Pedestrian Level Wind of the Existing Workers Villages in Shanghai

Xingzhao Zhang, Xinyu Wu, Luqiao Yang, Jiaqi Xu, Ruizhe Luo, and Jiawei Yao(✉)

College of Architecture and Urban Planning, Tongji University, 1239 Siping Road, Shanghai, China
jiawei.yao@tongji.edu.cn

Abstract. The workers villages are typical residential type during Shanghai's urbanization built from the 1950s to the 1980s. Due to changes in the urban environment and climatic circumstances, the workers villages have inadequate natural ventilation and difficulty in dispersing pollutants, putting residents' health at risk. In the context of urban renewal, it is necessary to clarify the effect of building morphological indicators on pedestrian level wind, especially in such old residential communities. In this paper, 100 workers villages representatives were gathered by GIS. Their summer ventilation conditions were simulated using the CFD solving the LES turbulence equation. The correlation between 9 morphological indicators and 2 pedestrian level wind indicators was obtained quantitatively by Pearson analysis and regression analysis. The result shows increasing the building coverage of 0.94% in the workers villages, the ratio of the area of the static wind in summer will increase subsequently by 10%. The results highlight the importance of considering morphological indicators to enhance the wind environment, and provide suggestions for the environmental transformation of communities with similar characteristic in the high-density city.

Keywords: The workers villages · Pedestrian level wind · Morphological indicators · Static wind · Linear regression

1 Introduction

The rising density and height of buildings within cities, along with the rapid development of urbanization, has had an impact on the original natural climate. The concentrated expansion of urban centers in coastal parts of China has exhibited a high degree of saturation, and dense building clusters in cities can decrease air flow and have a direct impact on urban wind speed [1]. Excessive wind speed and inadequate ventilation can both cause changes in the thermal environment, which can impair human comfort in outdoor settings and population health [2]. Uneven wind speed distribution also contributes to the accumulation of dangerous gases and air pollutants like sulfur dioxide [3], as well as major air pollution concerns like haze, which can cause respiratory ailments and reduce daily visibility [4].

P. F. Yuan et al. (eds.), *Hybrid Intelligence*, Computational Design and Robotic Fabrication,
https://doi.org/10.1007/978-981-19-8637-6_15

Because of the dense population of Shanghai's core city, convective winds are severely impeded, and significant sections of the building complex are windless [5]. Due to changes in construction requirements and changes in the surrounding environment, which is poorly ventilated and prone to heat and pollutant buildup, some old neighborhoods built in the last century have wind speeds of less than 1 m/s in parts of the interior of residential areas, causing problems for pedestrians. By the end of 2011, the existing workers villages in Shanghai had a total housing area of around 170 million square meters, accounting for 31% of the total area [6]. Workers villages have also been an important objective of Shanghai's urban renewal plan, which is focusing on urban regeneration and traditional old district renovation. In order to mitigate the heat island effect and promote sustainable urban growth, the rehabilitation of ventilation conditions in old neighborhoods in high-density cities has become a critical issue.

The goal of this research is to improve the outdoor wind environment in old communities, which are represented by the workers villages. Building coverage and height, for example, have been demonstrated to have a considerable impact on regional ventilation performance in studies [7], which will affect to some extent the comfort and safety of wind environment in outdoor pedestrian spaces [8]. The outdoor wind environment of the old communities represented by the workers villages are studied, quantitative indexes are introduced, and the relationship between building form and outdoor wind environment is presented qualitatively and quantitatively through the calculation of an idealized geometric form model and multiple regression analysis, and applied to the optimization of existing buildings.

2 Methodology

In the present study, a total of 100 neighborhood models of workers villages in different regions of Shanghai were collected via GIS and Rhino. The morphological indicators of the neighborhood models were calculated by the plug-in Grasshopper. Then, wind environment simulation of 100 workers villages and their surroundings were calculated, and the data obtained were filtered and processed by the plugin ladybug and eddy3D. It is an interactive interface on rhino/grasshopper platform based on openFOAM open source computational fluid dynamics (CFD) library. Solving LES equation has the advantages of fast and accurate. It has been proved to be adaptive in complex urban environment [9]. Eventually, correlation analysis and regression analysis were carried out between morphological and environmental factors to eliminate loosely related factors and clarify their correlations. Scatter diagrams were also created to verify the correlation. The overall process is shown in Fig. 1.

2.1 Geography and Climate of the Research Object

The rapid urbanization process is currently a big issue all over the planet. With such a large population, there is a great demand for natural resources, which stimulates city development at a high density. The high-density built-up area in Shanghai causes the wind speed to drop year after year, reducing the comfort of pedestrian wind environments

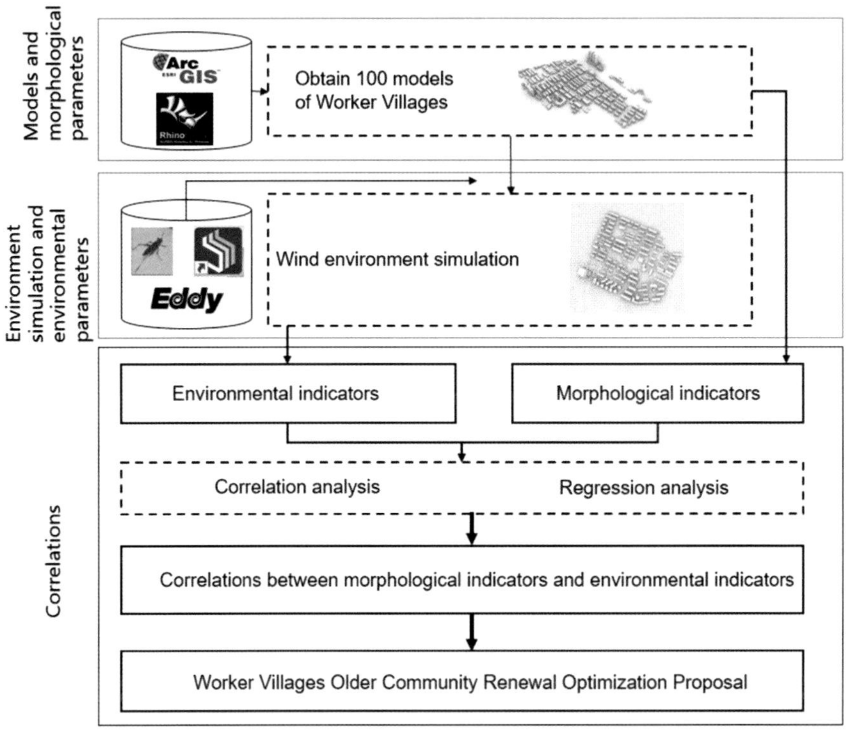

Fig. 1. Overall process

[10]. The workers villages were constructed earlier, and the wind environment was not taken into account during building, resulting in poor overall comfort.

In this study, about 100 existing old workers villages in Shanghai were selected as samples. The models were obtained and processed through software such as ArcGIS and Rhino to obtain a model library of the research objects.

2.2 Morphological Indicators

Eight morphological indicators were used in the study to quantitatively describe the varied spatial characteristics of workers villages: building coverage ratio, building volume density, floor area ratio, average building height, roughness length, height dispersion, frontal area index, and sky view factor.

Among these eight morphological indicators, the significant impact of average building height has been proven in numerical experiments [11], and its increment led to larger air change rates at roof and targeted canyon monotonously [12]. A study took place in Hong Kong [13] emphasis that frontal area index is to be a better morphological factor in depicting the wind environment at the pedestrian level. Illustration and equation of various indicators are showed in Fig. 2 and Table 1. Each indicator's abbreviations are included in the table and will be used later in the text.

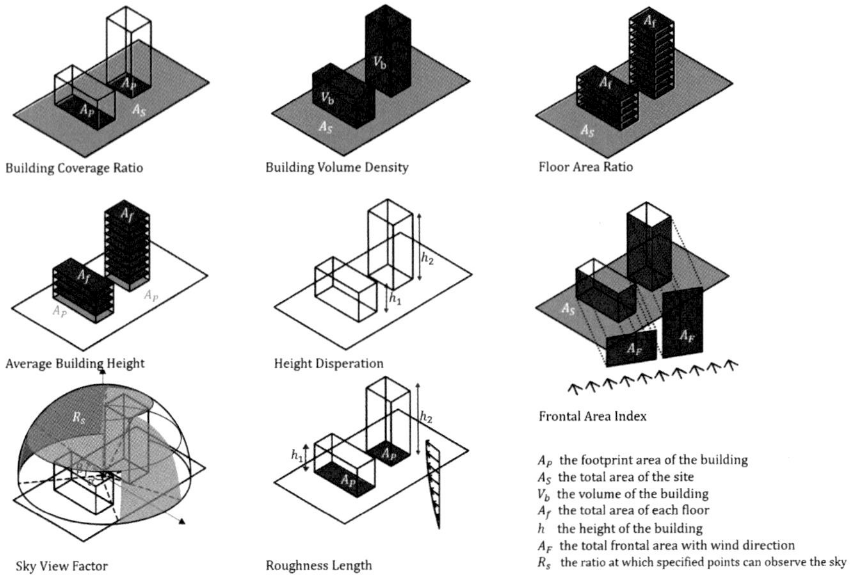

Fig. 2. Illustration of morphological indicators

2.3 Pedestrian Level Wind Indicators

In consideration of the characteristics of wind environment in the community, the geographical features of its location in high-density urban areas, where wind speeds greater than five meters rarely occur, and comfort under different pedestrian behavior, this paper puts forward an approximate division threshold for Shanghai workers villages. Static wind area is classified as areas where wind speed is smaller than 1 m/s, and areas with wind in 1–5 m/s and >5 m/s are defined as comfort wind area and strong wind area relatively. The area ratio of static wind area is used to describe the proportion of uncomfortable pedestrian areas in the community in the total area, especially in summer.

Other than static wind ratio, this paper also uses standard deviation to express the overall dispersion degree of each calculation point of wind speed in the whole test model site. Due to the construction of high density, the uniformity of air flow will be affected. The lower the wind speed dispersion, the more stable the overall wind speed distribution and the weaker the wind turbulence, resulting in a better wind environment. The unreasonable layout of building corners and building spacing will lead to a sudden increase in wind speed, leading to higher wind speed dispersion, resulting in unsafe or comfortable circumstances. Equation of various indicators are shown in Table 2.

2.4 Pearson Correlation Coefficient and Linear Regression

Pearson correlation coefficient is used to expose the correlation of indicators, whose output range from −1 to 1, and 0 means no correlation. Negative value means negative correlation and positive value means positive correlation. The larger the value to

Table 1. The equations used in the calculation of building morphological indicators

Morphological indicators	Unit	Calculation equation		Theoretical meaning
Building Coverage Ratio	%	$BCR = \frac{\sum_{i=1}^{n} A_p}{A_s}$	[14]	Building intensity in an area
Building Volume Density	m	$BVD = \frac{\sum_{i=1}^{n} V_b}{A_s}$	[15]	BVD indicates total volume of buildings divided by the total area
Floor Area Ratio		$FAR = \frac{\sum_{i=1}^{n} A_f}{A_s}$	[16]	Use intensity of construction land
Average Building Height		$\overline{h} = \frac{1}{n}\sum_{i=1}^{n}(\frac{A_f}{A_s})$	[17]	Vertical building intensity
Roughness Length	m	$z_0 = \overline{h} \cdot \frac{C_d}{0.74} \cdot (1 - \frac{z_d}{\overline{h}}) \cdot \frac{FAI}{BCR}$ $z_d = \overline{h} \cdot (1 + A^{-BCR} \cdot (\mathrm{BCR} - 1))$	[18]	the efficiency of transforming the energy of average wind speed into turbulent motion in the boundary layer [19]
Height Dispersion	m	$\sigma = \sqrt{\frac{1}{n}\sum_{i=1}^{n}(h_i - \overline{h})^2}$	[20]	discrepancy of building height in the site
Frontal Area Index		$FAI = \frac{A_F}{A_s}$	[21]	a building's frontal area over a site's area
Sky View Factor	[0–1]	$SVF = 1 - \sum_{i=1}^{n} sin^2 \beta_i (\frac{\alpha_i}{360^o})$	[14]	the ratio at which specified points can observe the sky

Table 2. The equations used in the calculation of building environmental indicators

Environmental indicators	Representative/calculation equation	References	Theoretical meaning
Static wind ratio	Ratio of areas with wind speed less than 1 m/s to the total area	[22]	Proportion of static wind accumulation area
Wind-speed standard deviation	$\sigma = \sqrt{\frac{1}{N}\sum_{i=1}^{N}(x_i - \mu)^2}$	[16]	Variation of wind speed

both sides, the stronger the correlation is. Pearson correlation coefficient's capability of achieving this is accomplished through centralization and cosine calculation.

Furthermore, linear regression explains the higher correlation and is able to generate the results of linear regression equation. Correlation analysis is the basis and premise of regression analysis, and regression analysis is the deepening and continuation of correlation analysis. Due to the existence of some samples with wide orders of magnitude, directly processing ignores some independent variables with extremely small value level may cause an inaccurate fitting result, normalization is introduced to avoid dimensional effects. Normal distribution of data is kept and zoomed to the range of 0 to 1.

3 Results and Analysis

After completing the wind environment simulation of 100 real models, this paper firstly makes a qualitative description of the overall and individual workers villages, along with description of data distribution, and then makes a quantitative analysis using Pearson correlation and linear regression.

3.1 Geographical Distribution and Analysis of Individual Cases

Figure 3 shows the geographical distribution of selected models, which display a circular distribution centered on urban area. It is obvious that most of the tested workers villages in Shanghai don't display an optimistic performance in comfortable wind area—in the figure, a large number of points is biased to orange and red.

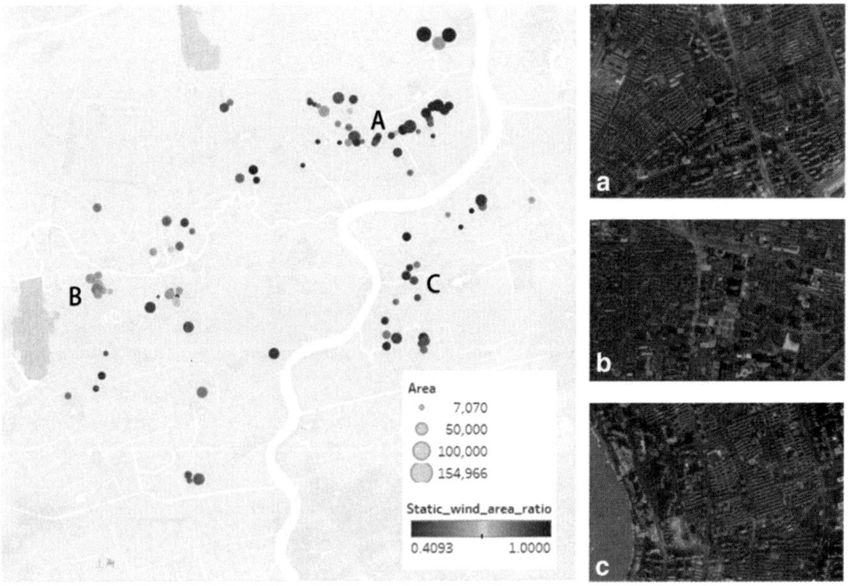

Fig. 3. Distribution of tested new workers villages and corresponding static wind area ratio and site area sizes, **a**–**c**, three typical areas with concentrated points are concerned.

The static wind ratio of the 3 areas (Fig. 4), on a mesoscale, is different from the overall condition. Lighter colored points are gathered in Area B than A and C, also showing the characteristics of homogenization, which means it has a better pedestrian level wind environment than area A and C in summer. Different geographical locations, preliminarily assumed, will alter specific morphological aspects, hence influencing the wind environment. Specifically, three areas have a varied overall tendency in terms of construction orientation. A and C are oriented primarily southeast, whereas B is oriented north–south. Besides, FAI and other morphological indicators will be affected by different orientations, which alter the wind environment. As the satellite image display, BD of surrounding sites of area B (Fig. 3b) is lower than which of the other plots (Fig. 3a, c), indicating that the overall building density of the two plots B is also much lower than that of the other plots.

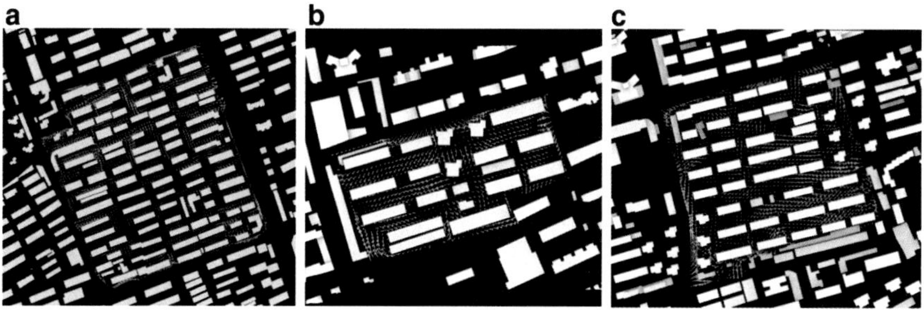

Fig. 4. Wind simulation of 3 typical cases corresponding 3 areas mentioned in Fig. 3

Some preliminary conclusions can also be obtained from the micro scale. In addition to the variances in morphological aspects induced by different geographical locations, the layout of workers villages has an impact on the wind environment. The determinant dominates most workers villages, although there are other combinations of determinant and point group or enclosed organization. Because the point group and determinant combination in area B (Fig. 4b) is more extensive, the area of static wind is comparatively small. As a result, it is clear that this arrangement is worthy of note and can be used as a component in the optimization and reconstruction of future workers villages.

3.2 Descriptive Analysis of Data

The morphological characteristics, BCR, and SVF have a more concentrated distribution with fewer outliers, as seen in Fig. 5. BVD, z_0, σ, and FAR are all right skewed, with more outliers. The distribution of static wind area ratio is more concentrated than the distribution of wind speed standard deviation among the environmental performance indexes, and the distribution of wind speed standard deviation is more scattered.

From the box-and Whisker plot for all indicator result distribution of wind simulated models (Fig. 5), it can be seen that the median of the static wind area ratio of the workers villages is about 0.87, and the vast majority of the values fall between 0.6 and 1, which means that the vast majority of the workers villages' summer static wind area ratios is above 60%. Thus the overall wind environment of Shanghai workers villages is unsatisfactory.

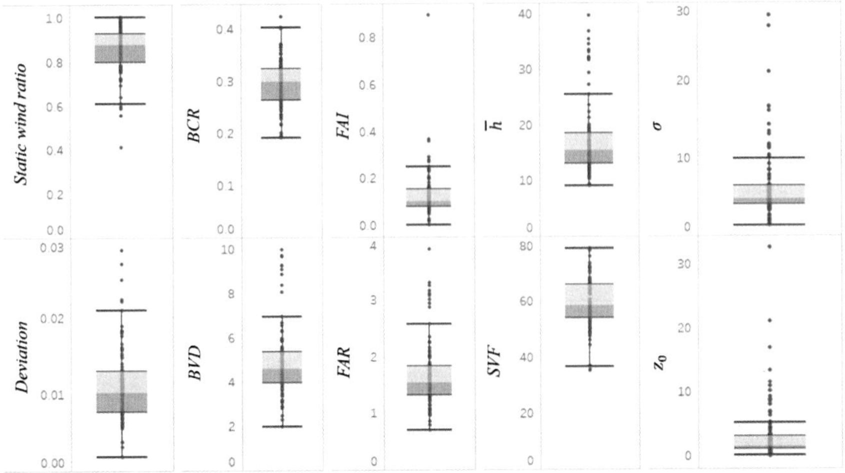

Fig. 5. Box-and Whisker plot for all indicator result distribution of wind simulated models, dots represent each number result, the left and right boundaries of each subgraph represent the lower and upper limit value respectively, and left and right boundaries mean the lower and upper quartile, and central dividing line represent median.

3.3 The Correlation Between Spatial Morphology and Wind Environment

Comprehensive scatterplots, data distribution of every indicator and relevant calculated Pearson correlation coefficient are shown in a matrix visualization (Fig. 6). In the matrix, the darker red the color is, the higher the positive correlation between the two indicators is explained, while the darker blue the color is, the higher the negative correlation between the two indicators is explained, corresponding correlation coefficients are also marked, and the range of which is 0–1. Through comprehensive analysis, it can be preliminarily determined that: (1) Within environmental indicators, Static wind ratio strongly correlated to wind deviation. (2) Between environmental and morphological indicators, **i.** Site area size strongly correlated to deviation, following a linear negative correlation. **ii.** BCR, BVD, FAR weakly correlated to static wind ratio, and all present positive correlation. **iii.** FAI and wind deviation show slight correlation, following a positive correlation. **iv.** z0 and deviation show weak correlation, following a positive correlation.

To further obtain quantitative relationships between morphological and environmental indicators, linear regression is conveyed. Based on its result, Fig. 7 shows the correlation and significance (P value) of coefficients. In general, the results are as follows: (1) Unfortunately, all data have a low degree of fitting. It illustrates that the real model's wind environment is more complex, as a result of numerous elements acting together, and that the link with specific morphological markers is weak. (2) Wind speed standard deviation and site area size show the highest regression coefficient among all indicators, which is close to 0.3 (Fig. 7a) and respond great significance ($P < 0.0001$) It remains to be explored why the site size is negatively related to wind-speed standard deviation. (3) Relatively speaking, BCR indicates higher correlation than other morphological indicators, but merely does it explain the regression relationship between the two indicators so far.

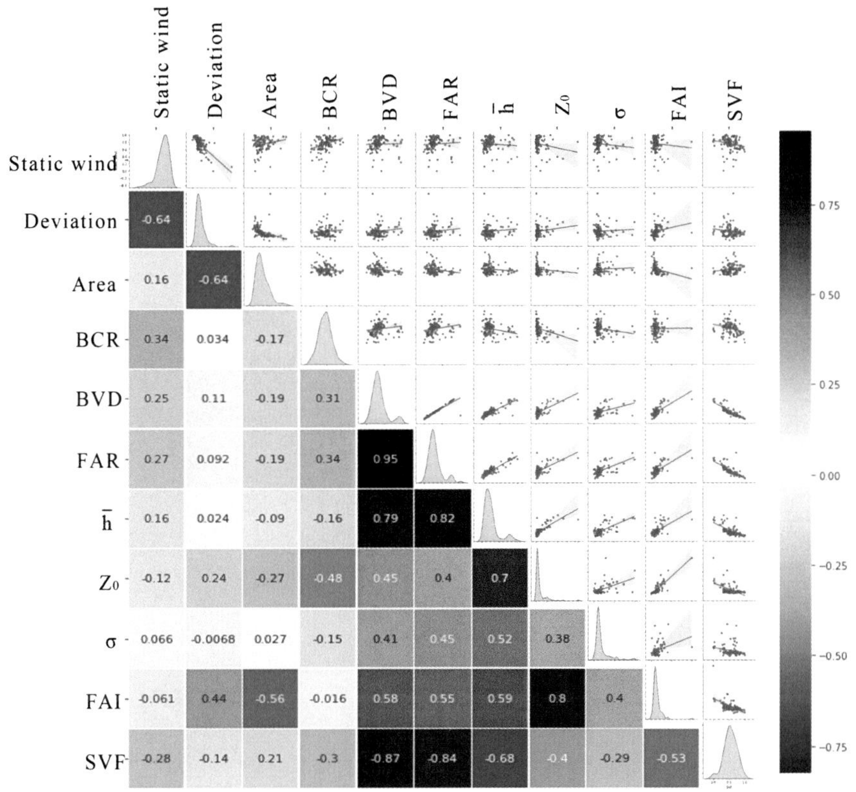

Fig. 6. Pearson correlation and scatterplots matrix

The regression analysis shows that increasing the building coverage of 0.94% in the workers villages, the ratio of the area of the static wind zone in summer will increase subsequently by 10%, with a site area of 1.4–18.1 hectares, a building volume density of 1.97–9.97, and an average building height of 8.9 to 39.6 m. In the workers villages with a site area of 1.4–18.1 hectares, a building volume density of 1.97–9,97, and an average building height of 8.9 to 39.6 m, the building site area is reduced by about 1.5 m^2 and the summer wind speed dispersion increases by 10%.

4 Conclusions

In this study, numerical simulations were used to analyze the wind environment and verify the influencing factors for 100 workers villages neighborhoods in Shanghai. Correlation analysis and regression analysis of wind environment indicators are conducted. The following findings provide researchers and planners with a deeper understanding of pedestrian level wind conditions in residential areas in Shanghai and similar climatic regions.

(1) The correlation analysis reveals that, among the morphological indicators, building coverage, building volume density and volume ratio correlate more significantly with

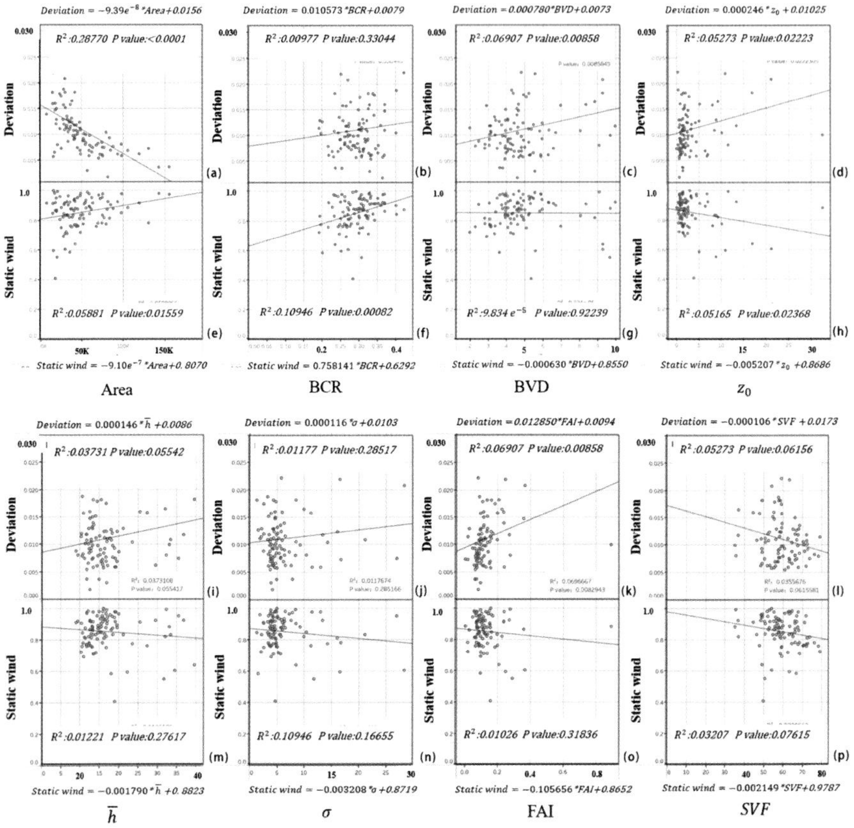

Fig. 7. Scatterplots of environmental and morphological indicators

the area ratio of the static wind zone. This conclusion may also be superimposed by multiple indicators. (2) Frontal area index, building coverage ratio, building volume density and floor area ratio are the four morphological indicators which affect the outdoor wind environment of Shanghai workers villages. These indicators can be used for urban renewal and renovation of old neighborhoods. (3) It can be seen that there is a correlation between morphological indicators and environmental indicators, but they don't have a high enough fitting precision. The main reason is conjectured to be the presence of more anomalies in the data analysis that affect the degree of fit. Furthermore, the measured data are based on real world model, which adds more uncertainties compared with idealized models.

In the subsequent study, more dimensions of environmental data such as wind speed ratio may be added and correlation analysis may be conducted. In addition, this study mainly uses linear regression equation to analyze the correlation between the indicators, but the correlation between wind environment and building morphology cannot be described by a simple linear equation. The N-S equation is a two-linear nonlinear equation, and should also be explored and analyzed in conjunction with the k-s turbulence model, etc. For further research, we hope to use black-box operation through machine learning to fit morphological and environmental indicators more intelligently and effectively, and better find the correlation between the data.

References

1. Yang Y, Gou A (2021) Research on wind environment simulation of commercial district based on Phoenics—taking shanghai central building group as an example. In IOP conference series: earth and environmental science (Vol 647, No 1, p 012193). IOP Publishing
2. Ng E, Cheng V (2012) Urban human thermal comfort in hot and humid Hong Kong. Energy Build 55:51–65
3. Tominaga Y, Stathopoulos T (2013) CFD simulation of near-field pollutant dispersion in the urban environment: a review of current modeling techniques. Atmos Environ 79:716–730
4. Ing-hui M, Zhuo-cheng M, Zhong-qi Y, Yuan-hao Q, Fu-hai G, Jian-ming X, Min C (2016) Analysis of the temporal and spatial distribution of haze and its influencing factors in Shanghai. Polish J Environ Studies 25(5)
5. Yang Y, Gou A (2021) Research on wind environment simulation of commercial district based on Phoenics—taking Shanghai central building group as an example. In: IOP conference series: earth and environmental science (Vol 647, No 1, p 012193). IOP Publishing
6. Peng Z, Jia L, Li L, Quan SJ, Yang PPJ (2017) How the roofing morphology and housing form affect energy performance of Shanghai's workers village in urban regeneration. Energy Procedia 142:3075–3082
7. Palusci O, Monti P, Cecere C, Montazeri H, Blocken B (2022) Impact of morphological parameters on urban ventilation in compact cities: the case of the Tuscolano-Don Bosco district in Rome. Sci Total Environ 807:150490
8. Yang F, Qian F, Lau SS (2013) Urban form and density as indicators for summertime outdoor ventilation potential: a case study on high-rise housing in Shanghai. Build Environ 70:122–137
9. Zheng X, Yang J (2021) CFD simulations of wind flow and pollutant dispersion in a street canyon with traffic flow: comparison between RANS and LES. Sustain Cities Soc 75:103307
10. Du Y, Mak CM, Tang BS (2018) Effects of building height and porosity on pedestrian level wind comfort in a high-density urban built environment. In: Building simulation (Vol 11, No 6, pp 1215–1228). Tsinghua University Press
11. Coceal O, Belcher SE (2005) Mean winds through an inhomogeneous urban canopy. Bound-Layer Meteorol 115(1):47–68
12. Chen G, Rong L, Zhang G (2021) Impacts of urban geometry on outdoor ventilation within idealized building arrays under unsteady diurnal cycles in summer. Build Environ 206:108344
13. Wong MS, Nichol JE, To PH, Wang J (2010) A simple method for designation of urban ventilation corridors and its application to urban heat island analysis. Build Environ 45(8):1880–1889
14. Xu Y, Ren C, Ma P, Ho J, Wang W, Lau KKL, ... Ng E (2017) Urban morphology detection and computation for urban climate research. Landscape Urban Plann 167:212–224
15. Shi Y, Xie X, Fung JCH, Ng E (2018) Identifying critical building morphological design factors of street-level air pollution dispersion in high-density built environment using mobile monitoring. Build Environ 128:248–259
16. Ma T, Chen T (2020) Classification and pedestrian-level wind environment assessment among Tianjin's residential area based on numerical simulation. Urban Climate 34:100702
17. Tian J, Xu S (2021) A morphology-based evaluation on block-scale solar potential for residential area in central China. Sol Energy 221:332–347
18. Macdonald RW, Griffiths RF, Hall DJ (1998) An improved method for the estimation of surface roughness of obstacle arrays. Atmos Environ 32(11):1857–1864

19. Gál T, Unger J (2009) Detection of ventilation paths using high-resolution roughness parameter mapping in a large urban area. Build Environ 44(1):198–206
20. Jiang D, Jiang W, Liu H, Sun J (2008) Systematic influence of different building spacing, height and layout on mean wind and turbulent characteristics within and over urban building arrays. Wind Struct 11(4):275–290
21. Shi Y, Lau KKL, Ng E (2017) Incorporating wind availability into land use regression modelling of air quality in mountainous high-density urban environment. Environ Res 157:17–29
22. Mittal H, Sharma A, Gairola A (2018) A review on the study of urban wind at the pedestrian level around buildings. J Build Eng 18:154–163

A Virtual Reality-Based Tool with Human Behavior Measurement and Analysis for Feedback Design of the Indoor Light Environment

Yunqin Li(✉), Nobuyoshi Yabuki, and Tomohiro Fukuda

Division of Sustainable Energy and Environmental Engineering, Graduate School of Engineering, Osaka University, Osaka 565-0871, Japan
li@it.see.eng.osaka-u.ac.jp

Abstract. Human behavior data provides essential feedback information for architects to improve a human-centered indoor light environment design. However, architects have difficulty capturing the complex, multidimensional, and unpredictable behavior of humans, often struggle to get users' feedback on time in the schematic phase. This paper proposes a new virtual reality-based behavioral measurement and assessment tool that quantitatively collects and analyzes individual behavioral data, including travel trajectory, travel time, and gaze points, to reveal user experience and interaction of light, aiming to better help architects get timely feedback from users and create human-centered indoor light environment designs in the scheme optimization phase. To showcase this tool, we utilize an exhibition hall of a museum design as an illustrative example. The experiment demonstrates the feasibility of the proposed tool, and its results suggest that different lighting schemes influence human behavior patterns and that the introduction of natural light usually stimulates more movement. The developed virtual reality tool prototype provides valuable visual information and statistics for analyzing human behavior and evaluating indoor light environment design schemes.

Keywords: Virtual reality · Human behavior · Behavior tracking · Feedback design · Scheme evaluation · Indoor light environment

1 Introduction

Human behavior data is essential feedback information for architects to provide and improve a human-centered indoor light environment design [9]. If the complex, multidimensional, and unpredictable human behavior in indoor spaces can be accurately captured before a building is constructed, it will provide architects with efficient context-based behavioral feedback information and alleviate the architects' labor for light environment design optimization [5]. Therefore, measuring and analyzing the interaction of user behavior with the light environment in building interior spaces has long been an important challenge affecting stakeholder engagement.

P. F. Yuan et al. (eds.), *Hybrid Intelligence*, Computational Design and Robotic Fabrication,
https://doi.org/10.1007/978-981-19-8637-6_16

Previous studies on human behavior analysis used multi-agent-based behavioral simulation and field observation using manual labor or monitoring techniques (e.g., camera, Wi-Fi, UWB, Bluetooth) [10]. They either fail to simulate realistic complex and fine-grained behavior or require strict field experiments, rely on a physically built environment, and are challenging to use in the schematic design phase. Recent emerging virtual reality technology has revealed the possibility of overcoming the previous limit, providing individuals an immersed interactive virtual environment that is highly similar to real [2]. For one, highly experimental controlled VR not only allows for rapid construction and modification of virtual scenes but also for the automatic collection of more accurate behavioral data [6]. For another, flexible control of experimental sites and time expands the VR experiment data sample and increases sampling heterogeneity [3]. Specifically, by controlling light factors related to human behavior in the virtual environment, designers can measure behavioral responses for quantitative analysis, an important source of data for the optimization stage of a -centered indoor light environment design.

Taking these advantages, this paper proposes a new VR-based behavioral measurement and assessment tool that quantitatively collects and analyzes individual behavioral data, including travel trajectory, travel time, and gaze points, to reveal user experience and interaction of light, aiming to better help architects create user-centered indoor light environment designs in the scheme of design and optimization phase. This new VR tool provides a platform for participants to freely roam through multiple virtual indoor light environment scenarios while recording their movements and views at high speed and accuracy. Combining real-time realistic rendering in this tool, designers could obtain accurate participants' immersed feedback and their behavioral interaction data in a close to the realistic indoor light environment.

2 Methodology

The workflow for feedback design using the proposed VR-based system is divided into five steps (Fig. 1). First, architectural models and lighting settings are prepared in 3D modeling software. Second, this design content is converted into a realistic-rendered VR environment using the proposed VR system. Then, volunteers are recruited to use head-mounted display (HMD) devices to experience the immersive virtual environment with the design content. Next, users provide their VR feedback through questionnaires, and the behavioral analysis results are obtained from the proposed VR system. Finally, with user feedback, architects can optimize their designs to accommodate user preferences and needs.

The experimental building is modeled in 3D modeling software (e.g., Revit). The model contains the basic structure of the architectural space, including walls, floors, ceilings, columns, windows, furniture, etc. The lighting setup is commissioned in 3D modeling software (e.g. 3dmax) to simulate a realistic light environment. All this information can be exported as an FBX file and then imported into the Game engine (e.g. Unity) [4].

The proposed VR system is designed to record information about the user's behavior in a virtual scene as close to reality as possible, including travel trajectory, travel time, and gaze points. To achieve this aim, first, when importing the 3d model into Unity,

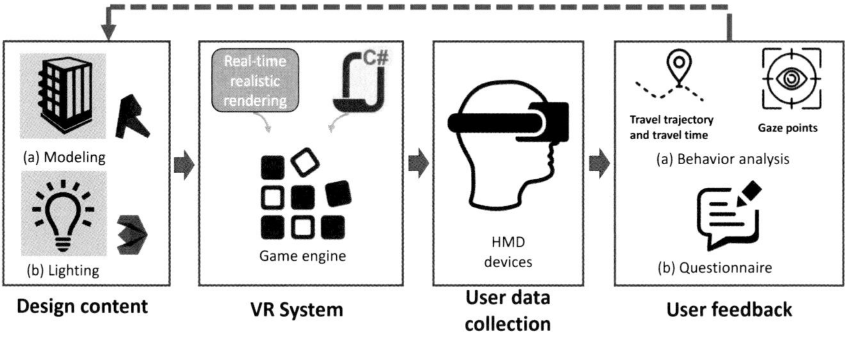

Fig. 1. Workflow of feedback design using the proposed VR-based system.

we use a rendering module (e.g., Universal Render Pipeline (URP)) to add missing information such as material types and properties (e.g., reflections, shadows, etc.) to create an immersive virtual space and achieve real-time realistic rendering. Second, we write code in Unity to record information about the participants' behavioral data and then write visualization scripts to show the results of the behaviors in the building layout.

The data collection contains behavior data from the VR system and subjective data from the questionnaire. For the former, as the participant roams through the virtual scene using the HMD device, the participant's position, head rotation, and timestamp in milliseconds are recorded in Unity. Also, the coordinates of the participant in space and the coordinates of the point where the participant's line of sight intersects with the spatial entity are obtained based on these data. Thus, the system collects three types of behavioral data: travel trajectory, travel time, and gaze point. By visualization scripts, these behavioral data can be displayed directly in the building layout from different views.

Since human feedback to the light environment is a complex cognitive process that requires a combination of multiple evaluation results to be predicted [7]. Therefore, in the proposed workflow, in addition to the measurement of human behavioral information by the proposed VR system, we also employ some traditional scales and questionnaires together. The proposed VR system can measure implicit visual physiological data, the questionnaire can assess the overall experience of the VR system, and the positive and negative affect scale (PANAS) can assess the emotional behavior of the participants [7].

The apparatus in this study include an HMD system (HTC-Vive 2016) and a desktop computer. In the experiment, participants will remain in a standing position and can move forward, backward, left, and right through the experimental scene at a speed of 1.25 m per second via the controller of the HTC-Vive (Fig. 2). The display resolution for both eyes was 2160*1200 pixels and the display screen refresh rate of 90 Hz.

The procedure of the experiment consists of five parts:

- Preparing experimental scenarios of the different indoor light environments in the same model.
- Introducing the overview of the experiment to the participants and making sure they are acquainted with the HMD device.

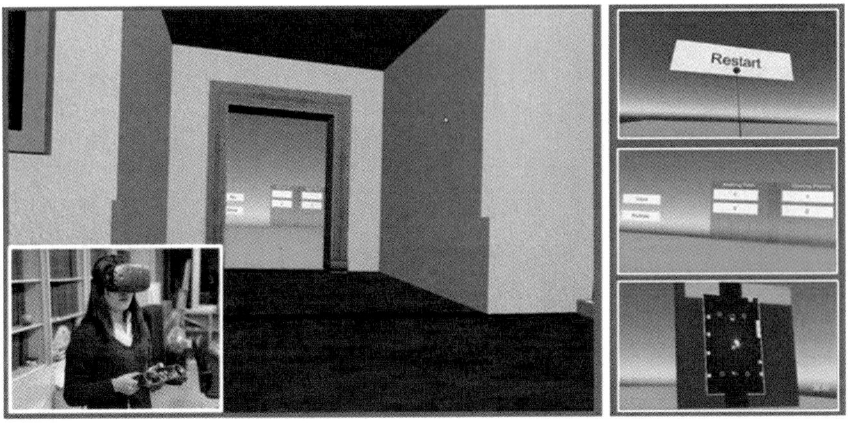

(a) (b)

Fig. 2. A participant using **a** the proposed system and **b** its interactive components roaming in an immersive virtual environment.

- Each participant will enter all experimental scenarios in turn, and there is a 5-min break for scene transitions. The order of the experimental scenes was randomized for each participant.
- Letting them roam freely in the experimental scene without their knowledge for a period determined by themselves.
- Letting the participants fill in the questionnaires including their basic demographic information, the simulator sickness, VR-based system evaluation and PANAS scale.

3 Experiments and Results

To showcase this tool, we utilize an exhibition hall of a museum design as an illustrative example. First, we prepared and constructed two contrast indoor light environment design schemes in 3D modeling software and then translated the 3D models into a VR environment with realistic rendering. Next, 10 participants were invited to roam in the spatial schemes and finish a questionnaire of their features and experiences afterward. And meanwhile, this VR system will record participants' positions, head rotations, and timestamps in milliseconds in the game engine. Then, by extracting three types of behavioral data, including travel trajectory, travel time, and gaze points, participants' path, area of interest, and gaze points were mapped of visualization. Finally, combined with participatory feedback from user questionnaires and visualized behavioral results, designers can compare and evaluate different indoor light environment design schemes for optimization.

We designed a virtual model of a museum gallery space 16 m long, 10 m wide, and 6 m high, with a total area of 160 m^2, with dimensions similar to the actual exhibition space. There are 11 virtual booths (with exhibits) in the space. The wall and ceiling materials simulate latex paint and the floor materials simulate a non-glossy wooden floor. The gallery includes a 25 m^2 top skylight and four sets of light fixtures, each consisting of three spotlights. Figure 3 shows the floor and profile of the designed virtual showroom. The artificial lighting is top directed with a color temperature of 3300 K and a color rendering index (cri) of 90. The natural lighting conditions are simulated as the sun's position at 14:00 at the same time of the year (the clear day on December 22) in the same location (Tokyo, Japan). In scenarios A and B, the exhibition halls were designed for artificial lighting, and mixed lighting (artificial lighting together with natural lighting), respectively (Fig. 4).

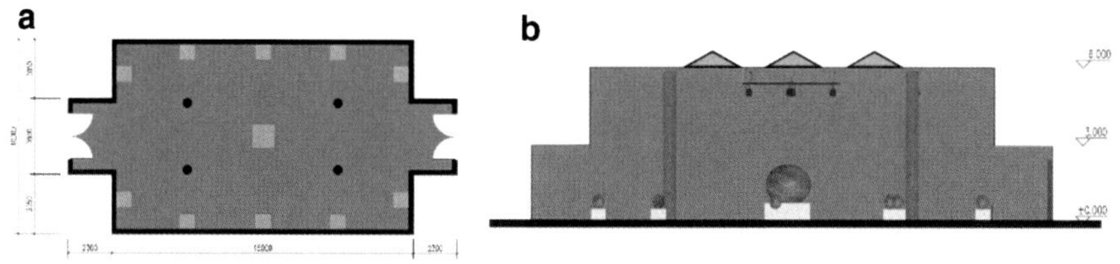

Fig. 3. **a** Floor and **b** section drawings of designed virtual exhibition space.

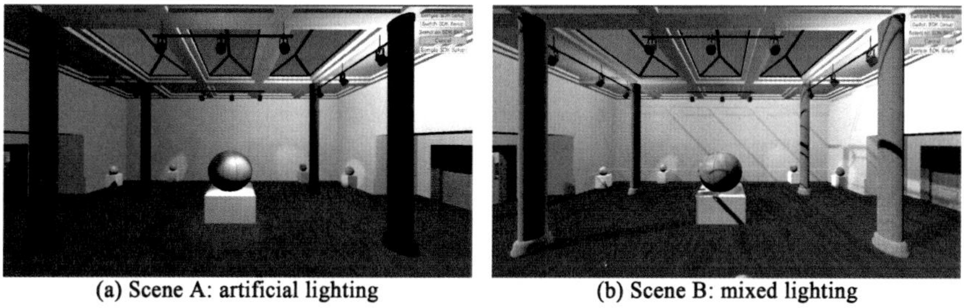

Fig. 4. Lighting scenes in the experimental exhibition hall: **a** scenario A with artificial lighting and **b** scenario B with mixed lighting.

A total of 30 participants (15 female and 15 male) with backgrounds in environmental design participated in this experiment, aged between 20 and 30 years old. Participants were unaware of the experimental scene in advance and were informed of the potential risks that could be present in the experiment.

All the participants completed the questionnaires at the end of each scenario, including their basic demographic information, the simulator sickness, VR-based system evaluation, and the positive and negative affect scale (PANAS). None of the subjects reported any discomfort or were asked to stop the experience. In the evaluation questionnaire for the VR-based system, 90% of the participants reported that the virtual architectural environment was realistic and immersive, while 73.3% reported that the lighting simulation in the virtual scenes was similar to the real scenes and 6.7% felt that the lighting rendering in the virtual scenes was not realistic enough. In addition, 93.3% of the participants found the interaction logic of the controllers in the system easy to understand, while

6.7% of the participants felt that the device took some time to get up to speed. Finally, 26.7%, 63.3%, and 10% of the participants believed that their behaviors performed with the VR system and HMD device in the virtual scene will be consistent, similar, and different, respectively, from those performed in the same scene in the real world.

The results of the PANAS questionnaire revealed that the majority of participants (63.3%) preferred mixed lighting (artificial and natural lighting). It is worth noting that 73.3% of the participants felt that artificial lighting was more conducive to focusing on the exhibit itself, and 86.7% of them felt more active, interested, and inspired in the artificial lighting scenario. In addition, 10.0% of the participants felt slightly nervous in the artificially lit scenes.

Travel trajectories are recorded at a frequency of one route point every 0.2 s. Each participant's route trajectory and travel time are recorded individually. Based on these data, we can analyze the trajectories and travel areas of interest individually or collectively. Figure 5a and b shows the travel trajectories and travel time of a participant in scenarios A and B. Figure 5c and d shows the summary of trajectories of all participants in scenarios A and B. As the dwell time increases, the transparency of participants' trajectory points gradually decreases. The average travel time for all participants in scenarios A and B were 489 and 378 s, respectively.

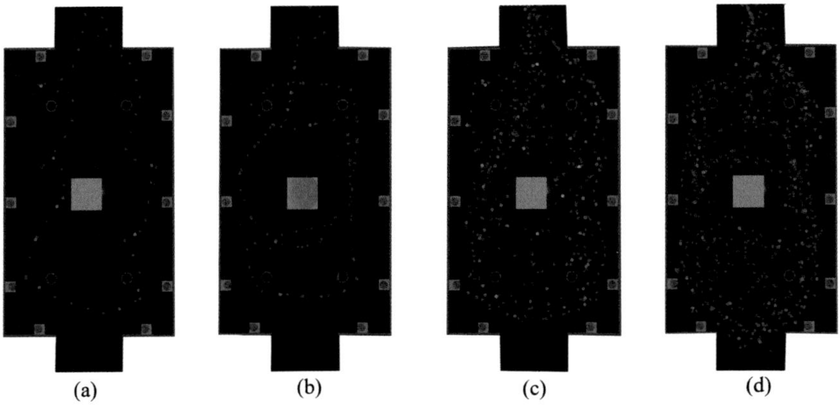

Fig. 5. Travel trajectories example of a participant in **a** scenario A and **b** scenario B and the cumulative trajectories of all participants in **c** scenario A and **d** scenario B.

The collected gaze point data can analyze which objects or information in the virtual environment attract the participants' attention. The frequency of gaze points was recorded once in 0.2 s. To better demonstrate the distribution of gaze points in space, we set up 3 different camera views. Figures 6 and 7 show the gaze points of one participant and the cumulative gaze results of all participants from three perspectives in scenarios A and B. As the dwell time increases, the transparency of participants' gaze points gradually decreases.

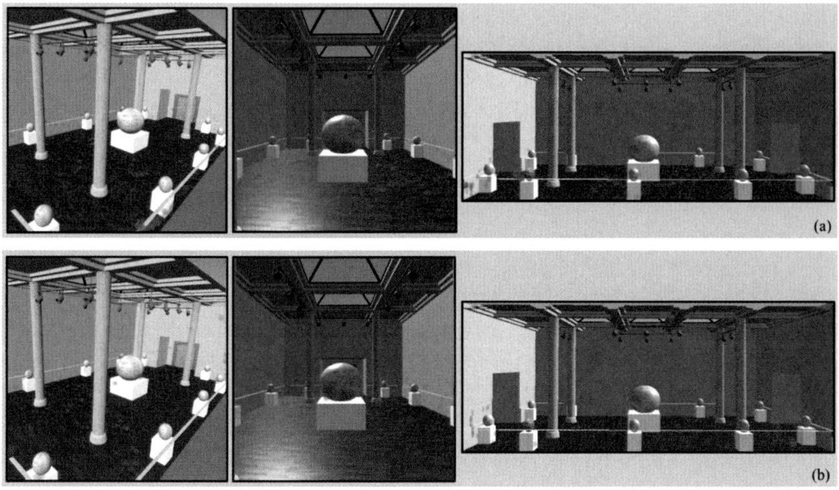

Fig. 6. A gaze points example of a participant from three perspectives in scenario A (**a**) and scenario B (**b**).

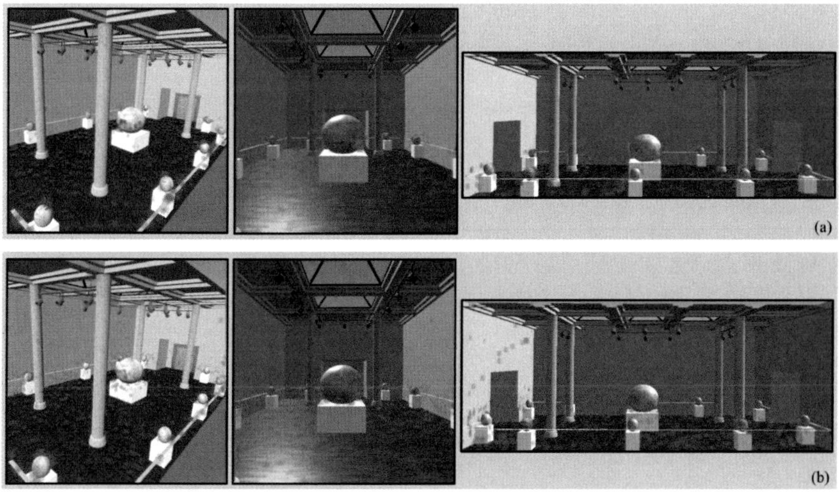

Fig. 7. Cumulative gaze results of all participants from three perspectives in (**a**) scenario A and (**b**) scenario B.

4 Discussion

This study uses the proposed VR system to record behavioral data while participants can walk and watch freely in the 3D space constructed by the game engine, which facilitates the analysis of participants' behavioral patterns under different light environment scenario simulations. Human behavior analysis based on VR technology applied to feedback-based design studies of indoor lighting environments is not common in previous related studies [8]. The results of the participant's questionnaire on VR-based system evaluation showed that the majority of people considered the system to be able to simulate virtual and real scenes and behaviors under the scenes more consistently, while being easy to operate. Therefore, the application of the proposed VR tool to the study of feedback-based interior light environment design is feasible, easy to use, and effective.

The results of the PANAS questionnaire on this and lighting preferences in the physical environment are consistent with previous studies [1]. Therefore, to meet the lighting preferences of users, design teams can provide more opportunities for natural light-harvesting in the experimental virtual environment. The results for the different emotional behaviors exhibited by participants in different light environments suggest that first, for some valuable and time-consuming exhibits, designers can use artificial lighting solutions to provide users with a focused atmosphere. Second, the partially dark light environment in the artificially lit room caused discomfort for a few users. Since to reduce the harm of light on the exhibits, the artificial lighting design guidelines suggest minimizing the illumination of the light source without affecting the normal activities of the occupants.

As shown in Fig. 5c, in the mixed lighting scene, the pedestrian trajectory points are more chaotic and relatively abundant in pedestrian trajectories. In Fig. 5d, in the artificial lighting scene, there are more highlighted stopping points which indicate that the fixed time of a single stopping point is significantly longer than in the mixed lighting scenes. In addition, we can observe some regions of interest, which are regions consisting of a high density of footprint points in Fig. 5c and d. In the artificial lighting scene, many of the regions of interest are near the exhibit in the center of the room, while the distribution of the regions of interest in the hybrid lighting scene is more uniform. These suggest that people in the hybrid lighting scenes prefer motion-based dynamic viewing behaviors, while the artificial lighting scenes tend to focus on stopping behaviors.

As shown in Fig. 7a, the gaze points in the artificial lighting scene were more focused on the exhibits, with almost no gaze points outside the illuminated range of the artificial light. In contrast, in Fig. 7b, the mixed lighting scene had a wider range of gaze points, with gaze points on the ceiling or wall caused by looking up or around. In addition, Fig. 7a has relatively more gaze points with low transparency, which implies that participants in the artificial lighting scene have longer single-point gaze time. Interestingly, the average travel time of all participants in the mixed lighting scenes was 29% higher than in the artificial lighting scenes. This suggests that natural light has a negative impact on participants' gaze duration, but a positive effect on pedestrians' willingness to stay in space.

Taken together, these results suggest that different lighting schemes have an effect on human behavior patterns and that the introduction of natural light usually stimulates more movement. Therefore, architects can design the light environment of the exhibition hall according to the exhibits' viewing characteristics and the desired viewing behavior.

The generic VR system presented in this study provides a systematic approach to collecting user behavioral information and preferences using immersive virtual environments, and such data collected can be used to improve the design of buildings based on user preference feedback. There are still some limitations. First, the proposed VR system makes it difficult to measure and evaluate the group behavior of humans. The current experiment only allows one person to roam in the venue, therefore, the method does not simulate the interactive behaviors, such as avoidance and gathering, that occur in a real environment with multiple people in the same space. Secondly, this experiment was only a control experiment for the lighting method (artificial and mixed lighting). In the complete light environment design, there are several light environment parameters (such as booth illumination uniformity, light color temperature, light source direction, etc.) that have not been further experimentally verified. It is possible to extend other light environment factors to collect user feedback-based behavioral information for designing detailed architectural lighting solutions suitable for users' needs and preferences. Third, the current tracking for the gaze point is based on the result of the center point of the line of sight, ignoring the situation of the gaze point shift due to the possible eye rotation in practice. Finally, the proposed VR system has only been experimented on fully virtual scenarios to verify its feasibility. In the future, controlled experiments with real sites and realistic rendered virtual scenes should be introduced to better evaluate the effectiveness of this VR system.

5 Conclusion

This paper presents a generic tool for collecting user behavior information (including travel trajectories, travel times, and gaze points) in an immersed indoor light environment based on virtual reality technology. By capturing and analyzing this information and user feedback, designers can translate this information into design boundaries promptly to ensure human-centered indoor light design. To demonstrate how to use our proposed VR system for feedback-based design in the schematic phase of a building, we showed an experiment in which we collected and analyzed behavioral data from 30 participants in artificial and mixed-light environments. The results showed that the VR system can collect and analyze human behavior data of walking, staying, and watching, and showed participants were willing to stay longer in the space under mixed lighting. The results of this study can help architects get timely feedback from users so that they can create spaces focused on the needs and preferences of end-users and improve user satisfaction, rather than just designing according to laws and codes. In future research, we will introduce eye-tracking as well as controlled experiments on real sites to further develop and validate the proposed VR tool.

References

1. Ai X, Wu Z, Guo T, Zhong J, Hu N, Fu C (2021) The effect of visual attention on stereoscopic lighting of museum ceramic exhibits: a virtual environment mixed with eye-tracking. Informatica 45. 10/gnhs8r
2. Chen H, Hou L, Zhang G, (Kevin) Moon S (2021) Development of BIM, IoT and AR/VR technologies for fire safety and upskilling. Autom Constr 125:103631. 10/gjn6jv
3. Feng Y, Duives DC, Hoogendoorn SP (2021) Using virtual reality to study pedestrian exit choice behaviour during evacuations. Safety Sci 137:105158. 10/gh4vjn

4. Heydarian A, Pantazis E, Wang A, Gerber D, Becerik-Gerber B (2017) Towards user centered building design: Identifying end-user lighting preferences via immersive virtual environments. Autom Constr 81:56–66. 10/gbpcn5
5. Li Y, Yabuki N, Fukuda T (2022) Exploring the association between street built environment and street vitality using deep learning methods. Sustain Cities Soc 79:103656. 10/gn6gmp
6. Li Y, Yabuki N, Fukuda T, Zhang J (2020) A big data evaluation of urban street walkability using deep learning and environmental sensors-a case study around Osaka University Suita campus. Presented at the proceedings of the 38th eCAADe conference, TU Berlin, Berlin, Germany, pp 319–328
7. Lin J, Cao L, Li N (2020) How the completeness of spatial knowledge influences the evacuation behavior of passengers in metro stations: a VR-based experimental study. Autom Constr 113:103136. 10/gnbnh3
8. Xiao H, Cai H, Li X (2021) Non-visual effects of indoor light environment on humans: a review☆. Physiol Behav 228:113195. 10/ghjp3v
9. Yan M, Tamke M (2021) Augmented reality for experience-centered spatial design—a quantitative assessment method for architectural space. Presented at the Towards a new, configurable architecture, Proceedings of the 39th eCAADe conference-Volume 1, Novi Sad, Serbia, pp 173–180
10. Zhang J, Fukuda T, Yabuki N (2021) Automatic object removal with obstructed façades completion using semantic segmentation and generative adversarial Inpainting. IEEE Access 9:117486–117495. 10/gm6hpm

Artificial Intelligence

Using Text Understanding to Create Formatted Semantic Web from BIM

Jingming Li[1,2](✉)

[1] Midea Building Technology, Midea Group, Foshan, China
jli@hnu.edu.cn
[2] College of Civil Engineering, Hunan University, Changsha, China

Abstract. The application of BIM in the building life cycle needs to be continuous. The information collected and accumulated in the early stages should flow to the subsequent phases. However, BIM applications currently focus on collision inspection, compliance inspection, and engineering calculation, few models can be successively used in the following stages. Remodeling is required in the operation and maintenance period, resulting in waste. Meanwhile, some of the information accumulated by BIM might be frequently used in the operation and maintenance stage, while some data are relatively rarely used. The semantic web can help manage building information at all stages. But the generation of a semantic web is mostly manually completed. It is necessary to standardize the repeated semantic description in the model and convert BIM into a standard semantic model for information indexing, reducing the resource consumption of model loading and optimizing the efficiency of the operation and maintenance system. When the existing research transforms from BIM to the semantic web, there will be a lack of information and descriptions of the ownership relationship between entities due to the limitation of formats. To realize the standard transformation from BIM to the semantic web, this work proposes a method of using Natural Language Processing (NLP) to understand the text and infer the relationship between entities according to the knowledge map. First, the entities are extracted from BIM, such as air conditioning unit, electric lamp, fan, etc., if the name of the extracted entity is irregular, the names are translated with the help of NLP and Ontology (such as brick or haystack) to obtain the standard definition. By comparing the complete knowledge graph (such as the knowledge graph of the air conditioning system), the relationships can be deduced, and then a standardized semantic model can be generated.

Keywords: BIM · Semantic web · Text understanding

1 Introduction

Global total energy consumption has reached 162 thousand terawatts in 2019. As the world's largest economies, China and the United States account for more than 40% of global energy consumption [1]. Among them, the building sector accounts for 18.35% of China's total energy consumption, and the ratio has reached 40% in the US [2, 3]. In

P. F. Yuan et al. (eds.), *Hybrid Intelligence*, Computational Design and Robotic Fabrication,
https://doi.org/10.1007/978-981-19-8637-6_17

the context of tackling climate change, how to effectively reduce building energy consumption has become an important topic of energy conservation and carbon reduction. Using Building Information Model (BIM), Semantic Web, Internet of Things and other information technologies to improve building operation is one of the key research areas.

BIM contains lots of details since early design stages in buildings. IFC is a common format for BIM. In addition to attribute information, IFC also contains a lot of 3D information, and the material and other information carried by IFC can also be used for rendering 3D effects. The model itself is a large dataset, meaning IFC is not an ideal index to manage the data in the Operation and Maintenance (OM) phase.

The Semantic Web is a network of data that includes dates, titles, part numbers, chemical properties, and any other types of data. Brick and Haystack [4] are two universal semantic schemas for defining entities in buildings. Haystack tags resolve data silos among various subsystems (HVAC, lighting, and enterprise scheduling). Brick aims to standardize the semantic description of buildings, including physical, logical, virtual assets and the relationships between them. Using the Semantic Web, Brick can coherently describe many special and custom functions, assets, and subsystems throughout the building life cycle.

Using Brick or Haystack as a description specification for a building reduces the cost of deploying analytics, energy efficiency measures, and smart controls across buildings, demonstrating the integration of numerous subsystems in a modern building: HVAC, lighting, fire protection, security, etc. Simplifies smart analytics and control applications development, as well as reduces reliance on non-standard, unstructured labels specific to building management systems. But the conversion process is still challenging.

1.1 Related Works

Currently, there are many types of research and applications in building compliance, with Ontology, Metadata, and Semantic Web, but most of them are used for building design and model detection. These studies usually extract semantics from BIM models to generate graphs, or directly use BIM models as query objects, and then carry out deductions such as cost budgeting, energy design, and construction hazard identification based on standards [5–7].

Through the integrated application of BIM and Semantic Web technology, the project [8] design conforms to the construction quality specifications. The project can automatically check the size and position of the BIM model components according to the requirements of the specification, thereby reducing the benchmarking workload of the relevant personnel during the construction process.

When extracting semantics from BIM and comparing them with standards, the process usually introduces ontology description technologies such as SPARQL and Web Ontology Language (OWL) or Unified Modeling Language (UML) to improve the standardization and efficiency of retrieving BIM semantics. For instance, McGibbney and Kumar [9] summarized BIM model detection based on semantic implementation. The process firstly converts specifications and models to semantics and then extracts the BIM model based on specification requirements, which improves the efficiency of specification and model checking. Bi et al. [10] used ontology for knowledge expression to promote the protection of ancient buildings.

The current applications of the Semantic Web are similar to labeling entities in Brick and Haystack. Labels produce good guidance for data, but cannot effectively represent relationships between entities. Also, the knowledge graph exported from IFC at this stage contains too much redundant data, which reduces the efficiency of data management. Although the complex types of data in the Operation and Maintenance (OM) phase are more suitable for the Semantic Web to exert its value, few studies have applied it to OM because of the difficulties in creating a semantic web. The application of BIM in the building life cycle is continuous. After accumulating information in BIM, only part of the data will be reused in the OM phase.

However, the applications of BIM are mainly concentrated in the application stages such as model collision check, compliance check, and engineering calculation, and rarely extend to the OM stage, causing massive remodeling works and waste of resources. It is necessary to standardize the repeated semantic description in the model and convert the BIM into a standard semantic model for information indexing, which can reduce model loading, reduce resource consumption and optimize the efficiency of data sharing in the building operating system. Currently, the semantic simplification of BIM models is still done manually, which consumes a lot of manpower and time.

1.2 Contributions

This work explores the scheme of building a standard semantic model based on BIM. First, the entities are extracted from the BIM model, and the names are standardized or transformed regarding the ontology dictionary. Then, the belonging and supply relationships between each entity will be reasoned by combining the knowledge graph. Finally, the custom entities in BIM are converted into standard semantic models and the transition of building information from early stages to OM.

2 Methodology

To create the standard conversion from BIM to Semantic Web, this chapter proposes a method of using NLP to understand entities and infer the interrelationships between entities according to the knowledge graph. The implementation framework of this method is shown in Fig. 1.

2.1 Entity Extraction

The first step is to extract entities from the data source, such as the BIM model.

When the existing research converts BIM information to the Semantic Web, due to the lack of information, the conversion in Industry Foundation Classes (IFC) is geographical rather than semantical. It cannot infer the relationship between entities when exporting ifcOWL.

There are options to extract entities from the BIM model, directly extracted through plug-ins, or converted into ifcOWL. The latter will cause information loss in the conversion process. In contrast, directly extracting entities from the model can avoid data loss during the conversion.

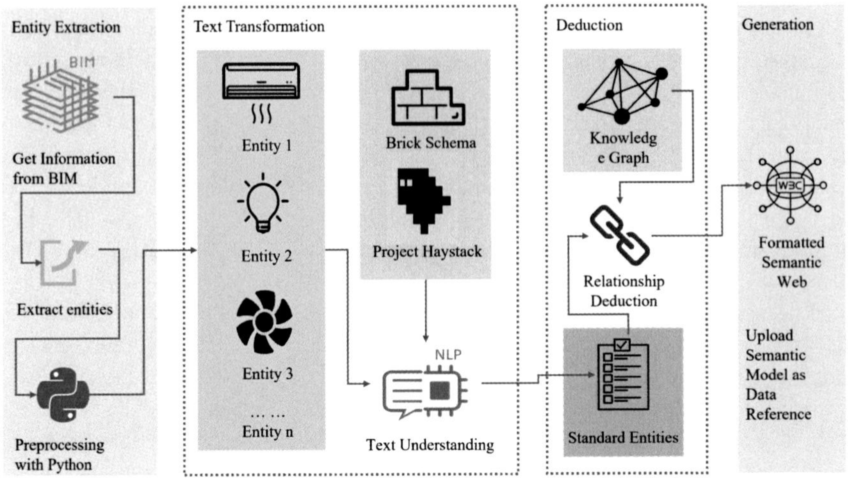

Fig. 1. The method of generating standard SW model from BIM

Direct data extraction from the BIM model can use plug-ins such as Dynamo or Grasshopper, as shown in Fig. 2. First, the script traverses all the categories in the file (such as piping, mechanical, electrical, etc.). Then, it traverses the entities under each category. Finally, the program returns all the entity names and family categories. The built-in conditional grouping module of Dynamo can group entities according to their spatial positions. After deduplication, entities and their positional relationships can be obtained.

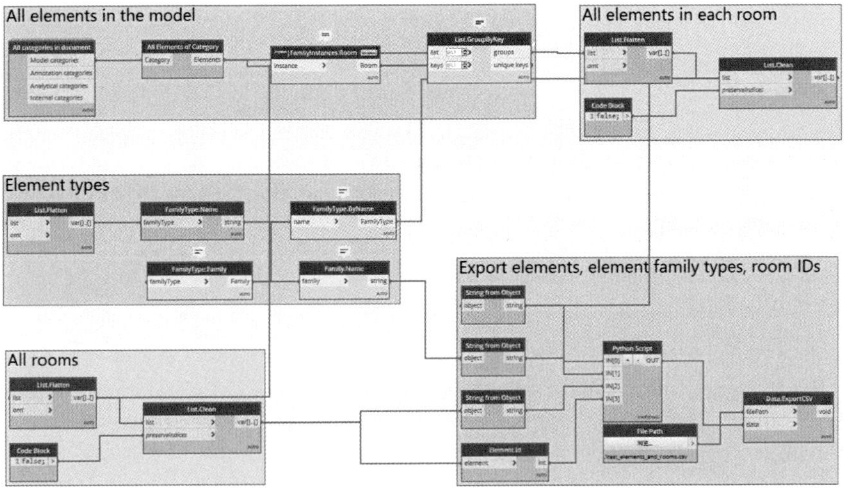

Fig. 2. Extracting entities based on location relationships from BIM

As illustrated in Fig. 3, following similar settings, the extraction of entities can also be carried out according to the supply relationship. The script uses the third-party plug-in (Spring Node) to obtain the associated entities within the view. By setting the category of the parent equipment and the category of the equipment being supplied respectively, and removing the repeating entities, the supply relationships can be obtained.

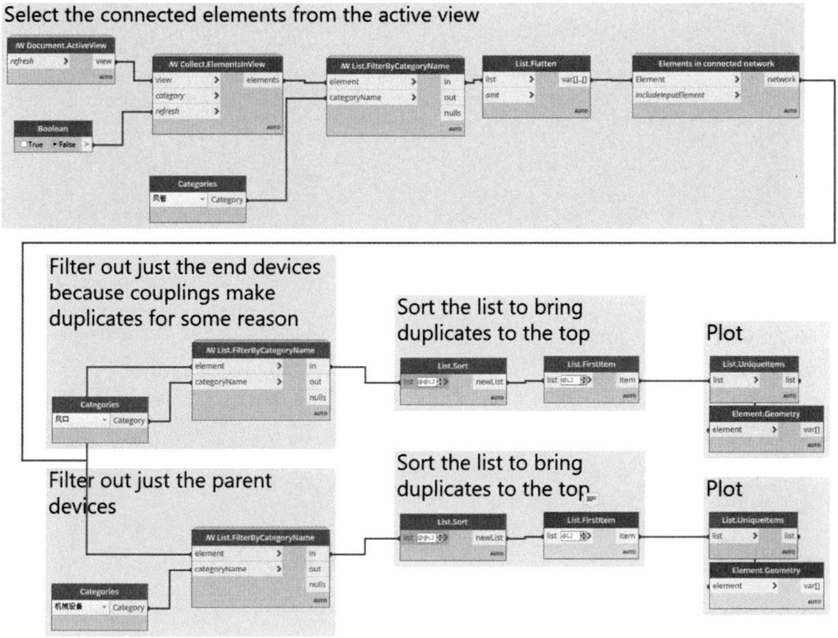

Fig. 3. Extracting entities based on feeding relationships from BIM

2.2 Text Transformation and Relationship Inference

The first step is to determine the entity name. If the names of the extracted entities are irregular, NLP is needed to understand its family name through cosine similarity, as shown in Fig. 4. Knowledge graphs use visualization techniques to describe knowledge resources, which can be used to display, analyze, and reason about the interconnections between entities. The relationship can be derived according to the knowledge graph. The translated standard entity can infer the relationships according to the location in the knowledge graph. The implementation of relational reasoning mainly relies on the SPARQL query on the complete knowledge graph and creates relationships for newly generated entities according to the query results.

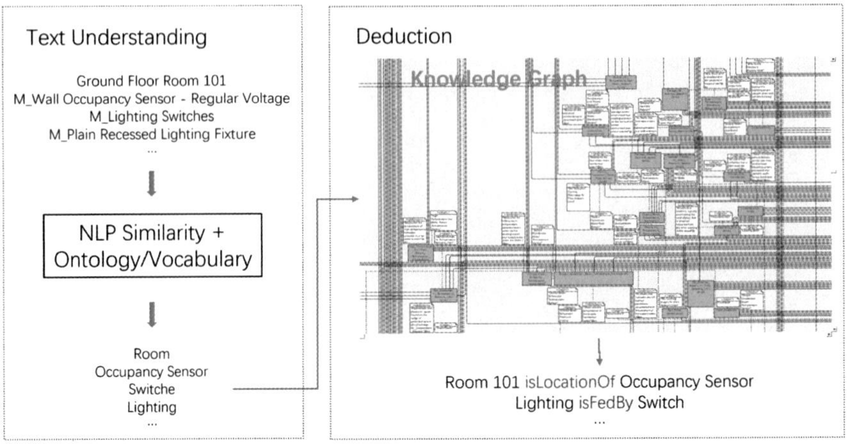

Fig. 4. The reasoning of relationships from custom entity names

Ai and Bi are the vectors transformed from the standard dictionary of name and ontology respectively, and the cosine similarity of the vectors is calculated in turn, as shown in Eq. 1. The closer the cosine similarity is to 1, the closer the two words are to each other, −1 means the words are opposite.

$$\text{similarity} = \frac{\sum_{\text{i}=1}^{n} A_i B_i}{\sqrt{\sum_{\text{i}=1}^{n} A_i^2}\sqrt{\sum_{\text{i}=1}^{n} B_i^2}} \tag{1}$$

Through the cosine similarity evaluation of the vector, the entity name can be transformed into the standard vocabulary of the ontology dictionary. The framework uses tools PyTorch and transformers [11]. BERT converts the text into entries with a length of 128. Each entry has a separate 768-digit vector. Pooling will extract the average of all tags and combine them into a unique 768 vector space to produce a "sentence vector". Based on the pre-trained models, the program converts the non-standard vocabulary and ontology vocabulary into vectors respectively, uses PyTorch to calculate the cosine similarity, and finally selects the closest standard word.

If the naming rules of the model are clear, such as RM_101_IDU_134_T1 refers to No. 134 indoor unit in Room 101, it only needs to be converted into the standard name of the body according to the naming rules. The naming rules may already include the location and supply relationships, and the relationship between entities can be directly disassembled from the naming. Figure 5 shows the process.

By comparing the inferred relationships with a complete knowledge graph (such as the knowledge graph of an air-conditioning water-cooling system), the interrelationships between entities can be deduced, and a standardized semantic model can be generated.

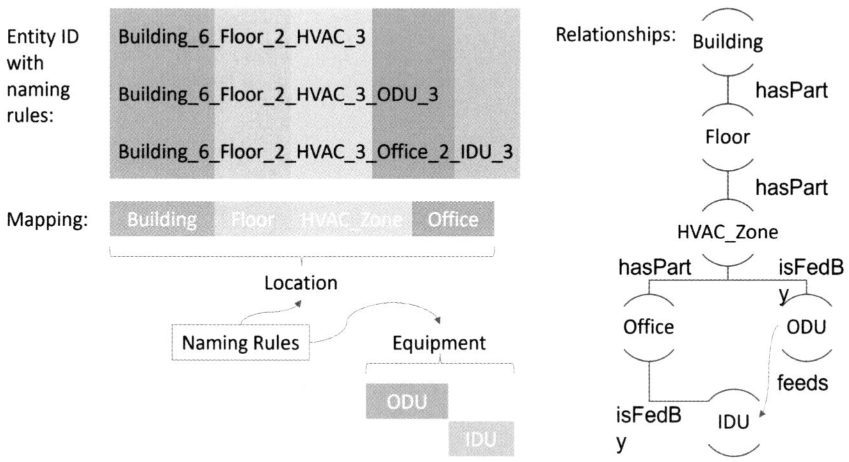

Fig. 5. The reasoning of entity relationships from formatted entity names

3 Experiment Results and Discussions

An experiment was conducted on the HVAC file within Revit. The program extracts examples from BIM files and classifies them into 42 categories. 25% of the categories, which are the data sources and control units, are included in Brick or Haystack. The classification of examples is carried out according to ontology standards, and the results are shown in Table 1. The extracted examples are mainly HVAC and electrical equipment. The entity names may not be standardized, for instance, Trck_BswySystms_Cooper_RSA_Profile Series_AR111 Closed Back Integral Xfmr is a lighting device.

Figure 6 shows the comparison results between the pre-trained language model and brick and haystack standards. The SentenceTransformer loads the pretrained models [12] and calculates the most similar word, respectively. From the results that even the model specially adjusted for similarity calculation has poor results. The accuracy of the first mock exam is only 60%, and the same model is not stable when dealing with different ontology standards. Paraphrase-mpnet-base-v2, which performs best in classifying and looking for similar texts, has a 40% difference in accuracy between brick standard and haystack standard. However, this problem will be improved with the strengthening of buildings and equipment by the pretrained model, which is also confirmed by the results of different pretrained models.

Compared with the results of brick vocabulary calculation, although the natural language model has been optimized in their respective training sets, its performance is still unstable in the process of practical application, especially in the fields involving specific professional knowledge.

Table 1. The words covered in Brick and Haystack

Model Entities	Entity Type	Brick	Haystack
M_Wall Occupancy Sensor—Regular Voltage	Occupancy sensor	√	√
M_Lighting Switches	Switch	√	√
M_Plain Recessed Lighting Fixture	Lighting	√	√
M_Pendant Light—Linear—2 Lamp	Lighting	√	√
M_Sconce Light—Flat Round	Lighting	√	√
Trck_BswySystms_Cooper_RSA_Profile Series_AR111 Closed Back Integral Xfmr	Lighting	√	√
M_Pile-Steel Pipe	Pipe	–	√
Water	Water	√	√
DC_Tankworks_PLM_5000L	Water tank	√	√
Miele MasterCool KF 1911 Vi	Freezer	√	–

For non-standard named entities, semantic understanding is required. Due to the ontology standards' low coverage of the BIM model and the fact that the pretrained model has not been strengthened for buildings and equipment, the accuracy of standardized text translation is low, and the automation level in the process of generating standardized semantic web from non-standard named entities in BIM file is low, so manual participation is still needed at present.

When using NLP to understand professional knowledge, it still needs to be constrained by a large amount of relevant professional knowledge. For example, when this chapter explores the use of NLP to transform the non-standard language description in BIM into the ontology standard in the field of HVAC, the implementation of this process requires HVAC professional knowledge to provide constraints for machine-reading and understanding and optimize the training set.

However, the pretrained model has not been optimized enough, and there is a large deviation between the results of machine understanding and the actual situation. The accuracy of BIM + NLP in converting non-standard entity names needs to be improved, and the current model also needs to strengthen the language recognition of professional vocabulary in the field of architecture and HVAC.

4 Conclusions

Based on BIM Technology, this work studies the method of standard semantic description of BIM entities, including the extraction of entities and their relationships from BIM and the entity name transformation of NLP semantic understanding.

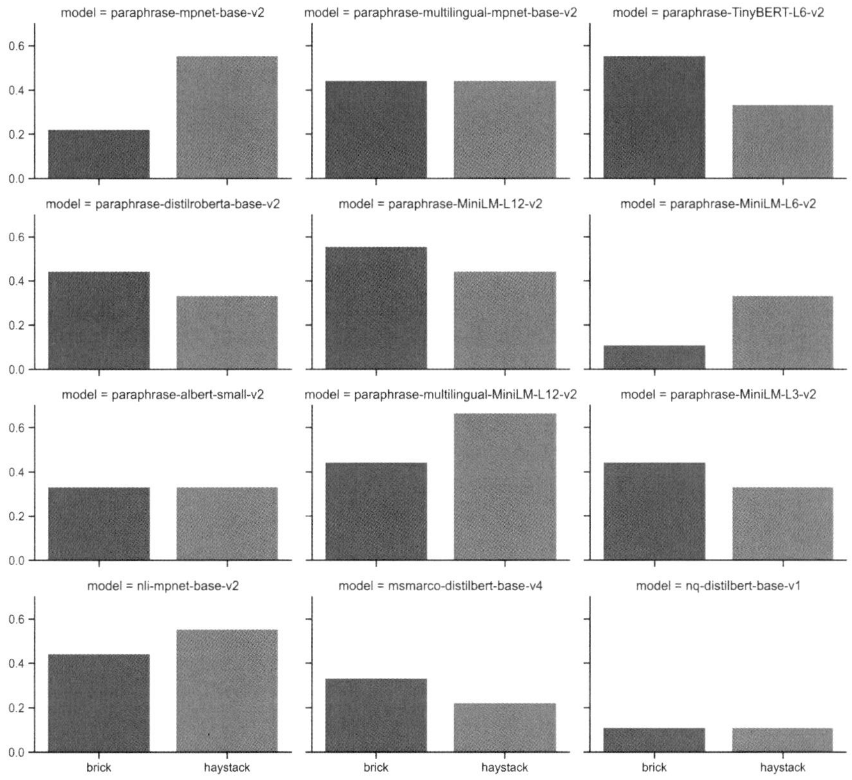

Fig. 6. The results of corrections

Currently, the standardization of BIM entity naming rules is an effective means. Due to the low accuracy in understanding professional HVAC knowledge, the conversion of BIM entities without clear naming rules is still dominated by manual work, and automation is limited. In future research, optimizing the pretrained model through professional knowledge will be the key to optimizing the NLP translation results. With the assistance of ERNIE 3.0 or GPT-3, semantic understanding should be more accurate during BIM to semantic web conversions.

References

1. Ritchie H, Roser M (2021) Energy production and consumption—our world in data. https://ourworldindata.org/energy-production-consumption#energy-production-and-consumption-by-source. Accessed 8 Mar 2022
2. Department of Energy (DOE) (2015) Chapter 5: increasing efficiency of building systems and technologies. Quadrenn Technol Rev An Assess Energy Technol Res Oppor, 143–181
3. Yan L (2018) China: energy efficiency report, 2018

4. Quinn C, McArthur JJ (2021) A case study comparing the completeness and expressiveness of two industry-recognized ontologies. Adv Eng Inf 47:101233. https://doi.org/10.1016/j.aei.2020.101233
5. Staub-French S, Fischer M, Kunz J, Ishii K, Paulson B (2003) A feature ontology to support construction cost estimating. Artif Intell Eng Des Anal Manuf AIEDAM.https://doi.org/10.1017/S0890060403172034
6. Lork C, Choudhary V, Ul Hassan N, Tushar W, Yuen C, Ng BKK, Wang X, Liu X (2019) An ontology-based framework for building energy management with IoT. Electron.https://doi.org/10.3390/electronics8050485
7. Zhang H, Gu M, Sun J (2017) The method and the device for BIM model specification detection based on semantic retrieval
8. Farghaly K, Soman RK, Collinge W, Mosleh MH, Manu P, Cheung CM (2022) Construction safety ontology development and alignment with industry foundation classes (IFC). Electron J Inf Technol Constr. https://www.research.manchester.ac.uk/portal/en/publications/construction-safety-ontology-development-and-alignment-with-industry-foundation-classes-ifc(67e547bd-c261-4c34-9460-afbeb103aedf).html. Accessed 8 Mar 2022
9. McGibbney LJ, Kumar B (2015) A framework for regulatory ontology construction within AEC domain. In: Ontology in the AEC industry: a decade of research and development in architecture, engineering, and construction
10. Bi Z, Wang H, Lu Y (2014) A construction model of ancient architecture protection domain ontology based on software engineering and CLT. J Softwhttps://doi.org/10.4304/jsw.9.11.2886-2894
11. Reimers N, Gurevych I (2019) Sentence-BERT: sentence embeddings using siamese BERT-networks. EMNLP-IJCNLP 2019–2019 Conf Empir Methods Nat Lang Process 9th Int Jt Conf Nat Lang Process Proc Conf, 3982–3992
12. Ubiquitous Knowledge Processing Lab (2020) Pretrained models—sentence-transformers documentation. https://www.sbert.net/docs/pretrained_models.html. Accessed 30 Aug 2021

Artificial Intelligence Prediction of Urban Spatial Risk Factors from an Epidemic Perspective

Yecheng Zhang[1(✉)], Qimin Zhang[2], Yuxuan Zhao[1], Yunjie Deng[1], Feiyang Liu[3], and Hao Zheng[4(✉)]

[1] College of Architecture and Art, Hefei University of Technology, Hefei, China
575860760@qq.com
[2] School of Mechanical Engineering, Hefei University of Technology, Hefei, China
[3] Wuhan Institute of Technology, Wuhan, China
[4] Stuart Weitzman School of Design, University of Pennsylvania, Philadelphia, USA
zhhao@design.upenn.edu

Abstract. From the epidemiological perspective, previous research methods of COVID-19 are generally based on classical statistical analysis. As a result, spatial information is often not used effectively. This paper uses image-based neural networks to explore the relationship between urban spatial risk and the distribution of infected populations, and the design of urban facilities. We take the Spatio-temporal data of people infected with new coronary pneumonia before February 28 in Wuhan in 2020 as the research object. We use kriging spatial interpolation technology and core density estimation technology to establish the epidemic heat distribution on fine grid units. We further examine the distribution of nine main spatial risk factors, including agencies, hospitals, park squares, sports fields, banks, hotels, Etc., which are tested for the significant positive correlation with the heat distribution of the epidemic. The weights of the spatial risk factors are used for training Generative Adversarial Network models, which predict the heat distribution of the outbreak in a given area. According to the trained model, optimizing the relevant environment design in urban areas to control risk factors effectively prevents and manages the epidemic from dispersing. The input image of the machine learning model is a city plan converted by public infrastructures, and the output image is a map of urban spatial risk factors in the given area.

Keywords: Coronavirus disease 2019 · Spatial risk factor · Machine learning · Incidence prediction

P. F. Yuan et al. (eds.), *Hybrid Intelligence*, Computational Design and Robotic Fabrication,
https://doi.org/10.1007/978-981-19-8637-6_18

1 Introduction

1.1 Research Background

In the post-epidemic era, scholars have conducted many relevant studies from the perspectives of pathology, molecular biology, and epidemiology. Still, most of them performed classical statistical analyses based on epidemiological surveys and classical statistical analyses. The exploration of the mechanism of association between disease transmission risk and spatial environmental factors has not been identified, since the types of data obtained from epidemiological surveys are primarily quantitative or qualitative indicators, such as numerical indicators used to measure the severity of the disease, the number of infected contacts, and the type of transmission of coronavirus. Furthermore, It is challenging to avoid covariance among multiple variables and quantify each factor's impact. Therefore, it is not easy to make accurate quantitative predictions based on epidemiological surveys.

1.2 Literature Review

With the growing popularity of machine learning in urban data [1, 3, 5, 10, 11], new technologies provide new solutions for epidemic prevention and control. Big-data analysis to accurately predict the risk of urban epidemic transmission provides the design improvement of urban-related facilities with new opportunities.

Machine learning in urban epidemic prevention research from an epidemic perspective focuses on the prediction of epidemic transmission risk and urban spatial optimization analysis. For the prediction of epidemic transmission risk, Cao Zhonghao et al. created an intelligent simulation model of the COVID-19 in Guangzhou city based on GIS technology, analyzed the transmission chain pathways by studying the Spatio-temporal trajectories of different individuals, and predicted the epidemic prevention and control trend, but lacked involving of urban environment and epidemic transmission mechanism [3], Li Zhao et al. proposed an epidemic transmission risk propagation model based on coupling LSTM algorithm and cloud model by the analysis of the June 2020 outbreak in Beijing, achieved the prediction of real-time and short-term epidemic transmission risk, but its data sample is exceptional and lack in universality [7]; Peng et al. studied the transmission trend of the COVID-19 by controlling and analyzing the real-time regeneration index of the epidemic in five countries, including China, Japan, and Italy, but the research granularity is large and fail to reveal profound mechanism behind transmission [9]. Analyzing optimizing urban space design is also an essential application in machine learning to prevent the epidemic. Xin Li et al. constructed an urban risk factor analysis model by analyzing the geographic location information of confirmed neighborhoods in Wuhan at the beginning of the epidemic outbreak, which provided an optimization plan for urban planning and architectural design [8, 12].

1.3 Problem Statement and Objectives

To sum up, most of the research granularity of machine learning models based on artificial intelligence technology and corresponding indicators at home and abroad are not suitable for exploring urban environmental risks' characteristics and can not predict and verify the interaction between urban areas' environmental risks and epidemic spread. In this paper, an image-based neural network is proposed to study the relationship between urban spatial morbidity and urban facilities design, find out the correlation mechanism between urban epidemic spread risk and spatial environment factors, and simulate and verify the epidemic outbreak in Wuhan. By using open-source data and generative antagonism networks (GANs) to experiment with various input and model types, we can easily compare the influences of different factors in the urban environment and reveal the interaction between urban environmental risks and urban design. Generating an antagonistic neural network is innovative in data types and learning methods.

2 Data and Spatialization

2.1 Study Area and Data Sources

In this paper, we selected the central area within the third ring of Wuhan city as the study object, covering about 591 km^2 and the seven most infected districts back to February 30. In this study, we used a multi-source data fusion approach. We modified data according to the Baidu map app and obtained the epidemic distribution data from various public websites and the official website of the Wuhan Health and Wellness Commission. The epidemic's increase in Wuhan tended to be stable after February 30 (Fig. 1), thus it could be argued that the data used in this paper are well interpreted and representative. As can be seen from the figure, the number of new infections fluctuated wildly between February 17 and 25; the spatial distribution of infected patients in the four days during this interval shows that there is an apparent spatial aggregation of infected patients, primarily in densely populated areas such as hospitals and shopping malls. The spatial migration process of patients was roughly gathered in the central district and showed a decreasing trend to the peripheral regions. Later after February 23, only the central section and scattered peripheral areas still had many infections.

2.2 Epidemic Distribution

The spatial analysis method needs to spatialize the information of infected patients based on table records in advance. The way of transmission for the disease is close contact and the spread of drop. Besides, hospitals and neighborhoods are two known places in transmission viruses. In this paper, we selected the neighborhoods address information as the geographic location of patients. After obsoleting POI data outside the third ring, a total of 30,617 data were involved as samples for this study. By combining the community-level population data released by the population census, we can improve the heat map (incidence rate), which better reflects the region's overall epidemic situation.

This paper used the kernel function (one of the most widely used methods for analyzing spatial patterns) to calculate the quantity per unit area based on the point elements to fit each outbreak distribution point, setting a smooth cone-shaped surface. The heat map in Fig. 2 shows the rough result of kernel density estimation, which indicates that

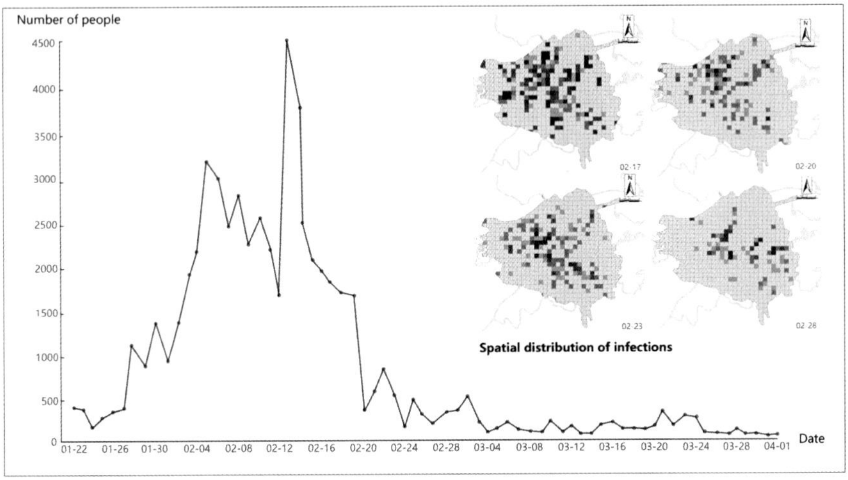

Fig. 1. Statistics of daily new infections in Wuhan.

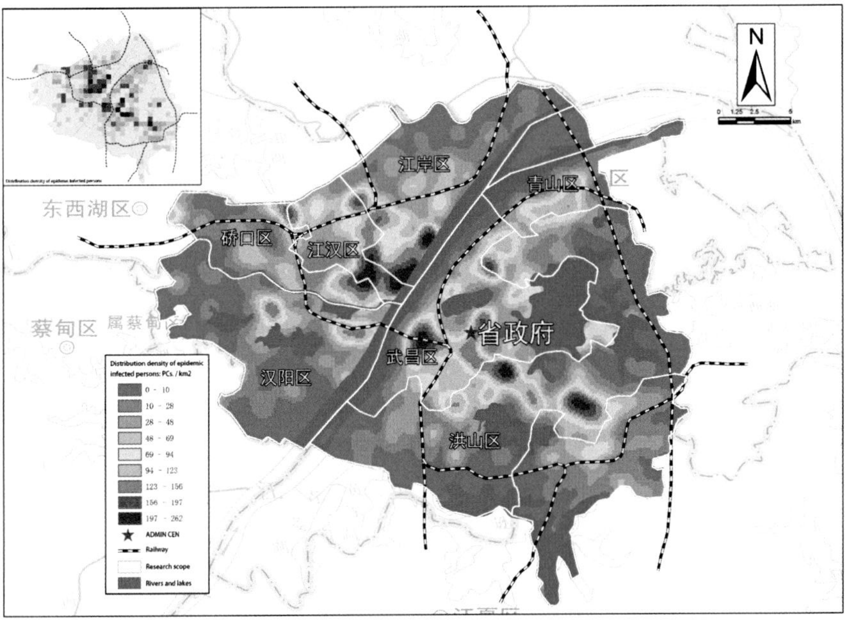

Fig. 2. Thermal and spatial risk factors of the epidemic in Wuhan.

the outbreak in Wuhan before February 30 was distributed in north–south along the river and aggregated in Wuchang, Hankou, and Hengyang. Due to the influence of natural factors such as water bodies and mountains, epidemic distribution in Wuhan is patchy; as the data in infected communities is limited by the people flow and activities frequency, it cannot accurately reflect the situation of the epidemic in the region, thus we introduce the definition of the incidence rate applied in the study of SARS transmission in Guangzhou [2]. Therefore, we can realize the spatial visualization from corresponding population census data.

2.3 Incidence Rate and Spatialization

Incidence Rate (IR) refers to the frequency of newly reported cases in a certain period, representing the regional distribution of cases. Furthermore, it can be visualized using the same criteria as the associated geographical factors. Since this paper focuses on the influence of various urban spatial risk factors at the microscopic scale, we pre-processed a grid cell-based incident rate map in the population spatialization with more presentable information.

Based on the 7th national census data, taking the kriging surface interpolation model to assign the demographic data in 1 km*1 km fine grid cells (591 in total). Compared with the beginning of 2020, the spatial pattern of population distribution has not changed much due to policy control. Therefore, after correction, it could be said that the population distribution is the population distribution map on the grid by the end of 2019. The spatial distribution of population density on the grid is close to the actual distribution. Kriging interpolation involved the spatial autocorrelation characteristics of Wuhan population density in data gridding, making the calculation results of population distribution more in line with the actual situation.

In this paper, we chose the core density estimation method to estimate the spatial distribution density of infected patients in Wuhan on 1 km*1 km grid cells. The core density function is the normal Gaussian curve function. In this paper, the size of the GIS grid network is 1 km, and the kernel radius is determined to be 2.5 km after several trials, which can retain enough details and meanwhile reflect the overall trend of spatial distribution. Dividing the spatial density value of infected data within the grid element by corresponding population density, the result shows the spatial incidence rate map of COVID-19. Since the grid cells do not contain enough information to support the features needed for network training, we densified the midpoint grid points of the incidence rate again according to the above parameters. Finally, we obtained the incidence rate label sample.

It can be seen from Figs. 3 and 4 that the epidemic of Wuhan's epidemic is mainly aggregated in the central district of the urban area with residential quarters as the core, which is also the most active commercial area of Wuhan's economic activities and complex transportation system. Regarding the study, it was found that spatial risk factors such as schools, shopping malls, subway stations, hospitals, and hotels were also important factors influencing the high incidence of the New Coronary Pneumonia outbreak, so a modeling analysis of spatial risk factors and incidence rates was conducted.

2.4 Spatial Risk Factors and Model Constructs

According to the first law of geography, the object of geographic space affects regional correlation, which is an essential basis for the spatial spread of epidemics. Scholars who studied the direction of Spatio-temporal modeling of infectious diseases have pointed out that the spatial spread of SARS and other infectious diseases is closely related to

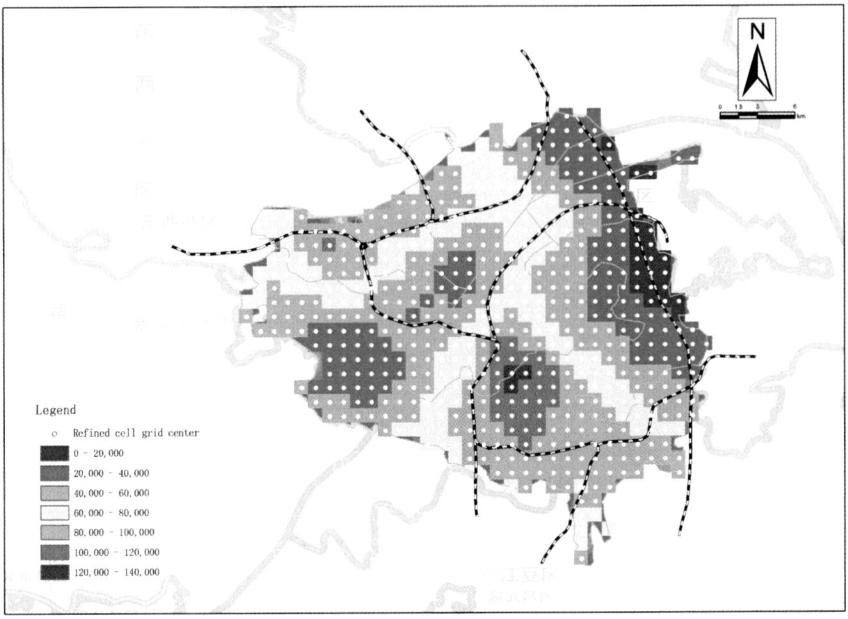

Fig. 3. Population density.

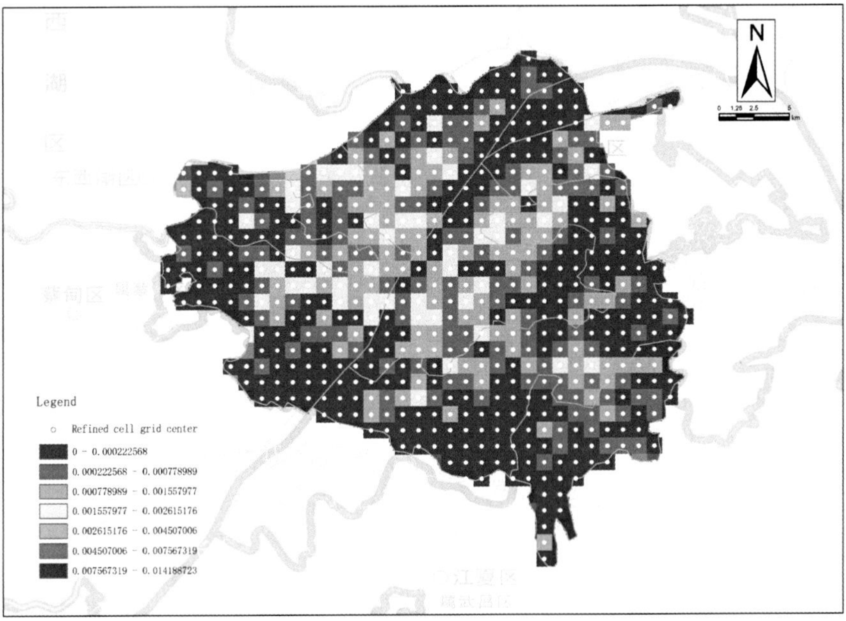

Fig. 4. Incidence Rate.

factors such as population, the environment and scope of human life, and the distribution of various other spatial influencing factors [7]. Here in this paper, with the help of Li Xin's team's definition of spatial risk factors for the spread of COVID-19 [7], we make an association test under normal distribution for its most critical urban facilities (Fig. 5).

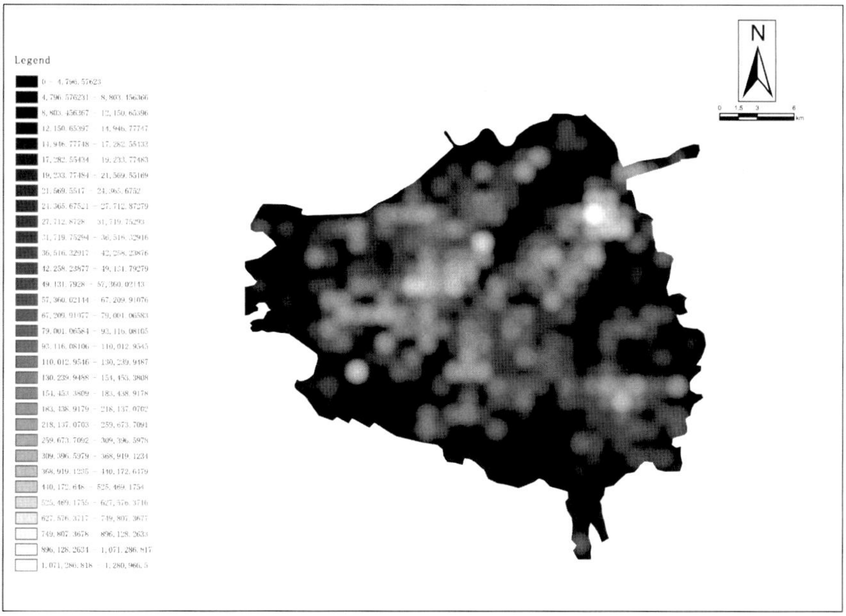

Fig. 5. Sample labels of incidence rate.

Comparing the incidence rate of COVID-19 and the density of various spatial risk factors (Fig. 6), it is evident to see a positive correspondence between them. We used 591 grid cells as samples and the Pearson index to the incidence rate and the density of various spatial risk factors using Statistical Product and Service Solutions. The results showed a significant positive correlation between each risk factor and the incidence rate of coronavirus in the two-tailed test with a significance level $P = 0.000$ (<0.05). The strongest associations are present in schools, supermarkets, subway stations, hospitals, parks and squares, and hospitals. These areas contained the following two characteristics: high population density and high population circulation; hospitals, as the main shelter for patients earlier, had a higher risk exposure for health workers on their way to treat patients, and thus became the areas with the highest incidence rates. In contrast, spots such as subway stations, government, and park squares have a high degree of overlap in the spatial distribution of outbreaks, particularly for the locations of the highest outbreak points in these areas that almost overlapped.

Therefore, this paper used the above correlation weights with poi data distribution to build an urban spatial risk factor map with a resolution of 11,871*12,630 as a feature sample for GAN training (Fig. 7). By image processing of the incidence rate label sample (Fig. 5), we used the PIL library in Python, cut the whole incidence rate map

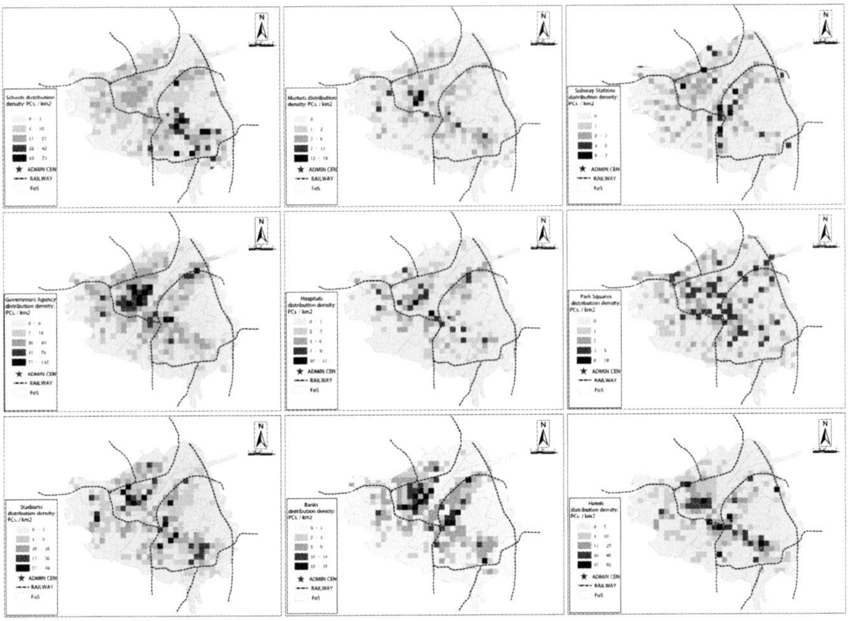

Fig. 6. Density distribution of COVID-19 spatial risk factors in Wuhan.

into small images with a resolution of 512 × 512 fragment. After the process, the size of each image became more suitable for machine learning. Meanwhile, the overlap the spatial risk factor feature samples with the incidence rate label samples to get the input of feature samples, acquiring 552 sets of slices of 23*24 in total. Since the validation scope used the area within the three-ring urban area, the regional boundary data were not highly reliable. After removing these data, we got 275 sets of feature and label slices. In addition, the uneven distribution of urban spatial elements results in few features and labels in rivers, greenery, and other areas. This paper manually eliminates unnecessary identification elements and obtains 225 sets of slices in total.

To describe the training model for forecasting incidence rate more conveniently, we used the value domain of incidence rate on fine spatial cells to establish its weights and implement data vectorization in rhino with the tools such as grasshopper (Fig. 7 right). The algorithm starts with the fine grid points (591 in total) as the center and the epidemic incidence value at this point as the search weight, traversing the remaining points within the connected range. The urban incidence rate grayscale map can be extracted from the grid values to achieve a better visual representation in grasshopper.

3 Neural Network Training and Data Analysis

3.1 Model Selection

The machine learning model is responsible for learning the relationship between the input urban facility layout with spatial risk information and the output of COVID-19

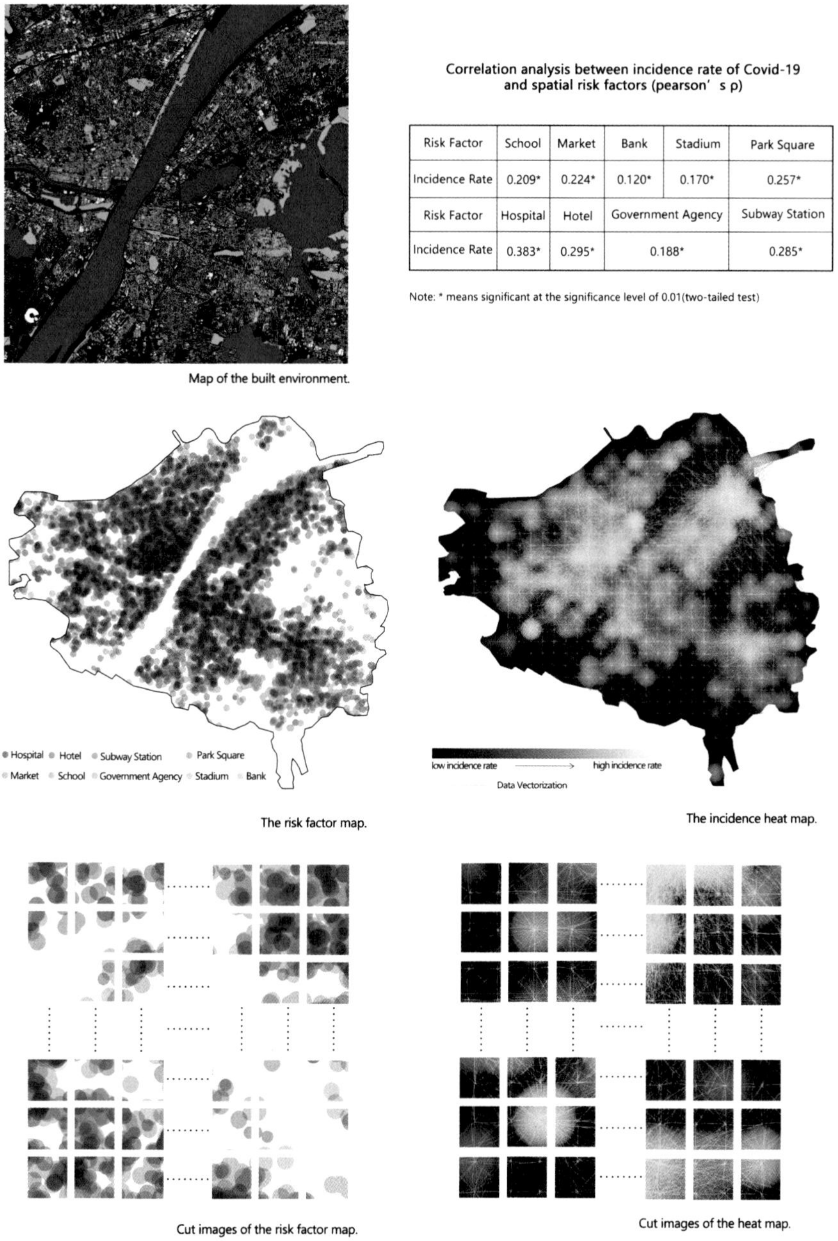

Correlation analysis between incidence rate of Covid-19 and spatial risk factors (pearson' s ρ)

Risk Factor	School	Market	Bank	Stadium	Park Square
Incidence Rate	0.209*	0.224*	0.120*	0.170*	0.257*
Risk Factor	Hospital	Hotel	Government Agency		Subway Station
Incidence Rate	0.383*	0.295*	0.188*		0.285*

Note: * means significant at the significance level of 0.01(two-tailed test)

Fig. 7. Processing of features and labeled samples.

epidemic incidence distribution, trained based on an image-based GAN framework with convolution and deconvolution kernels. Conditional GAN from Goodfellow et al. [4] and pix2pixHD (an open-source project) from Isola et al. [6] were used to develop the algorithms for this study. pix2pixHD implies a pixel-to-pixel transformation where the size of the input and output images remain constant.

In this training, the input feature sample is a spatially weighted distribution map of various facilities in the city with spatial risk information, and the output label sample is a COVID-19 onset distribution map generated by kriging interpolation. Since the data used is open source data of a single city in Wuhan, among the 275 sets of slices obtained, 80% (179 groups) were put into the training sample set and 20% (46 groups) into the test set to verify the accuracy of the model. After training, the model can predict the epidemic distribution with an unknown spatial factor map (Fig. 8).

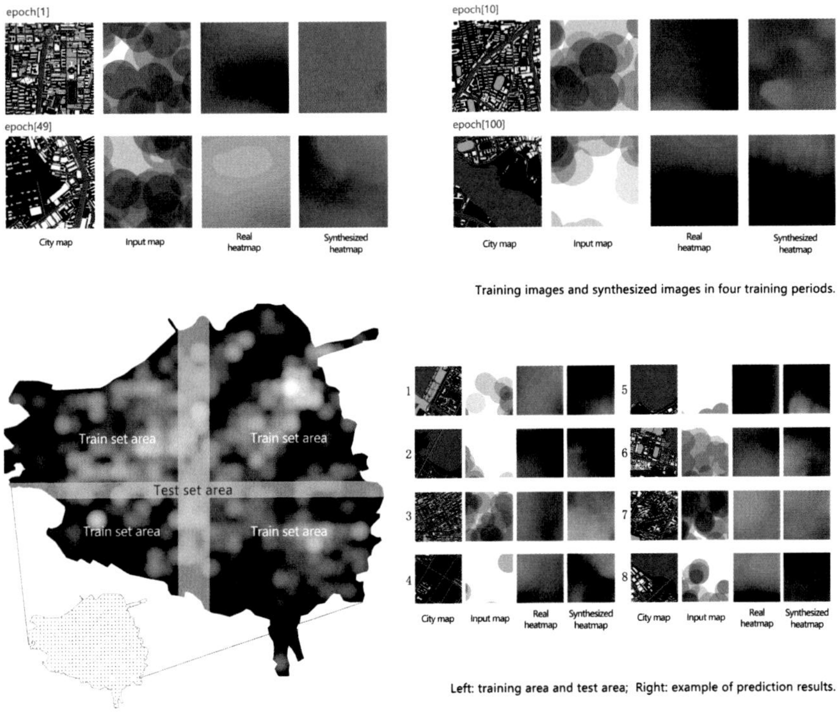

Fig. 8. Model training process.

3.2 Accuracy Measures

The training set should include all areas of the Third ring urban area of Wuhan. When the images of the test set are identified to represent different areas of the city, to reduce the influence of uneven spatial elements, we select two vertical columns and one horizontal column from the middle of the whole data area, totaling 46 groups of slices. This prediction result should be accurate because the training set includes 179 groups of screening slices in various areas of the Third Ring Road. To verify the accuracy of the model training, four ways are set to verify the fitting degree of the model. First, we

Traversed the pixel value m(m = r*0.299 + g*0.587 + b*0.114) of each pixel of 46 groups of samples (512*512 resolution) in the whole test set by getpixel method in the PIL library and obtained the average difference between the predicted value and the true value of all pixel points in each group of slices. The average Generation Accuracy of this training model test set is 0.7895. This judgment method can better explain that the absolute difference between the predicted value and the true value is small. Second, to test whether the model has formed a good fit to the spatial distribution, we set the randomly generated pixel value to compare with the true value and got the average Random Accuracy of 0.6959, which shows that the predicted value of the model is productively improved compared with the random situation. Third, by inverting the true pixel value, we calculated the difference between the inverted value and the true value and got the average Inversion Accuracy of 0.6401. It indicates the situation that the predicted values are entirely false compared to the real value. Last, we took the maximum difference between 0 and 255 and the true value to maximize the relative error rate, and get the average Lowest Accuracy of 0.3201. The comparison of the Generation Accuracy, Random Accuracy, Inversion Accuracy, and Lowest Accuracy is shown in Fig. 9, in which the accuracy of our generated values is much higher than the comparative groups. When the mathematical expectation of the Random Accuracy is regarded as 0.5, and the Inverse accuracy is regarded as 0, the predicted value of our model is increased by 167.74% compared to the random guess.

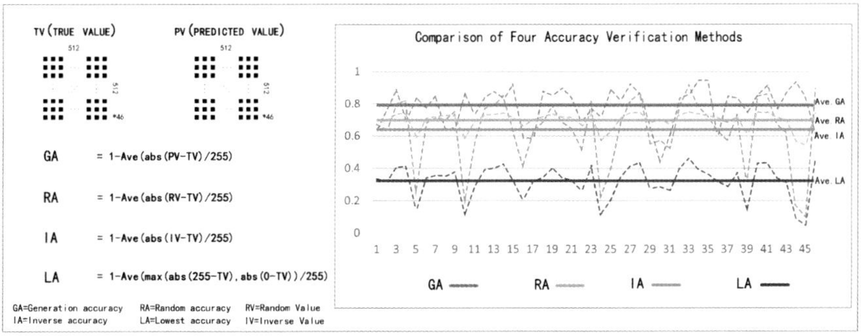

Fig. 9. Verification of model accuracy.

3.3 Layout Optimization and Prediction Application

With the trained model, we can continuously adjust design elements in urban planning to achieve the lowest public health risk by optimizing urban design plans for specific areas and predicting the spatial risk through neural networks (Fig. 10 left and middle). Changing the spatial attributes of the POI source data and testing implies that by increasing schools and hospitals and decreasing hotels and sports grounds, the predicted incidence rate has been changed evidently. In addition, assigning virtual negative correlation POI also corresponds to decreasing incidence rates, which indicates that this model learns

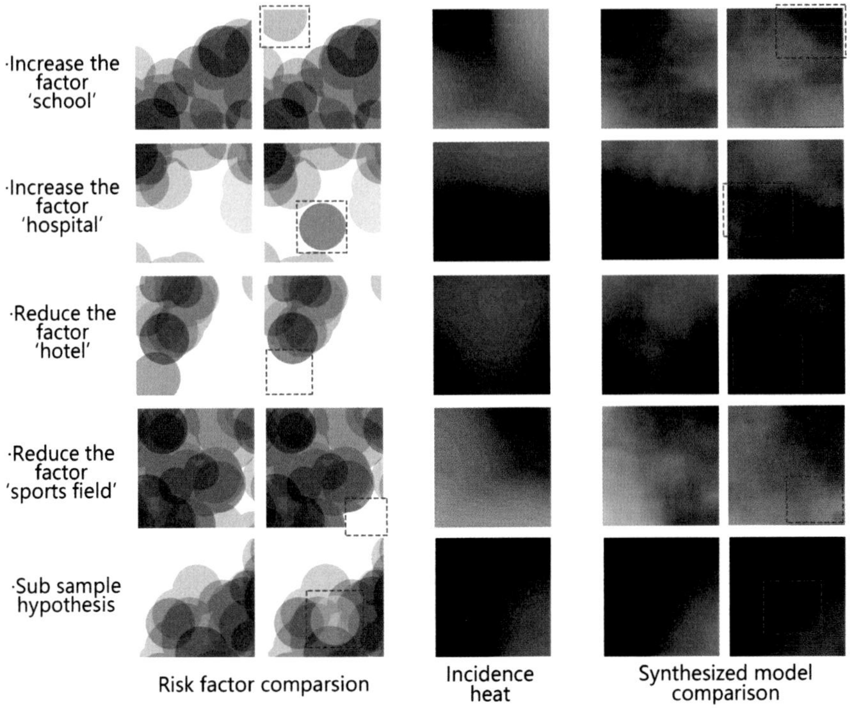

Fig. 10. The interplay between urban spatial risk and urban design.

better a particular mapping relationship between poi distribution and incidence rates and learns various risk factors weighting relationships between them.

Therefore, we can verify the interplay between urban environmental risk and urban design, revealing the potential optimization through combining multiple learning models in a logically closed loop, which quantitatively predict the epidemic incidence by modifying the figure-ground relationship and consequently iterating the urban design. It suggests that machine learning models can innovate advisors and supervisors to refine urban designers in planning and design and quantitatively achieve public health and safety.

3.4 Discussion

To more accurately visualize and predict morbidity in different urban design areas, research needs to further discuss the credibility of urban spatial risk factors. The more credible entropy correlation and ridge regression can be used to determine the correlation coefficient between urban space facilities and the epidemic distribution. At the same time, it is necessary to consider the precise quantification of the impact of various urban facilities on urban life by existing scholars. Under different circumstances The influence scope of various poi facilities is different. This model can be optimized by combining the existing quantitative analysis in the post-epidemic era and more accurate measurement research on the epidemic, and further inputting multi-dimensional natural environment elements such as ventilation environment and street built environment such as street openness. as a complement to space risk. In addition, the experience of epidemic prevention and control in countries around the world has proved that different social policy environments and human factors have a greater impact on the prevention and control of epidemic transmission. If we can start from spatial individuals, we can understand the process of epidemic transmission at the micro-geographical scale from the perspective of evolution and emergence., will be able to better explain the dynamic, sudden, self-organized and other complex characteristics of the epidemic. Therefore, it is necessary to supplement the explanatory variables for epidemic risk factors from local grid dynamic models such as urban constrained CA.

After the model is optimized above, it can be considered that a model for predicting the current generation of new coronary pneumonia has been obtained through GAN training. The adjustment process of streets and buildings in Fig. 10 is equivalent to simulating an urban planner using the training model in this paper to obtain feedback on the prediction of epidemic incidence and adjust the design solution to ultimately achieve a reduction in the distribution of the new crown in the city. In terms of accuracy, the optimization model of this paper, which further integrates multi-source data based on the characteristics of epidemic transmission, can quantitatively predict the incidence of new crowns and provide a usable model for urban infectious disease prevention and control planning.

4 Conclusion

This paper developed a suitable neural network model using Wuhan epidemic distribution data and census plot-level data to explore and validate the following. (a) The spread of an epidemic is due to the interaction of the infectious disease's driving forces and the multiple spatial factors and population movements in the environment. (b) The top three urban facility types in terms of increasing infection are hospitals, hotels, and subway stations, which address the shortcomings of subjective evaluation and the difficulties of data acquisition compared to previous methods. (c) The generated machine learning models could be extensively used for big data, as well as provide new capabilities for studying environmental behavior. The mapping relationship between urban facilities and epidemic onset can be learned by neural networks and used to predict the spread of epidemics in other cities. (d) By building multiple learning models, not only can the principles of interactions and determinants of behavior be derived. Nevertheless, real-time feedback on the assessment results can also support urban designers in improving designs to achieve some level of safe optimality in the urban environment. This study proposes new methods for risk assessment based on refined scales, which can provide fresh ideas for future disease risk assessment, decision support for epidemic prevention and control, and security for the people. Future research will base on a multi-scale epidemic prevention system. By incorporating other dimensional elements such as environmental elements, ventilation and sunlight, and social elements such as population movement, the research will be systematically developed for epidemic transmission.

References

1. Cao S, Zheng H (2021) A POI-based machine learning method for predicting residents' health status. Proceedings of the 3rd international conference on computational design and robotic fabrication (CDRF), Shanghai, China, 139–147
2. Cao Z, Wang J, Gao Y, Han W, Feng X, Zeng G (2008) Risk factors and autocorrelation characteristics on severe acute respiratory syndrome in guangzhou. Acta Ecol Sinica (9):981–993
3. Cao Z, Zhang J, Yang M, Jia L, Deng S (2021) The city agent model of COVID-19 based on GIS and application: a case study of Guangzhou. J Geo-Inf Sci 23(02):297–306
4. Goodfellow I, Pouget-Abadie J, Mirza M, Xu B, Warde-Farley D, Ozair S, Courville A, Bengio Y (2014) Generative adversarial nets. Adv Neural Inf Process Syst. MIT Press, Cambridge, MA, 2672–2680
5. He J, Zheng H (2021) Prediction of crime rate in urban neighborhoods based on machine learning. Eng Appl Artif Intell 106:104460
6. Isola P, Zhu J, Zhou T, Efros AA (2017) Image-to-image translation with conditional adversarial networks. arXiv preprint. https://arxiv.org/abs/1611.07004
7. Li X, Zhou L, Jia T, Liu F, Zou Y (2021) Decoding the impact of potential urban risk factors on the COVID-19 situation: a case study of Wuhan. City Planning Rev 45(08):78–86
8. Li, Gao H, Dai X, SunHai H (2021) Epidemic spread risk prediction model coupled with LSTM algorithm and cloud model. J Geo-Inf Sci 23; No.171(11):1924–1935
9. Peng ZH, Song WY, Ding ZX et al, Linking key intervention timings to rapid declining effective reproduction number to quantify lessons against COVID-19. Front Med.https://doi.org/10.1007/s11684-020-0788-3
10. Shou X, Chen P, Zheng H (2021) Predicting the heat map of street vendors from pedestrian flow through machine learning. Proceedings of the 26th international conference on computer-aided architectural design research in Asia (CAADRIA), Hong Kong, China, pp 2.569–578
11. Sun Y, Jiang L, Zheng H (2020) A machine learning method of predicting behavior vitality using open source data. Proceedings of the 40th annual conference of the association for computer aided design in architecture (ACADIA), Philadelphia, USA, pp 160–168
12. Yao Y, Yin H, Li X, Guo Z, Ren S, Wang R, Guan Q (2021) Fine-scale risk assessment of COVID-19 in Wuhan based on multisource geographical data. Acta Ecologica Sinica 41(19):7493–7508

The Analytical Workflow for Shifts in Post-COVID Living Preferences on Neighborhood Qualities

Zhiyi Dou[1], Waishan Qiu[2], Wenjing Li[3], and Dan Luo[4](✉)

[1] School of Geographic Information Science, University of Queensland, St. Lucia, QLD, Australia
[2] Department of City and Regional Planning, Cornell University, Ithaca, NY, USA
[3] Center for Spatial Information Science, The University of Tokyo, Tokyo, Japan
[4] School of Architecture, University of Queensland, St. Lucia, QLD, Australia
d.luo@uq.edu.au

Abstract. potential self-organization of work-life patterns and social profiles in the designated neighbourhood.

To evaluate the subjective perception of the urban residence, the study started with a comparative survey by asking residence to compare two randomly selected urban contexts in a data base of 398 contexts sampled across Hong Kong and state their living preference under the presumption of following scenarios: 1. working from home; 2. working in city centre offices. Core information influencing the spatial equilibrium are provided in the comparable urban context such as street views, housing price, housing space, travel time to city centre, adjacency to public transport and amenities, etc. Each context is given a preference score calculated with Microsoft TrueSkill Bayesian ranking algorithm based on the comparison survey of two scenarios.

The 398 contexts are further analysed via GIS and image processing, to be deconstructed into numerical values describing main features for each of the context that influence urban design strategies such as composition of spatial features, amenity allocation, adjacency to city centre and public transportations. Machine learning models are trained with the numerical values of urban features as input and two preference scores for the two working scenarios as the output. The correlation heat maps are used to identify main urban features and its p-value that influence residence's preference under two working scenarios in post–COVID era. The same model could also be applied to inform the direction of urban design strategies to construct a sustainable community for each type of working population and validate the design strategies via predicting its competitiveness in attracting residence and developing target industries.

Keywords: Urban data analysis · Sustainable communities · Post-COVID · Data driven urban design · Machine learning

P. F. Yuan et al. (eds.), *Hybrid Intelligence*, Computational Design and Robotic Fabrication,
https://doi.org/10.1007/978-981-19-8637-6_19

1 Introduction

The post-COVID era has influenced the way cities to operate. Workplace practices and modes of collaboration have subtly changed the preferences of the working population for urban existence. Modern cities are designed and developed to provide humans with a convenient and healthy environment. Cities are densely populated with commercial and industrial development, leading to vulnerability to epidemic outbreaks [5]. More companies are allowing their employees to work remotely permanently instead of in physical offices to cope with the effects of this global epidemic [2]. This change in work patterns has led to a mismatch between the spatial form of the city and the matching population in the post-COVID period. Residents will need to consider their work and lifestyle during Lockdown. When some work patterns shift to remote working, this has implications for mobility, transportation, and housing prices. The working population will face two scenarios. First, the convenience premium enjoyed by residents of neighbouring cities with short commutes and easy commutes will disappear for people working from home. Some workers may tolerate longer commutes from further afield if they only need to commute to the city center once or twice a week [6]. Fewer people coming to offices can dampen demand for office space, especially in expensive inner-city areas. Under the circumstance of ensuring the public health and safety of the city, the spatial distribution of urban resources will have two extremes of scarcity and excess. Therefore, urban design and construction will need to reconsider people's needs. Considering the interweaving of urban context at multiple scales and complex decision-making process based on subjective preference, the current methods in constructing modelled predicting the performance of urban development face the following key challenges:

1. The difficulties to evaluate the impact of urban features across multiple scales, such as urban planning qualities and local spatial experience.
2. Challenge to incorporating citizen's subjective opinions into the evaluation model.
3. How to expand a model trained with inputs from current context to predict the performance of future novel scenarios.
4. People's spontaneous activity may deviate from their stated preference from survey.

Based on survey and case study of Hong Kong, this research aims to address the key challenge by developing a comprehensive workflow with the following features:

1. Urban features from both planning data and urban street views are incorporated into the evaluation with learnable weight.
2. Intuitive subjective rating from users is applied to generate preference score.
3. A portion of the data for the sampled points are manipulated to include scenarios for future development via statistical analysis of data distribution in the design space.
4. Data mined from spontaneous activities such as real-estate rental and social media is used to validate the accuracy of the stated preference from the survey, and evaluate how faithful they are in reflecting people choose realistically.

2 Literature Review

As an essential part of urban public space, the street is not only the primary carrier of transportation, but also an important space for personal daily activities, including work and communication [7]. Street environment is a major factor influencing people's physical activity, the choice to move and willingness to pay [3]. Previous assessments of the street environment were mainly derived from two measures. The first was objective based data of city to measure street quality, such as building height, park area, number of surrounding transportation facilities, etc. However, a single objective measurement cannot represent the overall feeling of human residents. Another is a subjective measurement, such as interviews or surveys [8]. However, due to individual differences, the consistency and reliability of operations will influence the uncertainty of the results. Therefore, methods that combine subjective perception scores and field surveys to obtain objective measurement data have emerged. For example, Qiu et al. applied computer vision and machine learning to subjectively measure four perceptual qualities collected from SVI samples using Shanghai [3].

In the post-COVID period, changes in cities are mainly changes in the spatial distance of people, affecting living choices, travel, and commuting patterns. The link between COVID-19 prevalence and urban design features has sparked much debate in some studies. However, the existing literature does not detail which street environmental factors influence residents' preferences during the post-COVID period [6]. Therefore, our work aims to study the impact of work patterns on their preferences by targeting the working population of a company in a city center. Based on extracting the pixel ratio of objective street view elements using SVI data, adding urban spatial information to enrich urban perception to achieve subjective measurement. The real estate data is then used to validate the score. The results will show two work scenarios affecting residents' preferences in the post-COVID era. The same model will also inform post-pandemic urban design strategies.

3 Data and Methodology

This study attempts to take the current Hong Kong city as an example in response to the above issues. The street image of the study area is segmented to obtain the urban feature values, and the city's physical environment is perceived through the street view image. After adding objective urban spatial data, enriched investigators' subjective choice of residence preferences for workplaces in a randomly sampled Street View database. We use machine learning to predict future urban development design strategies in the post-COVID era. This research will construct a sustainable urban environment and future development direction for the working population by evaluating this model.

A key characteristic for the workflow of this methodology is that both urban context from planning scale such as the POI distribution, abundance of public transportation, parks and distance to city centre, as well as local urban quality such as street experience is taken into consideration altogether. Opinions of citizen in incorporated via a subjective rating process. The rating is based on manipulated data with improved distribution in design space to incorporate future scenarios of development. Lastly, data revealing behaviour and choice in real world, such as real-estate rental data is applied to validate the accuracy of the preference stated by the survey takers and evaluate the extend for them to forecast people's behaviours and opinions on current and future development (Fig. 1).

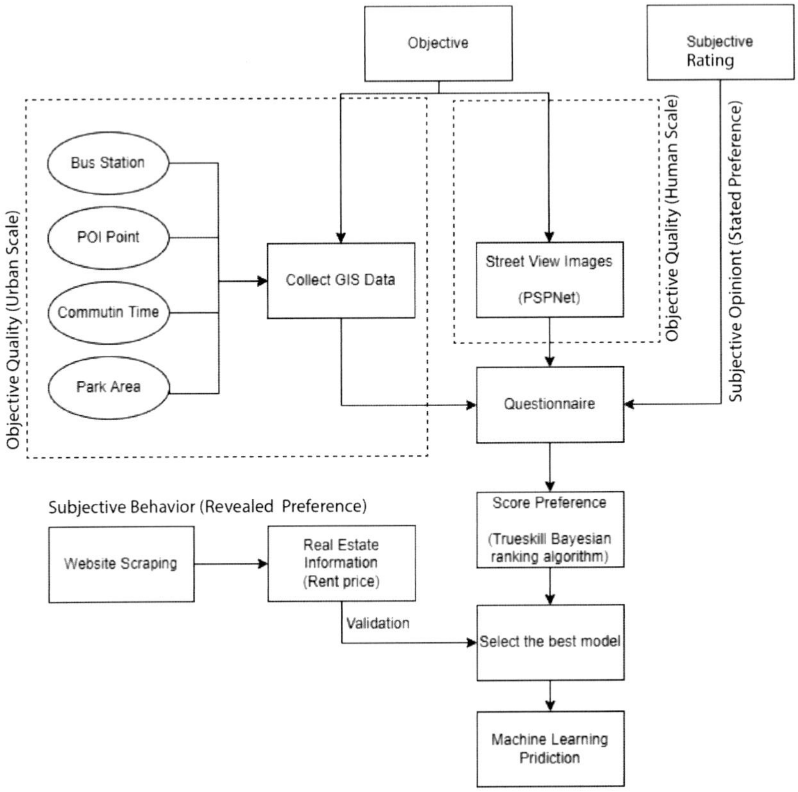

Fig. 1. Analytical framework

3.1 Study Area

The Hong Kong Special Administrative Region, located at the southern tip of mainland China, was selected as the study area. The whole territory consists of Hong Kong Island, Kowloon, and the New Territories. It is a highly prosperous free port and cosmopolitan city. With a total population of more than 7 million and an area of about 1,100 square kilometers, it is also the third-largest financial center in the world. Hong Kong has a unique geographical space environment. Due to its mountainous terrain, dense urban population and numerous high-rise buildings, it only accounts for about 20% of the general land area [9]. The analysis of the street quality in Hong Kong will provide inspiration for future urban renewal.

The objective spatial data of the study area include: (1) The distribution of Hong Kong bus stations, POI, and park areas from HONG KONG GEODATA STORE (https://geodata.gov.hk/gs/), (2) a shapefile of road networks from Open Street Map (https://www.openstreetmap.org/), (3) SVI collected from Google Maps Street View API, (4) property information from Centaline (https://hk.centanet.com/ info/index).

3.2 Data Collection Processing

The road network in the form of linear vector data is extracted from the open source of Open Street Map, then converted into shapefile by QGIS to obtain all road network data in Hong Kong. We are creating 1,000 points in the road network data for each district in Hong Kong with an interval of 50 m and getting 18,000 points. Ensure that our training images cover most urban area types and improve efficiency. We randomly selected 398 study sites. Then according to the geographic coordinates of the research point, download the corresponding SVI from the Google Street View static API (Fig. 2).

Objective factors are significant criteria to measure street quality. We selected "number of bus stops," "number of POIs," "size of park area," and "commute time" to present the perceived streetscape quality. Processing in the geographic information software ArcGIS Pro. A buffer zone of 1.5 km^2 is established with randomly selected research sites as the center. The number of bus stops in the buffer zone of each study site, the number of POIs, and the park area covered are counted. The commute time was then estimated based on the distance from the study site to the city center. The number of bus stops represents the degree of crucial urban infrastructure and is one of the criteria for people to judge the accessibility of a place of residence [10]. The number of POIs represents the urban spatial structure and provides scientific support for guiding regional spatial optimization and regulation. POI data records the spatial location information of various social and economic sectors, which can more clearly identify the types provided by urban functional areas [11]. The area of parks represents an essential indicator of the overall environmental level of the city and the quality of life of residents [12]. Commuting time represents the travel behavior of residents who live and work apart and is an essential factor affecting the residential experience of the working population.

After collecting context data based on existing scenario, a noise and variance is applied to data set, so that the context parameters are distributed across the design parameter space, enable the dataset to encompass future scenarios in addition to existing context (Fig. 3).

3.3 Physical Feature at Personal Scale

People's preference scores for Street View were collected through an online questionnaire platform developed (Fig. 4). The online survey system asked participants to click on one image in a pair of SVIs to answer the assessment questions. Street view quality is perceived through human vision, and provides urban spatial quality data to deepen judgment. If the company is in the city center, the participants will assume that they are in two different environments: choose the living area based on working from home or the office. The 398 SVIs mentioned above randomly sampled from the Hong Kong study area were used in the survey. A total of 1900 valid contributions were collected, and

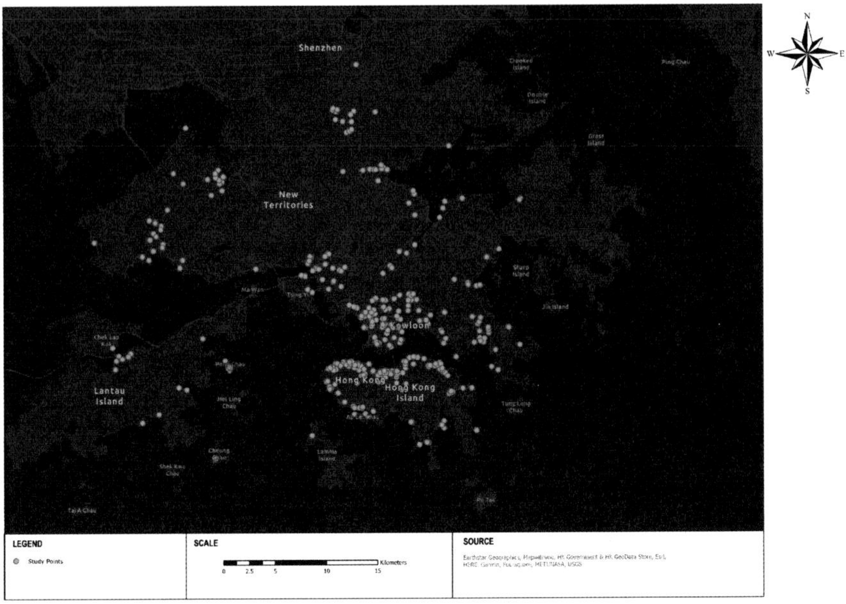

Fig. 2. 398 sampled study points

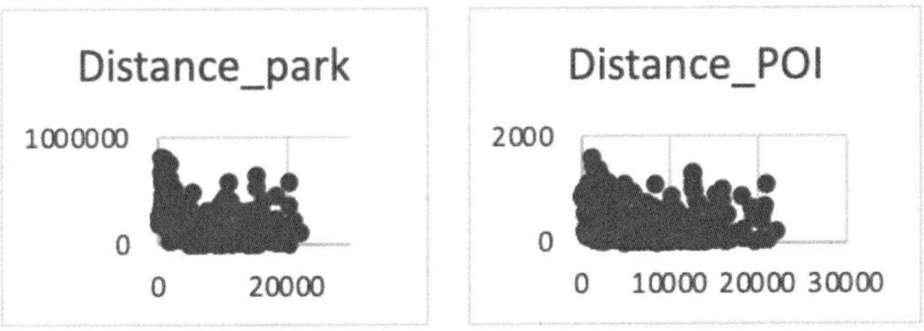

Fig. 3. Study points scatter plot

the participants were architecture students at the University of Queensland. Averaged to ensure that each image participates in more than 5 comparisons.

Part of the physical features in the street view is one of the essential factors for human perception of the quality of the street view [3]. The pixel ratio of a single feature in SVI is obtained through the Pyramid Scene Parsing Network (PSPNet). It is a pixel-level object recognition and classification algorithm with over 93.4% pixel-level accuracy. More than 30 Street View elements were detected in the study area. Since the uncertain number changes of pedestrians, cars and some facilities and the size of the covered image area have little significance to the pixel ratio, only eight meaningful street elements are selected as the primary reference standards.

To faithfully translate citizens opinions into statistical values, Microsoft True Skill ranking is applied to generate scores based on multiple comparison outcomes between randomly sampled urban scenarios. In order to generate the relative scores that reflects

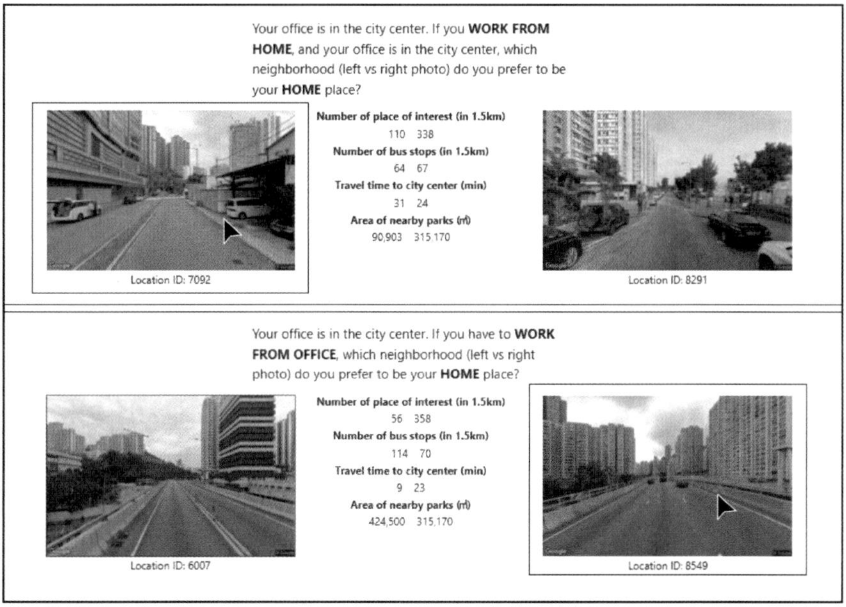

Fig. 4. Questionaire/Survey for Revealing Subjective Preference

the ranking of preference for individual scenarios, each scenario composed of a set of urban information and a street view photo should be compared with at least three other random scenarios. The benefit of using a dynamic comparison survey instead of a statice rating survey is to minimize human base bias in rating, such as some people tends to give higher rating comparing to others (Fig. 5).

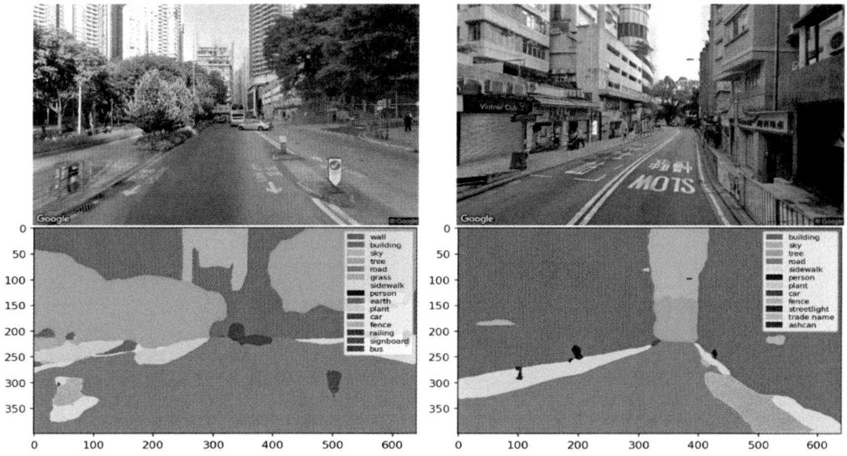

Fig. 5. PSPNet semantic segmentation results with raw input (study data 13 and 510)

3.4 Model Training and Validation

The target output variables of this study are two preference scores. The above-mentioned street view elements obtained from SVI are fed into the machine learning model. Refer to Qiu et al. for the Pudong Area related research using 8 ML algorithms [3]. They are Linear Regression, K-Nearest Neighbors (KNN), Support Vector Machine (SVM), Random Forest (RF), Decision Tree (DT), Voting Selection (VS), Gradient Boosting (GB) and Adaptive Boosting (ADAB), to Predict two preference scores. However, traditional machine learning method primarily working with higher lever features that require pre-processing of the street scene itself, such as extracted image segmentation value representing the percentage of urban element. This pre-processing of the image, though able to extract readable urban feature qualities, is in-efficient to translate the more abstract qualities such as style and atmosphere. Thus, in addition to applying traditional machine learning algorithm, the research has progressed toward applying Convolutional Neural Networks as the foundation of constructing the perdition model.

Set the Mean Square Error (MSE) as the loss function to evaluate the performance of different ML and Neural Network algorithm. It shows advantages in evaluating average models. The lower MSE value implies the higher accuracy of a prediction model. In addition, R-Square is a key parameter determine the proportion of the variance for a dependent variable that's explained by an independent variable or variables in a regression model.

Multiple models with traditional machine learning is applied including K-Nearest Neighbors (KNN), Support Vector Machine (SVM, Random Forest etc. as well as a customized Neural network with Convolutional layers (CNN). The better performing traditional machine learning method include KNN and SVM achieved RMSE of 0.21 and R2 of 0.24, while CNN is able to significantly improve the performance toward RMSE of 0.153 and R2 of 0.476. The performance of CNN has proven to be reaching the level of applicability in for predicting the human preference of living condition under the work from office scenario (Fig. 6).

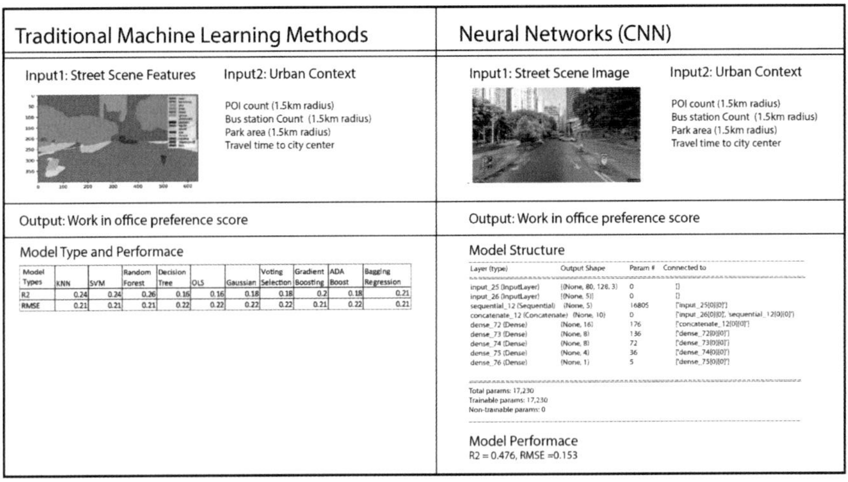

Fig. 6. Comparison between the performance of traditional machine learning algorithm and CNN

4 Conclusions and Feature Work

This study applies the trade-offs of urban residents' perception of urban spatial quality and urban public facilities in the prediction environment of ML models. The analysis was performed using QGIS, ArcGIS Pro and Python. The study areas were 398 study areas randomly selected from Hong Kong. From a theoretical point of view, we use the SVI segmented pixel ratio dataset to replace the subjective measurement of human eyes perception on the one hand, and then add objectively measured urban spatial quality data. The results will be more comprehensive and nuanced to reflect the working population's residency preferences.

The advantage of this evaluation model is that it combines subjective measurements of objective factors, which can more comprehensively predict and guide the design of the street environment in the post-COVID period. Since the dataset fits the pedestrian's perspective, the cost is meagre, and it does not require highly complex technology and can be used only in the area where the SVI dataset is available. According to the target population, such as the working population, relevant spatial elements (bus stops, parks, POI, commute time) are added to improve the accuracy of the prediction results.

Kindly note, this research is proof-of-concept research for the methodology and workflow, with preference data collected from design students. The preference data does not reflect the opinion of Hong Kong citizen and will not be published. In addition, The accuracy of PSPNet parsing SVI images is not 100%, though using NN and inputting realistic urban photo would improve this error. Then the environment of SVI can only be used in common scenes, and lack of analysis of specific environments, such as some private streets and interior scenes, cannot obtain images. For the 398 SVIs with a limited number of participants and randomly selected, the accuracy and reliability of the prediction results still need to be improved. These will also be the main directions of future research work, providing more complete and precise guidance for the sustainable development of urban construction. However, with the validation of the workflow, the current model has validated its performance statistically, thus proving this methodology can be applied toward multiple cities given a systematic survey with sampled targeted user group and additional urban samplings.

References

1. Gollin D, Kirchberger M, Lagakos D (2017) In search of a spatial equilibrium in the developing world. Nat Bureau Econ Res
2. Miranda AS, Fan Z, Duarte F, Ratti C (2021) Desirable streets: using deviations in pedestrian trajectories to measure the value of the built environment. Comput Environ Urban Syst
3. Qiu W, Li W, Liu X et al (2021) Subjectively measured streetscape perceptions to inform urban design strategies for Shanghai. ISPRS Int J Geo-Inf. https://doi.org/10.3390/ijgi10080493
4. Minka T et al. (2018) TrueSkill 2: an improved Bayesian skill rating system. Technical Report
5. Wang J (2021) Vision of China's future urban construction reform: In: the perspective of comprehensive prevention and control for multi disasters. Sustainable cities and society
6. Florida R, Rodríguez-Pose A, Storper M (2021) Cities in a post-COVID world. Urban Studies
7. Liu M, Jiang Y, He J (2021) Quantitative evaluation on street vitality: a case study of Zhoujiadu community in Shanghai. Sustainability

8. Lynch K (1964) The image of the city. MIT press
9. Kwok CYT, Wong MS, Chan KL, Kwan MP, Nichol JE, Liu CH, … Kan Z (2021) Spatial analysis of the impact of urban geometry and socio-demographic characteristics on COVID-19, a study in Hong Kong. Sci Total Environ
10. Rode P, Floater G, Thomopoulos N, Docherty J, Schwinger P, Mahendra A, Fang W (2017) Accessibility in cities: transport and urban form. Disrupting Mobility
11. Bing X, Bingyu Z, Xiao X, Jingzhong L, Xiao X, Wanxia R (2020) A POI data-based study on urban functional areas of the resourcesbased city: a case study of Benxi, Liaoning. Human Geogr
12. Yan L (2015) The progress of landscape planning: a brief discussion on urban parks. Chinese Science and technology periodical database engineering technology

Exploration on Diversity Generation of Campus Layout Based on GAN

Yubo Liu[1](✉), Zhilan Zhang[1], and Qiaoming Deng[2]

[1] State Key Laboratory of Subtropical Building Science, South China University of Technology, Guangzhou, China
liuyubo@scut.edu.cn

[2] School of Architecture, South China University of Technology, Guangzhou, China

Abstract. Previous studies have shown that GAN has made some progress in the generation of campus layout plan, but the result is single output for single input condition. This paper hopes to make some attempts and explorations on the diversity generation of campus planning layout design by machine learning. Based on Pix2Pix model, this paper proposes a method to divide image channels so that the campus function bubble diagram and the site boundary can both become the input conditions. There is a strong correspondence between the campus functional bubble diagram and the campus layout. The main idea of this study is to control the generated results by changing the input of the campus functional bubble diagram, so that we can have a diversity layout of campus according to the same site conditions. In the experiment, we train thirty samples of campus planning layout design, and finally evaluate the generated results in a qualitative and quantitative way, which proves that the generated results are relatively ideal. This research enables designers to participate in the process of machine learning generative design to control the generation results.

Keywords: GAN · Campus planning layout design · Campus function bubble diagram · Diverse generative design

1 Introduction

At present, the machine learning technology developed rapidly, brings new possibilities to the field of architectural design. Among the machine learning techniques, the generative against network (GAN) has certain potential in the architectural design layout generation research. GAN can generate a specific layout plan automatically according to the given site conditions by summarizing the potential composition rules of various elements in a large number of layout data. However, most of the current studies are focused on the apartment and block plan with relatively simple functions, and there are few studies on the more complex objects, and the generated results are unitary and uncontrollable.

Campus is large scale, with complex internal functions, clear zoning plan and has certain rules in element composition. It is a special community organization in the city,

P. F. Yuan et al. (eds.), *Hybrid Intelligence*, Computational Design and Robotic Fabrication,
https://doi.org/10.1007/978-981-19-8637-6_20

including not only educational buildings, but also business, entertainment, sports, and other functions, almost as a miniature society, so it has many possibilities of layout. The traditional campus planning and design is a major construction, which often needs a lot of manpower and material resources. The Architects need to spend a lot of time to repeatedly modify the design schemes and compare the multiple schemes to select the best one. Campus layout design is an important link in the early stage of campus planning. If multiple preliminary layout design can be proposed quickly in the early stage of design and multiple design options can be provided to designers, it will bring higher efficiency.

This research considered to apply machine learning technology in campus design layout generation. In this paper, we proposed a novel campus layout design method. According to the way of architect doing design by drawing the functional bubbles sketch, we took the campus site boundary and functional bubble diagram together as input. Different functional bubbles correspond to different functional zonings inside the campus. The machine can summarize the corresponding relationship between functional bubbles and campus building layout. Finally, by changing different functional bubble diagrams, we will have diverse campus layouts based on a single boundary. This research allows designers to control the generated design results, and they can adjust the results as needed.

2 Related Work in the Field of Architectural Layout

In recent years, machine learning technology has made some progress in the field of building layout generation [2, 4, 5, 7]. With more complex research objects and ever larger scales, some scholars begin to explore the possibility of combining machine learning with campus planning and layout design. Chang [1] proposed the use of reinforcement learning and parametric modeling for campus design generation. This method can provide many design choices based on different design parameters. However, it is based on artificial algorithms and limited optimization goals to guide the generation design, which cannot get a more organic layout form. Luo [3] used deep learning to discusses the idea and method of campus layout generation, and proposed a selecting rule and annotation method for small sample generation. With such a method, the computer can automatically generate the layout of the campus under the conditions of given campus boundaries and surrounding roads. This experiment demonstrates that machine learning techniques can generate more complex layouts like campus. However, this research is limited to the single output, and can only output a single result according to the condition of a site, which cannot meet the needs of comparing multiple schemes in campus design at the early conceptual stage. Therefore, the designer cannot control the final output result and adjust the generated results.

In terms of research on diversity generation of layouts. Pan [6] obtained the results of diverse generation by using GauGAN model to generate the layout of the neighborhood community in north of China. This study showed us the possibility of deep learning for diverse layout output, which affected the output results by changing the images of the input. But the final diverse output results did not change much, and it is difficult to see the direct influence of the input elements on the results. So far, the research on diversified output is not mature enough.

In conclusion, there has been few researches on the generation of campus planning and design combined with machine learning. This paper hopes to further explore the application of deep learning technology for diverse output of campus layout, and study how to establish a more direct connection between input elements and output elements, so that designers can control the direction of the generated results.

3 Methodology

The main process of diversified generation of campus layouts based on deep learning is as follows:

1. Database establishment. The main collection date is the central loop type campus planning and design plan.
2. Data labelling and extraction of campus functional bubble diagrams. Establish a data labeling method based on architectural knowledge.
3. Training and testing. Pix2Pix model is adopted for training and testing.
4. Evaluation and analysis. The results are evaluated and analyzed to verify the possibility of diversity generation.

3.1 Model Architecture

The Pix2Pix model is one of the conditional generative adversarial network models (cGAN). It can be used in image processing tasks to map the input image to the output image, and it can achieve the supervised image-to-image translation. The network structure of Pix2Pix contains a generator and a discriminator. The generator used the U-Net structure to encode the input image, and then decodes it into a fake image that resembles the real one. The discriminator used the conditional discriminator PatchGAN.

3.2 Training Method

The input and output of the original Pix2Pix model are three-channel images, which can only support single image input and output. In order to achieve multiple input, we modified the input image channel based on the original Pix2Pix model, changing the three channels to six channels, and finally the model can realize the simultaneous input of two images.

This research was based on the original experiment (Fig. 1) and then put forward a new training method (Fig. 2). Input the campus site boundary and campus functional bubble diagrams into the generator at the same time and it generated a layout plan of campus approaching the real sample. Then we put this plan and the corresponding real campus plan together into the discriminator. The discriminator evaluated both, and gamed with the generator until the model converged. Finally, we input different functional bubble diagrams to generate diversity campus layouts based on the same site conditions, so we can achieve the goal of controlling the result by inputting different bubble diagrams.

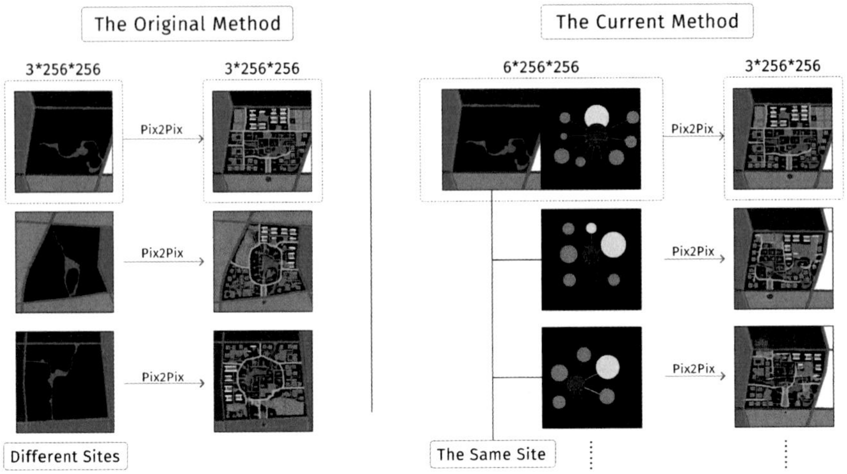

Fig. 1. The original method and the current method

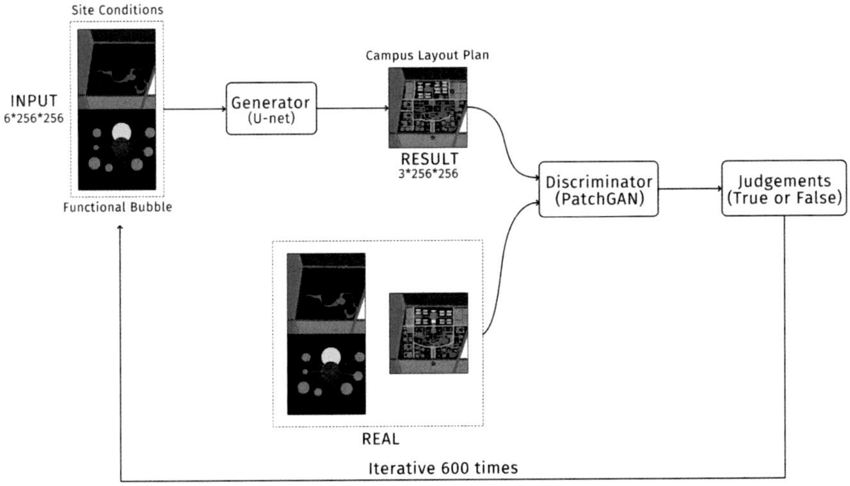

Fig. 2. Training method

3.3 Dataset

We have collected campus cases through portfolios of major design institutes, and papers, books and websites related to campus planning.

3.3.1 Selecting Rules

Considering the impact of samples on the generated results, we conducted the following selecting method for the data:

1. Appropriate Scale. Select the sites with scale between 50–100 hectares.
2. Square boundary. The boundary of the campus' site should be relatively square.
3. Consistent planning style. The layout plan used in this experiment is the central loop type.
4. Complete and clear functional zoning.

Finally, thirty campus samples (Fig. 3) were selected, four of which were used as test samples.

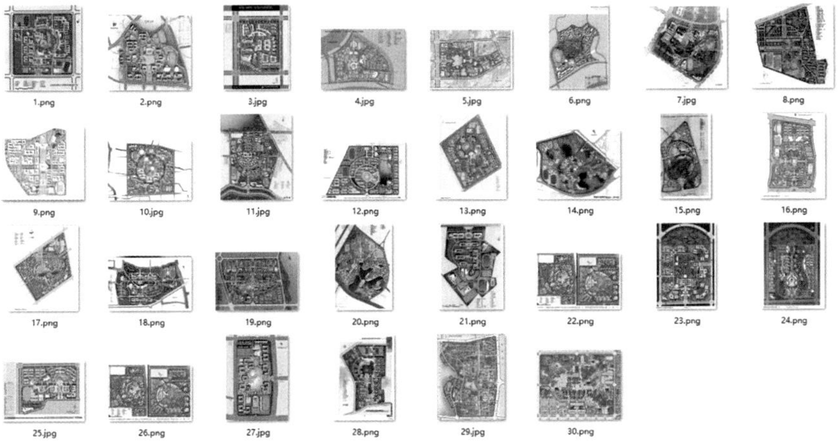

Fig. 3. Selected samples

3.3.2 Sample Labelling

We labelled the data of the campus plan with uniform form, proportion, and color block. By summarizing the main layout features of different campus building, such as teaching buildings and dormitories, we extracted the common forms and simplify some special building layouts.

We convert these functional zonings into the form of functional bubble diagrams (Fig. 4). Firstly, the color blocks of functional zoning are divided according to the original functions and roads. And then the size of the color blocks determines the size of the bubble. The larger the color block is, the larger the function bubble is. In addition, the functional bubble diagram can also reflect the proximity of different zonings and the total number of functional zonings.

3.3.3 Data Augmentation

In this experiment, a total of 120 data were obtained by expanding the data set through horizontal mirroring and vertical mirroring to improve the learning effect of the machine on the samples. Considering that rotation would affect the orientation of teaching buildings, dormitories, and other buildings, we did not rotate data in this experiment.

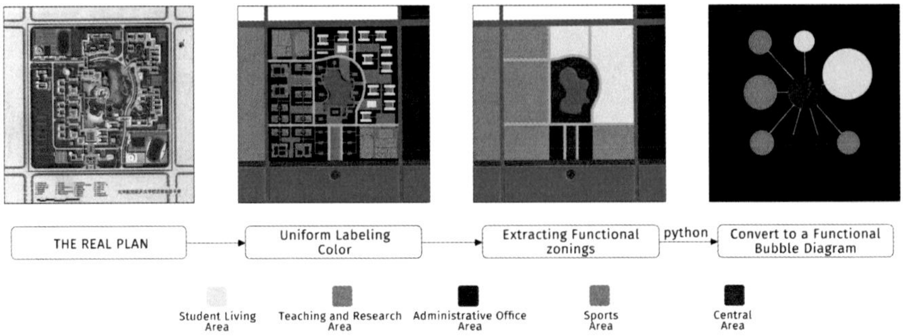

Fig. 4. Process of extracting campus functional bubble diagram

4 Training and Testing

4.1 Previous Experiment

Our research team has done an experiment of generating campus plan based on machine learning. The method of step-by-step training is adopted. The first training inputs site boundary and outputs functional zoning plan. The second training inputs functional zoning plan, and outputs campus layout plan. Through step-by-step training, the campus plans are generated (Fig. 5).

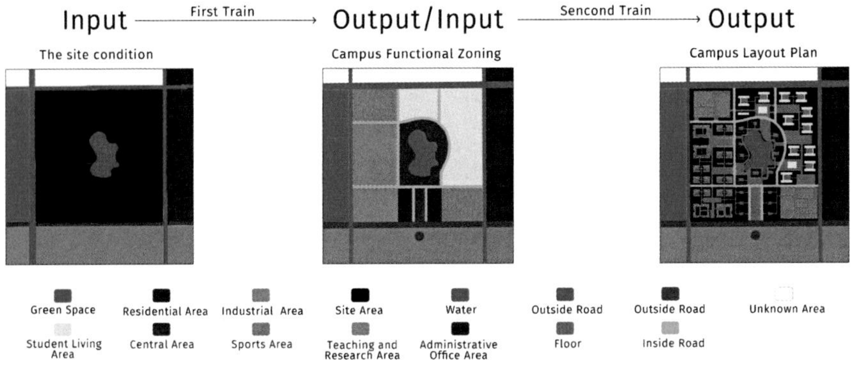

Fig. 5. Step by step training method

The experiment could generate a single result for each site, which cannot be changed and adjusted to the needs of designers to achieve diverse results. Therefore, we carried out the subsequent improved experiments.

4.2 Improved Experiments

In order to solve above problems, we make some adjustments and improvements in the following experiment:

1. Extract function bubble diagrams according to the original function layout.
2. The functional bubble diagram was proposed to be involved in the experiment. The site boundary condition and the functional bubble diagrams thus become the input together in order to realize the joint influence of multiple images on the generated results.
3. In the pre-experiment, the functional zoning plans of campus were generated.
4. In the final experiment, the layout plans of campus were generated.

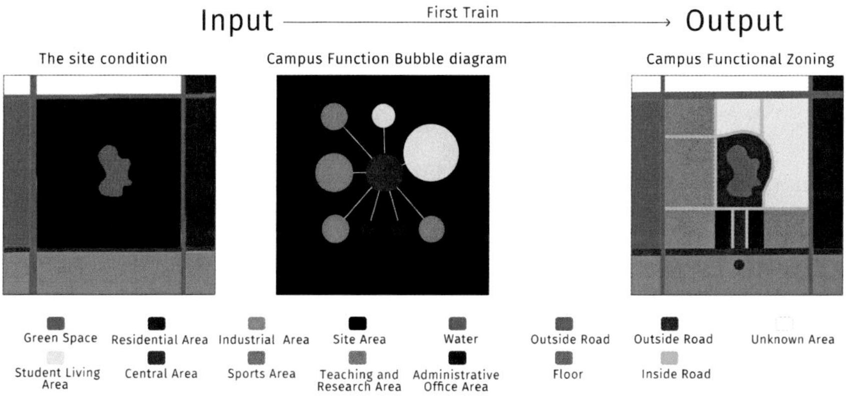

Fig. 6. Pre-experiment training data: input and output, labeling rules

In the pre-experiment (Fig. 6), we took the campus site boundary condition and the campus function bubble diagram as input, and the corresponding real campus function zoning plan as output. We input them into the machine for learning at the same time. The learning rate was adjusted to 0.002 and the number of iterations was 600.

The results of the above experiments (Fig. 7) were relatively good, and the generated functional zoning plans could change according to the change of functional bubbles diagrams. In order to explore the ability of the Pix2Pix model to generate complex plans, we tried to generate the layout of campus planning in one step in the following experiment (Fig. 8). The campus site condition and the campus function bubble diagrams were taken as input, and the corresponding real campus planning layout plan was taken as the output. The learning rate was adjusted to 0.002, and the number of iterations is 600 times, which could generate ideal effects.

Through the two experiments, the machine can directly learn the internal rules of the campus layout plans. Therefore, we will directly test the model of the final experiment and then analyze it.

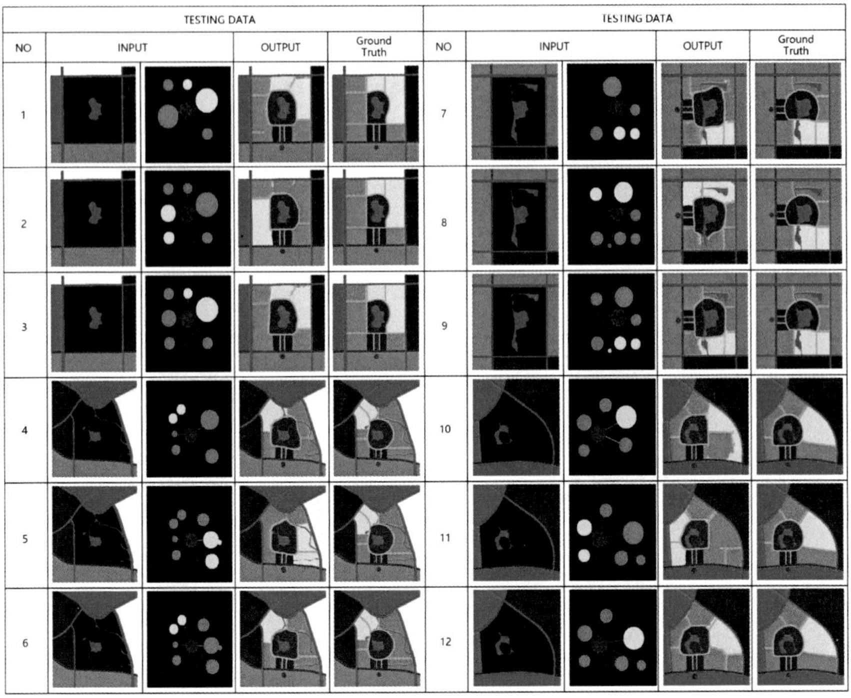

Fig. 7. Pre-experiment testing results

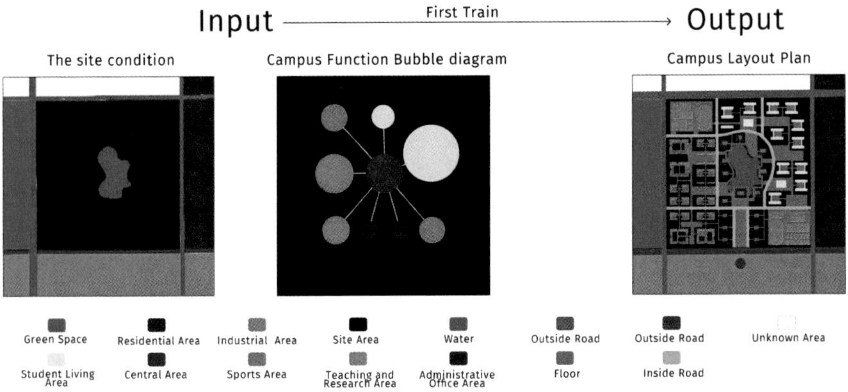

Fig. 8. The final experiment training data: input and output, labeling rules

4.3 Result Analysis

4.3.1 Qualitative Evaluation Criteria:

1. Adjust the position, size, and quantity of the functional bubble diagrams, to see if the results can show a variety of changes.

1.1. Modify the color of the bubble.
1.2 Modify the position of the bubble.
1.3. Add or subtract a bubble.
1.4. Modify the size of a bubble.
1.5. Customize a completely new layout.

2. Compared the result of the original functional layout with the real plan to see if it follows the common sense of architectural design, and observe whether the results of layout are reasonable. For example, it is advisable to have a north–south layout and a building distribution.
3. Observe whether it made some adjustments according to the surrounding environment, such as adjusting the overall direction of the building layout according to the inclined angle of the road, responding to the main landscape, and whether the changes of the surrounding site function will have an impact on the result.
4. Observe the sharpness of the generated images, including whether image pixels and building boundary are clear, and whether it can generate continuity of road.

4.3.2 Analysis

In this step, we used the method of controlling the variable (Fig. 9) to discuss the effect of functional bubble diagrams on the Generating effect and found the implicit correspondence relationship between functional bubbles and the site layout.

Through qualitative analysis, it can be found that changing the bubble diagram can affect the functional layout of the campus to a certain extent. The NO. 3–5 and NO. 4–5 illustrated that the arbitrarily placed bubble diagram can generate a new building layout, and can follow the campus layout rules well. The machine has learned the relationship between the administrative district and the main entrance, which shows the controllability of this approach. In terms of learning the layout rules of campus, the machine could basically master the layout of buildings in the central area around the central landscape, and the buildings are oriented towards the landscape, which can be a reasonable layout form. The buildings were able to give way to the road and some of them can be distributed along the trend of the road. But in the formation of image clarity, the architectural boundary is vague and the continuity of the road is poor.

Through quantitative analysis, functional elements in the generated images were relatively complete, and the machine basically learned the correspondence relationship between the building layout and the functional bubbles. In conclusion, this experiment demonstrates the ability of machine learning to learn complex objects.

5 Discussion

This paper showed a new idea of the diverse layout generation that users can control the results of campus layout by inputting different bubble diagrams. The results of the experiment reached our expected results. At the same time, the experiment showed that samples of common patterns can improve the efficiency of the machine learning complex building layouts, and demonstrated the huge potential of the machine to learn

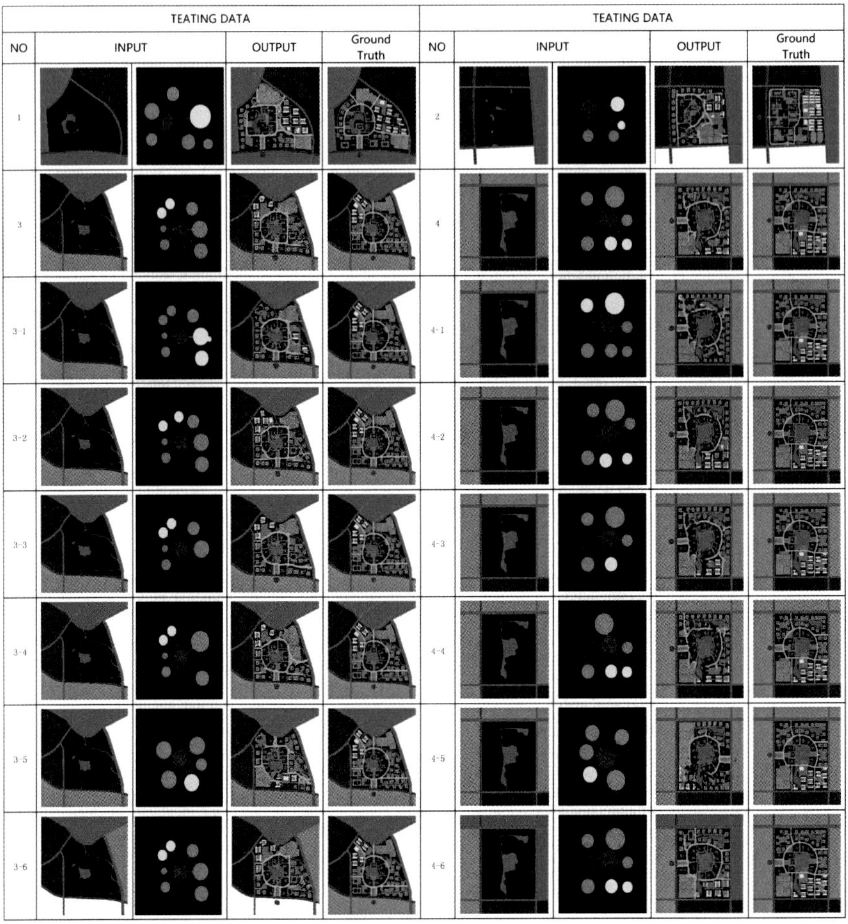

Fig. 9. The final experiment testing results

the complex layouts. In the future, on discussing the diverse generation, we hope to study different influencing factors: the impact of axis relationships on campus layout generation and another machine learning model which will generate diverse results of campus.

This study is only limited to the plan generation, without too much consideration of 3D generation. In the future research, we will build a three-dimensional model based on the generated results, and achieve a visual 3D model effect.

References

1. Chang S, Saha N, Castro-Lacouture D, Yang PPJ (2019) Multivariate relationships between campus design parameters and energy performance using reinforcement learning and parametric modeling. Appl Energy 249:253–264

2. Huang W, Zheng H (2018) Architectural drawings recognition and generation through machine learning
3. Liu Y, Luo Y, Deng Q, Zhou X (2020) Exploration of campus layout based on generative adversarial network. In: The international conference on computational design and robotic fabrication. Springer, Singapore, pp. 169–178
4. Liu Y, Fang C, Yang Z, Wang X, ZhouZ, Deng Q, Liang L (2021) Exploration on machine learning layout generation of Chinese private garden in Southern Yangtze. In: The international conference on computational design and robotic fabrication. Springer, Singapore, pp 35–44
5. Nauata N, Chang KH, Cheng CY, Mori G, Furukawa Y (2020) House-gan: Relational generative adversarial networks for graph-constrained house layout generation. In: European conference on computer vision. Springer, Cham, pp 162–177
6. Pan Y, Qian J, Hu Y (2020) A preliminary study on the formation of the general layouts on the northern neighborhood community based on GauGAN diversity output generator. In: The international conference on computational design and robotic fabrication. Springer, Singapore, pp 179–188
7. Zheng H, An K, Wei J, Ren Y (2020) Apartment floor plans generation via generative adversarial networks

Collective Intelligence and Effects of Anticipation

Ljubica Miric(✉)

CAUP - Tongji University, Shanghai 200070, China
miricl203@gmail.com

Abstract. Based on the extrapolation of contemporary theories that mind is a prediction machine, this paper points out the repetitive nature of the prediction parameters of collective intelligence anticipation scenarios and questions the hybridity of fears and desires regarding the evolution of artificial intelligence. Through the analysis of mnemonic principles of remembrance it combines the ancient technique of the art of memory with the contemporary views on the dynamics of perceptions in order to establish a link between the spatial constructs and their rapid expansion through the presence of the digital medium while in parallel suggesting an approach to the integration of the necessity for a more sophisticated systems of artificial intelligence into the collective intelligence. The main question of this paper is whether the anticipation of the future results in the creation of such a future, can a prediction-based interaction between man and machine govern its outcome? In order to attempt to develop potential new methods of integrating an idea of a different, more neutral outcome, the focus will be on the characteristics of perception that surpass the influenced, emotional response, and the observation of the general, innate human mechanisms of alignment. Through the study of the principles of memory the aim of this paper is to ask whether the individual comprehension of space and time as disengaged in given examples can in fact produce a system of ideas compatible in its nature to that of the machine itself. What are the main aspects of both human and machine that would stem their hybridity, and how should the collective intelligence adapt to enable the interlink?

Keywords: Collective hybrid intelligence · Mental imagery · Social imagery

1 Introduction

The pivotal aspect of intelligence, here defined as the ability to construct a model [8] gives the base through which this paper analyses the correlation between collective and artificial intelligence.

The following work explores aspects relating to memory, further than the contemporary *fears and hopes*—whose complementarity was defined by Descartes,[1] the mechanisms of generalized globally shared anticipation scenarios. The inquiry used for

[1] In his book the Passions of the Soul from 1649, Descartes describes both fear and hope as dispositions of the soul.

P. F. Yuan et al. (eds.), *Hybrid Intelligence*, Computational Design and Robotic Fabrication,
https://doi.org/10.1007/978-981-19-8637-6_21

proposing new kinds of understanding of the hybridity features will be centered around the shared aspects of spatial developments, as manifesting the use of AI on a social—urban scale, and the correlation between individual mental imagery with social imagery of anticipation.

The method of this paper is to propose an extrapolation of the individual brain's functional modes onto the social mechanics of reality lived, or in other terms: the application of the perceptual setting of an individual to the shared aspects perceived by the majority of the social group. Rather than standing by the idea of the collective consciousness, it explores the mechanism of perception which are inherently the same for each organism, the temporal dimension of the reality perceived in a space—virtual or material, as recollected, interpreted and acted upon. In proceeding, the terms: *cultural memory* [7], *collective intelligence* [2, 17] and the *extended mind* [5], will be used to refer to the social aspects through which the hybridity of intelligence between men and its digital environment of interaction is formed [13].

From the neuroscience standpoint our perception of reality is the brain's best guess of what reality is [15]. While locked inside a skull, the brain receives stimuli from the outside world, it interprets it and according to it makes the best probable supposition of what reality is, and somewhat a more accurate, yet rather unconscious, calculation of what are the conditions in which the rest of the organism that it is attached to finds itself in. How then can we define the parameters according to which this guessing takes place, or further so, time?

The established modes of how the prediction comes to be generated have been defined by Clark [6]. He applies the term *prediction error minimization* to refer to the process through which the guess is refined into a suitable inner representation of reality. Namely, in order to spend less energy on the calculations of the incoming stimuli, the stagnant aspects of the previous frame are kept and, if needed, corrected in accord with the incoming information about the temporal changes within the spatial context. Another aspect of this process is the constant back and forth between the past and the present, whereas previously collected information is updated, but still active in the calculated perception of the present frame. This interlinked relationship between the two temporal signifiers marks the key components of perceivable reality as repetitive.

In his 1908 ed. book titled *Matter and Memory*, Bergson writes: "perception is master of space in the exact measure in which action is master of time" [3, p. 24, translation]. Would it then be possible to interpret one of the causes of the present perception as the preparation for its future response? Taking this supposition further would mean interlinking the perceptual modes in a specific temporality. The linear passage of time is then varied within the perceived frame of reality. One might go as far as to assume that the perception of the present moment is a simultaneous calculation of the past retrieved and future predicted through the very process of *prediction error minimization*. The brain's best guess is thereby an action mastering time through the perception of space.

2 Mechanisms of Perception

The notion of *inference perception* as defined by Helmholtz, that can be traced back to Kant, is further analyzed by Seth through the idea of the *controlled hallucination*.

According to Seth [15] the three main components of it are: top-down predictions, bottom-up sensory signals providing prediction errors, and the hypothesis that one doesn't experience sensory signals but rather always their interpretation. He writes: "...what we actually perceive is a top-down, inside-out neuronal fantasy that is reined in by reality, not a transparent window onto whatever that reality may be" [15, p. 83]. This idea of *predictive processing*, hypothesis indicates that all conscious experiences are a form of appropriation of reality, even the one occurring in the present. Which begs the question whether the difference between past, present and imagined experience is in the quantity and quality of available stimuli.

Building on Peacocke's *Experiential Hypothesis*, [12] defines the Dependency Thesis, referring to the differences between two states that seemingly correspond, sensory experience and imagining, he writes: "imagery tends to be less determinate and replete in detail than sense experience, imagery can be subject to the will in ways that experience cannot be" [12, p. 405]. However, referring to the content, the dependency view attributes the perceptual experience to both the memory of an experience and imagination. A generic image, or the one that hasn't taken place takes the attribute of time within perception, while the experiential memory is an image which undertook place but lacks a fix temporality due to its potentially repetitive nature. Having in mind that the sensory experience of the present moment perpetually includes the capacity of a past prediction and the potential of the future action, a mental image is here as relevant as its correspondence to either dimension of space and time, or past and future.

In order to better comprehend the temporal aspect's repetitive nature of the mental image one can refer to the work of [19] on the *mnemotechnics*. With its origins in the ancient art of memory, this practice functions as an evocation of the place layered with detail or image. In order to memorize one is invited to visualize a spatial dimension to the temporal action. She writes: "We have to think of the ancient orator as moving in imagination through his memory building whilst he is making his speech, drawing from the memorised places the images he has placed on them" [19, p. 3]. Through this aggregation of temporality, past, present and future, while here action is defined as the continuum of speech, the orator expresses the spatial objectivity of reality through the virtualization of his actions, referring to the content. An essential aspect of this process is there by the layering itself of the imagined space with the factual image creating a virtual unity necessary to the perceptual process.

According to Bergson [3], "images themselves, they cannot create images, but they indicate at each moment, like a compass that is being moved about, the position of a certain given image, my body, in relation to the surrounding images" [3, p. 10]. Introducing the mnemotechnics to this process would signify a layering of space-image of the surroundings with the imagined place-image through which the content of the memory is recalled. It is thereby a simultaneous correspondence between the sensory image, or the neuronal fantasy, and the controlled imaginary of remembrance. Due to the overlap of control as action, one of the physical and other of the virtual movement, the virtual place would lack the three dimensional spatial aspects, and the interpretation of it would be achieved through the image-like, time-defined action of the virtual movement. Hence, the temporal aspect assimilates remembrance with imagination, indicating that the place of the mental image, past or potential is the distorted space within which the third dimension is that of time, or hereby action: the movement through it.

What this hypothesis indicates is that the process of remembrance is influenced by the degree of the distortion of place into image, or the individual interpretation of it depending on movement. Just as the memories are displaced through time, they are created through time, or place rather than space, allowing for the sequencing of images. However, it is the deformity in the expectation of the next image in the sequence based on the previous that induces the storage of information, and produces the possibility for the recollection of it. This distortion, enhanced through the joint interaction with the data collected, the constant application of the screen, stems the need for a further exploration of the mechanisms of shared hybridity.

How is then this temporal back and forth of the individual mind consistent with the collective memory and thus imaginary? A key notion to the understanding of the perceptual process of the social group is what has it been exposed to in terms of individual sensory experiences. It would seem evident then that the wider the distribution of the content the higher the percentage of its partaking in the collective's memory. The potential of the group to envision a future action, or the aggregation of individual responses, thus depends on its capacity to have retained stimuli which can potentially be influential of each individual on a larger scale. The agreement on what has been perceived can come through the universal understanding of what is objective, though this aspect falls outside of the scope of this paper, instead the focus will be on the mechanism of individual perception which is biologically universal. Thereby: instead of focusing on the fact that red means *stop* and green means *go*, this work is oriented around the fact that upon seeing red the majority will have understood the sensory input as the color red, regardless of the interpretations one might have regarding that experience.

According to Lévy [10], "the Internet represents the unmediated presence of humanity to itself" [10, p. 190] leading to the more evident conception of reality as collective, since "we are all in the process of thinking within the same network" [10, p. 191]. The argument for the virtualization of reality leading up to its collective generalization through the ubiquitous screen may be expressed through the theory of the *extended mind*. The Internet has become a valid component, an extension, of the individual mind—shared collectively. The consequences of this virtualization seems to be the exposure to the screen—an image, rather than space, leading up to the kind of distribution of the content's experience which hasn't been experienced directly in space. This resemblance to the mnemotechnics in terms of the virtual place has potentially led to the heightened memorization of the digital content, however the constant exposure to the incoming information might have produced an acceleration in the substitution of the remembered content with the recently registered, within the episodic memory.

Thereby, the focus is not on the collective memory as a whole, defined by Halbwachts, but rather the cultural memory as argued for by Assmann which is "exteriorized, objectified, and stored away in symbolic forms" [1, p. 110]. The digital objectification that takes place through the virtual experience of the sensory screen can accordingly be observed through its potential to be perceived as the actual[2] experience. The aspects of

[2] According to Levy [11, p. 23], the correlation is not between the virtual and the real but rather the virtual and actual, and the possible and real.

the virtual sensory input must then be in alignment with the spatial distortion in order to be assimilated into the perception of the pre-recorded as experienced within the present. Nevertheless, the lack of choice regarding the action of movement is reflected in the extension of choice to the realm of content, whereby the mastering of time is experienced through the attention given to the particular content individually, which creates an accumulation of the communally confirmed influence of such content.

The virtual representations of the world, in order to be integrated and influential, should correspond to the internal process of modeling, the intelligence as defined by Hawkins [8], or the neuronal ability to create a model that represents the world. The construct of the virtual image which has the status of the cultural memory and the mechanisms of the mental image through which the exterior is perceived are then in tight correlation in what concerns the intuitive human ability to integrate and actualize the observable stimuli. Therefore the argument for the collective memory here doesn't consist of collectively remembered sensory experience, but rather the individually generated imagery feedback to the distorted aspects of virtual space within the collective. It is this aspect of the distributed activity that then generates the future outcomes.

3 Anticipation of the Collective

Within the attempt to comprehend the interaction between humans and machines in the contemporary world, the accounts of the perceptual mechanisms are here confronted with the notion of anticipation as a process of action in virtual or material space.

According to Poli [14], "anticipation occurs when the future is used in action". He goes on to distinguish the components of anticipation as "a forward looking attitude and the use of the former's result for action" [14, p. 4]. While the study of the future consists of three levels, forecasting, foresight and anticipation, also called the design-based foresight [14, 18], the third level, anticipation is based on the results of former two and marked with the goal of "implementing them into decisions and actions" [14, p. 7].

Further than the predictive and spatial perceptual forecasting and the non-predictive temporal foresight which generates an exploration into the possible futures, the anticipation models evoke the enacting of both past and future within the present. In accord with the imbrication of the two, anticipation scenarios produce the action which directs towards a seemingly desirable outcome. Aligned with the view on fear and hope as having the same functional mechanism, the action of anticipation is here not defined according to the emotional response but rather it demonstrates how the mental image of action takes its form within both space and time. According to Bickhard [4], p. 15*, 2018, p. 328) *representation* and *motivation* are not "distinct subsystems" but "different aspects of one underlying dynamic of selecting interactions within a space of (anticipated) possible interactive trajectories".

In his 1969 paper *Social Research for Social Anticipation* Paul Smoker writes: "To date, the most desirable alternative futures have often been regarded as "utopian" and impossible, while those that are considered most probable have for the most part been mere extrapolations of dominant trends of the past" [16, p. 13]. In regard to Jamison's notion of the postmodern *perpetual present* (1991) this formulation of the overlap between temporal dimensions, socially resulting in the action which is most likely to correspond to the previous action, directs the comprehension of the envisioned outcome as an exaggeration of the social present. According to Poli's reading of Appaduarai's interpretation of Bloch, one can distinguish the *endpoint utopia* from the *everyday life utopia*: "the roots of the future are in the present, if only we learn to see them." p. 117 How is then the social anticipation of the future past shaping the perpetuity of the individual present?

The confrontation of the anticipation of the future based on the past which has never taken place, as is the case of fiction which exaggerates the contemporary fear and hope, with the seemingly evident potential of the future development which envelopes progress as such that has taken place, points towards the nature of origin of the mental image of action. The spaceless past which is socially distributed is then further than the individual experience, an image embedded within the collective. "Imagination, and especially collective imagination, *produces* reality" [11, p. 189]. An arguable take would then be that the fiction which has been experienced as a mental image is equally relevant as an experiential memory. Generating a mental image of action is thereby equally dependent on the past that took place and the past that only took time.

The repetitive nature of the temporal recreating of the mental image, and the unique distortion of place generated within each sequence, influences a response to that which has been integrated—as a uniform potential contemporary outcome action. On a larger scale, the controlled hallucination of the individual addresses the exposure to the shared stimuli as integrated in different manners, but acted upon in accord with the collective reality of the present. In other words: regardless of the individual angle at which the green light was perceived to be the color green, there is a consensus on the fact that the experience of the color green was in fact green, which within the collective that has perceived it individually stems a decision routing of action as: action, inaction or no regard. An accumulation of varied responses thus produces the socially constructed response to the past experience collectively shared. If the past experience is that of the collective response in the form of a mental image of future action, the collective itself is the driving force of such an action.

4 Artificial Intelligence on a Social Scale

The notion of AI formed through the social conception of what is artificial intelligence is in many ways different to the mechanisms of its functioning.

Instead of focusing on the social perception of what artificial intelligence could become, or further more the social aspects of functioning of the artificial intelligence, the interest falls within the domain of the operational hybridity between the society and numerous systems of artificial intelligence which govern it.

An evident example of this comes through the extended mind theory. Knowing the route has unarguably become a possibility to know the best route at a given time. The human capacity to anticipate has been enhanced through the computational power of calculation. The argument is thus for the hybridity of collective action of social anticipation and the enhancement of it through the shared, distributed algorithms of AI. If the potential of a future action depends on the past memory, obtained through a direct experience or an indirect impression of an experience, and the recalculating of it through the present circumstances of perception mechanism, then an evident addition to that calculation is the prediction proposed by artificial algorithms.

Where the movie *Matrix*[3] depicts the new generation of the governing artificial system as the result of it being challenged by an individual, the reality of contemporary AI is that each individual contributes to the constant re-enhancement of the perpetually regenerating system. A dystopian prediction, here viewed as a *dark* utopia, focuses in many aspects on the everyday life rather than the final outcome, seen as such, one is prompted to acknowledge the shared computer simulation. The comprehension of virtuality comes most evident in its manifestation as augmented reality, where the historical backdrop gives spatial characteristics to the physically elusive digital medium. Hence hybrid urbanism in the context of the individual on a social scale has the potential to stem new modes of approach to the urban reconfiguration of the already built landscapes.

Just as the cultural memory finds its actuality in the monumental structures which landmark the urban fabric, so can a digital expression of style signify a virtual landmark of the said fabric. Upon classifying the depictions of the urban hotspots an algorithm can be used to identify different aspects which mark the urban zone as of particular interest, intensity, brightness, colorimetry, flow, etc. These data, collected and interpreted could potentially signify a means of classification of the urban zones and their further redevelopment in regard to the necessity to lessen or heighten the particular relationship between the inhabitant and their environment.

Instead of proposing a new manner of shaping the urban tissue, the focus in the particular examples falls on the capacity of the individual to retain and reproduce an idea of urban landscape in accord with the socially constructed consensus on what the built landscape itself should resemble. By creating an urban project which focuses on the distribution of an idea of a city, one which can be achieved as an accumulation of individual actions, it is arguably possible to achieve a socially appealing representation towards which the inhabitants alongside developers will strive towards. Such project would then combine the existing systems of algorithms with the reconfiguration of their application in the attempt to create a more balanced functionality of the built landscape. This would stand for the merger between the individual actions resulting from the shared anticipation scenarios and the governance of those scenarios by urban planners.

5 Conclusion

Rather than opposing artificial intelligence to human intelligence, the line is drawn between the multiplicity of algorithms generating the outcome action of what is known as the artificial intelligence and the social mechanisms of producing a collective response as an aggregation of individual processes.

The argument is thus for the hybridity of the collective with the algorithms which govern it, rather than an attempt to invoke differences and similarities between the individual brain and the artificial software. Just as artificial intelligence could not achieve the

[3] The *Matrix* (1999) directed by Wachowskis, Distributed by: Warner Bros. Pictures, Village Roadshow Pictures, Roadshow Entertainment.

type of individual human consciousness, the collective itself lacks a uniform consciousness. But do they lack intelligence? If intelligence is the ability to generate a model, with an awareness of what that model is, then the answer for both would be negative due to the absence of consciousness, however, if intelligence is the ability to generate a model and act upon it, then the hybridity of the collective with the artificial systems of operating within the world becomes more evident.

A mechanism of the response generated by the AI is in many ways similar to that of the collective, it is based on an accumulation of individual processes of learning, and though a prediction is possible it is not based on an individual emotion but a systematic organization of individual responses. The most influential systems of AI in the contemporary world operate in relation to the collective of individuals, and though there are examples of individually controlled algorithms with an attempt to resemble human consciousness, the majority of the AI systems interact with- and are based on- the collective. An example of such a system at an urban scale is the initiative called the city brain.[4] The way in which it operates deserves further investigation from a philosophical stance, or the modules in which it interacts with- and is influential of- the society it enhances.

References

1. Assmann J, Nünning A, Erll A (2008) Communicative and cultural memory. In: Cultural memory studies: an international and interdisciplinary handbook (Media and Cultural Memory/ Medien Und Kulturelle Erinnerung), 1st ed. Walter de Gruyter, pp 109–118
2. Bastiaens TJ, Baumöl U, Krämer B (2010) On Collective intelligence (Advances in intelligent and soft computing, 76). Springer
3. Bergson H (1911) Matter and memory. The Macmillan Co
4. Bickhard MH (2003) An integration of motivation and cognition. In: Smith L, Rogers C, Tomlinson P (eds), Development and motivation: joint perspectives, monograph series II(Leicester: British Psychological Society), pp 41–56
5. Clark A, Chalmers D (1998) The extended mind. Analysis 58(1):7–19. https://doi.org/10.1093/analys/58.1.7
6. Clark A (2016) Surfing uncertainty: prediction, action, and the embodied mind (Reprint ed.). Oxford University Press
7. Erll A, Nünning A (2008) Cultural memory studies: an international and interdisciplinary handbook (Media and Cultural Memory/Medien Und Kulturelle Erinnerung) (1st ed.). Walter de Gruyter
8. Hawkins J, Dawkins R (2021) A thousand brains: a new theory of intelligence (First Edition). Basic Books
9. Leach N (2022) Architecture in the age of artificial intelligence: an introduction to ai for architects. Bloomsbury Visual Arts
10. Lévy P (2005) Collective intelligence, a Civilisation: towards a method of positive interpretation. Int J Polit Culture Soc 18(3–4):189–198. https://doi.org/10.1007/s10767-006-9003-z
11. Lévy P (1998) Becoming virtual: reality in the digital age. Plenum Trade
12. Martin M (2002) The transparency of experience. Mind Lang 17(4):376–425. https://doi.org/10.1111/1468-0017.00205

[4] According to Leach [9] City Brain is "Perhaps the most extensive exploration application of AI to the city" [9, p. 120].

13. Miorandi D, Maltese V, Rovatsos M, Nijholt A, Stewart J (2014) Social collective intelligence: combining the powers of humans and machines to build a smarter society (Computational Social Sciences) (2014th ed.). Springer
14. Poli R (2019) Handbook of anticipation: theoretical and applied aspects of the use of future in decision making (1st ed. 2019 ed.). Springer
15. Seth A (2021) Being you: a new science of consciousness. Dutton
16. Smoker P (1969) Social research for social anticipation. Am Behav Sci 12(6):7–13. https://doi.org/10.1177/000276426901200603
17. Tovey M (2008) Collective intelligence: creating a prosperous world at peace. Earth Intelligence Network
18. Tuomi I (2013) Next-generation foresight in anticipatory organizations. Background study for the european forum on forward-looking activities (EFFLA), European Commission
19. Yates FA (1966) Art of memory (Selected Works, Vol 3) (Revised 1999 ed.). Routledge

A Rapid Wind Velocity Prediction Method in Built Environment Based on CycleGAN Model

Chuheng Tan[1](✉) and Ximing Zhong[2](✉)

[1] The Bartlett School of Architecture, University College London, 22 Gordon Street, London, UK
chuheng.tan.21@ucl.ac.uk
[2] Aalto University Finland Espoo, 02150 Espoo, Finland
ximing.zhong@aalto.fi

Abstract. Although the wind microclimate and wind environment play important roles in urban prediction, the time-consuming and complicated setup and process of wind simulation are widely regarded as challenges. There are several methods to use deep learning (DL) models for wind speed prediction by labeling pairs of wind simulation dataset samples. However, many wind simulation experiments are needed to obtain paired datasets, which is still time-consuming and cumbersome. Compared with previous studies, we propose a method to train a DL model without labelling paired data, which is based on Cycle Generative Adversarial Network (cycleGAN). To verify our hypothesis, we evaluate the results and process of the pix2pix model (requires paired datasets) and cycleGAN (does not requires paired datasets), and explore the difference of results between these two DL models and professional CFD software. The result shows that cycleGAN can perform as well as pix2pix in accuracy, indicating that some random city plans image samples and random wind simulation samples can train surrogate models as accurate as labelled DL methods. Although the DL method has similar results to the professional CFD method, the details of the wind flow results still need improvement. This study can help designers and policymakers to make informed decisions to choose Dl methods for real-time wind speed prediction for early-stage design exploration.

Keywords: Deep learning · Wind velocity prediction · Pix2pix · CycleGAN

1 Introduction

The Wind environment is an important factor in urban microclimate studies [1]. Since 1930s, the Navier–Stokes equations have been used for calculating fluid dynamics problems, which enable people to use mathematical equations to simulate the principles of fluid mechanics [2]. The development of computer technology increased the development of three-dimensional methods. Until now, Computational Fluid Dynamics (CFD) simulation is regarded as one of the most popular and useful methods for urban wind

Zhong: Contributed equally to this work and should be considered the co-first author.

P. F. Yuan et al. (eds.), *Hybrid Intelligence*, Computational Design and Robotic Fabrication,
https://doi.org/10.1007/978-981-19-8637-6_22

environment studies but is usually complicated and time-consuming, which is widely regarded as a challenge [3]. In 1999, the advent of Fast Fluid Dynamics (FFD) proposed by Jos Stam [4] addresses this challenge to some extent [5]. FFD is a technique for solving the incompressible Navier–Stokes equations and it is developed for real-time fluid visualization in the video and gaming industry, consequently, which has the advantage of being very fast to run and simple to code [6]. Since the 2000s, Deep Learning (DL) methods have been implemented in predicting the wind environment, providing the possibility to analyze or predict the urban wind environment much faster [7]. Although some research on simulating and predicting the urban wind environment based on GAN and FFD has been proposed [5], a large number of wind simulation experiments are needed to obtain paired data sets, which is a complex and time-consuming process for data preparation and labeling [8]. CycleGAN is a new technique that involves the automatic training of image-to-image translation models, without paired examples. It has the advantage of training models in an unsupervised manner using a collection of images from both source and target domains that do not need to be related in any way [9]. How to achieve a workflow that can produce wind speed predictions without preparing paired datasets remains a challenge.

The main purpose of this article is to find a more convenient DL workflow, rather than improve the accuracy of existing simulation software. Compared with the previous study, we choose the cycleGAN model, which doesn't require paired and labelled datasets. Designers input only random city samples, and random wind simulation samples, which are not matched. To verify our hypothesis, we evaluate the results accuracy of two different GAN models, pix2pix and cycleGAN, aiming to find a fast and easy workflow to generate high-quality results for city block wind prediction. Besides, we compare the results with CFD simulation results concurrently to discover the advantages and limitations of this method.

2 Related Work

Many studies successfully used machine learning and the DL method to investigate wind characteristics and wind effects on buildings and the urban environment (e.g., wind flow, wind pressure, and wind-induced responses) [8]. The large full flow fields data of CFD simulations has sparked interest in applying deep neural networks (DNNs) for fast approximations to computational fluid dynamics. DNNs were firstly applied to surrogate modelling and realize design exploration [10]. Convolutional neural networks (CNNs) can perform better than DNNs in the design exploration domain because they represent non-linear input and output functions while extracting spatial relationship [5], such as Jin's prediction of the velocities filled around a circuit cylinder [11].

The advent of Generative Adversarial Networks (GANs) greatly reduces the calculation cost and time and produces minor error results by learning the characteristics of paired samples [12, 13]. Kim develops a kind of GAN model, called Generative Adversarial Imputation Network (GAIN) to impute the unmeasured velocities around buildings [8]. He Yi develops a hybrid framework for rapid evaluation of wind velocity around buildings through parametric design, CFD simulation and machine learning on Pix2pix [14]. Mokhtar used cGAN and Pix2pix for pedestrian wind comfort estimation [15].

The studies above highlight the potential of deep learning networks in learning and predicting the wind environment. As discussed in the introduction section, labeling datasets is a cumbersome process, a large number of paired maps of urban block prototype plans and corresponding wind simulation results sometimes are hard to prepare. CycleGAN model is a kind of unsupervised model for image-to-image translation that doesn't require paired datasets, which makes it possible for faster dataset collection [9].

In contrast with previous work, our focus is to verify whether unsupervised methods that don't need labeled datasets can also realize wind prediction. In this study, we compare the performance of two different GAN models, a supervised model (Pix2pix) and an unsupervised one (cycleGAN) in the DL method for wind velocity simulation. The aim is to find a faster workflow to generate datasets without labelling and obtain better training results in a shorter time. Compare with the supervised model, this workflow enables designers to train the unsupervised model for wind environment prediction faster and easier.

3 Method

The research workflow of this paper is shown in Fig. 1. The process can be divided into 3 steps: (1) dataset preparation, (2) pix2pix model and cycleGAN model training, (3) comparison of the results from these two models. We use Houdini to generate city block plans and perform wind simulation in step2, and train the current DL model in Google Colab in step2. Finally, we observe the training loss and FID value of two models in step 3, the focus of our study, aims to evaluate the performance of these two models and to find a faster workflow for city wind velocity prediction.

3.1 Data Preparation for the Pix2pix Model and CycleGAN Model

The main difference between the pix2pix model and the cycleGAN model is the use of labelled datasets. The former uses labelled input and output data, while the latter does not [16]. Pix2pix is a kind of supervised model that's defined by its use of labelled input and output datasets, it is composed of a generator and a discriminator. These datasets are designed to train or "supervise" algorithms into classifying data or predicting outcomes accurately [17]. A cycleGAN model works on its own to discover the hidden pattern in the dataset without the need for human intervention. It is trained in an unsupervised manner using a collection of images from the source and target domain without a one-to-one mapping between the source and target domain that does not need to be related or paired in any way [9]. Compared to supervised models such as pix2pix, it is more convenient and faster to create a large dataset, and we don't need to make paired datasets samples and label them.

There are two sets of datasets for these two models, datasets 1 (D1) is for pix2pix model training, which contains input (train A) and output (train B) datasets. Train A are city block plans and train B are their corresponding wind simulation results, the A and B images are paired and matched. Datasets 2 (D2) is for cycleGAN model training, which also contains train A and train B, but they are random and not matched.

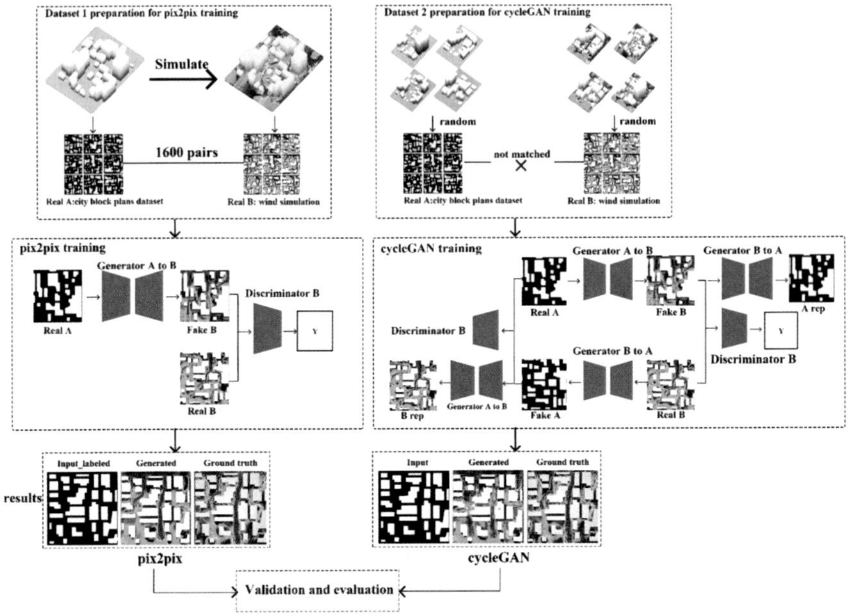

Fig. 1. The whole workflow of comparing two DL models for wind prediction in this study

In this study, we choose Houdini parametric modelling to prepare city blocks and generate wind velocity simulation, as Houdini has a good synergy in modelling and wind simulation for our comparative experiment. Firstly, we use Houdini parametric modelling method attached to the PDG system to generate city blocks prototype plans. The PDG system provides a rich set of stock nodes to enhance productivity, which can generate a great number of datasets automatically in a short time [18]. Every generated block plan was limited to a 200 m × 200 m square, which was based on typical dimensions and densities of urban contexts [15]. And in each block, the buildings were set up with heights ranging from 9 to 100 m, and lengths and widths ranging from 12 to 60 m.

Secondly, each city block is entered into the Houdini wind simulation system automatically, which has a good synergy. This simulation method is a kind of FFD method, which has been verified as a relatively fast and high-accuracy method in 2015 [19] and it is successfully applied in much parametric design research, such as Nodado et al.'s wind-induced architectural systematization [20]. It can build different models and generate corresponding wind environment analysis charts in real-time without manually exporting the model for wind simulation in other software. To set up the wind test environment, the minimum meshing cell size is defined as 3 m for all the models, which can take a better balance between computational feasibility and a low average error. The wind velocity was set as 5 m/s at a 10 m reference height for all plans. This wind velocity was commonly observed in most urban area [15]. As for the datasets for the pix2pix model, the input images (city blocks plans) and output images (wind speed simulation) need to be sorted consistently and set in the same size for matching into pairs in the machine learning process. For architects, this process may be tedious and error-prone.

Finally, for training the Pix2pix model, we generated 1600 paired city block plans and corresponding wind speed simulation images, the colormap of wind analysis can be converted to label information. And for training the cycleGAN model, we generate 1600 city block plans and 1600 wind simulation images that aren't matched. Figure 2 datasets samples for pix2pix model and cycleGAN model separately. The dataset samples are shown. For curating the datasets, we divided the dataset into 90% training and 10% testing sets, thus we used 1440 in training and 160 in testing, according to the sample density and proportion of the dataset.

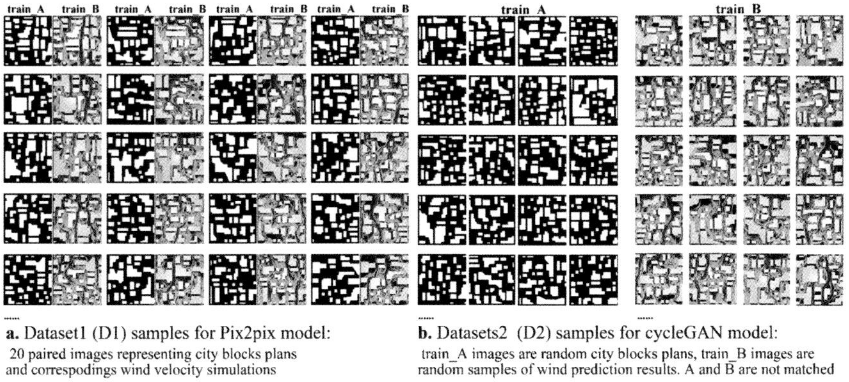
train_A
train_B
train_A
train_B
train_A
train_B
train_A
train_B
train_A
train_B
a. Dataset1 (D1) samples for Pix2pix model:
20 paired images representing city blocks plans and correspodings wind velocity simulations
b. Datasets2 (D2) samples for cycleGAN model:
train_A images are random city blocks plans, train_B images are random samples of wind prediction results. A and B are not matched

Fig. 2. Datasets samples for the pix2pix model and cycleGAN model separately

3.2 DL Model Training

In this section, we train the pix2pix model and cycleGAN to find an accurate and faster wind simulation and prediction workflow. The two processes are set up and implemented in Google Colab, which provides a runtime fully configured for deep learning and free-of-charge access to a robust GPU, and it is convenient to write and execute code [21].

Pix2pix model is developed based on a conditional generative adversarial network (cGAN) to learn a function to map from an input image to an output image. The network is made up of two main pieces, the Generator, and the Discriminator. The Generator transforms the input image to get the output image [22]. In this study, we use the pix2pix proposed by Phillip Isola in 2017 [22]. the input and output resolution are set to 256 × 256 pixels, a 50 m receptive field for the discriminator, with a learning rate of 0.0002 and generator adversarial to L1 loss ratio of 1 to 100, and all experiments were trained for a total of 200 epochs. After every epoch, the trained weights were saved to monitor the progress of training. Figure 3a shows the process of pix2pix model training.

In this study, we use the cycleGAN model proposed by Jun-Yan Zhu and Taesung Park in 2017 [17]. A cycleGAN model is composed of 2 GANs, making it a total of 2 generators and 2 discriminators. One generator transforms city block plans into city blocks with wind speed simulation result, and the other transform wind results into city block plans. In the case of cycleGAN, a generator gets additional feedback from the

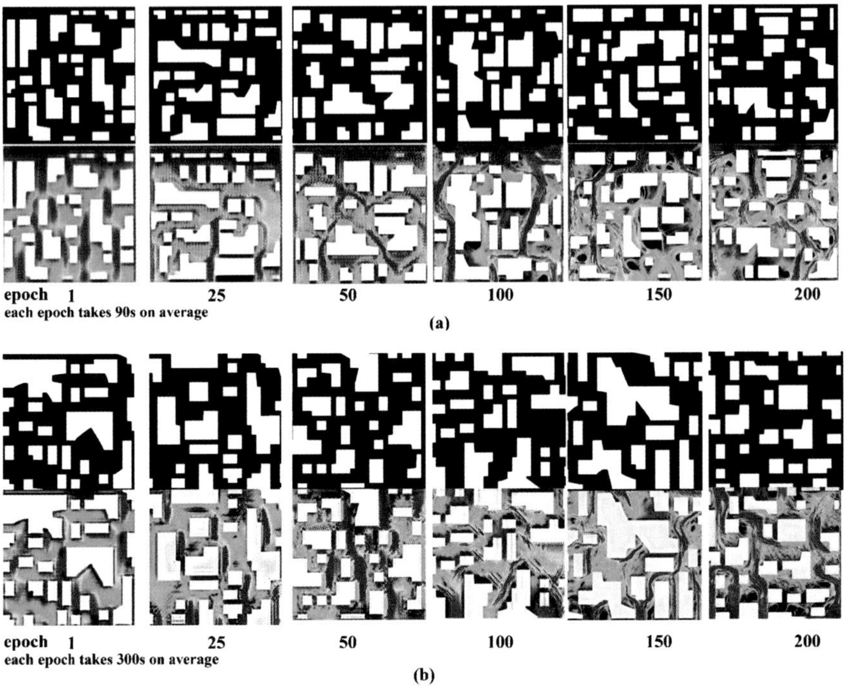

Fig. 3. **a** pix2pix training process **b** cycleGAN training process during epoch 200

other generator. This feedback ensures that an image generated by a generator is cycle consistent, meaning that applying consecutively both generators on an image can yield a similar image.

To ensure the same experimental conditions, in this study, we set the same environment and parameters for the cycleGAN model as for the pix2pix model. The image of training dataset A and B are converted to 256 × 256 pixels, with the learning rate of 0.0002 for a total of 200 epochs. When training the discriminator, the loss is divided by 2. The weights are initialized with a Gaussian distribution with a mean of 0 and a standard deviation of 0.02. Every epoch the training set is shuffled and partitioned into subsets the size of the minibatch. After every epoch, the trained weights were saved to monitor the progress of training. Figure 3b shows the process of cycleGAN training.

3.3 Performance Evaluation Method and Criteria

Firstly, we evaluate the stability and accuracy of these two models by mainly observing the Training loss and the Fréchet inception distance (FID). We compare the training loss graphs to observe the loss decreasing and convergence of the two models as shown in Fig. 5a. For clearer image contrast, we reparametrize the value of the diagram. The FID is a metric used to assess the quality of images created by a generative model, especially in GAN [23]. It is the current standard metric for assessing the quality of GANs as of 2020, which has been used to measure the quality of many recent GANs. The FID is improved on the IS by actually comparing the statistics of generated samples to real samples. It compares the distribution of generated images with the distribution of real images that were used to train the generator [24]. The FID score is then calculated using the following equation in this study:

$$FID = ||\mu - \mu_w||_2^2 + \mathrm{Tr}\left(\Sigma + \Sigma_w - 2\left(\Sigma^{\frac{1}{2}}\Sigma_w\Sigma^{\frac{1}{2}}\right)^{\frac{1}{2}}\right) \tag{1}$$

The FID metric is the squared Wasserstein metric between two multidimensional Gaussian distribution: $\mathcal{N}(\mu_w, \Sigma_w)$, the distribution of the neural network features of the images generated by the GAN and $\mathcal{N}(\mu_w, \Sigma_w)$ the distribution of the same neural network features from the real images used to train the GAN [24].

A lower FID indicates better-quality images, conversely, a higher score indicates a lower-quality image and the relationship may be linear.

Secondly, although the FFD simulation method in Houdini is a practical and rapid way to generate large datasets, the accuracy may not be as well as the professional CFD simulation software. Considering more convincing results, we compare the results of DL models, Houdini wind simulation and ANSYS Fluent (a popular and professional CFD software). The environment settings in ANSYS are the same as in Houdini.

4 Results and Discussion

Figure 4 shows the comparison of results generated from two models respectively and the real wind simulation from Houdini and ANSYS fluent. Figure 5 shows the performance comparison in training loss and FID value of the pix2pix model and the cycleGAN. In Fig. 4, the results indicate that both cycleGAN and pix2pix can perform well in generation images, which are very close to the output dataset. And FFD can predict wind environment with reasonable accuracy. The flow direction is similar between FFD and CFD, however, discrepancies are present in flow speed.

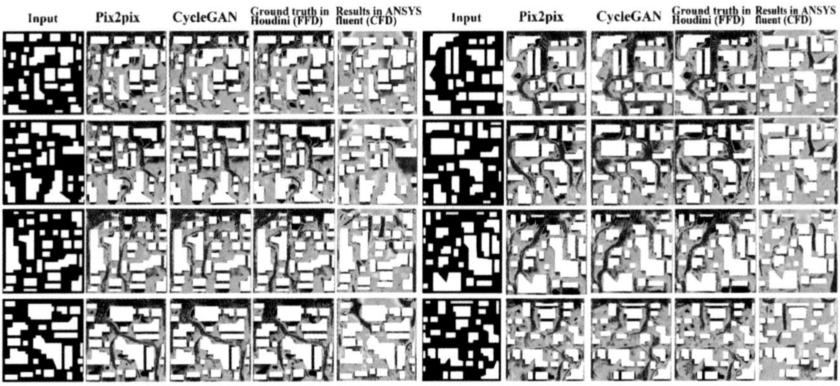

Fig. 4. Testing results comparison of two models and ground truth simulation

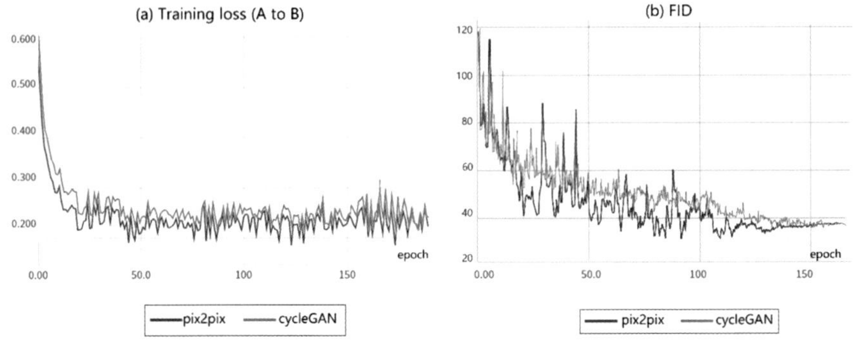

Fig. 5. Training loss diagram and FID of pix2pix and cycleGAN model

In the training loss diagram of Fig. 5, we can see that at the beginning of the training process, the accuracy and stability of cycle GAN are not as good as that of the pix2pix model, but after epoch 50, the two loss of models decreases and gradually converges. In the FID diagram, both values decline from 120 to lower than 20 after epoch 100, meaning that both of them can generate higher-quality images, the cycleGAN shows a steady downward trend, while the fluctuation of pix2pix is relatively larger. From the perspective of training time, for each training epoch, pix2pix only takes the 90–100 s on average, while cycleGAN takes 300 s on average. As a result, cycleGAN can generate images as high-quality as the pix2pix model, although it requires more training time, it saves a lot of time performing simulation to prepare paired datasets.

However, this study still has some limitations. Although DL prediction results are similar to CFD results in wind direction and speed, which can be used for preliminary design decisions, wind speed results are still not as accurate as CFD results in details. Future research could aim at using CFD samples to train a DL model for more convincing methods. The above methods are the results of software simulation. Wind simulation of the real environment can help us to correct a more accurate DL wind simulation workflow. Moreover, cycleGAN can only be used for preliminary and simple wind simulation judgment at present. The application at different scales has not been tested. Further research will utilize a larger amount of data to evaluate whether its accuracy is related to the amount of data that achieves better performance than the pix2pix model.

5 Conclusions

In this study, with the comparison of the results from the cycleGAN model and pix2pix model, we discover that cycleGAN can predict wind speed as accurate as pix2pix and it has a relatively stable state in the training process but only requires more training time. Besides, from the perspective of dataset preparation, it can greatly save time because it enables designers to input unpaired datasets of city block and wind simulation images, meaning that they don't need to perform complex and repetitive wind simulations. This study offers new thinking that enables designers to choose a faster and more accurate workflow in deep learning for the wind environment.

References

1. Toparlar Y, Blocken B, Maiheu B, Van Heijst GJF (2017) A review on the CFD analysis of urban microclimate. Renew Sustain Energy Rev 80:1613–1640
2. Milne-Thomson LM (1973) Theoretical aerodynamics. Courier Corporation
3. Blocken B, Janssen WD, van Hooff T (2012) CFD simulation for pedestrian wind comfort and wind safety in urban areas: general decision framework and case study for the Eindhoven University Campus. Environ Model Softw 30:15–34
4. Stam J (1999). Stable fluids. In Proceedings of the 26th annual conference on Computer graphics and interactive techniques, pp 121–128. (1999, July)
5. Mokhtar S, Beveridge M, Cao Y, Drori I (2021) Pedestrian wind factor estimation in complex urban environments. In: Asian conference on machine learning, pp 486–501. PMLR. (2021, Nov)
6. Waibel C, Bystricky L, Kubilay A, Evins R, Carmeliet J (2017) Validation of Grasshopper-based fast fluid dynamics for air flow around buildings in early design stage. Building Simulation, pp 7–9
7. Calzolari G, Liu W (2021) Deep learning to replace, improve, or aid CFD analysis in built environment applications: a review. Build Environ 206:108315
8. Kim B, Lee DE, Preethaa KS, Hu G, Natarajan Y, Kwok KCS (2021) Predicting wind flow around buildings using deep learning. J Wind Eng Ind Aerodyn 219:104820
9. Brownlee J (2020) A gentle introduction to cyclegan for image translation. Machine Learning Mastery.
10. Kutz JN (2017) Deep learning in fluid dynamics. J Fluid Mech 814:1–4
11. Jin X, Cheng P, Chen WL, Li H (2018) Prediction model of velocity field around circular cylinder over various Reynolds numbers by fusion convolutional neural networks based on pressure on the cylinder. Phys Fluids 30(4):047105
12. Guo Y, Liu Y, Oerlemans A, Lao S, Wu S, Lew MS (2016) Deep learning for visual understanding: a review. Neurocomputing 187:27–48
13. Farimani AB, Gomes J, Pande VS (2017) Deep learning the physics of transport phenomena. arXiv:1709.02432
14. He Y, Liu XH, Zhang HL, Zheng W, Zhao FY, Schnabel MA, Mei Y (2021) Hybrid framework for rapid evaluation of wind environment around buildings through parametric design, CFD simulation, image processing and machine learning. Sustain Cities Soc 73:103092
15. Mokhtar S, Sojka A, Davila CC (2020) Conditional generative adversarial networks for pedestrian wind flow approximation. In: Proceedings of the 11th annual symposium on simulation for architecture and urban design, pp 1–8. (2020, May)
16. Delua J (2021) Supervised vs. unsupervised learning: what's the difference. Artif Intell Retrieved 5(09)
17. Zhu JY, Park T, Isola P, Efros AA (2017) Unpaired image-to-image translation using cycle-consistent adversarial networks. In: Proceedings of the IEEE international conference on computer vision, pp 2223–2232
18. Sidefx.com. (2022). PDG|SideFX. https://www.sidefx.com/products/pdg/. Accessed 12 Mar 2022
19. Kaushik V, Janssen P (2015) Urban Windflow: Investigating the use of animation software for simulating windflow around buildings
20. Nodado CD, Yogiaman C, Tracy K (2021) Towards wind-induced architectural systematization-demonstrating the collective behaviour of urban blocks as a design asset
21. Carneiro T, Da Nóbrega RVM, Nepomuceno T, Bian GB, De Albuquerque VHC, Reboucas Filho PP (2018) Performance analysis of google colaboratory as a tool for accelerating deep learning applications. IEEE Access 6:61677–61685

22. Isola P, Zhu JY, Zhou T, Efros AA (2017) Image-to-image translation with conditional adversarial networks. In: Proceedings of the IEEE conference on computer vision and pattern recognition, pp 1125–1134
23. Heusel M, Ramsauer H, Unterthiner T, Nessler B, Hochreiter S (2017) Gans trained by a two time-scale update rule converge to a local nash equilibrium. Adv Neural Inf Process Syst 30
24. Dowson DC, Landau B (1982) The Fréchet distance between multivariate normal distributions. J Multivar Anal 12(3):450–455

Demand-Driven Distributed Adaptive Space Planning Based on Reinforcement Learning

Jiaqi Wang(✉) and Wanzhu Jiang(✉)

School of Architecture, South China University of Technology, Guangzhou, China
{ucbqwj0,ucbq121}@ucl.ac.uk

Abstract. In the second digital turn, the architecture driven by big data logic is gradually shifting from a traditional static entity to an intellective living organism. This paper explores a space planning algorithm that applies reinforcement learning to the multi-agent system to achieve condition adaptability. This algorithm contains an inclusive environment and programmable agents that represent independent spaces. Through reinforcement learning, personalized space needs are quantified as the agent's Space Schema, which can provide adaptive behavior strategies to adjust volumetric room boundaries. The spatial organization emerges in multi-agent competition, guided by the Negotiation Schema, realizing the dynamic equilibrium of spatial relations and the stable maximization of collective interests. Through real-time interaction and distributed decision-making, this bottom-up method defines a new architectural paradigm that continuously changes based on demands with its high degree of variability, adaptability and evolvability.

Keywords: Demand adaptability · Space planning · Multi-agent system · Distributed decision-making · Deep reinforcement learning

1 Introduction

We live in an era with excess data, characterized by dynamism and complexity. Our environment is no longer seen as fixed, or shaped by forces beyond our control, but as in constant and noticeable change [2]. On the one hand, the changes in natural and social environments lead to unstable site conditions. On the other hand, the trends of diversification and customization in spatial needs result in various design requirements. However, architects always regard the environment and users as constant factors and acquiesce to the static nature of spaces. To cope with this stereotype, we need to consider the real-time adaptability of architectural design. While realizing automatic generation, architecture also needs to interact with designers, users, and environments to achieve a state of autonomy.

It is possible to realize the above goals in the second digital turn, especially by applying generative design methods and artificial intelligence in architecture. Under this influence, the architecture system can form intelligence to cope with changes in external

These authors contributed equally to this work and should be considered co-first authors.

P. F. Yuan et al. (eds.), *Hybrid Intelligence*, Computational Design and Robotic Fabrication,
https://doi.org/10.1007/978-981-19-8637-6_23

conditions, realizing the transition from small data logic based on deterministic formulas to big data logic according to mass computation.

This research is part of the frontier study on autonomous architecture systems to cope with the ever-changing environment and complex design requirements. It aims to combine multi-agent system generation methods and machine learning technology to endow the digital architecture system with autonomy, establishing the Distributed Adaptive Space Planning Algorithm (DASP) that can respond to changes in diverse design problems in real-time and adaptively reconfigure space conditions and organizations. As an intelligent architectural framework, this bottom-up responsive design generation method includes the needs and wishes of users, the coordination and integration of AI, and the thoughts and notions of designers, forming a hybrid Intelligence (Fig. 1). Through distributed decision-making, it can maximize group goals while meeting individual needs, defining a new paradigm of architecture that adapts to human willingness continuously.

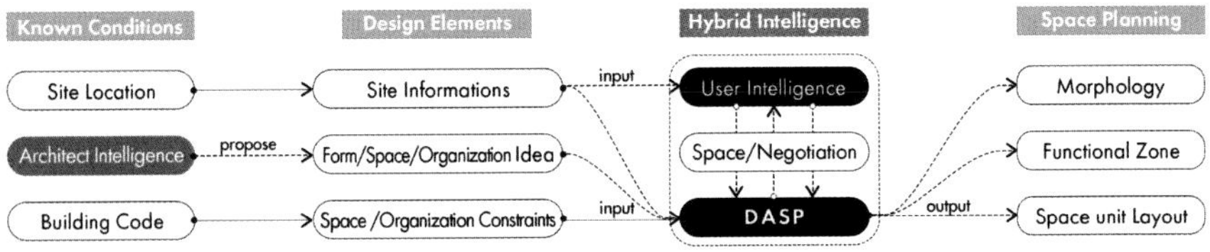

Fig. 1. The space planning process based on the DASP method

2 Literature Review

Space planning is a main task of architectural design: organizing space elements appropriately to meet a set of standards or achieve certain purposes, whose automation is what computational design hopes to achieve initially. However, the non-determinism of space planning is one of its main contradictions with computational methods, mainly manifested in (1) Design elements are discrete, unstable, and unstructured; (2) Incomplete design constraints prevent a single solution; (3) Conflicting design goals indicate there is no perfect solution; (4) Design evaluations are influenced by subjective styles and preferences.

The multi-agent system is an appropriate computational model to deal with indeterministic space planning problems. Its application can be divided into two types: the physical system in which agents represent space units, and the living system in which agents occupy space voxels. Guo [1] adopted three shapes of agents to represent spaces with different functions, whose positions are organized by attraction and repulsion forces, guiding the multi-story building generation. Meyboom [4] controlled the agents to occupy or release voxels through the principle of stigmergy. The agents interact indirectly with pheromone to arrange the position and shape of each space and form the building's spatial layout. For these two algorithm models, the former has a simple framework and better controllability, but its insufficient organizational accuracy leads to the low information capacity and distributed decision-making ability. In contrast, the advantage of the latter model lies in the interactivity and extensibility of its framework, which can add diverse spatial and environmental requirements. But due to the weak controllability of its agent behaviors, the planning results are difficult to echo with traditional spatial forms. Thus, this research tends to explore more operating rules and organizational forms of agents based on the latter model, focusing on agents' control strategies for diverse spatial requirements.

Faced with the intractable problems of indeterministic algorithms, the implantation of reinforcement learning is effective. As one of the three basic machine learning paradigms, reinforcement learning is a process of training the agent to continuously adjust its own strategy to take the optimal action to maximize the accumulated reward through interactions with the environment. The current application of reinforcement learning in CAAD focuses on topics of intelligent control [3, 5] and machine feedback [8], which also shows great potential in space planning research. NoMAS [6], a project in UCL, trained a library of space modules as agents which can be reconfigured according to user needs, developing an algorithm model to generate the residential complex. Pedro Veloso [7] trained a set of spatial agents to autonomously complete custom spatial configurations and interact with designers in real-time to generate floor plans. In the above work, reinforcement learning endows the agent with specific behavior patterns when dealing with the design target. This study also implants it into the multi-agent system to enhance their control, training agents to coordinate different spatial demands and respond in real-time.

3 Methodology

Distributed Adaptive Space Planning (DASP) takes the multi-agent system as a framework, adopting various methods including reinforcement learning and stigmergy principle, comprehensively combining bottom-up local agents with top-down global constraints. It connects users and the environment to accommodate dynamic activity needs, diverse design specifications, and critical site conditions, becoming an interactive, adaptable, and expandable space planning method (Fig. 2). The process can be divided into four steps: (1) the construction of the planning environment; (2) the representation of the space agent; (3) the implantation of the Space Schema; (4) the implementation of the Negotiation Schema.

3.1 Planning Environment

DASP uses a 3D grid formed by discrete spatial voxels as an inclusive planning environment. Each node's size is determined by the spatial scale of the agent and the resolution required for planning. In this experiment, 0.3 m is selected as the side length of the basic unit to respond to the ergonomic scale and acquire the refined planning results. Each vectorized point in this matrix has two basic information of position and color, where the initial color is ($r = 0$, $g = 0$, $b = 0$), which can be dyed by the agent and its pheromone. In addition, each node can also be assigned multi-level environmental information, imported in the form of formulas or bitmaps, and superimposed on each voxel.

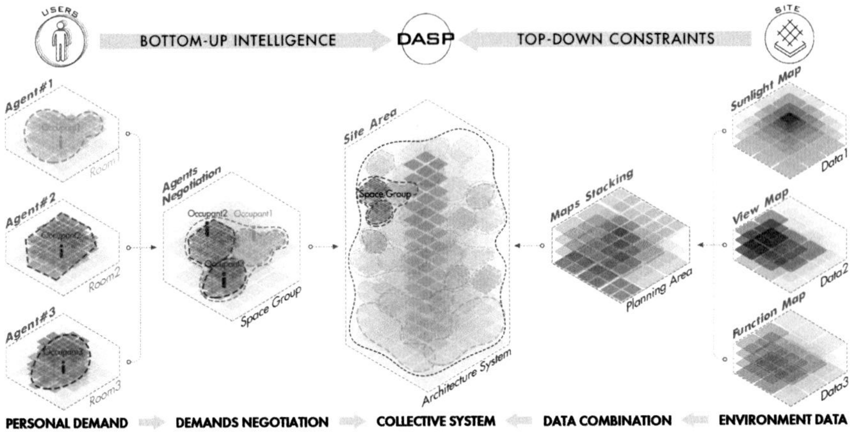

Fig. 2. The basic framework of DASP: integration of local agents and global constraints

This preliminary pheromone that guides the activities of agents can be transformed into a data set of [0, 1] in proportion as the top-down global constraints.

Different global constraints are divided into two categories in DASP, including masking-stack and mapping-stack (Fig. 3). The data in the masking-stack demarcates the areas that do not participate in the planning, including terrains and obstacles, etc. The mapping-stack contains physical environment maps (sunlight maps, wind maps, noise maps) and social environment maps. The conversion ratio (T_E) of different types of data (E) can be set according to empirical research or local requirements and unified into the demand degree of the same measurement ($E_D = E \times T_E$). These global constraints can be used as a manifestation of group consciousness, setting the collective goals of the multi-agent system and preventing its autonomous behavior from falling into chaos.

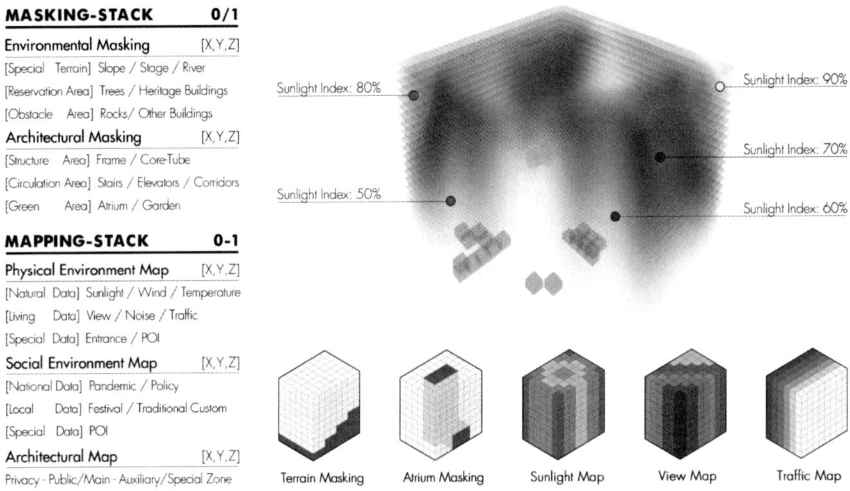

Fig. 3. Global constraints, left: detail parameters of two stacks, right: some examples

3.2 Space Agents

An agent representing a user space is embodied as a point cloud collection of nodes with a particular shape in the environment. In the beginning, the agent will be initialized as a cell at a specified position, following preset rules to occupy or release voxels in the environment and gradually forming a closed geometry, that is, the expected space. When the algorithm runs, the agent will emit pheromone and receive pheromone from adjacent agents and the environment to perceive the surrounding state and determine the expansion behaviors. In addition, two agents cannot occupy the same node simultaneously but can negotiate and reconfigure the occupied area. In this framework, if the design requirements and specifications are properly translated into the operating rules of the agent, it is possible to simulate any spatial state and its combination in real-time (Fig. 4).

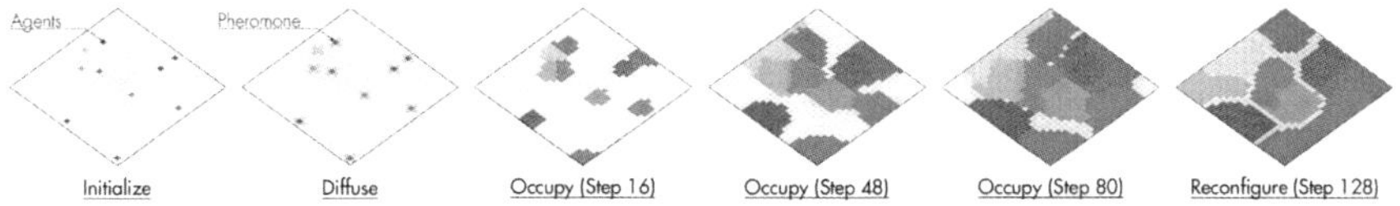

Fig. 4. Planning process of the agents

Each agent in DASP has the following basic parameters.

1. The color (R, G, B) is defined as a user hue, which can link the user characteristics. These parameters are obtained from questionnaires, which can provide personalized information and help organize the agent relationships.
2. The pheromone carries color information, starting from the center of the occupied shape, picking up the three-dimensional Von Neumann Neighborhood iteratively, and diffusing outwards to realize the gradual contagion of the nodes.
3. The capacity represents the upper limit of the node number that the agent can accommodate, corresponding to the space volume. For example, a standard room of 6 m × 3 m × 3 m contains 2000 voxels with a side length of 0.3 m. When the occupied nodes reach capacity, the calculation will continue, and the agent can update the occupied area by releasing internal voxels and occupying new ones.
4. The expansion rules guide the agent to select suitable occupation voxels, controlling the growth mode and the generation result. Each adjacent node (P_N) of the occupied area P is substituted into the customizable function F_t (x, y, z), and the node (P_{Ni}) with the smallest or largest result (F_t (P_{Ni}) = F_t (P_N)min/max) is chosen as the next occupied one. Based on the different design objectives, various calculation methods can be set to promote diverse growth behaviors.

3.3 Space Schema

Each agent is assigned local identities to develop a real-time adaptive connection between abstract agent behaviors and specific user needs. The concept of Schema is introduced into DASP as an agent attribute, which reflects its instinctive behaviors in response to condition changes or external disturbances.

The significance of the Space Schema (Fig. 5) is to enable the agent to generate a specified space according to personalized requirements. It employs three control elements of volume, proportion, and form. The volume determines the spatial scale, the proportion indicates the aspect ratio, and the form describes the particular shape. The setting of the Space Schema realizes a comprehensive description of a space and ensures that the user's demand can be transformed into a combination of several parameters to direct the agents' planning behaviors.

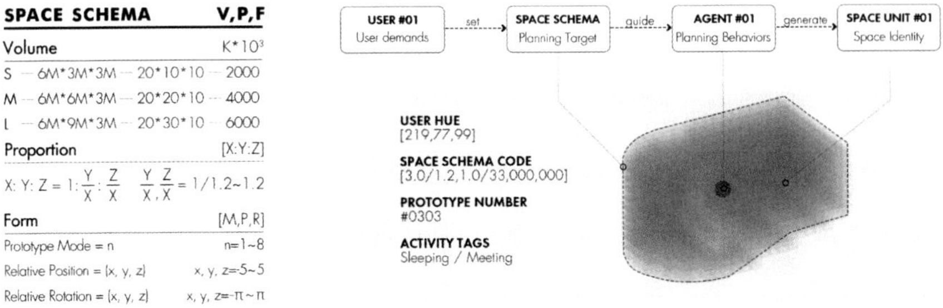

Fig. 5. Space schema, left: detail parameters, right: principle and example

This study innovatively introduces reinforcement learning as a decision-making mechanism to embed the Space Schema into the agent (Fig. 6). In the experiment, the randomness of V, P, and F is limited to preliminarily verify the method's feasibility. The space volume is set to 2–4 ($\times 10^3$) voxels, the rate(y/x) and rate(z/x) choose 5 values from 1.2^{-1} to 1.2, and the form takes 8 geometries of different themes, generating a total of 600 parameter combinations. Before each episode starts, the system will initialize a target schema in the planning environment according to a random parameter set.

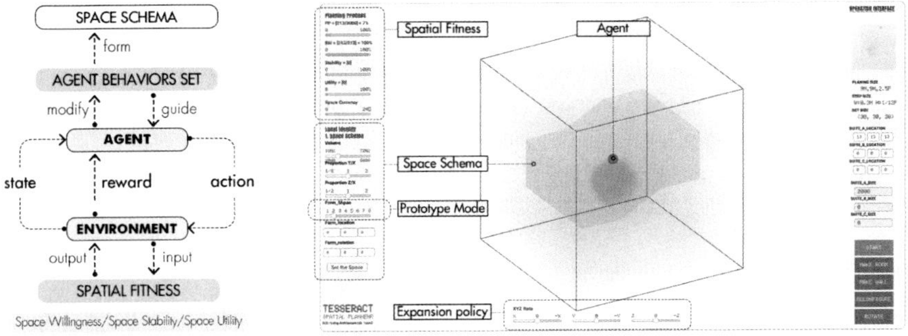

Fig. 6. Left: the principle of reinforcement learning, right: the computational model

In the training framework, the discrete actions were set to 7 to separately increase or decrease the values in the policy (x, y, z) to decide the agent's expansion function and adjust its growth direction and intensity in steps.

The observation contains the information for planning goals and current status. In this experiment, it consists of these parameters: (1) the index of Space Schema as the target status. (2) the agent current state, including its centroid position, the last occupied point position, and the expansion rules (x, y, z). (3) the current planning state, including all the points in the occupied area.

There are three types of rewards in the experiment: (1) real-time rewards, each time the expansion points are inside the target schema, a reward will be added. (2) Periodic rewards, when the number of points that meet the requirements in the agent-occupied area accounts for a certain value (30%, 60%) of the agent capacity (this value is called space willingness), real-time rewards will increase. (3) Target rewards, when the agent reaches capacity and the space willingness exceeds 90%, the planning is completed, and the maximum reward is obtained. We also set up rewards based on structural stability and spatial practicality analyses.

After 6×10^6 times of training, the training curve is stable at the maximum value of 36. In this case, the agent can substantially adapt to the changes in the Space Schema in real-time and complete the corresponding space planning tasks (Fig. 7).

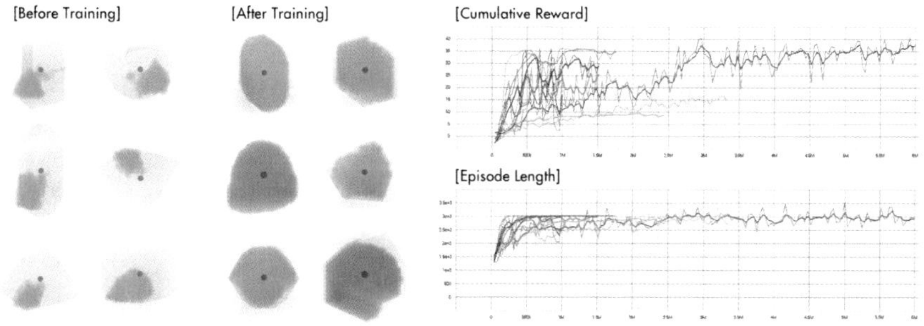

Fig. 7. Left: The behaviors of the agents after training, right: graphs (pink one success) with the cumulative reward and the episode length of the agents (y-axis) over the episode (x-axis)

3.4 Negotiation Schema

Negotiation Schema is another local identity of agents, representing the tendencies in the space organization. It is mainly responsible for regulating the agent's relationships with its neighbors and environment to endow the whole system with adaptability. The setting of the Negotiation Schema (Fig. 8) aims to realize the agent's response to the user's organizational willingness, completing the collective space arrangements through deformation or displacement.

In DASP, the Negotiation Schema consists of three sets of parameters, which describe the organizational relationship at different scales. Among them, the interaction tendency represents the adjacency willingness of the agent with surrounding agents. According to the propensity ranking of "away–reject–none–accept–adjacency", the set of weights K_I contains a set of numbers $\{k_{i1}, k_{i2}, k_{i3}\ldots\}$ in the range $[-2, 2]$ for each surrounding agent. In the algorithm, this relationship is encoded in a group of attractors, which generate attractive or repulsive forces according to the positive or negative values of K_I. These forces make agents away from or close to each other and weaken as the distance

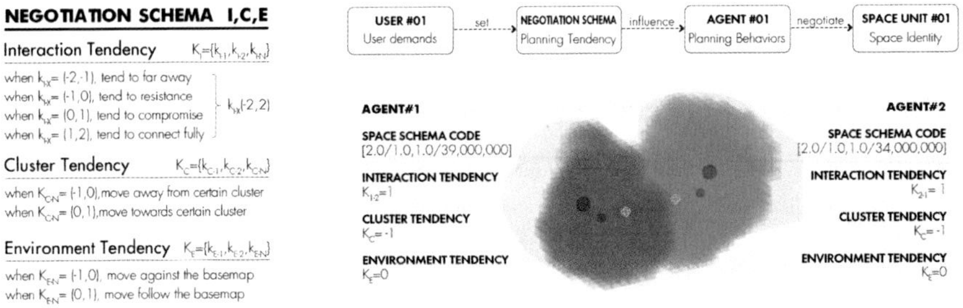

Fig. 8. Negotiation Schema, left: detail parameters, right: principle and example

increases. Each agent will experience a combined force from neighbors to adjust its morphology and steer it toward a suitable social position.

Cluster tendency is more inclined to describe the grouping willingness based on the agent's color. The user can set the tendency weight K_C to similar color groups in the range of $[-1, 1]$, which is multiplied by the pheromone value in the planning area and directly acts on the expansion rule of the agent, to lead the agent to gradually move toward or away from the group in a specified color.

Environment tendency quantifies the agent's willingness for environmental resources. The set K_E, which is influenced by space function and user preferences, includes multiple weights between $[-1, 1]$ corresponding to indexes in global constraints. In this way, the agent's comprehensive environmental satisfaction degree for a certain node can be obtained quantitatively through a weighted average method. Therefore, the environmental resources of the whole system can be reasonably allocated to maximize the utilization rate.

The Negotiation Schema coordinates the agents' relational network and further optimizes the spatial form generated by agents, which greatly expands the adaptability of DASP when dealing with complex demands.

4 Experiment and Application

We validated the space planning potential of DASP when dealing with complex real-time demands in practice through several sets of experiments. First, the simulated daily schedule of an occupant guides the Space Schema for parameter modification. Figure 9 shows the different spatial forms generated by the agent in this continuous planning. When the Space Schema is changed in this experiment, the DASP algorithm can autonomously remove the unnecessary nodes and perform the space planning according to the adjusted parameters. However, for some shapes with acute angles, such as the space at 19:30, the error is still relatively large, related to the precision setting of the agent's expansion rule.

The second experiment was carried out in a range of 12 m × 12 m × 6 m. By placing five agents in this area, the possible negotiation activities of two users in this system were simulated to generate a space group. By adjusting the interaction tendency, we simulated the gradual process of the organizational relationship between two users from no tendency to mutual resistance and mutual acceptance. The parameter changes of the Negotiation Schema trigger agents' various behaviors such as deformation, attraction,

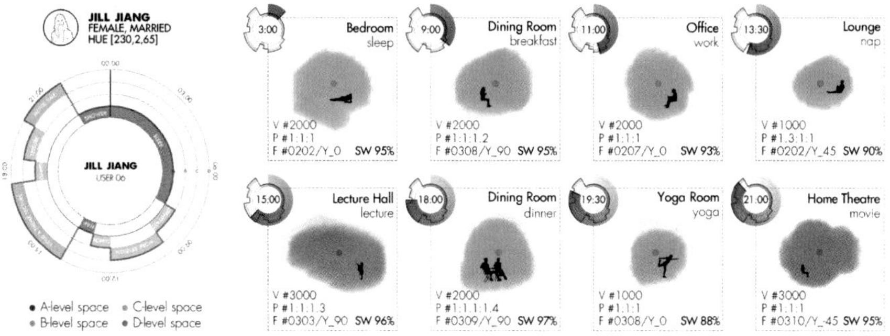

Fig. 9. Experiment 1, left: simulated timetable, right: the space sequence and its parameters

and repulsion. Thus, the space occupancy is negotiated, and different space forms emerge (Fig. 10). Obviously, the growth of the five agents in the first stage does not affect each other, tends to be far away in the second stage, and begins to approach in the third stage.

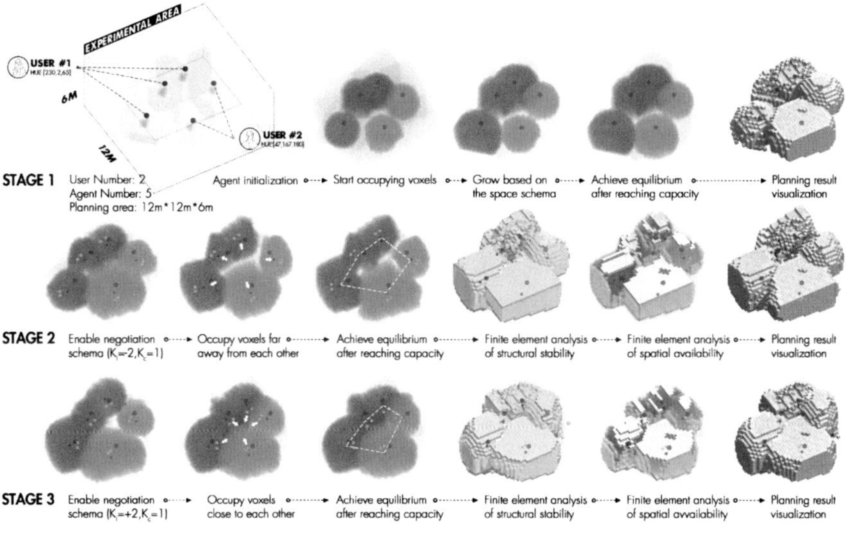

Fig. 10. Experiment 2, the planning process of three stages and their planning results

Finally, this algorithm is applied to developing the adaptive living complex, TESSERACT (Fig. 11). This project uses DASP as the core space design and generation tool, cooperating with several well-designed subsystems like the interactive platform and robotic material. Relying on the DASP's adaptability, interactivity, and expandability, TESSERACT defines a new paradigm of living architecture that adapts to the human will and changes continuously.

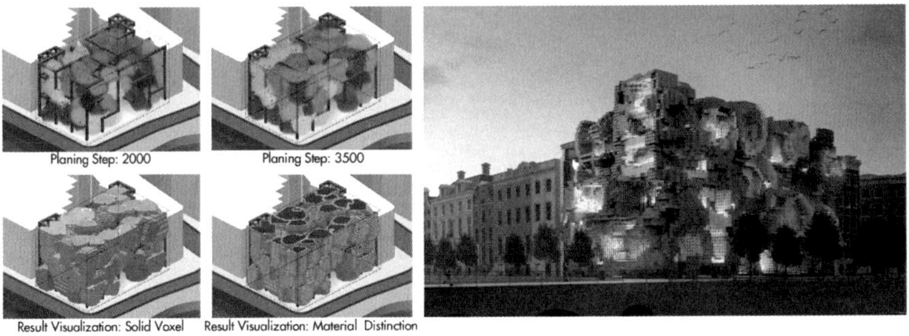

Fig. 11. Project TESSERACT generation, visualization and rendering

5 Conclusion and Future Work

This research proposes an interactive algorithm DASP with demand adaptability, using reinforcement learning to establish a spatial solution that can respond to changes in user needs and environmental conditions in real-time. DASP implants Space Schema, Negotiation Schema, and global constraints to the multi-agent system as linkages to users, communities, and environments, forming a distributed decision-making method. The experiments show that trained agents can respond well to the parameter adjustment and perform adaptive behaviors, revealing the potential of DASP in autonomous space planning. The future work will focus on the limitations of DASP. First, the spatial organization purely based on social relations and user regulation may lead to space configuration problems. Other architectural constraints like topological relations should be considered. Second, the adaptability now totally depends on input prototypes. It is necessary to continuously modify preset prototypes based on user interaction, and guide the transfer of agent's behavioral patterns, realizing the evolution of the Space Schema. Also, using it as a design or analysis tool for architects is a new direction.

Acknowledgements. Project TESSERACT is conducted in UCL B-Pro AD RC3, supervised by Tyson Hosmer, Octavian Gheorghiu, Philipp Siedler, and Ziming He, and developed by Jiaqi Wang, Wanzhu Jiang, Ying Lin, and Zongliang Yu.

References

1. Guo Z, Li B (2017) Evolutionary approach for spatial architecture layout design enhanced by an agent-based topology finding system. Front Archit Res 6:53–62
2. Hanna S (2020) Architecture as agent. Georgia Institute of Technology School of Architecture
3. Hosmer T, Tigas P (2019) Deep reinforcement learning for autonomous robotic tensegrity (ART). In: Proceedings of the 39th annual conference of the ACADIA, Austin, Texas
4. Meyboom A, Reeves D (2013) Stigmergic space. In Proceedings of the 33rd annual conference of the ACADIA, Cambridge
5. Smith S, Lasch C (2016) Machine learning integration for adaptive building envelopes. In: Proceedings of the 36th annual conference of the ACADIA, Ann Arbor

6. Tigas P, Hosmer T (2021) Spatial assembly: generative architecture with reinforcement learning, self play and tree search. arXiv:2101.07579
7. Veloso P, Krishnamurti R (2020) An academy of spatial agents: generating spatial configurations with deep reinforcement learning. In Proceedings of the 38th eCAADe conference, vol 2, Berlin, Germany
8. Xu T, Wang D, Yang M, et al (2018) An evolving built environment prototype. In: Proceedings of the 23rd CAADRIA conference learning, adapting and prototyping, vol 2, pp 207–215

Nolli Map: Interpretation of Urban Morphology Based on Machine Learning

Zhiyong Dong[1(✉)] and Jinru Lin[2(✉)]

[1] School of Architecture, South China University of Technology, Guangzhou, China
zhiyong_Dong@foxmail.com, ar_dongzhiyong@mail.scut.edu.cn
[2] College of Architecture and Urban Planning, Tongji University, Shanghai, China
jadya.lin@gmail.com

Abstract. Nolli map is the earliest diagram tool to simplify and quantify urban form, which most intuitively reflects the spatial layout of tangible elements in the city. The urban morphology contains its inherent evolutionary laws. Exploring the inner rules of cities is helpful for people to conduct urban research and design. Unlike the traditional research methods of urban morphology, the neural network algorithm provides us with new ideas for understanding urban morphology. In this experiment, we label 136 European cities samples in the rules of Nolli map as a training set for machine learning. We use Generative Adversarial Networks (GAN) for multiple mapping experiments. The generated images present recognizable and plausible images of the urban fabric. The results show that the machine can learn the inherent laws of complex urban fabrics, which expands a new applied method for the study of urban morphology.

Keywords: Nolli map · Urban morphology · Deep learning · GAN

1 Introduction

In 1736, Giambattista Nolli, the Italian architect and surveyor, was commissioned by Pope Clement XII to draw a plan of Rome (Fig. 1). Nolli recorded every part of the space of Rome, using complex iconographic schema, illustrative cartographic symbols to depict urban fabric, which comprehensively and effectively display the spatial elements of the city, reveal the urban structure composed of streets, squares, buildings, and show other tangible elements of the city. Nolli map now is the most widely used diagram method of urban morphology [7].

The urban morphology is the external presentation of the city which has developed in periods of years. Thus, the old fabric and the new one are intertwined [13]. In recent years, the development of open data platforms has provided numerous geographic information data resources for the quantitative investigation of urban morphology [18]. However, urban morphology is so complex that it is influenced by factors such as geography, socioeconomics, politics, etc. under nonlinear laws. The urban morphology research based on traditional statistical analysis fails to completely integrate all elements and to grasp the subtle characteristics of urban evolutionary laws.

P. F. Yuan et al. (eds.), *Hybrid Intelligence*, Computational Design and Robotic Fabrication,
https://doi.org/10.1007/978-981-19-8637-6_24

Fig. 1. Nolli map in 1748 and its sections *Source* https://nolli.stanford.edu/

With the rise of artificial intelligence especially deep learning technology, these problems may be solved. An important feature of deep learning is that machines can automatically extract general features from data through data learning, instead of extracting data features in a manual way and inputting them into machines like traditional research methods. It will provide a new way of urban morphology study. In this way, cities can be more accurately understood and represented.

2 Background

Before starting the experiments, we summarize the feature of Nolli map, propose the research implications, and analyse the existing researches on deep learning for city form.

2.1 Nolli Map and Urban Morphology

Nolli chose the traditional Roman scale of 1:2750 to draw. The information of the whole map can be summarized according to different parts of the city and divided into the following three different spatial categories (Table 1). The first category is urban built-up area, which is based on the figure-ground plan concept: (1) public spaces such as streets and squares are depicted with a white background; (2) inaccessible private buildings are indicated with black blocks; (3) open artificial green lands are illustrated with light grey. The second category comprised natural environment elements like different types of hills, rivers, plants, which are distinguished through different fabrics. In addition, Nolli also made sketch drawings to describe municipal infrastructures such as bridges and street furniture in details.

From the perspective of urban design research, Nolli map can more comprehensively depict urban form. Nolli map distinguishes the urban fabric by figure-ground plan, fully indicating the relationship between the private building and the public space. Comprehensive information and clear morphological features would help the machine to learn the organic fabric of the city under machine learning method.

Other diagram methods for urban morphology may be not suitable for the rapid learning by machines. Conzen divided urban plans into buildings, plots and streets,

Table 1. Categories of diagram theory in Nolli map

Urban built-up area				Natural environment elements			Municipal infrastructure and street furniture		
Street	Square	Buildings	Artificial green area	Hill	River	Plants	Bridge	Fountain	Drain
			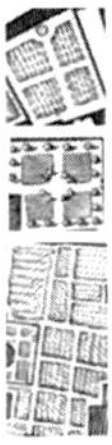		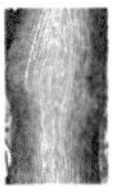		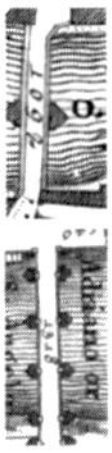	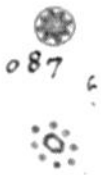	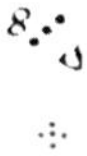

marked them with linear elements and different fabrics [1], whose description of urban structure ignores other city elements, urban planners mark various macro quantitative indicators such as plot ratio, functional density and height in different colours, which do not reflect spatial characteristics. The research on the relationship between spatial policy and urban form use geometric elements of point, line and surface, which cannot show a tangible form of the urban fabric [9].

Research in the past enriches the diagram of Nolli map and proposes future research trends in digitization and machine learning. Venturi created a Nolli map of the Las Vegas Strip, integrating road elements into the map. Huimin Ji analyzed and proposed a new urban public space mapping method, and pointed out that Nolli-type maps can be efficiently created by using the data of urban maps and indoor maps in various digital maps [6]. Hwang and Koile [4] proposed an illustration of the public space of Boston's main streets, and discussed the role of machine learning techniques in that diagram.

2.2 Applications of Deep Learning in Urban Design

At the urban level, the initial applications of deep learning mainly focus on the generation of urban streetscape maps [15] and quick previews of urban block renovations [17]. Subsequent related research gradually emerges in the generation of floor plans at different scales. Liu et al. [8] used cGAN models trained from a database containing 85 pairs of samples to generate building layouts at different building densities within a plot. Shen et al. [14] applied GAN to create urban design solutions in small plots. Pasquero and Poletto [11] used cycleGAN to explore "non-human" urban forms by transferring biomorphic styles to the urban fabric. Pan et al. [10] trained a large sample of northern neighborhoods to generate diverse layouts within a plot [10]. Fedorova [2] used the model trained by five existing urban environments to observe the possibility of style transfer between different cities, and presented quantitative and qualitative evaluations of the results (Table 2).

Through the analysis of the above existing studies, it is easy to find that: (1) Most of the studies are not from an urban morphology perspective, but from image fitting

Table 2. Comparison of various studies in the research

Paper	Authors	Model	Application scenario	Sample size	City range	Image resolution
Urban Design process with conditional generative adversarial networks [8]	Yuezhong Liu, Stouffs Rudi	cGAN	Small plots: road boundaries generate building rows; not clearly imaged	85	~0.5 km (Image metering)	512 × 512
Machine learning assisted urban filling [14]	Jiaqi Shen, Chuan Liu, Yue Ren, Hao Zheng	Pix2PixHD	Small plots: road network generates building rows	/	~0.5–0.7 km (Image metering)	256 × 256
DeepGreen–coupling biological and artificial intelligence in urban design [11]	Claudia Pasquero, Marco Poletto	cycleGAN	An attempt to urban style transfer	/	/	256 × 256
Suggestive site planning with conditional GAN and urban GIS data [16]	Runjia Tian	Pix2Pix	Single plot: road boundaries generate building layouts	4400	~0.5 km (Image metering)	256 × 256
A Preliminary study on the formation of the general layouts on the northern neighborhood community based on GauGAN diversity output generator [10]	Yuzhe Pan, Jin Qian, Yingdong Hu	GauGAN Pix2PixHD	Single plot: road boundary generates building row; single function: residential area	167	0.11–0.33 km	512 × 512
Generative design of urban fabrics using deep learning [12]	Jinmo Rhee, Pedro Veloso	WGAN	Small plots: random urban fabric generation from noises; unsupervised learning	45,852	~0.15 km (Image metering)	512 × 512
Generative adversarial networks for urban block design [2]	Stanislava Fedorova	pix2pix	Small plots: road boundaries generate building layouts	/	~0.4 km (Image metering)	256 × 256

in computer graphics; the studies ignore the laws of urban evolutionary development. (2) Most of the studies stay in the relationship between boundaries and building layout, ignoring other elements which influence the development of the city, such as water system, green space. (3) The urban scope of the experimental sample is small, mostly within 0.5 km × 0.5 km, which is more like the internal generation of land parcels rather than urban-level studies; it is also because of the limitation of the image processing capability of deep learning models, that the image resolution is too small to show more details of the city.

2.3 Objective

The research is to explore the strategy of using GAN model to realize the automatic generation design of urban form. According to the evolution laws, the corresponding relationship between the pairs of data sets can be sorted out. The urban fabric is automatically generated based on the city's underlying structure. We adjust the model framework of GAN to fit high-resolution images, which enables a wide range of Nolli map images to be learned. The machine cognition of the laws of the overall shape of the city has made a breakthrough on a large scale, which proves that deep learning has great potential in urban morphological researches and design applications.

3 Methodology

The main process of exploration on machine learning generation of urban morphology is as follows: (1) Database establishment: Select city data which meet the standard and collect their relevant information. (2) Sample processing and labelling: Redraw and label samples on basis of Nolli Map. (3) Training and testing: Input one-to-one corresponding sample sets to train and test the machine learning model. (4) Evaluation and modification: Evaluating the results and putting forward further adjustment of labeling method to improve the final generation.

3.1 Model Architecture

In this experiment, based on the data type and the characteristics of neural networks, Image-to-Image Translation with conditional Generative Adversarial Networks [5] are used as the main learning model. Taking this model as a basis, we adapt its model architecture, such as the number of neural network layers, to compute images at higher resolutions (1024 $\times$ 1024).

The concept of Generative Adversarial Networks (GAN) was first proposed in 2014 by Ian J. Goodfellow, Jean Pouget-Abadie, Mehdi Mirza, and other scholars [3]. GAN consists of a Generator and a Discriminator. In the training process, the Generator first generates an alternative image from the latent space and passes it to the Discriminator. The Discriminator takes either the real image or the alternative image as input and tries to distinguish whether the current input is the real data or the alternative data. After the mutual game between Generator and Discriminator, the GAN model completes the effective learning of the data and finally synthetize the high-quality fake images.

3.2 Dataset Construction

According to the experimental objectives and the requirement of the deep learning model, we collect the urban data that meet the requirements; then sort, analyze and label them.

The original city data source is the open data from OpenStreetMap, and we use QGIS and python to download the OSM city data. We select 136 European city samples, which can represent the typical urban morphological feature (Fig. 2).

After the city data is collected, we complete the data analysis and graphical labelling of the data set (including training set A and training set B) and the export of images in the data processing stage. We make the cleaning of the city elements in the OpenStreetMap attribute table, and label the main elements of the city in the rules of Nolli Map.

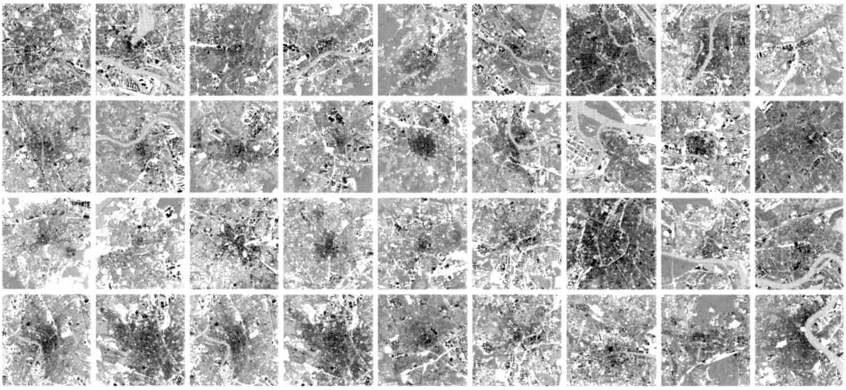

Fig. 2. Part of city dataset

In training set A, the basic data of roads, green areas, and water systems that affect the development of the city are retained in the OSM data attributes, and the artificial parts of green areas and water systems are deleted and the natural green areas are dissolved and expanded; in training set B, the data of existing buildings, green areas and water systems are retained in the OSM data attributes. Through the learning from training set A to training set B, the influence of the underlying information on urban morphology during urban evolution is explored.

The labeling method of the data is transformed from the Nolli map. In the labeled image (Fig. 3), different elements are represented by different labeling symbols: black blocks are buildings; diagonal filled and gray areas are green lands; dotted areas are water bodies; and blank areas are the places that are accessible to citizens.

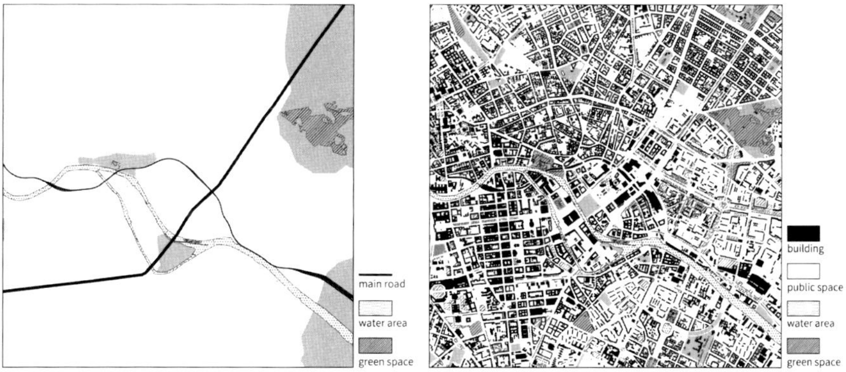

Fig. 3. Training set A &Training set B

3.3 Training and Testing

After the construction of training sets A and B, the images of the training dataset are fed into the optimized GAN separately. To achieve the goal of learning large-scale urban fabric, the model is adjusted to handle and run 1024 × 1024 pixels images. In the optimizing process, we adjust the number of net layers and the number of neurons in each layer of the neural network, to help the computer power to meet with requirements of high-resolution images and avoid the frequent overfitting phenomenon in machine training.

During the training process, the generator and the discriminator play against each other. A higher discriminator loss value and a lower generator loss value means that the training process tends to succeed. The training gradient is set to a constant learning rate before 200 epochs, and a decaying learning rate is used after 200 epochs, which facilitates better fitting of the data. From the loss images, it can be found that the losses of the generator and discriminator stabilize after 680 epochs. The loss values of the generator and discriminator during the training process are recorded in Fig. 4.

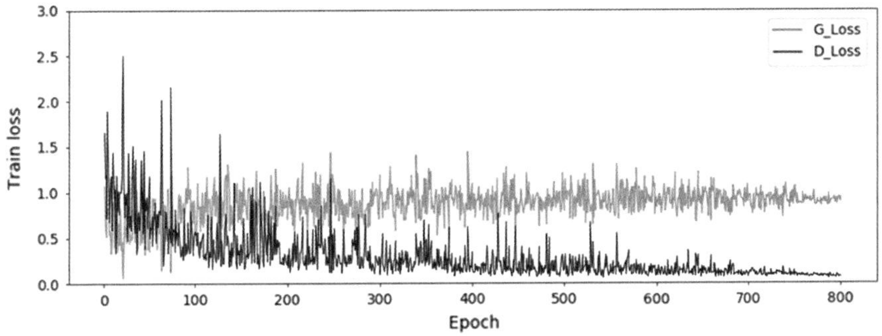

Fig. 4. Generator loss (G_LOSS) and discriminator loss (D_LOSS) during training

In addition, we set a monitor web to record and display the process of the generated images during the training. Figure 5 shows that the generated images are blurred before the 70th epoch; the generated images at the 70–470th epoch recorded are filled with the repeating blocks, which lacks the diversity of urban fabric; the generated images from the 470–770th epoch recorded are more reasonable and the city streets are more obvious. In the 770th epoch, the machine has grasped the more obvious urban morphological laws. Finally, we choose the 800th epoch trained model as the urban generated experimental model. After the model training is completed, inputting the testing data, the new urban form image responding to the base conditions can be quickly output.

4 Analysis of Results

Meantime, we make labelled maps that contain urban basic data from 25 European cities as the testing data inputting into the model. Then the synthesized city image is generated.

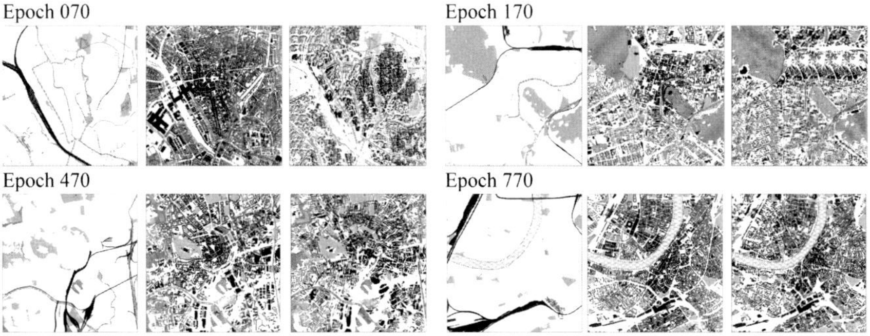

Fig. 5. Output results for each training epoch (input image; real image; synthesized image)

Compared with the real city's image, though the synthesized urban fabric is different, it retains the typological characteristics of the urban fabric (Fig. 6). In Frauenkirche's fabric comparison, the baroque radial street pattern in the real city is replaced by the natural organic pattern, while the synthetic fabric retains the morphological structure with a ring center and several radial axes as the skeleton. In the synthesized image of Leeds, the shape of the blocks and the shape of the secondary road network are quite different from the real ones, but the buildings layout is the same: the east and south parts are arranged with regular plots and large-scale buildings, which reveals the industrial area; the north part is arranged with residential areas with dot-like and strip-like buildings. In the comparison of the three pairs of images, the scales of buildings and street blocks are reasonable and the variation patterns are similar. The urban center is extremely in high density, with recognizable medieval city boundary patterns.

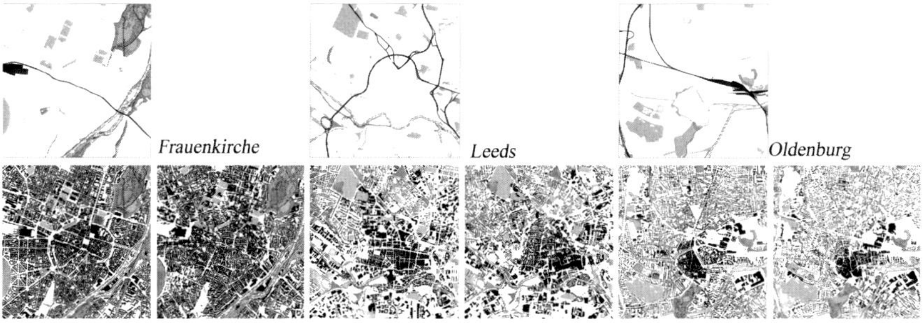

Fig. 6. Comparison between the real images and the generated images

We also input the urban initial condition to get the virtual city image (Fig. 7). The layout of the images maintains the main characteristics of the European cities: the street pattern conforms to the skeleton structure of the city. Most of the streets inside the skeleton have natural and organic morphological features with small radial structures inside; the street block scale is 50–100 m; high-density blocks exist in the center and low-density blocks are in the suburbs; buildings in the central area are dominated by enclosure groups, and more independent buildings appear in the suburbs. The simulated urban fabric also effectively responds to natural information such as water and green space: the building layout responds to the "Fixation line" elements (arterial roads, rivers) of the urban form; the edges of natural green spaces are eroded by buildings; more green space appears nearby the water bodies; buildings in the center enclose clear patterns of plazas.

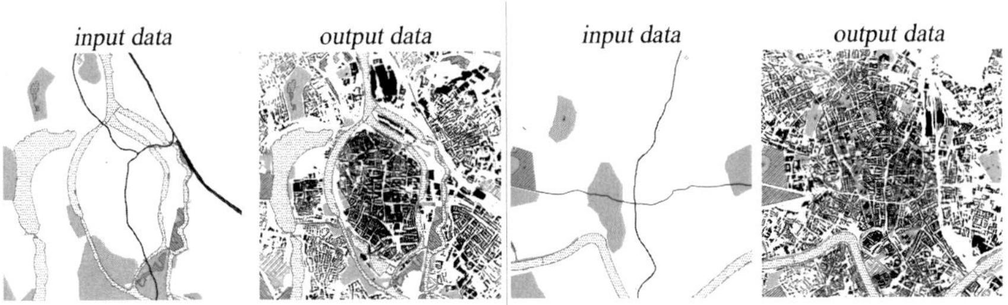

Fig. 7. Generated urban form

5 Conclusion

This experiment basically completes the generation of city form based on machine learning, and proves that the data-driven generative method is a new way of the cognitive research of cities. The diagram representation of Nolli Map fits well with the data training process of deep learning, and the clarity of figure-ground diagram facilitates the training of the city plan pattern and the features of the elements by the machine. Labeled city data is fed into GAN for training and rapidly predicts the city form based on the initial conditions. This experiment proves that Nolli Map as a classical urban research method still has important value in the era of artificial intelligence; machine learning will inject new vitality into urban morphology research.

References

1. Conzen MRG (1960) Alnwick, Northumberland: a study in town-plan analysis. Trans Pap (Inst Br Geogr) 27:111–122
2. Fedorova, S. (2021). Generative adversarial networks for urban block design. arXiv:2105.01727arXiv:2105.01727
3. Goodfellow I, Pouget-Abadie J, Mirza M, Xu B, Warde-Farley D, Ozair S, Courville A, Bengio Y (2014) Generative adversarial nets. Adv Neural Inf Process Syst 2014:2672–2680
4. Hwang JE, Koile K (2005) Heuristic Nolli Map: representing the public domain in urban space. In: Proceedings of CUPUM05, vol 1, p 16
5. Isola P, Zhu JY, Zhou T, Efros AA (2017) Image-to-image translation with conditional adversarial networks. In: Proceedings of the IEEE conference on computer vision and pattern recognition, pp 1125–1134
6. Ji H, Ding W (2021) Mapping urban public spaces based on the Nolli map method. Front Archit Res 10(3):540–554

7. Li M, Feng J (2017) An analysis of the value of the Nolli Map's Analytical methods. New Archit 11–16
8. Liu Y, Rudi S, Yang Y (2018) Urban design process with conditional generative adversarial networks. Archit J 09:108–113
9. Liu Q, Ding W (2012) The graphical method and its significance in the study of urban fabric form. Architect 5–12
10. Pan Y, Qian J, Hu Y (2021) A preliminary study on the formation of the general layouts on the northern neighborhood community based on GauGAN diversity output generator. Springer Singapore, Singapore, pp 179–188
11. Pasquero C, Poletto M (2020) DeepGreen-coupling biological and artificial intelligence in urban design. In: 40th annual conference of the association of computer aided design in architecture (ACADIA), pp 668–677
12. Rhee J, Veloso P (2021) Generative Design of Urban Fabrics Using Deep Learning. In: 26th international conference on computer-aided architectural design research in Asia (CAADRIA), Hong Kong, pp 31-40
13. Saint A (1993) The city shaped: urban patterns and meanings through history
14. Shen J, Liu C, Ren Y, Zheng H (2020) Machine learning assisted urban filling. In: 25th international conference on computer-aided architectural design research in Asia (CAADRIA), Bangkok, Thailand, pp 679–688
15. Steinfeld K (2019) GAN Loci: imaging place using generative adversarial networks, Austin, pp 392–403
16. Tian R (2021) Suggestive Site Planning with Conditional GAN and Urban GIS Data. In: Proceedings of the 2020 DigitalFUTURES-The 2nd International Conference on Computational Design and Robotic Fabrication (CDRF 2020), Shanghai, pp 103–113
17. Xu H, Li L (2021) Research on analysis and generation of street view color based on deep learning. In: 2021 national conference on architecture's digital technologies in education research, Wuhan, China, pp 120–126
18. Ye Y, Zhuang Y (2016) The raising of quantitative morphological tools in urban morphology. Urban Design 4(006):56–65

Developing a Hybrid Intelligence Through Hacking the Machine Learning Neural Style Transfer Process for Possible Futures

Ralph Spencer Steenblik(✉)

Indiana University Bloomington, 333 2nd St, Columbus, IN 47201, USA
steenblik@phi.archi

Abstract. This article highlights work using machine learning in collaboration with designers for speculative world building. The process is unique because of the feedback loop, between the designer and the computational process. World-building is a speculative practice and requires vision and courage on the part of the designer. Working with machine learning neural style transfer (NST) allows the designers to consider possibilities humanity may not otherwise allow ourselves to imagine. This is important because human imagination paves the path for the future of humankind. Imagining a sustainable future requires considering unconventional solutions. Imagining non-probable futures allows humanity to glean desirable aspects to strive for. Even if a conceived future is impossible within the built environment, there are many opportunities for people to inhabit these environments virtually. Letting yourself get lost in these places is a form of travel, even when conditions limit one's ability to physically do so.

Keywords: Hybrid intelligence · Cognitive prosthetic · Machine learning · Schematic design · Possible futures · Sublime · Neural style transfer · Neural networks · AI in design · World building · Future · Film · Virtual reality · Urban planning · Science · Science fiction · Climate change · Humanity · Design workflow · Design exploration · Early design stage

See (Fig. 1).

P. F. Yuan et al. (eds.), *Hybrid Intelligence*, Computational Design and Robotic Fabrication,
https://doi.org/10.1007/978-981-19-8637-6_25

Fig. 1. "Quantum Clouds" by the author, through collaging several NST along with post processing inspired by Archigram.[1] See Fig. 5 for more detail about the NST process for this image

1 Introduction

This work outlines an approach to using neural style transfer (NST) in a novel way as a part of the early architectural design process. We seek to use the tool in a way unintended by its creators, for results different from many examples or use cases.[2,3,4] Additionally we seek to composite the results of multiple NST iterations to create a composition that is more directed by the author (Figs. 3 and 5). The hypothesis of utilizing this procedure is that the results become hybrid between the tool and the designer. A feedback loop: both providing input and responding to the other.

The work presented herein attempts to blur the lines between architectural design, storytelling, digital media, computer science and other disciplines. The objective of the effort is to inspire humanity toward a brighter future through imagining and instantiating possible futures. Divergent from many other documented use cases within the design process, this use case of NST focuses on a more intuitive, iterative, feedback loop between designer and the computational process (Fig. 3). We attempt to achieve a guided vision, possibly beyond the potential conceptions of the designer in unaided circumstances to conceive of the outcomes, or potential design schemes. Built on several speculative architectural workshops, art exhibitions, and design studios at Tongji DigitalFUTURES, FutureLab Shanghai, Longxi Art Museum, and Wenzhou-Kean University, respectively, this paper seeks ontological uncertainty between the artist and a cognitive prosthetic using the machine learning technologies of a convolutional neural network (CNN), and more specifically NST in the early design phase for novel design solutions.

The collection of NST compositions represented in this article are primarily *possible futures*. One might ask about the merit of compositions that do not seem to be feasible or viable as constructed architecture. Yet these blue-sky explorations can exist only

[1] Cook [2].

[2] Liu et al. [7].

[3] Liu et al. [9].

[4] Wang [12].

digitally, or possibly become a part of humanity's-built reality at some point in the future. **Regardless of their final instantiation, by very conception, these possibilities become exactly that, a possibility, otherwise impossible.** This paper will first outline our process. Afterward consider some of the lineage from which this work springs and the potential power of the speculative world building process.[5] Next we will consider the low hanging fruit of instantiating these places at least virtually in the metaverse and/or through VR, animation to solidify their somewhat timid present reality (Fig. 2).

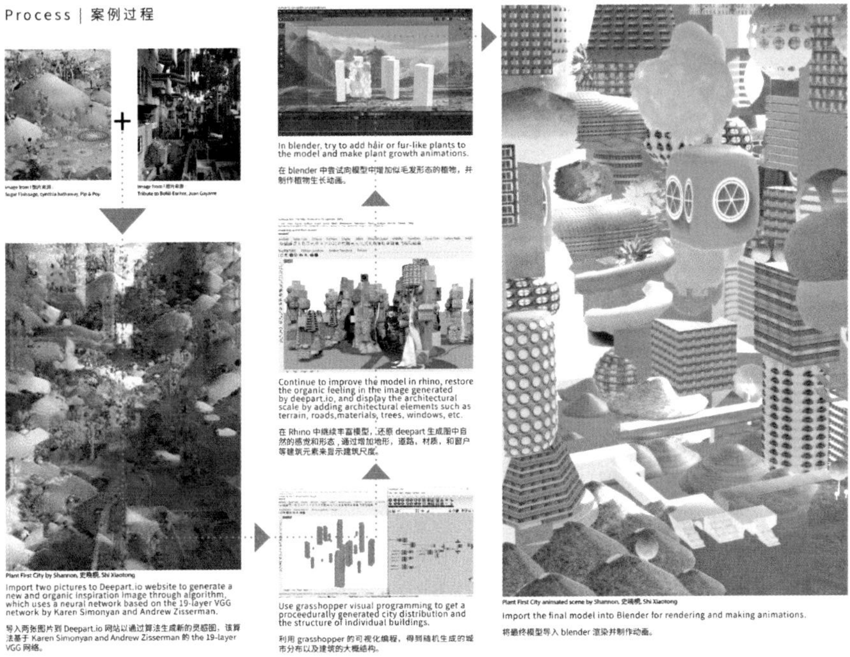

Fig. 2. This diagram outlines the artistic process, of "Plant First City" by Shannon Xiaotong Shi 施晓彤. A straightforward subversion of the NST tool by combining two "non-style" images to conceive of and create a novel urban environment

2 Novel Methods Toward Future Possibilities

This paper reflects on a collection of works by the author and collaborators that seek to subvert the intended use case of NST. The NST process is built on VGG16 and subsequently VGG19 conceived by Oxford's Visual Geometry Group (VGG, 19 denoting the number of layers in the network).[6,7] The technology is designed to apply the style

[5] Young [16].
[6] Gatys et al. [3].
[7] Hassan [4].

from, for example, a famous painting such as "Starry Night" by Van Gough and apply it to a base image such as a photo taken with your phone.

Our unconventional approach to the process involves intuition on the part of the designer as a part of the style transfer and compositing process. We outline a use case for NST in the early design phase of an architectural project where the designer goes back and forth between the NST process and other inputs toward optimized results which meets the subjective, intuitive exploratory criteria of the designer. Our process encourages mis-use of the NST process by subverting the labeling mechanisms inherent to the NST process; encouraging a style image to be used as a base image and what may be considered a base image as a style, in other cases we forgo the style and instead opt to only feed the system base images; starving the system for a style. We encourage the compositing of multiple more traditional NST images into a single composition. After several NST exercises, the design can then respond to the collected results through collaging to create outcomes that otherwise would prove impractical (Fig. 2).

Several NST use cases within the framework have resulted in the output weighing significantly on final outcomes of the architectural design project. The aesthetic of an NST often approaches something akin to computational impressionism.[8] This style inherently lends itself to the potential for an emotional response from a viewer. Paired with unexpected, uncanny, or emotionally vivid elements the compositions often approach the sublime.[9] These results can then be taken by the designer as a prompt, for further speculation and development based on a set of constraints. Our process maintains that NST output commonly provides a small peek into a potential world. It is up to the designer to take the prompt further and develop what they see in the image. After the designers develop their interpretations and speculations as a more flushed out image; often literally expanding the frame and providing elements of scale and context, a more traditional modeling process can follow, with the outcomes moving from not possible to conceived and on to a navigable experience, and possibly built. Although, there are several developing technologies with increasing capabilities able to translate single images into 3D digital models. Exploring these tools will be at the heart of future work in this research (Figs. 3 and 4).

Three of the architectural design studios that have used this process have ended with results which have been directly inspired by the early visioning produced with the designer and the NST process. In the spring of 2022, the author ran an architectural design studio with co-instructor David Vardy, where NST was a core aspect to the early design phase. Later in the semester the results were curated in the "Computational Impressionism 无极之地" exhibition at the Longxi Art Museum in Zeya, Zhejiang Province (Fig. 5).

The students began with site research. Collecting material samples, natural artifacts, photographs, etc. They also began processing the programmatic requirements, and in turn begin thinking about precedent projects. All these inputs inform the subject matter input into the NST process. Yet with the distinct interest in subverting the *style transfer* in favor of a mixing of ideas, the outcomes demand new subjective interpretations of the original object, not allowing the observer (including designers) to be passive about

[8] Lee [6].

[9] Kant and Bernard [5].

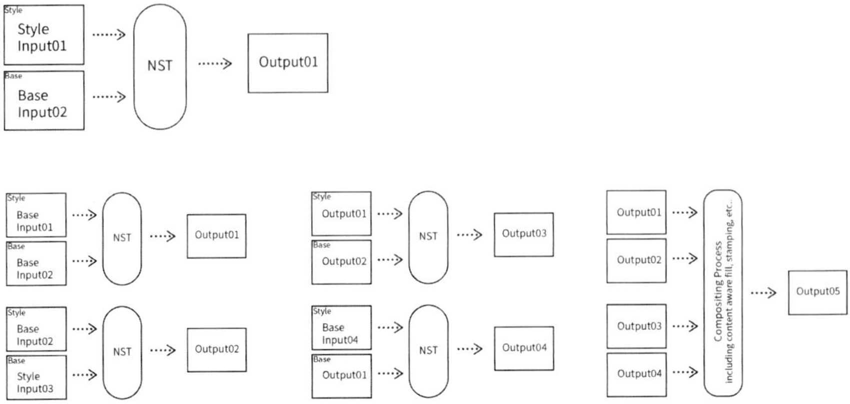

Fig. 3. Comparing a typical NST process (top) to the hybrid approach outlined in this paper (bottom)

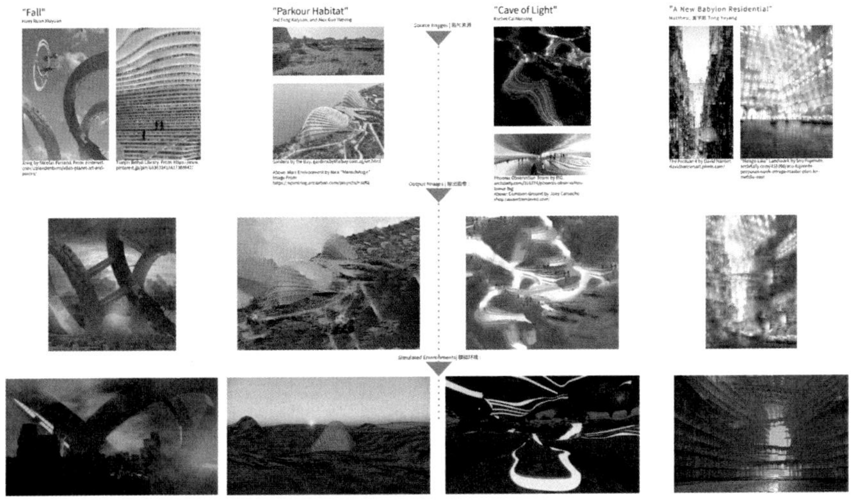

Fig. 4. Student processes example from image selection through to virtual environment

their viewing experience of the NST outcomes. The outcomes of this superficial process generally lack scale and context. The designer is called to engage with the art. The original subject is often no longer readily recognizable. Instead, the designer must layer their own psychological imprint on the piece to interpret it. It is the opportunity of the designer to define both. As mentioned earlier this happens through an expanding of the frame to define elements such as horizon, orientation, scale, and provide some reference indicators. The results of all these efforts are both inspiring and important in the development of hybrid intelligence within creative disciplines.

Fig. 5. On the left is a photograph of the style transfer work in the Computational Impressionism 无极之地 exhibition at Longxi Art Museum. On the right is the collection of process work used to derive the "Quantum Clouds" composition

3 Imagination: Pioneering Effort for Humanity

Much of what drives design are performative aspects of the outcomes, yet if we are only driven by the utilitarian requirements, we will never allow ourselves the opportunity to imagine alternatives. By pushing beyond what is conceivable we enable new possibilities for the future.[10] This can be referred to as the development of possible futures. It is a significant part of guiding the future. Hybrid intelligent processes such as style transfer can play a key role in this and other problems humanity faces.

This research project builds on the heritage of the digital studios pioneered by Bernard Tschumi, and before him the work of Syd Mead (1933–2019), Lebbeus Woods (1940–2012), Roger Dean (1944), and others. We see the legacy of the Columbia University digital studio effort going on to create such projects as the digital Guggenheim, New City, and has even influenced Hollywood world building see examples from Alex McDowell and the World Building Media Lab[11] at the University of Southern California. Yet this is not the only lineage of computation influencing the future of urban conceptions. We have seen, particularly in the last ten years, the influence of computational methods such as parametric, procedural, and machine learning rework the possibilities of creativity to incorporate computation as a collaborator (Fig. 5).

Beyond those technical aspects driving creativity and innovation, there are other stirrings including more cultural ones such as the post-digital movement which has worked to flatten and colorize predominant conceptions of our future. One could argue that the digital screen is the ultimate flattening, and yet the digital screen offers so much in terms of dimensionality. The imagery produced together with the artist and the computer using NST arguably creates a new aesthetic, a new style entirely, obviously grounded in

[10] Rhees [8].

[11] Unknown [11].

a strong foundation briefly explored above. Yet this aesthetic quite possibly may have been previously out of reach without computational collaboration.

The work that this group has undertaken is inherently optimistic and is rooted in questions such as: What kind of a world would you like to exist in if some limitation was removed? What possible futures could embody more sustainable circumstances for humanity? What is the ideal community for an individual? For industry? For community? What elements are needed to cultivate the most conducive environments? What methods of digital representation are optimal for expressing these ideas? How can one question traditional typologies and outdated organizational systems for the benefit of humanity? How do diverse family structures, demographics, and living/working styles affect the city?

Consider the work of Shannon Xiaotong Shi 施晓彤 with her work highlighted in Fig. 2 entitled Plant First City. In this work she considers the absurd possibility of a city where plants receive the primary and first consideration. What would such a reality feel like? In her extra-planetary conception, plants are the dominant species and in order to survive humans must live symbiotically. Even the conception of such a reality is quite intensive, let along the instantiation of it as a navigable environment, which is exactly what she was able to accomplish with her partner Luna Yuwen Wang 王宇文.[12] Of course this fabricated virtual environment is not built, but regardless this may be the fiction that inspires the next generation to do something truly historically altering, in the same way that the pacemaker was inspired by Frankenstein.[13] Regardless, this effort by the artist has impacted if no one else, her future. Hopefully she will reflect on this part of her architectural training as she is considering her career path. Content creation is a growing sector with promise. Regardless, her ability to conceive of realities outside of conventional human conception has been enhanced, and her confidence entertaining computational collaborations has grown.

4 Virtual Habitation

The technologies necessary to create complex digital worlds using automated computational methods seems to still be an elusive matter. NST and related technologies by themselves fall short of being a creative force in 3D world building alone.[14] Yet NST can be a part of a collaborative process with designers to directly inform and participate in "world building"[15] minimally for virtual spaces. As virtual spaces are not limited by the constraints outside of computational environments, the NST results can then be taken by the designer as a prompt, for further speculation and development based on a set of constraints.

NST seems to allow designers/architects to envision realities outside of the constraints of currently understood physics. This allows those who experience the place to engage in potentially impossible phenomena under any other circumstances. For example, "Parkour in Mars City", an Unreal Engine simulation by Alex Guo Yupeng 郭玉鹏

[12] Xiaotong 施晓彤 Shannon Shi [13].

[13] Rhees [8].

[14] Zhang and Blasetti [17].

[15] Young [15].

and Ted Kaiyuan Feng 冯开元, based upon a series of NST compositions.[16] again deals with an extra-planetary parkour experience designed around the divergence in gravitational properties between our own Earth and what might be; elsewhere. How does our design of space change based on this parameter? This example lets you virtually realize those differences. These landscapes become a "digital shadow"[17] of a physical landscape that has not yet been realized. These shadows have a life of their own outside of any physical reality. The ambition is that through these virtual prototypes we can iterate faster through possibilities toward a better future.

5 Conclusion

We have explored some pedagogical case studies, strengthening the argument that this process can yield compelling results that inspire and demand questioning, and quite possibly yield unconventional approaches to real world problems. NST has many use cases, but the most interesting may be the misuse of the tool by not simply taking the transfer at face value, but by subverting the labeling methods employed in the training set of the model to create unexpected results that take NST beyond just transferring a style toward honest unique creation. With this use case it is not difficult for the imagery to help in the process of speculating about future possibilities. Synthetic intelligence[18] allow us to move past human conceptions considering realities we may not deem feasible. Human imagination is always a precursor to what becomes reality. Thinking beyond our current conceptions of reality is essential to move past barriers that currently hold humanity. Non-probable futures allow humanity to identify aspects worth striving for. Even if a world building exercise is not probable for instantiation within the built environment, people can inhabit these environments virtually. What future do you want? Maybe working with NST can help us to create it.

References

1. Bratton BH, Stack T (2016) On software and sovereignty. The MIT Press, Cambridge, MA, p 213
2. Cook P (2020) Instant city in a field, typical set-up. Archigram Archives, Archigram. https://www.archigram.net/portfolio.html
3. Gatys LA, Ecker AS, Bethge M (2016) Image style transfer using convolutional neural networks. In: 2016 IEEE conference on computer vision and pattern recognition (CVPR), pp 2414–2423. https://doi.org/10.1109/CVPR.2016.265
4. Hassan MU (2021) Evolution of style transfer techniques. VGG16-convolutional network for classification and detection, 24 Feb 2021. https://www.neurohive.io/en/popular-networks/vgg16/
5. Kant I, Bernard JH (1914) Kant's critique of judgement. Macmillan, London

[16] Yupeng [14], https://www.yupengguo.wordpress.com/2021/05/05/example-post-3/.

[17] Bratton [1].

[18] Ibid.

6. Lee K (2019) Computational impressionism: aesthetic transference between impressionism and emerging media arts Kyungho Lee Illinois informatics institute. In: ISEA2019 Gwangju LUX AETERNA proceedings of the 25th international symposium on electronic art, Art Center Nabi, Seoul, Korea, pp 668–669
7. Liu C, Shen J, Ren Y, Zheng H (2021) Pipes of AI–machine learning assisted 3D modeling design. In: Yuan PF, Yao J, Yan C, Wang X, Leach N (eds) Proceedings of the 2020 DigitalFUTURES. CDRF 2020. Springer, Singapore. https://doi.org/10.1007/978-981-33-4400-6_2
8. Rhees DJ (2009) From Frankenstein to the pacemaker a profile of the Bakken museum. IEEE Eng Med Biol Mag 78–79
9. Liu S, Bo Y, Huang L (2021) Application of image style transfer technology in interior decoration design based on ecological environment. J Sens 2021, Article ID 9699110, 7. https://doi.org/10.1155/2021/9699110
10. Steenblik RS, Wang X, Dall'Asta JC, Saarinen E (2020) DigitalFUTURES Tongji University. https://www.digitalfutures.world/workshops/108.html, https://www.youtube.com/playlist?list=PLBtPB9RpflEa1g1NLn70EPrxnoRVNKIBo
11. Unknown Author (2022) The amazing 5th dimension. World Building Institute|The Future of Narrative Media. https://worldbuilding.institute/videos/the-amazing-5th-dimension
12. Wang L (2022) Workflow for applying optimization-based design exploration to early-stage architectural design - Case study based on EvoMass. Int J Archit Comput 20:41–60. https://doi.org/10.1177/14780771221082254
13. Xiaotong 施晓彤 Shannon Shi (2022) Plant first city. Shannon Xiaotong Shi 施晓彤 Personal Portfolio, Cargo Collective. https://www.xiaotongshishannon.cargo.site/
14. Yupeng Alex 作者 Guo, and Kaiyuan Ted 冯开元 Feng (2021) Parkour in mars city. Yupeng Guo Alex Portfolio, 7 June 2021
15. Young L et al (2020) Planet city. Australia, Uro Publications
16. Young L (2021) Planet city-a sci-fi vision of an astonishing regenerative future. ted.com/talks/liam_young_planet_city_a_sci_fi_vision_of_an_astonishing_regenerative_future.
17. Zhang H, Blasetti E (2020) 3D architectural form style transfer through machine learning (Full Version). https://doi.org/10.13140/RG.2.2.16791.52645
18. Zhang X et al (2015) Accelerating very deep convolutional networks for classification and detection. IEEE Trans Pattern Anal Mach Intell 38(10):1943–1955
19. "The Amazing 5th Dimension". World Building Institute|The Future of Narrative Media. https://www.worldbuilding.institute/videos/the-amazing-5th-dimension

AI Urban Voids: A Data-Driven Approach to Urban Activation

Amal Algamdey(✉), Aleksander Mastalski, Angelos Chronis, Amar Gurung, Felipe Romero Vargas(✉), German Bodenbender, and Lea Khairallah

IAAC-Institute for Advanced Architecture of Catalonia, Barcelona, Spain
{amal.algamdey,felipe.romero}@students.iaac.net,
algamdeyamal@gmail.com, romerovargas.f@gmail.com

Abstract. With the development of digital technologies, big urban data is now readily available online. This opens the opportunity to utilize new data and create new relationships within multiple urban features for cities. Moreover, new computational design techniques open a new portal for architects and designers to reinterpret this urban data and provide much better-informed design decisions. The "AI Urban Voids" project is defined as a data-driven approach to analyze and predict the strategic location for urban uses in the addition of amenities within the city. The location of these urban amenities is evaluated based on predictions and scores followed by a series of urban analyses and simulations using K-Means clustering. Furthermore, these results are then visualized in a web-based platform; likewise, the aim is to create a tool that will work on a feedback loop system that constantly updates the information. This paper explains the use of different datasets from Five cities including Melbourne, Sydney, Berlin, Warsaw, and Sao Paulo. Python, Osmx libraries and K-means clustering open the way to manipulate large data sets by introducing a collection of computational processes that can override traditional urban analysis.

Keywords: Computational urban design · Artificial intelligence · Machine learning · Urban data · KMeans clustering · Data visualization

1 Introduction

By 2050, 68% of the world's population will be living in cities. In order to adapt to this rapid urbanization growth while making cities more sustainable, livable, and equitable, designers must utilize qualitative and quantitative tools to make better-informed decisions about future cities [1]. In addition, big urban data is now readily available online, allowing the opportunity to utilize this information to generate new urban analyses between various features within the urban fabric [2]. A new digital layer can be added toward urban complexity through the novel perspective of data accumulation. However, despite these advancements, urban analysis and planning processes still follow the most static models that do not fit the development of today's cities (Jordan and Mitchell 2015; Al-Garadi et al. 2020).

P. F. Yuan et al. (eds.), *Hybrid Intelligence*, Computational Design and Robotic Fabrication,
https://doi.org/10.1007/978-981-19-8637-6_26

Moreover, the available urban data is vast and beyond human capability to handle; consequently, building new tools to understand and manage such substantial information is needed. Recently, cities are increasingly incorporating Machine Learning(ML) applications, primarily to meet economic and sustainability goals, etc. (Li et al. 2020; Choung and Kim 2019; Liu et al. 2017). Therefore, several approaches are presented to model the dynamics of urban drivers as a function of the different features of the urban form. The latest research results show that ML methods have significantly exceeded the conventional prescriptive modelling methods of urban indicators to evolve into an essential tool for urban planning decision-making (Ma et al. 2020; Hecht et al. 2013). To meet the challenges of the current urban complexity of emerging big data, the modelling of urban indicators increasingly exploits intelligent automatic methods using ML algorithms which can override traditional methods.

This paper aims to investigate the role of ML in analyzing and predicting the accessibility and proximity to urban features of the city of Melbourne. In addition, it demonstrates the advantages of building new tools for prediction-based planning to accurately define the interconnections within cities. With the availability of massive datasets, this study integrates a Machine Learning process using (K-Means clustering) to improve capabilities for urban planners to understand cities better. Also, It demonstrates how technological advances in data gathering and processing can depict more aspects of urban complexity and dynamic urban nature. However, due to the complexity of computation, these technological advances are still not accessible and lack connectivity to real-world scenarios and design practices [3]. For this reason, there is a need for urban planners and urban designers to operate and integrate computational design processes. Moreover, it is essential to develop new tools to have a deeper understanding of cities and take advantage of computational design and machine learning to improve design decisions for better cities.

2 Background

Urban indicators analysis using ML has been growing with great success to address urban planning challenges in recent years, but some are still in a very early stage of development or have been developed as part of research without implementation on real-case scenarios [4]. This method previously addressed some urban issues, including smart cities, mobility, climate, density, and energy. However, the evolution of that research indicates a promising future and outcome for ML application in this area [2]. In contemporary research, the ML algorithms applied are deep learning, artificial neural networks (ANN), support vector machines, neuro-fuzzy, and decision trees. These methods are usually used to classify information and create predictive models [5]. The reviewed papers highlight applications of ML algorithms to understand trade-offs between the city, its indicators, and urban planning. Also, to support the discussion on the role of these methods, the implementation of ML strategies on a city scale, and the possibility of integrating ML with other technologies.

The design of digital, smart, and connected cities is at the core of modern urban planning (Ma et al. 2020; Middel et al. 2019; He et al. 2018). Cities must demonstrate digital transformation and integration initiatives by creating intelligent and smart cities to become more influential. The Internet of Things is at the centre of the smart city and enables the enrichment of collected urban data into Urban Big Data. These big data capture information on all urban activities and are also the ideal source for ML algorithms. Indeed, from this big data, ML will allow the creation of AIs, allowing to make the city more intelligent (Jordan and Mitchell 2015). But also, and most importantly, the processing of big data by the rising power of ML algorithms will allow urban planners to predict the city's evolutionary trends and regulate them by orienting the city's shape towards the most sustainable, intelligent, digital, and connected form possible. Thus, urban data enriches the ML to create AI for IoT and intelligent urban planning, making the city more and more intelligent, digital, and connected, which helps enhance big urban data. Therefore, the power of ML algorithms, IoT applications, the engine of AI, and intelligent urban planning are crucial for addressing the challenges of the smart city (Al-Garadi et al. 2020; Jordan and Mitchell 2015).

Therefore, to develop our project AI Urban Voids, some projects were studied to understand existing computational design processes applied to urban environments. The Hive, for instance, is a web-based urban map that displays the building properties in conjunction with pedestrian movements at different times of the week (https://comotionla.com/more-la-parking-to-places). In addition, Hive proposes a 3D urban data visualization tool that helps find urban relationships between business opening hours by integrating civic open data sets. In this case, this project was the primary reference as an interface for urban indicators metrics and for understanding the potential for urban tools to analyze urban activation and economy (Fig. 1).

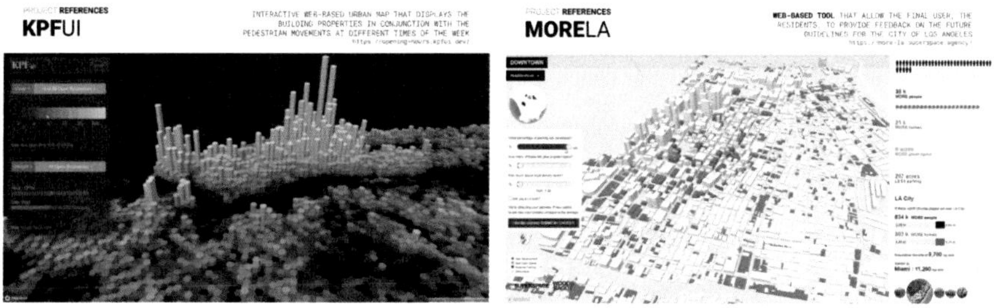

Fig. 1. Simulation Platforms: Hive, and more LA.

Secondly, in another study, RE{CODE} is an urban simulation platform developed at IAAC (Institute of Advanced Architecture of Catalonia) that modifies & understands the existing urban fabric and proposes methods to minimize the social and economic imbalances. The tools suggest a new rearranged urban pattern based on performance relationships and urban dynamics, this tool supports the urban analysis for large scale urban centers by identifying urban discrepancies and future opportunities.

Finally, MoreLA is a project developed by Superspace-Woods Bagot in the mobility framework (https://comotionla.com/more-la-parking-to-places). This project was developed as an interactive survey and a web-based tool created to allow the final user, the residents, to provide feedback on the future guidelines for the city of Los Angeles. MoreLA introduces the concept of participatory design by integrating users' opinions for large urban centers; this opens the opportunity to understand people's choices in early urban planning phases and identify urban solutions for local communities.

The overall analyses of these three projects conclude that computational design processes & machine learning applications have recently grown in urban studies to solve new challenges. Furthermore, data availability and accessibility of computational tools helped enhance this growth. However, some of these urban dynamics analyses approaches lack processes to identify the potential location for future urban amenities based on proximity and score criteria; this is to draw a relationship between urban nodes.

There is a necessity to develop computational urban design tools to manage and understand extensive data sets that cities are generating from time to time; this could lead to a more dynamic process between designers, users and urban communities [6].

3 Methodology

The project is conceived as a data-driven approach to analyzing and unveiling the urban fabric's hidden opportunities by introducing a computational design process that calculates distances and walkability using urban points of interest and existing pedestrian networks (OSM Data & Python NetworkX). This process allows the evaluation of the results based on clustering performance, comparing distances/areas, and introducing and analyzing new urban amenities.

This research starts with gathering and cleaning from Open Street Maps (OSM) data (Fig. 2). OSM Data plays an important role based on its availability aspects; however, it also opens the questions of how reliable open-source data sets are. Therefore, the first step is collecting and cleaning data from Open Street Maps OSM; this process opens the possibility of bringing a structured OSM dataset for further computation.

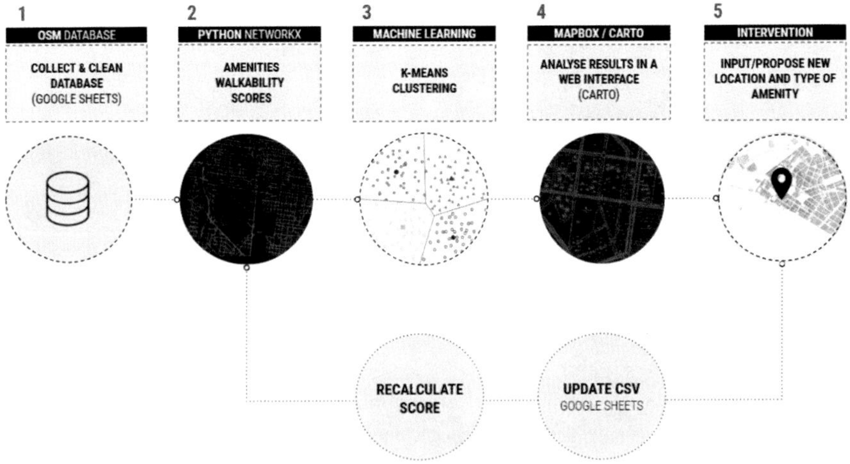

Fig. 2. Project methodology diagram.

For the intention of this project, the data set was narrowed to three key features that include:

- Points of Interest (Urban Nodes)
- Pedestrian Network (Urban)
- Locations with addresses (Urban Nodes)

Points of interest were selected based on the daily usage and importance of the services provided to cities. Some of the critical points of interest include education facilities (primary and higher), healthcare and hospitals, nightlife and entertainment facilities (Fig. 2). The second step began with Python NetworkX calculating distances between amenities and ranking them based on score criteria. NetworkX is a python library used to create, manipulate, and study the structure, dynamics, and functions of complex networks; in this case, cities.

The third step was to cluster all scores gathered from Python and NetworkX using K-means. K- means is an unsupervised machine learning process that helps identify data clusters in large datasets. The identification of these clusters was later visualized within a web interface. As a final stage, the project takes advantage of Carto as a data visualization by linking the result of the previous computation process; furthermore, this tool enables user interaction and data manipulation by a series of filters that display results instantly. Finally, for later stages, Urban Voids open the opportunity to introduce a new input where the user can introduce new urban features, whether a location or amenity type, that will feed into the database to recalculate the scores resulting in a new score and clustering based on the new input.

3.1 Dataset Creation (Extracting, Compiling and Clustering)

For the reliability and efficiency of the process, Python and OSMX libraries are used to extract the data from Open Street Maps to provide the script with the city-CRS, which extracts three main data frames: points of interest, pedestrian network, and address points. After this process, the data used for K-Means clustering were cleaned and adjusted (Fig. 3). The objective of the data cleaning was defined in three steps:

1. Build a Pandana Network–"Pandana is a Python library for network analysis that uses contraction hierarchies to calculate super-fast travel accessibility metrics and shortest paths." Pandana v0.6.1 (2021).
2. Generate tags for different Points of interest.
3. Clean and provide a clear structure for different addresses in a city.

The purpose of these three objectives is to prepare the definition of processing data:

- To calculate the shortest distance from every point to an "n" number of closest points of interest.
- To evaluate a based scoring system on the average walking time of 1.2 m/s as a normal pedestrian pace.
- To introduce an optimized computation process to calculate distances between multiple points of interest tags. In this case, the calculation achieved for a tag: 'bus stop', for the address data frame of 320 000 rows took less than 3 s.

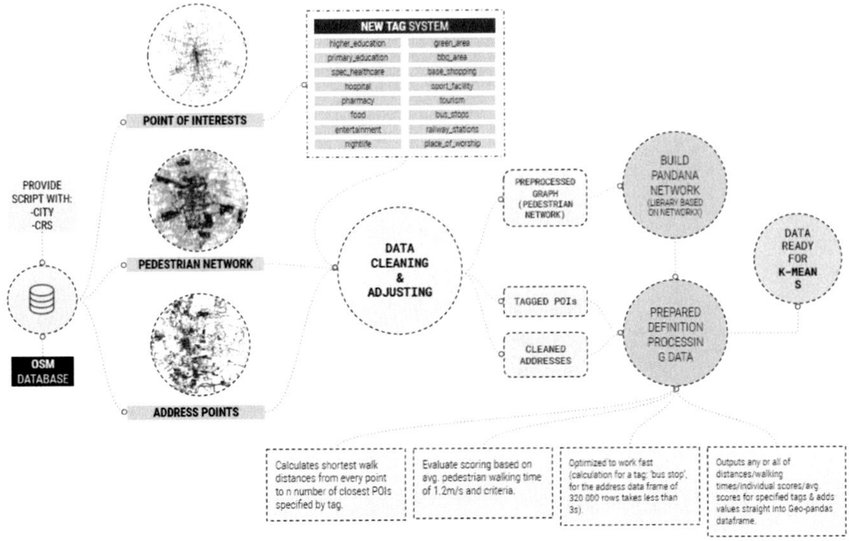

Fig. 3. Workflow of dataset creation.

- To achieve an output of any or all distances/walking times/individual scores/avg. & scores for specified tags and add values straight into GeoPandas data frame.

Building scoring and walkability analysis were developed using a scoring system as a result of the walking time from one point to another. Walking times vary between 0 min (score of 100 pts) and 70 min (score of 0 pts); using this method helped the project to identify & classify amenities for multiple addresses. As a result of this analysis, the script provided the number of X amenities closest to the subject address, based on multiple criteria, as shown in Fig. 4. Traditional Urban Design analysis strategies include using a Ped Shed Analysis that understands the proximity of urban nodes within a radius of 400, 800 and 1,200 m (5, 10 and 15 min walking distances). However, this technique can be considered a high-level analysis tool that most of the time does not consider the existing urban network and does not specify the relationship between multiple uses to the specific node.

3.2 Machine Learning (K-means Clustering)

"Plot pairwise relationships in a dataset. By default, this function will create a grid of Axes such that each numeric variable in data will be shared across the y-axes across a single row and the x-axes across a single column. The diagonal plots are treated differently: a univariate distribution plot is drawn to show the marginal distribution of the data in each column." Seaborn (2021).

For this project, a pair plot was used to understand the pairwise bivariate distribution of multiple datasets and their relationships within the subject city (Fig. 5).

In addition, the Pearson correlation coefficient was used to understand which attributes are linearly related to the predicted set (Fig. 6). Furthermore, a biplot overlays

BUILDING **SCORING AND WALKABILITY ANALYSIS**

WALK TIME	SCORES	DESCRIPTION
0	100	Destination spot
5	90	Highly walkable
10	80	Pedestrian friendly
15	70	Very Comfortable
20	60	Comfortable
25	50	Somewhat Comfortable
30	40	Somewhat walkable
40	30	Difficult
50	20	Very difficult
60	10	Extremely difficult
70	0	Too far to walk

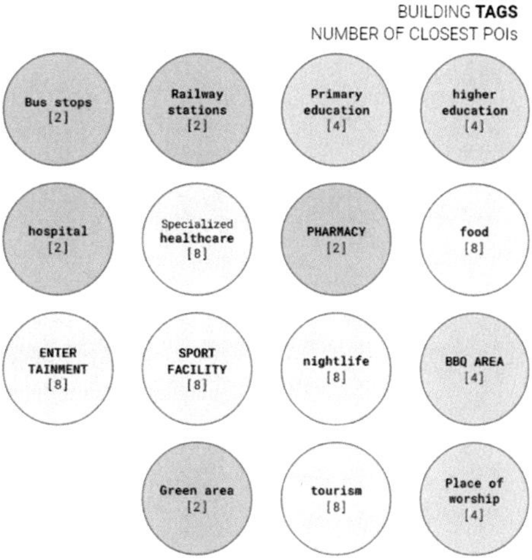

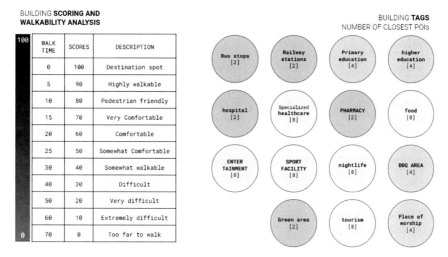

Fig. 4. Building scoring and walkability analysis.

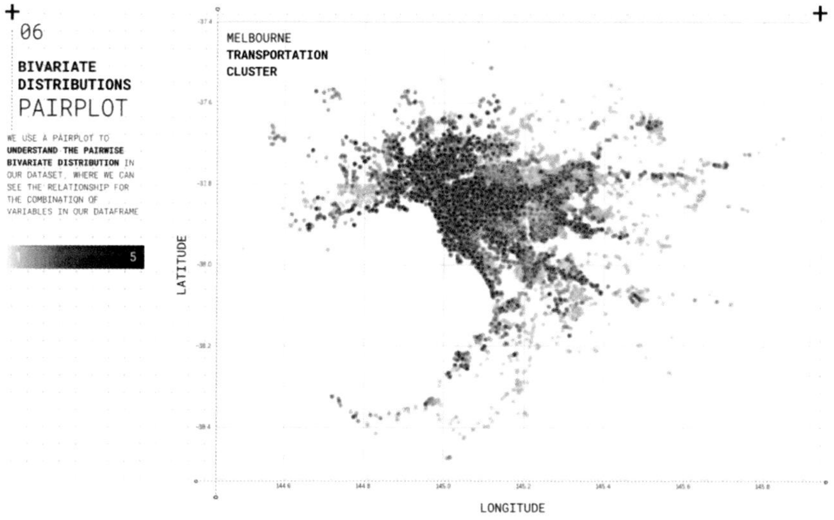

Fig. 5. Bivariate Distributions–Pair plot for Melbourne (Transportation Cluster).

both a score plot and a loading plot onto a single graph to visualize high-dimensional data onto a two-dimensional graph. Finally, the elbow method was used to understand the ideal number of clusters that should be used for the clustering based on the shape and features of the data. Then, the final clustering was plotted on a two-dimensional plot for each category based on overall performance.

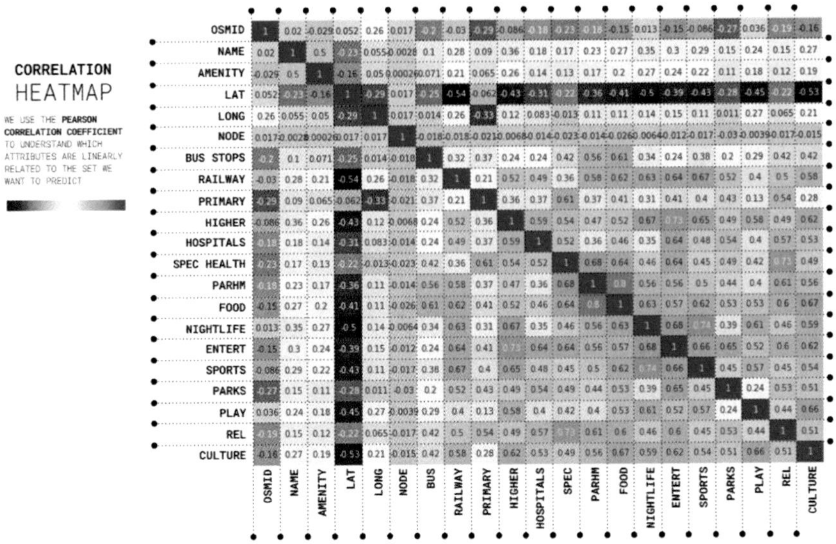

Fig. 6. Correlation Heatmap.

4 Case Study

Melbourne city was used as a case study to test the different clusters and their relationships to examine the described methodology. For the purpose of this project, Carto was used to create a web-based application to analyze and unveil the correlations of urban features and their opportunities for multiple cities. In this case, opportunities are measured based on the scoring system that provides a classification process based on the accessibility and direct relationship within urban areas. Five cities with different urban conditions were used as case studies for the analysis and comparison: Melbourne, Sydney, Berlin, Warsaw, and Sao Paulo. The data visualization tool allows the user to filter & compare the information based on the project criteria and display the number of clusters based on the multiple uses that are part of the data set (Fig. 7).

5 Results

As a result, this study can be divided into three categories, including statistical analysis, visualization, and web application. As shown in Fig. 8, this tool allows users to compare five different cities and visualize K-means values, this step demonstrates the ability of the tool to integrate different urban data sets and provide the results. However, it is essential to highlight that the results are driven by the quality & accessibility of the data. Figure 9, the web application, shows the AI Urban Voids as an interactive app. The overall idea is to provide a platform where the user can manipulate, filter, and understand the geographical relationship and connections within the city by using K-means clustering. This web interface can help policymakers, architects, and urban designers to evaluate and understand the city's behaviour and provide a better design response. It is essential

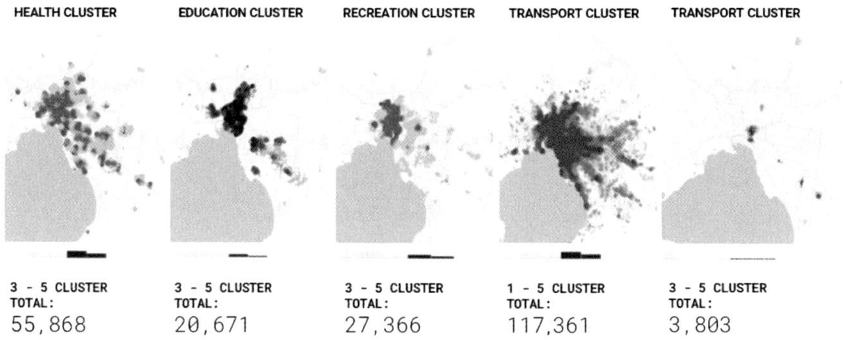

Fig. 7. Data visualization and analysis.

to highlight that the results can vary based on the quality and the data input; this is what is essential to improving and maintaining open data sources. Finally, this web interface can be used in strategic urban Planning & Urban Design phases by private or government entities to understand current urban conditions better and provide strategic projects for multiple urban contexts.

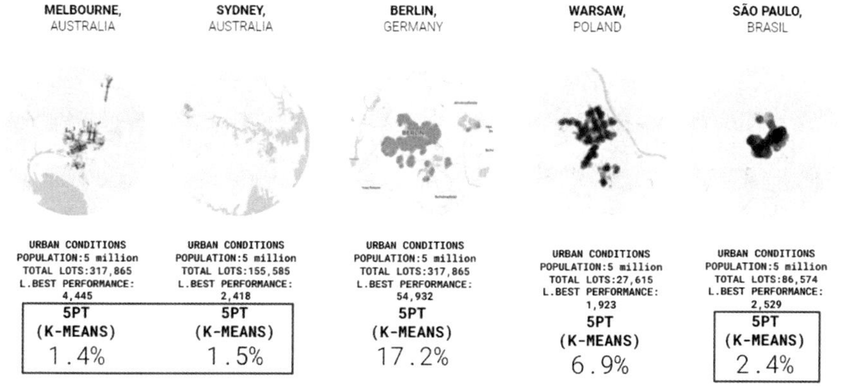

Fig. 8. K-means results from comparison and visualization.

6 Future Research Direction

With the evolution of today's cities and with Urban big data, there is a need for new efficient models more than the traditional static models. Moreover, more efficient ML techniques allow for more helpful data processing. Thus, intelligent urban planning could benefit from creating a smart city. The future work coming out of this research applies to the K-means clustering method of proximity to healthcare. For instance, the correlation approach considering road networks and trips from each park can correlate parks' mutual attractiveness rather than generic distance. Additionally, developments might come from creating 3D urban forms based on the results of clusters.

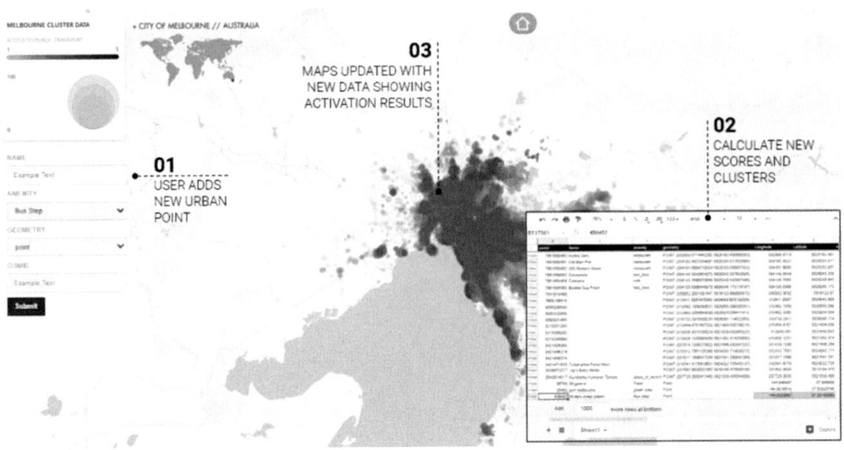

Fig. 9. Web application AI urban voids. (https://city-lab.wixsite.com/urbanvoids).

7 Conclusion

This research has presented very positive results in breaking down the workflow between urban indicators analysis and machine learning. Moreover, this study can be considered a workflow for further exploration by linking open data sources to a web interface to unveil hidden urban relationships that can improve urban accessibility for certain uses.

Open data sources and computational skills are the major limitations to developing urban tools that can integrate large data sets. Nevertheless, despite the disadvantages above, understanding urban centres and open data sources provide a benefit that opens the possibility to integrate Computational Design & Artificial Intelligence processes.

Moreover, this workflow demonstrates the different limitations tools can have when handling large data sets. Python and Osmx libraries open the way to manipulate large data sets that benefit multiple urban communities. There is an opportunity to complement, improve and create new/existing large open-source datasets that can test and inform design processes. Google places + OSM datasets are often driven by commercial applications, neglecting non-marketable areas and spaces still important for the city.

The process allows users a new series of opportunities, but the question about real-case applications in the urban area remains, who could benefit from this, and how can we make better cities with it? The deployment of the application provides an understanding of a new possibility of making AI accessible for urban designers, planners, developers, and policymakers. However, sufficient back-end work has to be done to integrate the urban data, machine learning processes and its definition in data visualization platforms. Since the data used now is entirely open source and the workflow completely established, the next step is to make this available for every city and minimize the computation to be accessible. Finally, this study has demonstrated the importance of using computational design methods in urban matters; this allows the creation of new tools that can improve urban analysis phases that will result in better design decisions & urban policies that can improve urban communities. The resulting data from this project can be used as a base to identify & allocate future urban density by providing an interconnected city; moreover, it can be used as a planning tool to re-interpret land zoning by local governments.

References

1. Moustaka V, Vakali A, Anthopoulos LG (2018) A systematic review for smart city data analytics, vol 51, no 5
2. Ulusar UD, Ozcan DG, Al-Turjman F (2019) Open source tools for machine learning with big data in smart cities. Smart cities performability. Cogn Secur 153–168
3. Haldorai A, Ramu A, Murugan S (2019) Machine learning and big data for smart generation. Comput Commun Syst Urban Dev 185–203
4. Nosratabadi S, Mosavi A, Keivani R, Ardabili S, Aram F (2019) State of the art survey of deep learning and machine learning models for smart cities and urban sustainability, no Aug 2019
5. Nikolaeva A, Adey P, Cresswell T, Lee JY, Nóvoa A, Temenos C (2019) Commoning mobility: towards a new politics of mobility transitions. Trans Inst Br Geogr 44(2):346–360
6. Gao X, Lee GM (2019) Computers & Industrial Engineering Moment-based rental prediction for bicycle-sharing transportation systems using a hybrid genetic algorithm and machine learning. Comput Ind Eng 128:60–69. (no Dec 2018)
7. Psyllidis A, Bozzon A, Bocconi S, Bolivar CT (2015) A Platform for urban analytics and semantic data integration in city planning, 21–36
8. Derix C (2019) Paradigm reversal–connectionist technologies for linear environments. In: Research culture in architecture, vol 21
9. Woods B (nd) Woods Bagot-sponsored study More LA anticipates major transportation changes for Los Angeles. https://www.woodsbagot.com/global-studio/news/ideas/woods-bagot-sponsored-study-more-la-anticipates-major-transportation-changes-for-los-angeles/. Accessed 21 Mar 2022. (re of modern genomics. Blackwell, London)
10. Podrasa, D., Zeile, P. and Neppl, M., 2021, September. Machine Learning for Land Use Scenarios and Urban Design. In CITIES 20.50–Creating Habitats for the 3rd Millennium: Smart–Sustainable–Climate Neutral. Proceedings of REAL CORP 2021, 26th International Conference on Urban Development, Regional Planning and Information Society (pp. 489-498). CORP–Competence Center of Urban and Regional Planning.

Co-creation: Space Reconfiguration by Architect and Agent Simulation Based Machine Learning

Anni Dai(✉)

University of Applied Arts Vienna, Studio Greg Lynn, Oskar-Kokoschka-Platz 2, 1010 Wien, Austria
annidai0202@gmail.com

Abstract. This research is a manifestation of architectural co-creation between agent simulation based machine learning and an architect's tacit knowledge. Instead of applying machine learning brains to agents, the author reversed the idea and applied machine learning to buildings. The project used agent simulation as a database, and trained the space to reconfigure itself based on its distance to the nearest agents. To overcome the limitations of machine learning model's simplified solutions to complicated architectural environments, the author introduced a co-creation method, where an architect uses tacit knowledge to overwatch and have real-time control over the space reconfiguration process. This research combines both the strength of machine learning's data-processing ability and an architect's tacit knowledge. Through exploration of emerging technologies such as machine learning and agent simulation, the author highlights limitations in design automation. By combining an architect's tacit knowledge with a new generation design method of agent simulation based machine learning, the author hopes to explore a new way for architects to co-create with machines.

Keywords: Machine learning · Agent simulation · Co-creation · Artificial intelligence · Space reconfiguration

1 Background

Agent simulation based design methodology has been around for decades, studies such as space syntax, agent-based semiology have been investigating, simulating and predicting spatial occupation patterns. However, the results are restricted to analysis and evaluation, which are yet to allow the space to become responsive. The hypothesis of the research is, having a space able to reconfigure itself based on agent behaviors.

1.1 Agent Simulation in Architectural Design Methodology

What is the medium of architectural design? Here I'm referring to Archer [2], 'The essential language of Design is modeling. A model is a representation of something' ([2], p. 20). 'Thus design activity is not only a distinctive process, comparable with but

P. F. Yuan et al. (eds.), *Hybrid Intelligence*, Computational Design and Robotic Fabrication,
https://doi.org/10.1007/978-981-19-8637-6_27

different from scientific and scholarly processes, but also operates through a medium, called modeling' ([2], p. 18).

Simulation works as one of the mediums of architectural design. Banks et al. [6] explained a simulation is the imitation of the operation of a real-world process or system over time. Simulations require the use of models; the model represents the key characteristics or behaviors of the selected system or process, whereas the simulation represents the evolution of the model over time. Simulation, as an evolution of the models of ideas over time with execution from computers, works as a 'new' medium in the architectural design process.

The technological development enables simulation of agent, and the combination between design experience with agent simulation. Penn and Turner [8] studied space syntax based agent simulation, giving many agents simultaneous access to the same pre-processed information about the configuration of a space layout. Today, there is also agent based parametric semiology whose goal 'is to design a new, coherent system of signification, a new artificial architectural language, without relying on the familiar codes found in the existing built environments' [1].

1.2 Machine Learning in Design Scientific Research

What is design science? Bayazit [4] referred to Vladimir Hubka and Ernst Eder's definition, the term 'design science' is to be understood as a system of logically related knowledge, which should contain and organize the complete knowledge about and for designing.

Looking back at the origin and history of scientific development in design research, Norbert Wiener played an important role. His idea 'Cybernetics' 'became the model for rational behavior employed in economics, and obtaining information and making decisions using computer systems' ([4], p. 23). 'There was a close relationship between design research and the developments in the IT field, especially in cognitive sciences, and "artificial intelligence" (AI) and expert systems' ([4], p. 27).

Machine learning as a part of artificial intelligence, has recently been used in architectural design. Machine learning (ML) is the study of computer algorithms that can improve automatically through experience and by the use of data [7]. By employing machine learning in the architectural design process, design automation is implied and thus brings both opportunity and threat to architects and the architectural design creative process.

1.3 Agent Simulation Based Machine Learning as a New Generation Design Method

Human behavior study has played a significant and continuous role in design methodology. Baudrillard claims that our current society has replaced all reality and meaning with symbols and signs, and that human experience is a simulation of reality [3]. With technological development, simulation, which is a 'new' medium in architectural design, has been widely adopted and applied to human behavior, which is agent simulation. Agent simulation is an architectural design medium and a design method.

Machine learning as part of contemporary design sciences, has in its nature relation with design research. 'Mutual influences of information technologies and design research were the requirements of the era' ([4], p. 28). It makes sense to combine machine learning with agent simulation. It is worth noting that in this case, different from the conventional routine of design sciences that goes from design to analysis to redesign, such as shadow analysis or Karamba analysis, machine learning breaks the routine and adapts design according to analysis results automatically.

1.4 Co-creation with Architects

Adoption of machine learning in architecture design pushes forward the idea of design automation, which puts the role of the architect in question. It can relate back to Donald Schön's study on designer behavior, which did not seem to relate to computer science back then, but here can be compared under the design automation discussion. Can machines replace architects by learning designer behaviors? And is machine learning, or design automation exempt from human/architect intervention?

The author would like to briefly answer the question from one aspect, without forming a comprehensive opinion. Tacit knowledge—as opposed to formal, codified or explicit knowledge—is knowledge that is difficult to express or extract, and thus more difficult to transfer to others by means of writing it down or verbalizing it. This can include personal wisdom, experience, insight, and intuition [9]. There is tacit knowledge in humans or architects that can hardly be conveyed, which is an important characteristic of architects. Architect design not only through accumulation of architectural knowledge, but all the other experiences form over the years and transforms to one's unique tacit knowledge.

One can argue that machines can learn and therefore form a type of tacit knowledge as well, as machine learning learns through past experiences. Here it is important to note that the database that machine learning from, and is based on is created and structured by humans, which exempt machines from gaining unwanted information from humans. The unwanted information might become part of the contribution to the machine's tacit knowledge. However, with unstructured or unwanted information in the database, the learning curve can be unnecessarily long, and might end in unsuccessful training. The database has human intervention, therefore the experience the machine gains is limited and structured, and differs itself from human knowledge.

The author wants to highlight the importance of co-creation between machine learning and an architect. By adopting the data processing strength of machine learning and the tacit knowledge of an architect, an architectural project can develop its most potential.

2 Application

The application part of this project mainly consists of four parts: Space, Agent, Machine Learning and Co-creation. Firstly, a space is needed to practice the design method within.; Secondly, an agent's behavior in this space, which is the agent simulation that is used as a database for machine learning; Thirdly, the machine learning training principle and outcomes; Fourthly, An architect's co-creation with the trained machine learning model.

2.1 Housing Precedents as Study Subjects (Space)

Even though modern buildings are generally different from ancient buildings, because of shared human behaviors in a living space, housing is a historically consistent, architecturally diverse typology, which allows a possible comparison between a modern building and an ancient building. Therefore, housing precedents are ideal for the study purpose of this agent simulation based project.

2.1.1 Precedents Principles

Twenty housing precedents were chosen as the study subject as they are historically consistent, architecturally diverse. Precedents chosen mostly have only one floor, which gives controllable parameters to study and compare among the precedents. Otherwise it will give unnecessary and uncontrollable parameters, such as how the stairs and lifts as a different agent behavior consideration for machine learning. However, It is difficult to find many representative one-floor housing precedents in classical architecture periods. In this case, the ground floor, where most active behaviors happen, is considered for the study (Fig. 1).

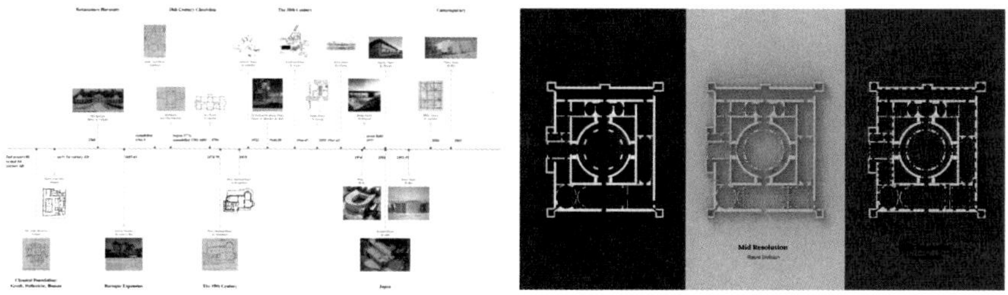

Fig. 1. Housing precedents timeline and three resolution illustrations. Images by Anni Dai

2.1.2 3D Model Preparation and Three Resolutions

In the process of transforming these precedents into 3d models for training, only the elements that are useful for machine learning and architectural representation are preserved. All the precedents are represented by their walls, the roofs and floors are hidden for the better observation and presentation of the project.

During the modeling process, the author observed that the building's room size, room density and room distance varies. The author decides to experiment the project on three different resolutions of these twenty housing precedents. Low resolution, where the walls are divided by structural integrity; Mid resolution, where the walls are divided by the size of the room; High resolution, where the walls are divided in 1.5 m blocks.

2.2 Agent Simulation (Agent)

2.2.1 Agent Simulation Rules

There are three typical agent types: Master, who is the owner of the building; Staff, who is serving the master and the guest, while responsible for cleaning the house; Guest, who visits the master and occasionally stays. Most ancient precedents have three agent types, while some contemporary ones which are highly private, therefore the guest agent is not considered as it is not the main intention of the buildings.

The author tries to make the agent's behavior comply with each building's specific situation. There is randomness in agent behavior that allows certain unpredictable developments that imitates the realistic scenarios (Fig. 2).

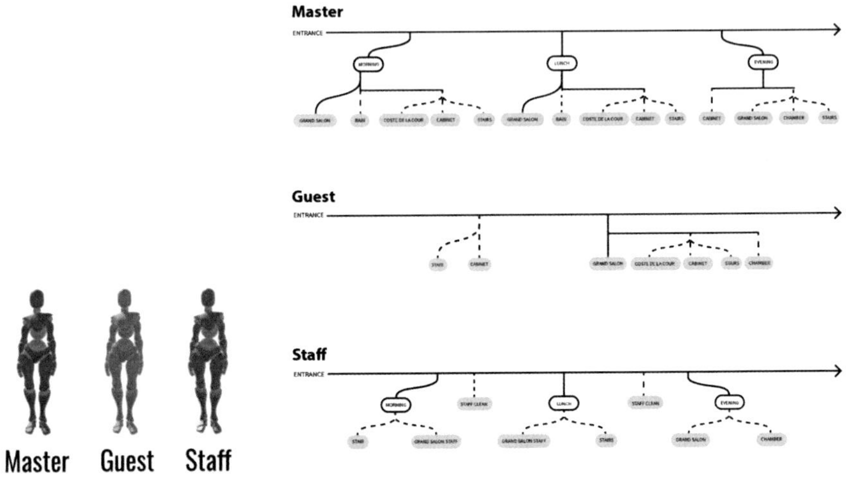

Fig. 2. Typical agent behavior. Image by Anni Dai

2.2.2 Agent Simulation Results

The complexity of the behavior is intentionally developed to use as many rooms as possible to gain the ideal training results. Each agent has different tendencies of space usage. For example, guest and master do not use the kitchen as much as the staff; guests have very limited access to rooms; Staff tend to access most of the rooms as they need to undertake a cleaning routine, while not spending as much time inside as the master, etc. The trail of agents movement is a direct representation of the room usage frequency (Fig. 3).

2.3 Machine Learning (ML)

2.3.1 Machine Learning Method

This project uses 'ML Agent', a Unity programme, as the Machine Learning Tool. Aas it was compatible with the 3d training environment, and Unity is compatible with agent simulation.

'ML Agent' uses reinforcement learning. In this project, each wall (gameobject) was given a ML brain. The author set the distance between agents and a wall as a goal. When an agent's distance to a wall is less than 1 m or exceeds 5 m, it is given a penalty

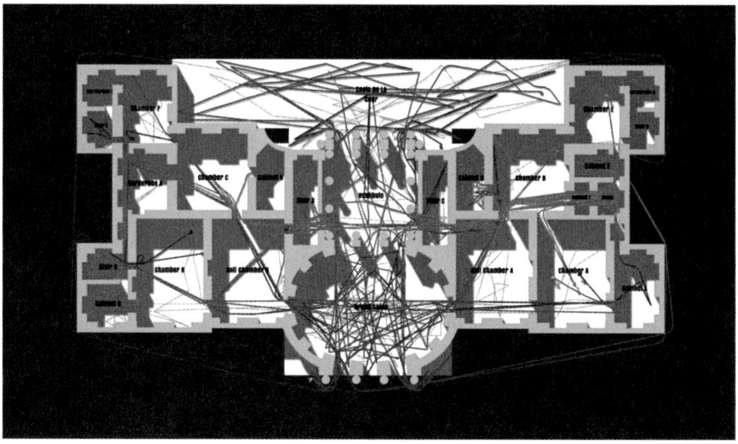

Fig. 3. Typical agent trail. Image by Anni Dai

(red); when an agent's distance to a wall is more than 1 m and less than 5 m, it is given a reward (green). Therefore, this project's training goal is to make the wall achieve a 'good' distance to the agents (Fig. 4).

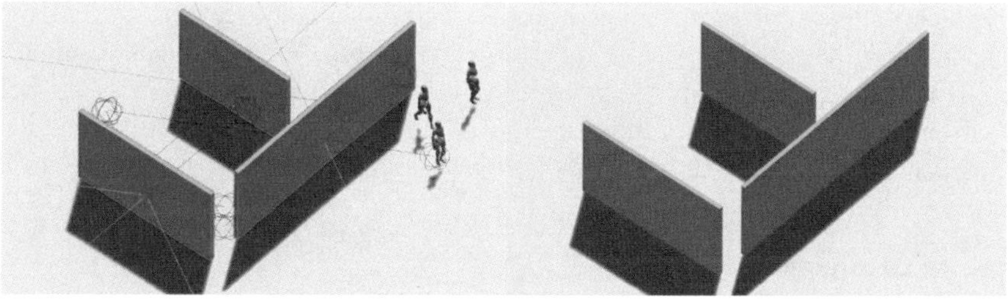

Fig. 4. Machine learning training diagram. Animated GIF by Anni Dai

The detection of the distance between agents and a wall used the Raycast 3D component. The wall is detecting agents within 10 m radius, it is tested as a good distance that will use less time to train. 10 m for a typical housing architecture is a big enough radius for a room. Using the Raycast 3D component has proven to be efficient to train a large amount of walls.

This project used the high resolution of twenty housing precedents as a training dataset, as it has the most walls for training. The result was the most comprehensive compared to other resolutions while not overbearing the computer. After 50,000,000 (50 million) steps, the trained result turned out satisfying. This trained model was then applied to each building again as an inference behavior, for further development (Fig. 5).

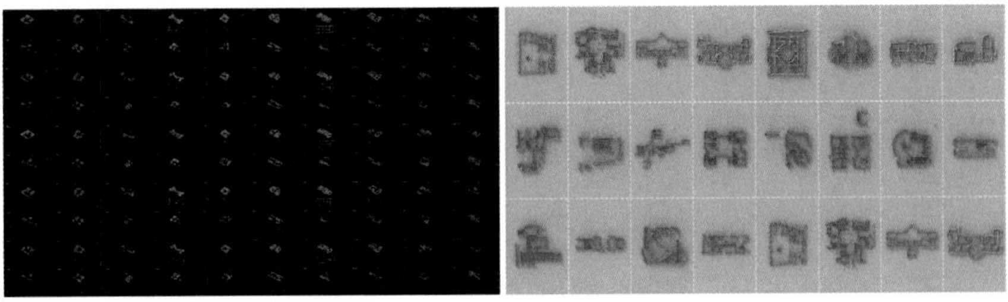

Fig. 5. Machine learning training overview and trained outcome. Image by Anni Dai

2.3.2 Machine Learning Training Outcomes

The training results typical developments are: entry creation, area increase, area reduction, room creation and new connection.

In general, the conclusion from this training outcome is, the room size increases as more agents use it, and reduces in size the less it is used. Machine learning only recognizes the frequency access of the room, and ignores the actual parameters, such as agent's time spent and functional differences between rooms.

2.4 Co-creation and AI Compliments (Co-creation)

As we have noticed from the machine learning training results above, there are limitations in this training. The other aspects of a housing architecture, for example, the amount of time an agent spent in a room, the activeness of different functional rooms, etc. should have been considered as part of the space reconfiguration parameter, however cannot be comprehended in this machine learning training outcome.

2.4.1 AI Compliments

Room Size and Room Height

To make each room size not only respond to agents' visiting frequency, but the realistic functional needs as well, the author assigned each room with a different **time weight** and **agent quantity weight**, which allows specific control over the room size and height. Room size is decided by measuring the distance between the wall and the center of the room constantly. Room height is associated with the agent quantity and amount of time spent in the room. The programme accumulates and averages the results (Fig. 6).

$$\begin{aligned} Ideal\ Room\ Size/Height = {} & Current\ Room\ Area/Height \\ & \left(\frac{Agent\ Quantity + 1}{Current\ Room\ Area} \times Agent\ Quantity\ Weight\right) \\ & \frac{Agent\ Room\ Time}{Time\ Weight} \end{aligned}$$

Fig. 6. Room size and height adjusting diagram. Animated GIF by Anni Dai

2.4.2 Co-creation

The human interaction parts come in where the weights are adjusted. This project created a UI where the **agent quantity weight** and **time weight** (mentioned above) of each room are presented and to be used by the architect. This is where the co-creation happens. While the programme still runs a machine learning model, the user or the architect can have real-time observation of the building. By tweaking the slider on the UI, Architects use their tacit knowledge to decide on how big the rooms should be, while the machine learning model is still processing and suggesting potential connections and entries. Machine Learning reads agent behaviors and makes suggestions on space layout, while architects use their tacit knowledge controlling and correcting mistakes the machine cannot recognize. Co-creation uses both sides advantages and makes the most of architectural design (Fig. 7).

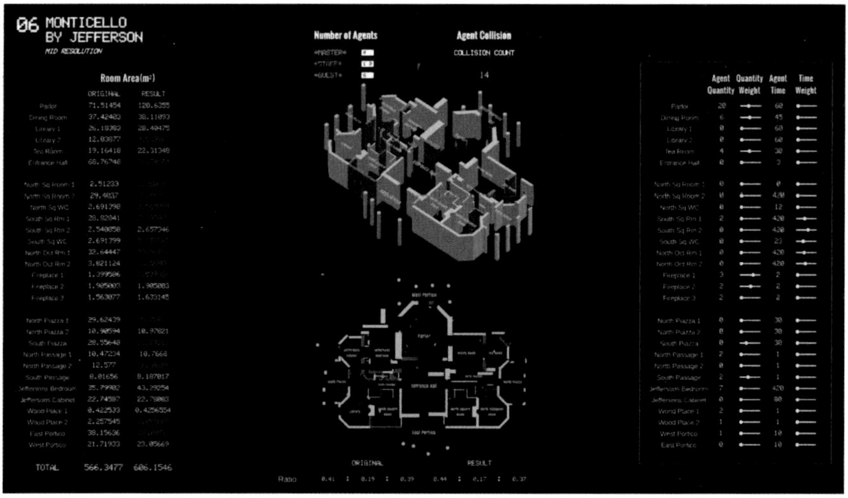

Fig. 7. Co-creation indication diagram. Image by Anni Dai

3 Summary

3.1 Results Comparison

Three Resolutions

The high resolution, in overall, varies the most from the original building. The example below shows representative similarities and differences among the 3 resolutions. All three resolutions create connections between rooms. However, the low resolution often eliminates and simplifies rooms; mid resolution creates more room size variations; high resolution tends to create spaces and creates openings at unprecedented places (Fig. 8).

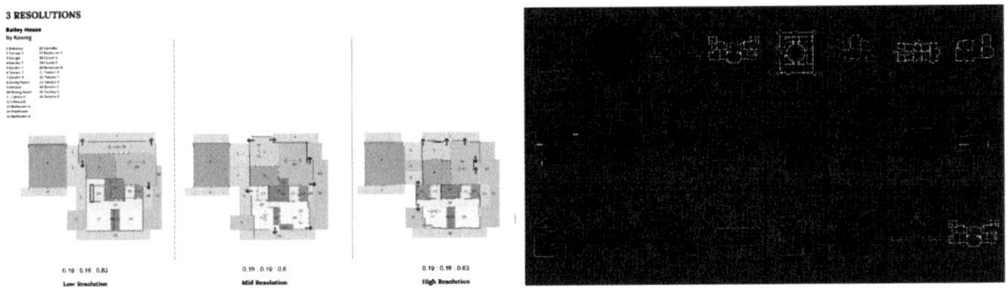

Fig. 8. Three resolutions and twenty precedents comparison diagrams. Image by Anni Dai

Twenty Precedents

Among the twenty housing precedents, the author finds modern buildings tend to have less variation compared to the ancient and classical buildings. Influenced by minimalism, most modern houses simplify room functions to only what was necessary. While in ancient houses such as 'Villa of the Mysteries', rooms such as pharmacy, furnace, are not used as often as any functions in modern architecture. While in classical architecture, rooms such as 'anti chamber', 'rotunda' were included because of their ritual value or aesthetics. Circulation was an important part of the modern house design, which is another reason why this research has less influence on modern buildings than on the ancient and classical buildings.

4 Conclusions

This paper first looked at the theory background on simulation and agent simulation's role in architectural design methodology, and machine learning's role in design scientific research. Then the author argued the concept of design automation, and proposed agent simulation based machine learning as a new generation design method, along with the importance of co-creation between machine's data processing ability and an architect's tacit knowledge.

This research was an attempt on creating a co-creation between machine learning and an architect. The result achieved the research purpose. However there is room for improvement. The co-creation is restricted within room size and height, which can be developed into specifying wall positions. The flexibility of agent simulation can be further explored to where agent positions can be adjusted in real time, and further explore how it will reflect on space reconfiguration.

References

1. Anon (n.d.) ABSTRACT–agent based semiology. http://www.parametricsemiology.com/?page_id=204#:~:text=The%20aim%20of%20agent%20based. Accessed 22 Nov 2021
2. Archer B (1979) Design as a discipline. Des Stud 1(1):17–20
3. Baudrillard J (1994) Simulacra and simulation. University of Michigan Press
4. Bayazit N (2004) Investigating design: a review of forty years of design research. Des Issues 20(1):16–29
5. Banks J, Carson J, Nelson B, Nicol D (2001). Discrete-Event System Simulation. Prentice Hall, p 3
6. Mitchell T (1997) Machine learning. McGraw Hill, New York
7. Penn A, Turner A (2002) Space syntax based agent simulation. Springer
8. Polanyi M (1958) Personal knowledge: towards a post-critical philosophy

Research on Architectural Generation Design of Specific Architect's Sketch Based on Image-To-Image Translation

Yuqian Li, Weiguo Xu(✉), and Xingchen Liu

School of Architecture, Tsinghua University, Beijing, China
{liyuqian17,liuxc20}@mails.tsinghua.edu.cn,
xwg@mail.tsinghua.edu.cn

Abstract. Sketch is a way for architects to communicate with others. Architects record their own ideas through rapid drawing. However, sketches are abstract, vague, and even ambiguous. To this end, architects need to spend a lot of time, through modeling and other means, to present the architectural plan that can be understood by people. However, this method is time-consuming and laborious. Due to the development of deep learning technology, especially convolutional neural networks (CNN) and generative adversarial networks (GAN), they have shown great advantages in the field of image recognition and generation. With the help of these technologies, ambiguous architectural sketches can be directly transformed into architectural scheme drawings, and architects' creative intentions can be continuously improved and developed, It will be very convenient and efficient. Therefore, based on the image-to-image translation, this paper realizes the mapping from architectural sketches to architectural scheme drawings with the help of CycleGAN. Through the analysis of the architectural generation design results of Frank Gehry's and Alberto Campo Baeza's architectural sketches, firstly, the feasibility of this method is verified. Secondly, it is found that this method can well complete the identification of sketch boundaries. In the generated scheme drawings, it can not only reflect the volume and lighting changes of the building, but also reflect the architect's creative intention and style to a large extent, The side reflects the cognitive ability of this method to architectural design.

Keywords: Architectural sketch · Architectural scheme drawings · Image-to-Image translation · CycleGAN

1 Introduction

With the development of computer technology, especially computer graphics, the use of 3D modeling, rendering and other means of design has become very convenient and efficient, but the concept sketch for architects in the process of architectural creation, always has a unique irreplaceable. A sketch has a definite goal and is a picture with a specific intention. Sketch is an action process in which architects record their own ideas through rapid drawing.

P. F. Yuan et al. (eds.), *Hybrid Intelligence*, Computational Design and Robotic Fabrication,
https://doi.org/10.1007/978-981-19-8637-6_28

Le Corbusier said: I like to use painting to express design ideas. Painting can be faster and more realistic. In addition to Corbusier, famous architects such as Mies van der Rohe, Frank Gehry and Zaha Hadid all like to express their design ideas and intentions through sketches. To them, the sketch is like an unfinished architectural work of an architect, a way of communication between the architect and others, which is full of the complex mental process of the architect and shows the architectural ideal of the architect.

However, sketches are often abstract, fuzzy and even ambiguous. Take Zaha Hadid as an example. Her sketches are very abstract. Zaha once said that she hopes to use abstract expression to break the thinking of traditional architecture. For this kind of intentional design expression, people's interpretation of it is often not clear, even sometimes, for the architect himself can not imagine such a fuzzy sketch, its development into a building will be. So, generally, architects and their teams need to spend a lot of time, through modeling to present architectural scheme drawings that can be understood by people. If the output is not satisfactory, they also need much time to modify, and constantly repeat this process.

Because of the development of deep learning, especially convolutional neural networks (CNNs) and generative adversarial networks (GANs), they have shown great advantages in the field of image recognition and generation. If we develop these technologies with architectural design, ambiguous sketches can be directly transformed into scheme drawings, and architects' creative intentions can be continuously improved and developed, then the work will be very convenient and efficient.

It is worth mentioning that from modernism, post-modernism to deconstructionism, architects of various schools emerge in endlessly, and their design styles are quite different. Extensive research increases the difficulty of understanding sketches. It is very necessary to study from individual to general.

Therefore, with the help of deep learning, this paper will extract the architectural sketch features of a specific architect, and produce the corresponding architectural scheme drawings, so as to realize the translation between the sketch and the building image, which could help the process of architectural design and achieve the purpose of convenience and efficiency.

2 Related Work

In the early stage of AI, machine learning has become a cutting-edge technology and has been applied in many researches. The black box process of machine learning is very similar to the process of people's cognition of the world. Through the learning and training of a large number of data, we can find its inherent law and map this law to the generated results. At the beginning of the twenty-first century, deep learning has shown greater advantages in big data training by increasing the depth of hidden layer of neural network in machine learning, and has successfully made a breakthrough in speech, image and other processing. This also makes the field of image translation as a branch of deep learning research develop rapidly.

In the deep neural network, convolution neural network is the most effective network for image processing. CNN can extract the specific features in the image, that is, a group of computational elements can process the visual information hierarchically through forward feedback. When using CNN for model recognition, the representation information learned by CNN will change with the change of network level, that is, with the increase of network level, the focus of image representation will change from the initial specific pixel to the image content. Therefore, the low-level CNN will capture the style representation of the image, while the high-level network will capture the content representation of the image. On the basis of this theory, scholars such as Gatys et al. [2, 3] proposed the style transfer and applied it to the creation of artists' style paintings.

Although the style transfer can achieve image translation successfully, because this method needs deep neural network in training, and the trained model can only be applied to specific image migration, the image translation algorithm based on GAN shows better performance.

In 2014, Goodflow et al. [4] and others proposed the GAN, which is composed of two models: generator and discriminator. Since the discriminator will constantly judge the similarity between the model generated by the generator and the original model, Isola [5] and others proposed Pix2Pix, an image translation framework based on GAN. After that, Pix2PixHD solves the problem of high-resolution image translation on the basis of Pix2Pix, and Vid2Vid solves the problem of high-resolution video image translation on the basis of Pix2PixHD. Since then, many image-to-image algorithms based on GAN have emerged.

Whether it is based on CNN or GAN, its goal is to learn the mapping between the input image and the output image, that is, image-to-image translation requires not only the generation of dual image in the target domain according to the image in the source domain, but also the consistency of the image in the translation process.

3 Methodology

3.1 Network Architecture

As mentioned above, GAN has obvious advantages in the field of image generation. Pix2Pix, Pix2PixHD and Vid2Vid are known as the trilogy in the field of image translation, which can achieve most of the work of image-to-image translation. Scholar Huang and Zheng [6] and Zheng et al. [7] have done a lot of research on the generation of room types through Pix2Pix. These results have inspired architects' design ideas to a certain extent.

However, in the process of architectural design, as a way to express the design intention, sketch is often not achieved overnight, it needs to be continuously developed and improved, it is a relatively circular process, rather than one-way. Therefore, this paper chooses the Cyclegan based on Pix2Pix. Cyclegan has a continuous cycle of generation process, in order to simulate the design process of sketch implementation scheme.

CycleGAN is a technology that uses unpaired image collections from two different domains to train an unsupervised image conversion model through the GAN architecture.

The CycleGAN (Fig. 1) presents an approach for learning to translate an image from a source domain X to a target domain Y in the absence of paired examples. The goal is to learn a mapping G: X → Y, such that the distribution of images from G(X) is indistinguishable from the distribution Y using an adversarial loss. Because this mapping

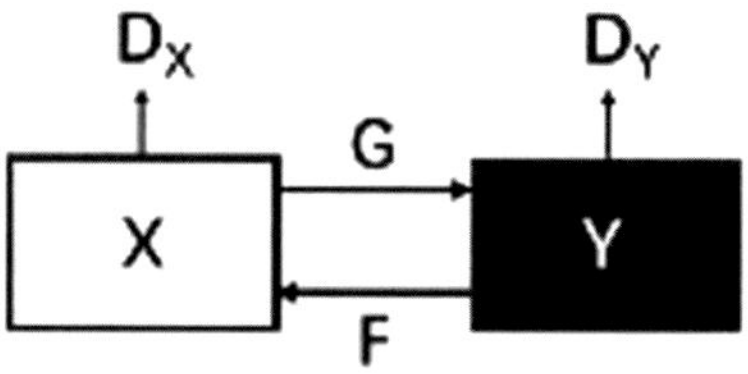

Fig. 1. The network architecture of CycleGAN

is highly under-constrained, CycleGAN couple it with an inverse mapping F: Y → X and introduce a cycle consistency loss to enforce F(G(X))≈X (and vice versa).

There are many creative applications of CycleGAN. It was first used in the transfer of photographs to an artist's painting such as converting a photograph into a Van Gogh painting. Jack Clark used this algorithm to convert ancient maps of Babylon, Jerusalem, and London into modern Google Maps and satellite views. Mario Klingemann used the code to translate portraits into dollface. Besides, CycleGAN is also used in medical fields, such as translating MRI to CT data.

3.2 Data Preparation

3.2.1 Sample Selection

The purpose of this paper is to extract sketch features and generate architectural scheme drawings through cyclegan algorithm. As mentioned above, the sketch itself has ambiguity, and each architect has his own design style. Collecting mixed architectural style data will increase the difficulty of computer cognition.

Frank Gehry is a master of deconstruction. His design is characterized by peculiar and irregular curves, and his sketches are relatively abstract. Alberto Campo Baeza is a Spanish modernist architect. His architecture is pure, elegant and poetic. He is good at using the combination of flowing space and light. His sketches are simple and readable. Therefore, this paper selects these two famous architects who are good at sketching as learning samples, and because of their different sketching styles, we can get a closer understanding of computer cognition through comparison.

3.2.2 Data Collection

Due to the limitation of the number of projects, architects can not collect a large number of data, and to achieve one-to-one correspondence also increases the difficulty of data collection. In this paper, through the collection of different perspectives of the same building, different scenes of the same building, for the two architects, we collected 100, that is, 50 pairs, 80, that is, 40 pairs of data as the research samples, 80% as training data, 20% as test data.

3.3 Training Process

Take the cycle generation from Gehry's sketch to scheme drawing as an example (Fig. 2).

Fig. 2. The data of Gehry's work

Gehry's sketches and building images are taken as two groups of pictures and they are unpaired, while the sketch and building images are one-to-one correspondence.

The CycleGAN will develop an architecture of two GANs, and each GAN has a discriminator and a generator model, meaning there are four models in total in the architecture.

The first GAN will generate pictures of sketches given pictures of building images, and the second GAN will generate scheme drawings given pictures of sketch.

Each GAN has a conditional generator model that will synthesize an image given an input image. And each GAN has a discriminator model to predict how likely the generated image is to have come from the target image collection. The discriminator and generator models for a GAN are trained under normal adversarial loss like a standard GAN model.

So far, the models are sufficient for generating plausible images in the target domain but are not translations of the input image.

Each of the GANs are also updated using cycle consistency loss. This is designed to encourage the synthesized images in the target domain that are translations of the input image.

Cycle consistency loss compares an input picture to the Cycle GAN to the generated picture and calculates the difference between the two, e.g. using the L1 norm or summed absolute difference in pixel values.

There are two ways in which cycle consistency loss is calculated and used to update the generator models each training iteration.

The first GAN (GAN 1) will take an image of sketches, generate image of scheme drawings, which is provided as input to the second GAN (GAN 2), which in turn will generate an image of sketches. The cycle consistency loss calculates the difference between the image input to GAN 1 and the image output by GAN 2 and the generator models are updated accordingly to reduce the difference in the images.

This is a forward-cycle for cycle consistency loss. The same process is related in reverse for a backward cycle consistency loss from generator 2 to generator 1 and comparing the original pictures of the buildings to the generated picture of scheme drawings.

4 Results

See (Fig. 3).

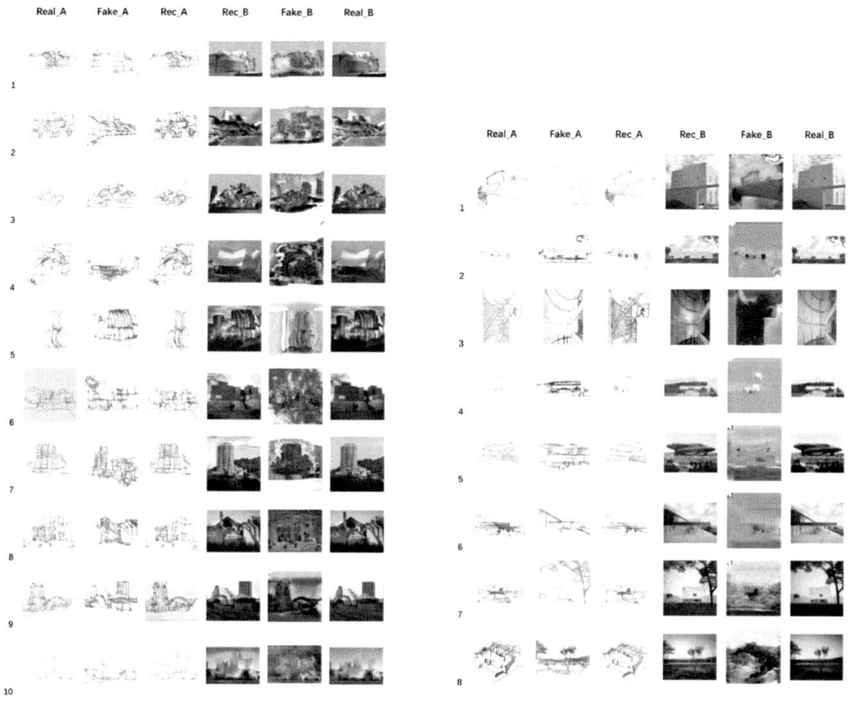

a. The output b. The output of Alberto's work

Fig. 3. The results of the test training

4.1 Data Preparation

Whether it is Gary's work (Fig. 4) or Alberto's work (Fig. 5), in the generated scheme drawing, the sketch boundary recognition is well completed. In the scheme drawing, the architecture, sky and other environmental factors are clearly expressed.

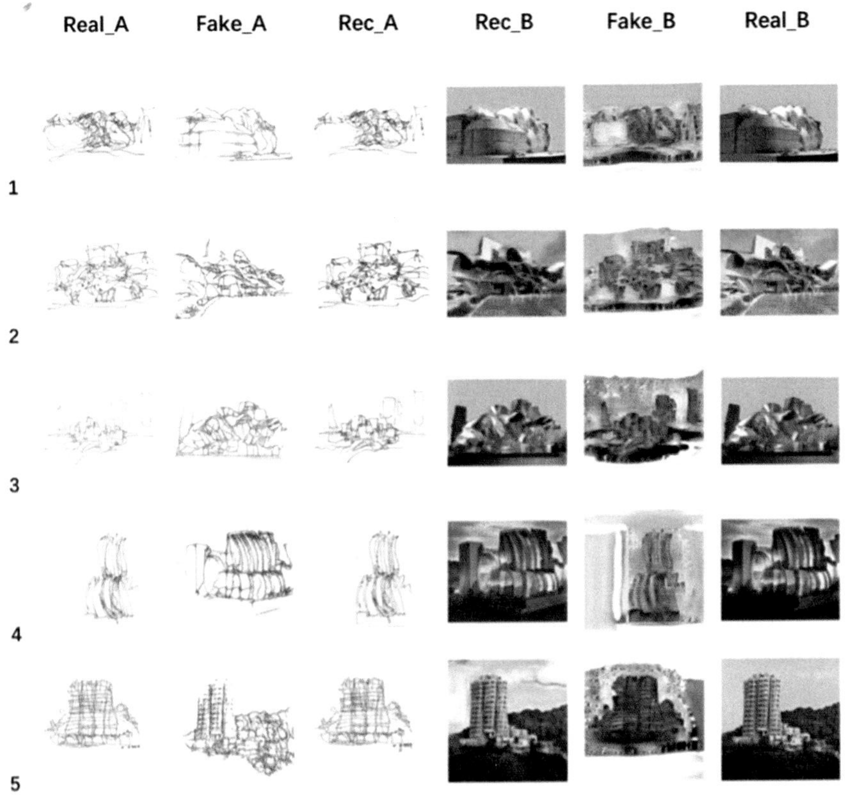

Fig. 4. xxx

4.2 Certain Cognitive Ability for Different Perspectives of the Same Building

In order to expand the sample size, this study collected different perspectives of the same building. It can be seen in the Fig. 6 that due to the continuity of design elements from different perspectives, the generated scheme drawings have similar color attributes. This may be because the computer recognizes the similarity between images to a certain extent.

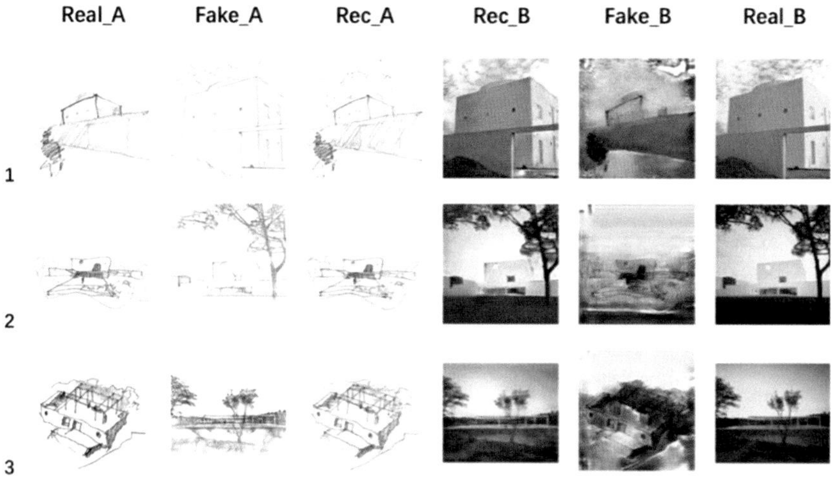

Fig. 5. xxx

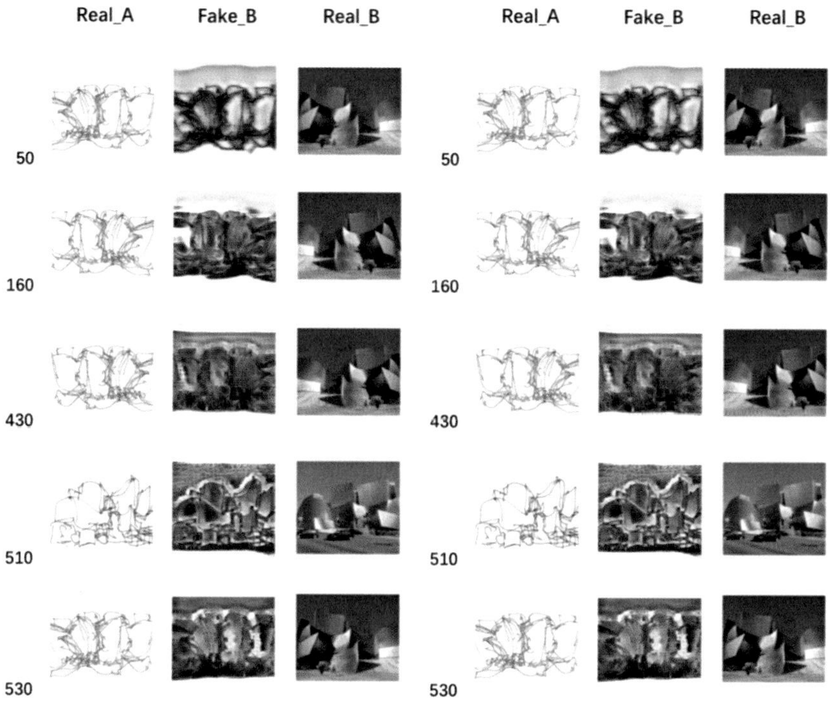

Fig. 6. xxx

4.3 The Generated Scheme Drawings Can Reflect the architect's Creative Style

In Alberto's design, he is good at using natures and hopes to integrate architecture and environment. It can be seen from the Fig. 7 that the generated architectural scheme drawing has a very similar architectural scene, and the natural elements, light and environment are fully expressed.

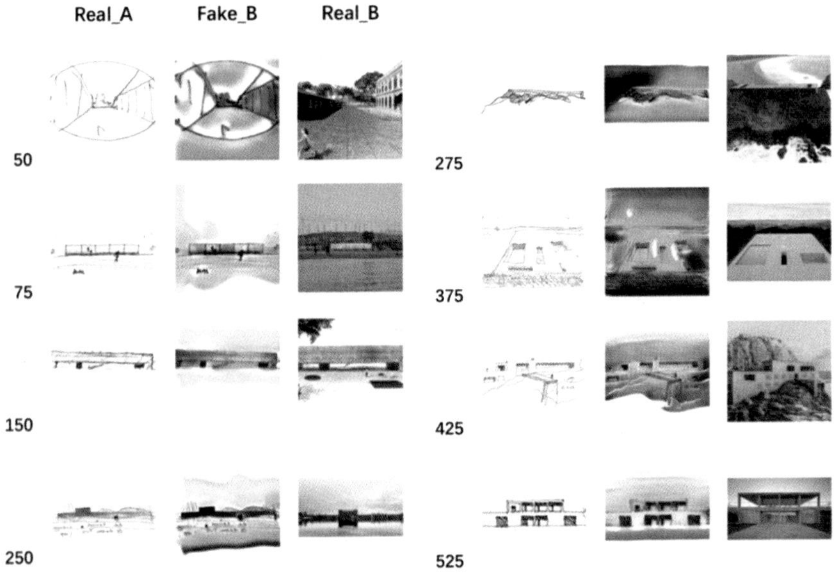

Fig. 7. xxx

4.4 The Generated Scheme Drawings Can Reflect Good Shadow Changes

In the Fig. 8, the architectural light and shadow expressed by Alberto in real buildings are well interpreted in the generated renderings. It creates the same interior atmosphere as the architect's intention.

It can also be seen from the Fig. 9 that in the design of Gehry, the light and shadow between the building volumes are also expressed, which interprets the relationship between the building volumes.

5 Conclusion and Discussion

In this study, based on image to image translation, with the help of GycleGAN, the mapping from architectural sketches to building images is realized. Through the analysis of the architectural generation design results of Gehry's and Alberto's architectural sketches, the feasibility of this method is verified. Secondly, it is found that this method can well complete the identification of sketch boundaries. In the generated scheme

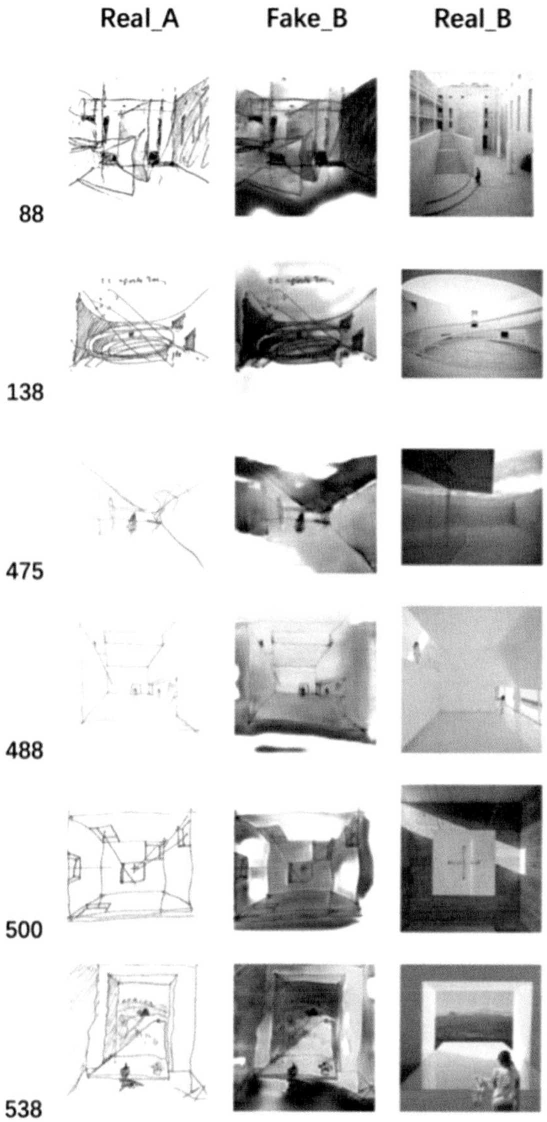

Fig. 8. The results of the test training

drawings, it can not only reflect the volume and lighting changes of the building, but also reflect the architect's creative intention and style to a large extent, the side reflects the cognitive ability of this method to architectural design.

Through the horizontal comparison of the sketch generation results of the two architects, we can see that the more abstract the sketch is, the more difficult it is to identify, the clearer and simpler the sketch is, the better the effect of the generated scheme drawing will be.

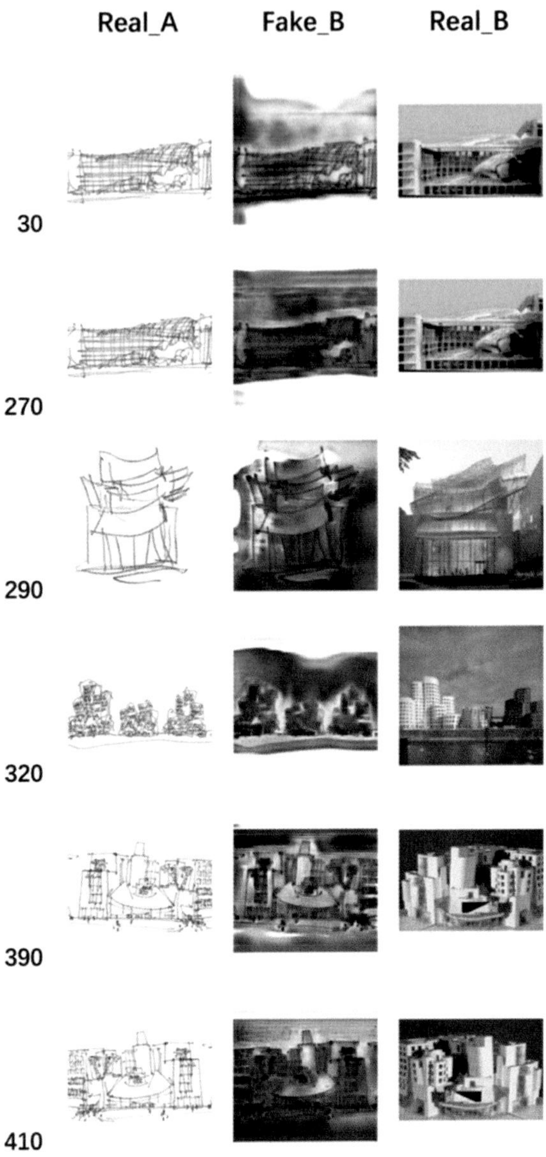

Fig. 9. The results of the test training

Of course, this also shows that the cognitive ability of computer needs to be further strengthened. This is also one of the important tasks of the next research.

References

1. Simonyan K, Zisserman A (2014). Very deep convolutional networks for large-scale image recognition. arXiv
2. Gatys LA, Ecker AS, Bethge M (2015) Texture synthesis using convolutional neural networks. MIT Press
3. Gatys LA, Ecker AS, Bethge M (2015) A neural algorithm of artistic style. J Vis
4. Goodfellow IJ, Pouget-Abadie J, Mirza M, Xu B, Warde-Farley D, Ozair S et al (2014) Generative adversarial networks. In: Advances in neural information processing systems, vol 3, pp 2672–2680
5. Isola P, Zhu JY, Zhou T, Efros AA (2016) Image-to-image translation with conditional adversarial networks. In: IEEE conference on computer vision & pattern recognition. IEEE
6. Huang W, Zheng H (2018) Architectural drawings recognition and generation through machine learning. In: Proceedings of the 38th annual conference of the association for computer aided design in architecture (ACADIA). Mexico City, Mexico 18–20 Oct 2018, pp 156–165. ISBN 978-0-692-17729-7
7. Zheng H, An K, Wei J, Ren Y (2020) Apartment floor plans generation via generative adversarial networks. In: Anthropocene, design in the age of humans-proceedings of the 25th CAADRIA conference, vol 2. Chulalongkorn University, Bangkok, Thailand, 5–6 Aug 2020, pp 599–608
8. Zhu JY, Park T, Isola P, Efros AA (2017) Unpaired image-to-image translation using cycle-consistent adversarial networks

Predicting the Vitality of Stores Along the Street Based on Business Type Sequence via Recurrent Neural Network

Zidong Liu[1], Yan Li[2], and Xiao Xiao[3](✉)

[1] The University of Texas at Austin, 310 Inner Campus Drive, Austin TX 78712, USA
zidong.liu.22@utexas.edu

[2] University of Sydney, Camperdown, NSW 2006, Australia
yali3816@uni.sydney.edu.au

[3] Politecnico di Torino, Corso Duca degli Abruzzi, 24, 10129 Torino, TO, Italy
xiao.xiao@polito.it

Abstract. The rational planning of store types and locations to maximize street vitality is essential in real estate planning. Traditional business planning relies heavily on the subjective experience of developers. Currently, developers have access to low-resolution urban data to support their decision making, and researchers have done much image-based machine learning research from the scale of urban texture. However, there is still a lack of research on the functional layout with shop-level accuracy. This paper uses a sequence-based neural network (RNN) to explore the relationship between the sequence of store types along a street and its commercial vitality. Currently, the use of RNNs in the architectural and urban fields is very rare. We use customer review data of 80streets from O2O platforms to represent the store vitality degree. In the machine learning model, the input is the sequence of store types on the street, and the output is the corresponding sequence of business vitality indexes. After training and evaluation, the model was shown to have acceptable accuracy. We further combined this evaluation model with a genetic algorithm to develop a business planning optimization tool to maximize the overall street business value, thus guiding real estate business planning at a high resolution.

Keywords: Machine learning · Big data analysis · Vitality prediction · Recurrent neural network · Genetic algorithm

1 Introduction

Business management experience shows that the order in which stores are located on the street has a significant effect on business. Stores at the street corner tend to have higher popularity and therefore higher rents. People without specific goals are more likely to shop at the first supermarket they see. Store's neighbours may also have a complex impact on its operation, depending on the types of stores. For example, two supermarkets that are located close to each other may have a vicious competition, but

P. F. Yuan et al. (eds.), *Hybrid Intelligence*, Computational Design and Robotic Fabrication,
https://doi.org/10.1007/978-981-19-8637-6_29

for McDonald's and KFC, putting them together may help enhance their visibility on the street. A bank placed next to a luxury store may help to increase the sales of the luxury store.

1.1 Problem Statement

The current business planning model still relies heavily on the subjective experience of real estate developers, which leads to the uncertainty in planning results and adversely affects the profitability of businesses. There has been increasingly data analysis such as customer base analysis and regional vitality analysis to support low-resolution issues like the proportion of store types [5]. However, the resolution of these data is still not sufficient to guide shop-level planning. There is still a lack of research for more precise planning such as the location sequence of business types along a street. Therefore, this paper has a strong practical research significance.

1.2 Literature Review

There have been many studies on region vitality through machine learning. However, most of them are image-based (GAN-based) and do not achieve store-level accuracy. Among these studies, GAN models are predominant. A study transforms the citizens' cycling route data into an urban heat map to represent community vitality and explores its relationship with urban fabric [8]. Similar approaches can be used to predict other urban metrics, such as urban crime rate [3] and commercial value [7]. However, due to the limitation of computation and data resolution, the generated results always have ambiguous areas. This is the reason why some studies have attempted to vectorize images before performing machine learning [9].

In this study, we choose RNN as the basic neural network model. RNN is based on sequential data, widely used in natural language processing, advertising recommendations and so on. Compared with other models, RNN's features are highly compatible with our research object and goal. Here are the reasons:

1. RNN uses sequential data as input and output.
2. In RNN models, the order of data has a decisive influence on the results.
3. The input and output in the RNN training set can be of different lengths.

Among the sparse RNN-based studies in the architectural and urban fields, there is one relevant to the topic on business optimization [4]. Using the behaviour of pedestrian inside a mall as data, the researchers trained a behavioural predictor that can infer the pedestrian's walking direction. This model in turn guides the design of the mall, leading to higher commercial value on the pedestrian's expected route. In addition, some researchers have tried to use RNNs from the perspective of software operation.

1.3 Project Goal

The paper aims to explore the relationship between the order of store business types along the street and their commercial vitality by a sequence-based neural network (RNN). The machine learning model simulates the behaviour of people walking down the street and passing through stores. In the model, the input is the sequence of store types and the output is the sequence of vitality indexes. After training, this machine learning model can predict the vitality of each store, thus guiding real estate business planning at a high resolution.

2 Methodology

The research process is divided into three parts: data collection, model training and model evaluation. We collected data of stores along the streets from O2O platforms including Gaode Map, Meituan and Dianping and transformed these data into sequences that can represent the types of stores and their sales status. After that, the sequence data are entered into the seq2seq model and trained in the LSTM layers. Then the model outputs the sequence of letters that can represent the vitality level. Finally, we use Cross Entropy Loss Function and the prediction accuracy function to evaluate the effectiveness of this prediction model (Fig. 1).

After obtaining the prediction model, a street outside the training set is used to verify the effectiveness of the model. Furthermore, we can combine this prediction model with a genetic algorithm to develop a business planning optimization tool: it automatically gives the best ranking order based on the input store types to maximize the business value of the whole street.

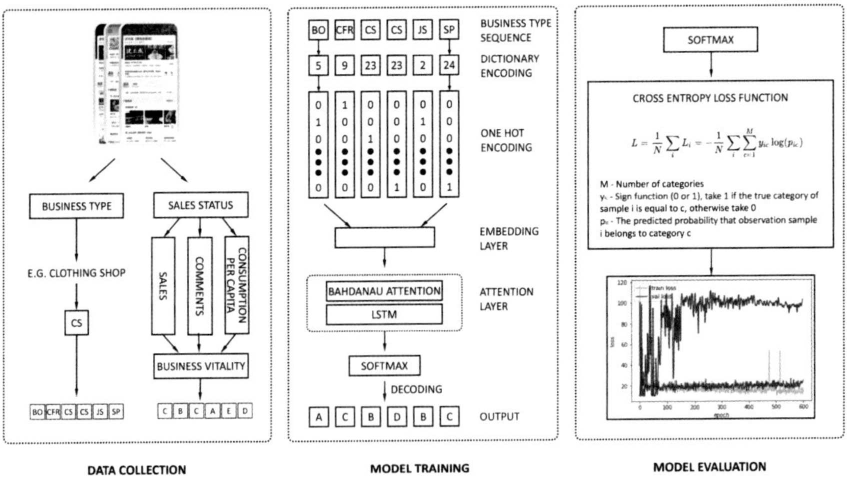

Fig. 1. Research framework

2.1 Data Collection

We selected 80 streets, 1261 stores, and 29 store types from 8 representative cities in China from O2O platforms (Fig. 2). As the main O2O platforms vary from city to city and different merchants on the same street might choose different platforms, it is necessary to collate data from multiple mainstream platforms. In this research, the commercial data was comprehensively collected on Meituan, Dianping and Gaode Map. In this way, we collect as complete data as possible for every store on each of the 80 streets. Regarding a tiny number of shops with missing data, we take the average of the nearby shops of the same type as a replacement. O2O platforms provide a variety of information: shop type, number of reviews, per capita spending. There is also information on sales volume (some semi-annual, some monthly).

Fig. 2. POI data statistics

2.2 Data Processing

Quantitative assessment of business vitality is very complex since no platform provides direct information on the sales of every shop in the street. Based on the assumption that all shops have the same review rate, we can use the number of reviews multiplied by the per capita spend to estimate the sales of each shop. However, after research, we found that the type of shop significantly impacts the number of reviews. For example, milk tea shops and fast-food restaurants tend to have very high review rates. In contrast, some support facilities such as banks and bicycle repair points have low review rates though their existence can have a significant impact on the surrounding stores.

In order to provide a more objective assessment of the commercial viability of shops, a relative quantity approach is applied here. For these 1261 shops, we compare the number of reviews multiplied by the value of per capita consumption within each type of shop, and then classify their relative vitality into five classes: ABCDE. For example, there are 75 pastry shops, so we rank their vitality, then the top 10 are ranked A, 11–25 are ranked B, and so on (Fig. 3). For those supporting facilities with few reviews like banks, we unify their vitality value C. After calculating the vitality values of the stores in these 80 streets, we can get some interesting statistical conclusions. Shanghai, Nanjing, Wuhan and Suzhou have higher average store vitality than Kunming and Changsha, which is in line with daily experience: store vitality is positively correlated with the economic development of a city (Fig. 4).

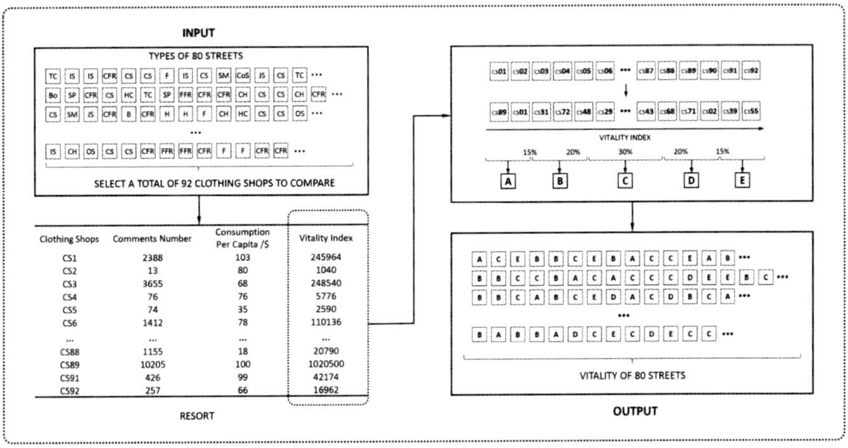

Fig. 3. Translating business data into relative vitality values

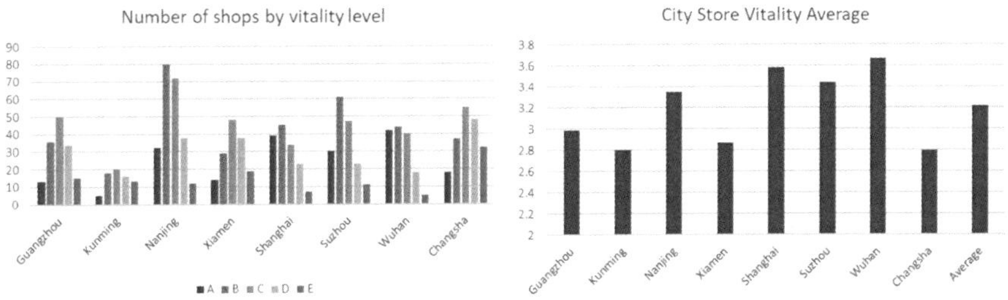

Fig. 4. Comparison of city store vitality

2.3 Training Set Expansion

The machine learning model simulates the behaviour of people walking down the street and passing through stores that is a one-way experience. However, since both ends of the street can be the starting points, the sequences can all be trained in reverse, so the dataset was expanded from 80 streets to 160. To expand the sample size further, we extracted all the subsequences whose length are greater than five from the beginning of these

160 sequences (Fig. 5). This is reasonable because we may not go through the whole street in daily shopping but finish shopping after passing several stores. By this method, we obtained a total of 1820 sequential data. This method of expanding the database is inspired by the research of Weixin Huang's team on the modelling operation, in which they also applied a similar subsequence approach [2].

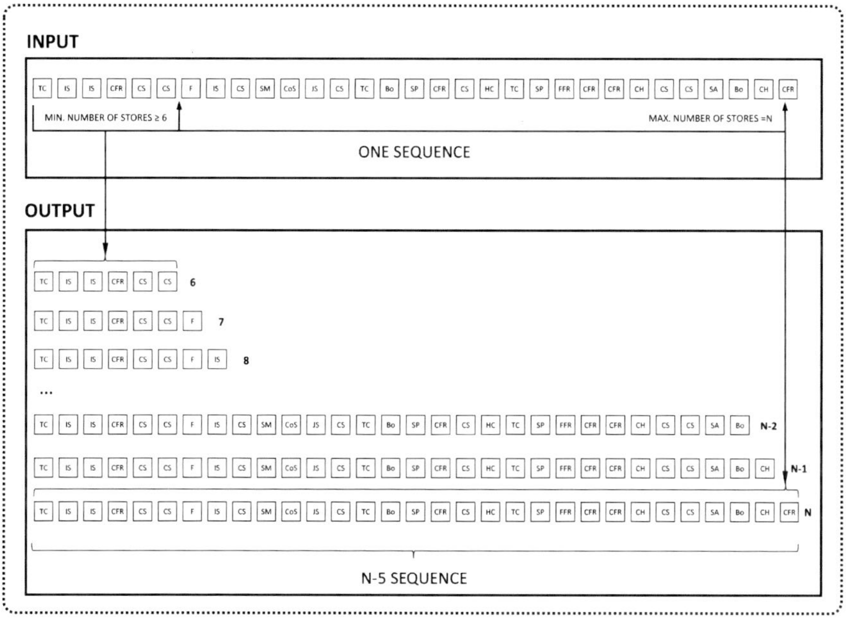

Fig. 5. Sub-sequences generation for the training set expansion

2.4 Machine Learning

Machine training is based on the Seq2Seq attention model (Fig. 1). Data set is divided into the training set, validation set and test set according to the ratio of 7:2:1. We evaluate the effectiveness of this model by two functions: Cross Entropy Loss Function (Eq. 1) and the Prediction Accuracy Function (Eq. 2). The Prediction Accuracy Function is formulated by the specific issue of this paper. The difference between the predicted value and the target value varies depending on the predicted value (Table 1). The accuracy of random guess is the sum of all the values in Table 3 divided by 25 equals 46.56%.

$$L = \frac{1}{N}\sum_{i} L_i = -\frac{1}{N}\sum_{i} L_i \sum_{c=1}^{M} y_{ic}\log(P_{ic}) \tag{1}$$

Table 1. Accuracy calculation table

Accuracy		Target vitality				
		A	B	C	D	E
Perdicted vitality	A	1	0.66	0	0	0
	B	0.75	1	0.5	0.33	0.25
	C	0.5	0.66	1	0.66	0.5
	D	0.25	0.33	0.5	1	0.75
	E	0	0	0	0	1

M: Number of categories y_{ic}: Sign function (0 or 1)

P_{ic}: The predicted probability that$i$th item belongs to category c

$$P = \frac{1}{n_t}\sum_{i=0}^{\min(n_t,n_p)} 1 - \frac{\Delta r_i}{\max(R - r_t, r_t - 1)} \times 100\%, \quad \Delta r_i = \left|r_{ip} - r_{it}\right| \tag{2}$$

R: Range of vitality level

n_t: Target sequence length n_p: Predicted sequence length

r_{it}: Target vitality of the i th term r_{ip}: Predicted vitality of the i th term

The training results after 600 epochs with 15 batches per epoch are shown in Fig. 6. The training effect is good. The model never enters the overfitting state since the training loss curve and the validation loss curve remain stable and the accuracy curve keeps increasing.

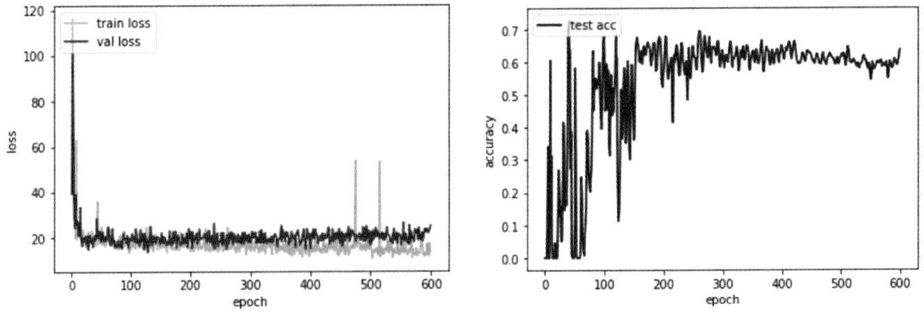

Fig. 6. Training results

3 Case Study

We chose Gungyuan West Street in Nanjing, outside the training set, to apply our trained evaluation model. Gongyuan West Street is in the historical centre area of Nanjing, with a wide variety of businesses and high popularity. The commercial situation of the site is shown in Fig. 7.

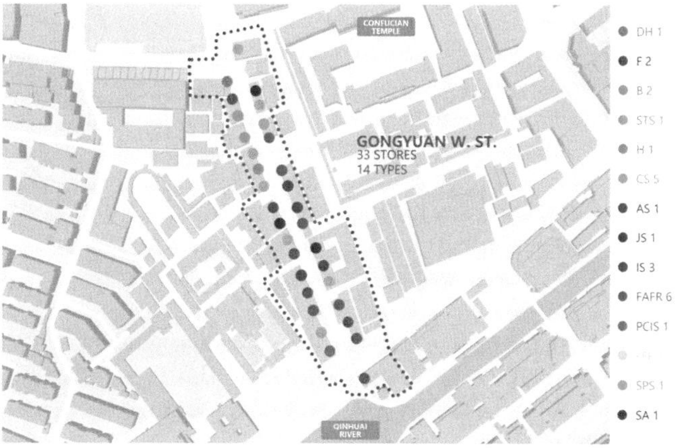

East		West	
Type	Vitality	Type	Vitality
CS	D	DH	B
FFR	D	F	C
JS	D	B	A
CS	B	B	B
CS	D	StS	A
FAFR	E	H	D
FAFR	D	CS	C
IS	D	AS	B
IS	E	JS	C
FAFR	A	CS	B
JS	C	F	C
SA	B	IS	D
CS	D	B	A
FAFR	C	FAFR	B
B	B	StS	B
FAFR	C	HC	E
B	C		

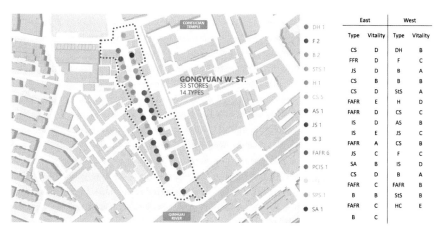

Fig. 7. Vitality of stores in Gongyuan West Street

The types of stores in West Street were input into the trained model, and the output vitality prediction was "b c b c b c b c b c b c b c b c b c", with an accuracy of 77% according to Formula (2) (Table 1). Experiment 1 adds a movie theatre at the beginning of the street, and the model had a higher expectation of street vitality (Fig. 8). Experiment 2 arranges the same kinds of stores together. The model also has a higher expectation of the overall vitality of the street (Fig. 9 and Table 2).

```
testSeq = ["TC","DH","F","B","B","StS","H","CS","AS","JS","CS","F","IS","B","FAFR","StS","HC"]
testSeq_str=" ".join(str(i) for i in testSeq).lower()
print(m.infer(testSeq_str)[0])
[101]  ✓ 0.7s
...  b c b b c b b c b b c b b c b c b
```

Fig. 8. Vitality of stores in Gongyuan West Street after Experiment 1

3.1 Vitality Optimization Based on Genetic Algorithm

Further, we combined this evaluation model with a genetic algorithm to develop a reference tool that can provide suggestions for optimizing the location of stores. The vitality levels correspond to specific numbers: A scores 5, B scores 4, C scores 3, D scores 2,

```
testSeq = ["DH","CS","CS","FAFR","F","F","StS","StS","B","B","H","AS","JS","IS","B","HC"]
testSeq_str=" ".join(str(i) for i in testSeq).lower()
print(m.infer(testSeq_str)[0])
[226]  ✓ 0.6s
... b c b c b c b b c b c b a a a a
```

Fig. 9. Vitality of stores in Gongyuan West Street after Experiment 2

Table 2. Expected street vitality after location adjustment

Target	B	C	A	B	A	D	C	B	C	B	C	D	A	B	B	E	
Prediction	B	C	B	C	B	C	B	C	B	C	B	C	B	C	B	C	
Accuracy	100%	100%	80%	75%	80%	75%	66%	75%	66%	75%	66%	75%	80%	75%	100%	50%	77%

and E scores 1. The genetic algorithm takes the total score of vitality as the optimization target. At each iteration, the genetic algorithm randomly swaps two store locations. Through continuous iterations, the genetic algorithm then gives the optimal solution of this prediction model.

After hundreds of iterations, the system did find a solution with a high vitality index: "CS F STS CS AS JS IS STS B HC B DH B H F FAFR". The vitality prediction for this sequence is: "B C B A A A A A A A A A A A A A" with a score of 76. Figure 10 records an evolutionary process.

```
First Generation Street Vatility is : 56
b c b c b c b c b c b c b c b c
10's Generation Street Vatility is : 58
b h b sts cs is cs as js dh fafr f b f sts hc
b c b b c b b c b b c b c b c b
20's Generation Street Vatility is : 59
hc b dh cs is cs sts as js b fafr f h f sts b
b c b b c b b c b b c b b c b b
30's Generation Street Vatility is : 61
cs b sts cs is js dh b hc b sts f h f fafr as
b c b b c b b c b b b b b b b b
40's Generation Street Vatility is : 67
cs f sts cs is js sts hc b as dh b h f fafr b
b c b b c b b c b a a a a b a a
50's Generation Street Vatility is : 73
cs f f cs as js sts hc b is b dh fafr h sts b
b c a a a a a a a a a a b c b a
60's Generation Street Vatility is : 75
cs cs f f as js sts is b hc b dh b h fafr sts
b c b a a a a a a a a a a a b a
70's Generation Street Vatility is : 76
f f sts cs as js is sts b hc b dh b h fafr cs
b c b a a a a a a a a a a a a a
80's Generation Street Vatility is : 76
cs f sts cs as js is sts b hc b dh b h f fafr
b c b a a a a a a a a a a a a a
```

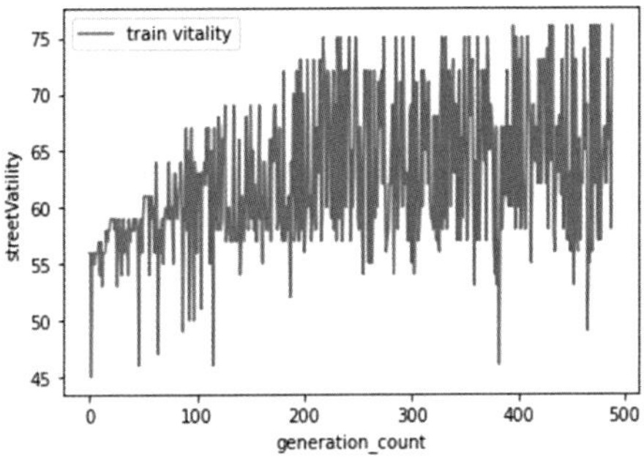

Fig. 10. Overall street vitality optimization by genetic algorithm

4 Conclusion

This paper presents a method that uses machine learning to predict commercial vitality along streets and provide optimization advice. This study has important practical value for high-precision business planning. Although there have been many machine learning studies based on urban texture images, few studies are accurate to the prediction of vitality of stores. Compared with previous studies, this study creatively interpreted people's walking and shopping behaviour in the street as a linear sequence. It converted POI data collected from the O2O platform into a sequence format to train the RNN model.

In the future, this study still has much room for improvement. The accuracy of the current model is still not high enough. In the data collection stage, a larger data set is needed in the future. Since the information accuracy requirement is very high (relative location of each store), the automatic POI data collection method based on geographic coordinates is not applicable. Currently, we use manual methods to collect data one by one along the street. In the future, however, automated data collection algorithms will have to be developed to replace the current manual methods to remarkably expand the scale of the training set. In the data processing stage, there are many noise points in the data set due to many factors affecting the vitality of the real-world stores. In the future, homogenized data algorithms will be used to eliminate the effect of noise [6]. In the model training phase, we will use more RNN models such as Transformer, GRU, BiLSTM to compare which model is more suitable for this research in the future.

References

1. Gao W et al (2021) A data structure for studying 3D modeling design behavior based on event logs. Autom Constr 132(103967):103967. https://doi.org/10.1016/j.autcon.2021.103967
2. Gao W et al (2022) Command prediction based on early 3D modeling design logs by deep neural networks. Autom Constr 133(104026):104026. https://doi.org/10.1016/j.autcon.2021.104026
3. He J, Zheng H (2021) Prediction of crime rate in urban neighborhoods based on machine learning. Eng Appl Artif Intell 106(104460):104460. https://doi.org/10.1016/j.engappai.2021.104460
4. Karoji G, Hotta K, Hotta A, Ikeda Y (2019) Pedestrian dynamic behaviour modeling. In: Proceedings of the 24th international conference on computer-aided architectural design research in Asia: intelligent and informed, CAADRIA 2019, pp 281–290
5. Schlegel A, Birkel HS, Hartmann E (2021) Enabling integrated business planning through big data analytics: a case study on sales and operations planning. Int J Phys Distrib Logist Manag 51(6):607–633. https://doi.org/10.1108/ijpdlm-05-2019-0156
6. Sheng Q et al (2018) The application of space syntax modeling in data-based urban design—an example of Chaoyang square renewal in Jilin city. Lands Archit Front 6(2):102. https://doi.org/10.15302/j-laf-20180211
7. Shou X, Chen P, Zheng H (2021) Predicting the heat map of street vendors from pedestrian flow through machine learning. In: Proceedings of the 26th international conference on computer-aided architectural design research in Asia: projections, CAADRIA 2021, pp 569–578
8. Sun YJ, Jiang L, Zheng H (2021) A machine learning method of predicting behavior vitality via urban forms. In: Proceedings of the 40th international conference on computer aided design in architecture: distributed proximities, ACADIA 2021, pp 160–168

9. Xia X, Tong Z (2020) A machine learning-based method for predicting urban land use. In: Proceedings of the 25th international conference on computer-aided architectural design research in Asia: anthropocene, CAADRIA 2021, pp 21–30

Study on Optimization of Building Climate Adaptive Morphology in Cold Regions of China: Case of U-Shaped College Building

Ping Chen[1,2], Chang Liu[1], and Hsin-Hsien Chiu[2,3](✉)

[1] School of Architecture and Urban Planning, Shandong Jianzhu University, Jinan 250101, China

[2] School of Architecture, Harbin Institute of Technology, Harbin 150001, China
hhchiu@berkeley.edu

[3] Key Laboratory of Interactive Media Design and Equipment Service Innovation, Ministry of Culture and Tourism, Harbin Institute of Technology, Harbin, China

Abstract. Proper design of building form will facilitate the use of climate environment in order to reduce the reliance of buildings on active equipment. This study takes the cold region of China as the research area, and Jinan city of Shandong province as a typical city in the cold region for specific research. The multi-objective optimization tool based on NSGA-II algorithm is used to optimize the opening angle, length of both sides and floor height of the building, and finally the optimal size range of the university teaching building under the influence of solar radiation heat gain in winter and summer is obtained, and the results show that for the U-shaped university teaching building, the parameters that affect the building performance more in the case of the east side opening are the length of the north side building and the rotation angle of the south side building, and the parameters that affect the performance more in the case of the west side opening are the length of the building on the south side.

Keywords: Solar radiation heat gain · Multi-objective optimization · NSGA-II algorithm

1 Introduction

The cold region is one of the five climate zones in China. It's climate index is: the average temperature of the coldest month 0–10 °C as well as the average daily temperature ≤5 °C for 90–145 days [10]. Cold regions in winter temperatures are low and last for a long time, the lowest temperature can reach minus 20 °C, the summer temperature is generally higher up to 32 °C, the duration of 2 months [9], the sunshine time varies greatly with the seasons, in this case the cold regions need to consume a lot of energy in the winter heating, and the summer need to use air conditioning and other mechanical equipment for cooling, because of the continuous transformation of the climate environment, cold regions each year need to consume a lot of energy to improve the thermal environment, how to improve the status quo has become a problem that needs to be solved. Jinan City,

P. F. Yuan et al. (eds.), *Hybrid Intelligence*, Computational Design and Robotic Fabrication,
https://doi.org/10.1007/978-981-19-8637-6_30

Shandong Province, located at 36°40′N, 117°00′E, in the eastern part of China, belongs to the cold region and has more obvious cold region climate characteristics. In this paper, Jinan City will be selected as a representative city in the cold region for the study.

In Chinese cities, the number and scale of colleges and universities are large (Figs. 1 and 2), which makes the college buildings rely too much on equipment to regulate the building environment, and their energy consumption accounts for a relatively large amount [4]. In view of these, it is urgent to find the best way to reduce the dependence of teaching buildings on equipment and use their own architectural advantages to obtain a suitable building environment.

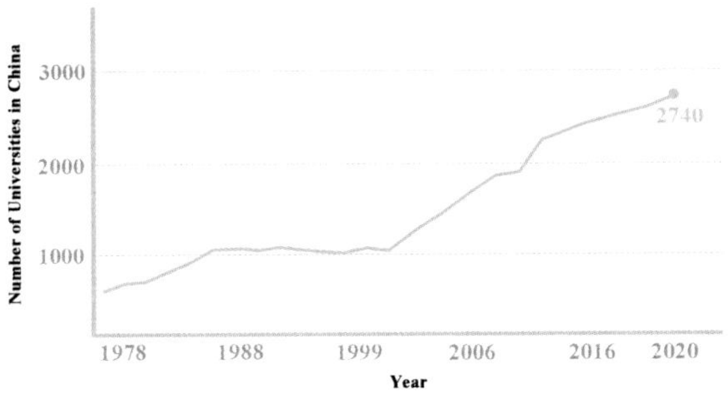

Fig. 1. Number of universities in China

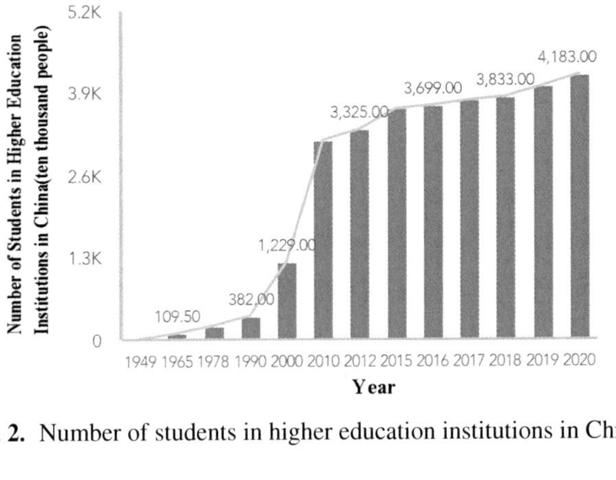

Fig. 2. Number of students in higher education institutions in China

The process of changing the thermal environment of the building due to seasonal changes plays a key role in the solar heat gain of the building, if the building solar heat gain is reduced in summer and increased in winter, it can effectively improve the building thermal environment and reduce the energy consumption required in the later stage. As the interface between the building surface and the outside world, the specific form of the building has a direct impact on the solar heat gain of the building, if we consider the coupling relationship between the solar heat gain of the building and the building form, we can take into account the changes of the building heat gain caused by seasonal changes at the early stage of the building design to maximize the use or avoid the solar heat gain, so as to reduce the energy consumption caused by improving the thermal environment of the building at a later stage and realize the sustainable development of the building.

Many researchers have paid attention to this problem and conducted a series of experiments and explorations. Mustafa did a study on the correlation between building form and solar radiation heat gain for a standard building form of rectangular shape [14]. Kampf Jerome Henri and other scholars applied a new method of predicting daylight radiation to study the morphological and energy-saving design of building forms and neighborhood forms [7]. In 2012, Zerefos and others quantified the difference in energy consumption between different forms of prismatic buildings [18]. After 2015, related scholars optimized the general building layout [15], building morphology [1, 12], and building details [13] with the goal of building solar radiation heat gain, and obtained a better building thermal environment. From past studies, it can be seen that the building form is closely related to the solar heat gain of the building, and by optimizing the design of the building form, the solar heat gain of the building can be changed, thus improving the thermal environment of the building.

How to quantitatively optimize the design of building form according to the external environment has become the core problem. In the early days, due to the limitation of technical conditions, researchers could only carry out quantitative design by "trial and error method". This approach is not only inaccurate, random and inefficient, but also requires a lot of time and cost. With the development of computer technology and optimization algorithms, the optimization of buildings can be considered directly from the optimization target, and then the parameters to be optimized can be set and the optimization design can be completed by iterative operation of optimization tools (Fig. 3).

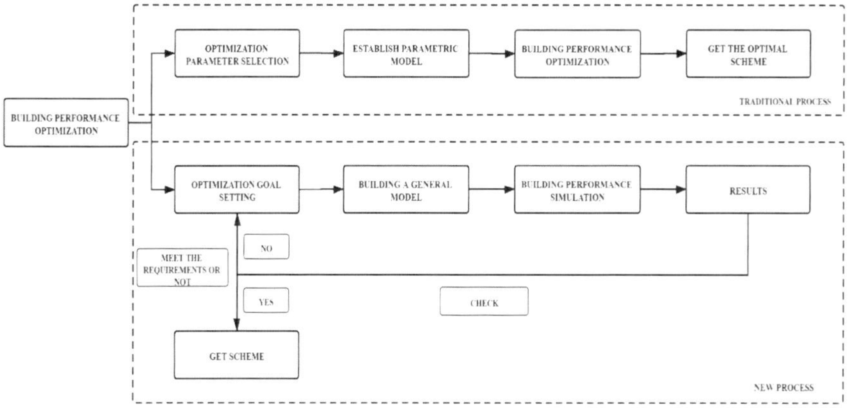

Fig. 3. Comparison of traditional methods and new methods

For this aspect of research, most of the early studies were explored on technical framework [17] and building construction [3], and after 2010, studies were focused on building group form and building form optimization, such as Martins selected typical building layout features in five Brazilian cities and conducted a multi-objective optimization study on the morphology of building clusters based on solar radiation heat gain [8], Longwei Zhang adjusted the morphology for freeform buildings in cold and severe regions represented by Shenyang, driven by several optimization objectives such as daylight radiant heat gain, volume coefficient, and space efficiency [19].

However, most of the existing studies are focused on single-objective optimization, and the optimization of multi-objective mainly considers the influence of different environmental elements, lacking the consideration of the influence of environmental elements on the research subject due to time change, which leads to the existing research subject can only consider the situation in a certain time period, and it is difficult to take into account the environment of other time periods in the region. For buildings that are used by a large number of people for a long period of time, it is necessary to consider the differences in solar heat gain of the buildings due to seasonal changes, and to make full use of or avoid the solar heat gain of different seasons, so as to reduce the energy required to improve the thermal environment of the buildings in the later period. The purpose of this paper is to construct an optimal design method for U-shaped college buildings in cold regions of China, explore the relationship between the optimization objectives, and make specific suggestions.

2 Materials and Methods

2.1 Intelligent Optimization Methods

This paper proposes an intelligent design method for building form optimization based on solar heat gain in an attempt to improve the thermal environment of university buildings in cold regions (Fig. 4). The goal of the optimization is to optimize the solar heat gain of the building so that the building has more solar heat gain in winter and less solar heat gain in summer. This process includes several steps such as determining morphological parameters, model construction, morphological optimization, and analysis and processing.

2.2 Optimized Objects

This paper selects the teaching building in the cold region of China as the research object. In the cold regions of China, due to the constraints of climate conditions, design specifications and functional requirements, there are three main types of college buildings in terms of overall layout, namely: one-way type, combined type and centralized type, and the one-way type can be divided into one-way type, U type, L type and H type (Fig. 5).

U-shaped teaching buildings are widely used, highly applicable and influenced by solar radiation, so this paper chooses one type of U-shaped teaching buildings for in-depth study. Through the research of U-shaped college buildings in cold areas, it is found that the main function distribution is linear, and the building blocks facing the opening of the U-shaped building are mostly facing east or west because of the number of classrooms, and the function is mostly small classrooms for small group teaching or

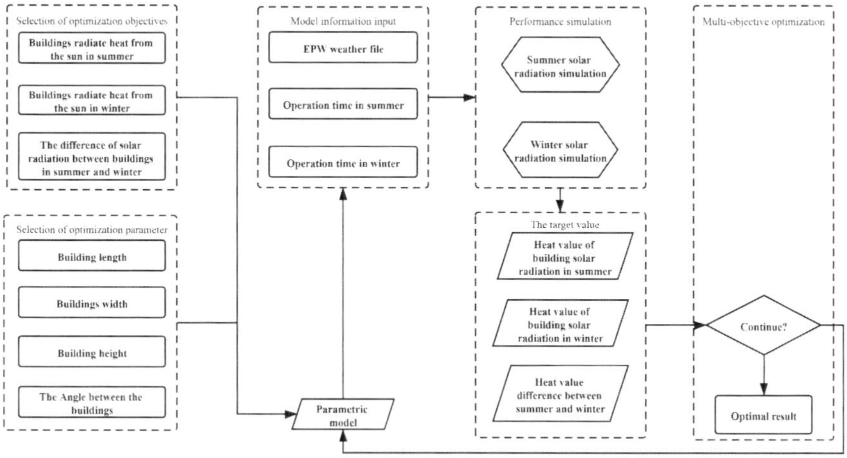

Fig. 4. Intelligent design method of building form optimization

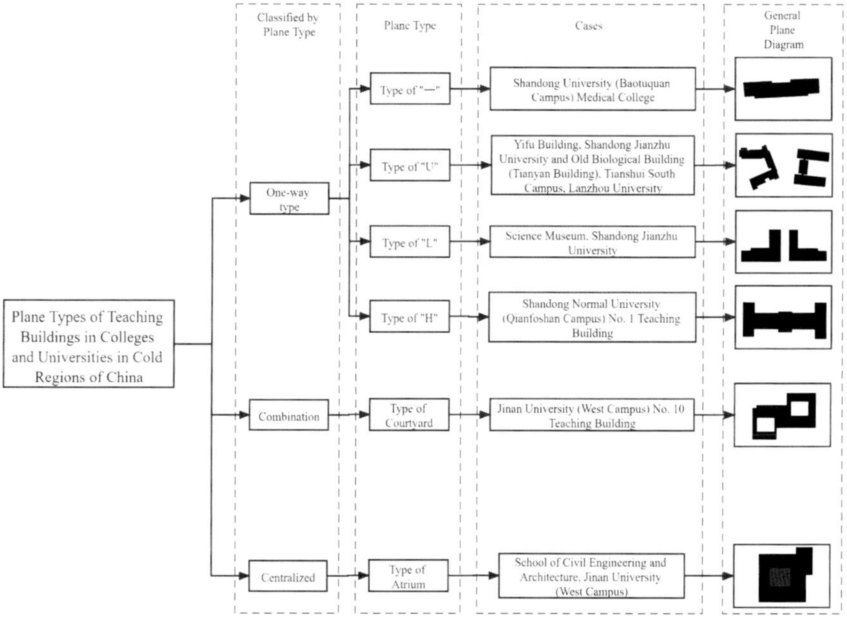

Fig. 5. Classification of colleges and universities in cold regions of China

students' self-study, and the building blocks on both sides of the opening are mostly arranged in the south and north as the main classroom. Besides, the multi-functional classrooms are usually set at the end of the corridor, and the number of symmetrical arrangement is usually two (Table 1).

Table 1. Cases of academic building plan

Old biological building (Tianyan building), Tianshui South Campus, Lanzhou University	Yifu Building, Shandong Jianzhu University	Molecular Science Innovation Platform, Institute of Chemistry, Chinese Academy of Sciences
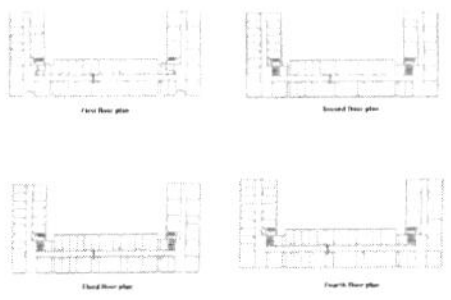	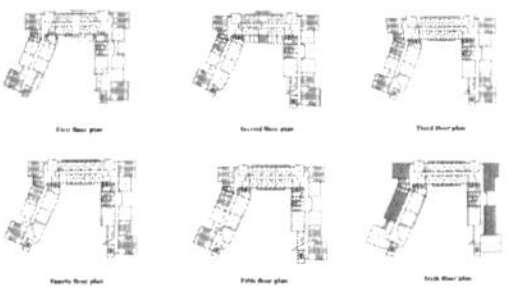	

2.3 Optimized Goals

In this paper, we focus on the problem of solar radiation heat gain of buildings caused by seasonal changes, so the optimization objectives are solar radiation heat gain of buildings in winter and solar radiation heat gain of buildings in summer, and in order to consider the influence of the two objectives on the optimization results and avoid the situation that the optimization results are excessively biased to one side, we need to add a third objective: the difference between solar radiation heat gain of buildings in summer and solar radiation heat gain of buildings in winter, and by adding this objective, we can consider the influence of the previous two on the results.

2.4 Optimizing Variables and Constraint Ranges

For U-shaped buildings, the length and height of the building and the angle of the building openings determine the building form and also affect the solar heat gain of the building, so these variables need to be set specifically (Fig. 6).

2.4.1 Building Length

According to the results of the above research and related data collection, the main functions of U-shaped college buildings in China are usually arranged in the building blocks on both sides of the openings, while the blocks facing the openings are often small classrooms with some auxiliary functions. The lengths of each side of the building are added up by each classroom, so the total length can be set by setting and adding up the lengths of individual classrooms. Classrooms are divided into three main categories according to their functions, namely, regular classrooms, multifunctional classrooms and auxiliary classrooms. According to the requirements of the teaching building and the building modulus, the length of a single general classroom is generally 8.1–9 m with a step length of 0.3 m, and the length of a single multifunctional classroom is 11.1–12 m with a step length of 0.3 m (Table 2), while auxiliary classrooms are not specifically set because they are not within the optimization range (Fig. 7).

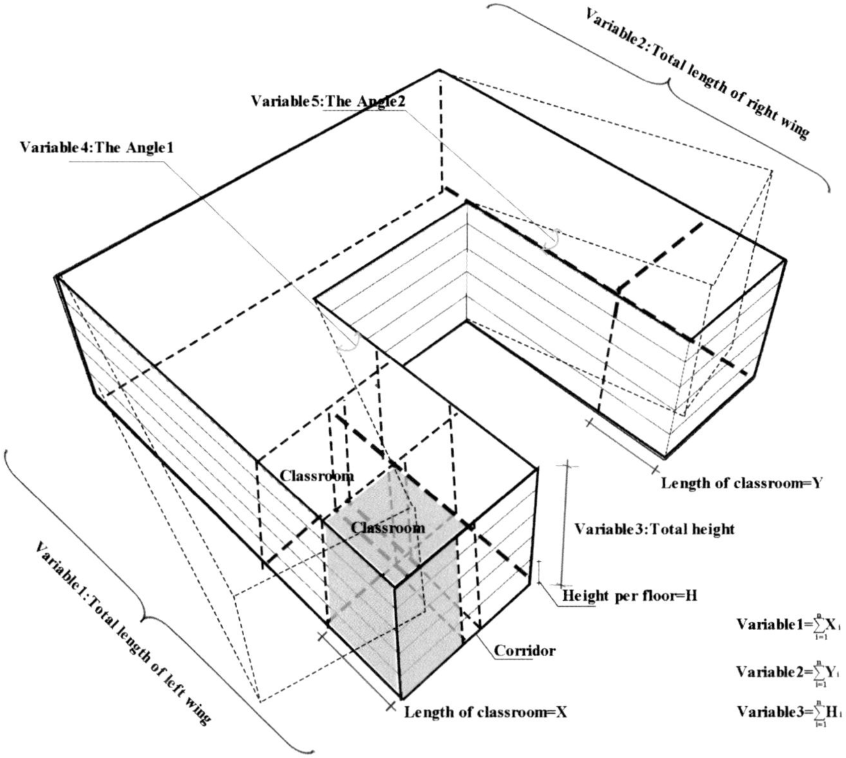

Fig. 6. Graphical representation of building optimization variables

Table 2. The value ranges and the steps of variables

Variables	Unit	Value ranges	Steps
Length of a single general classroom	m	8.1–9.0	0.3
Length of a single multi-purpose classroom	m	11.1–12.0	0.3
Height of each floor of the building	m	3.6–4.5	0.1
North side rotation angle	°	20	1
South side rotation angle	°	20	1

2.4.2 Building Height

The height of each floor is set with reference to the relevant regulations of China for teaching buildings and combined with the current function of classrooms, with 0.1 m as the step, and the range is chosen from 3.6 to 4.5 m.

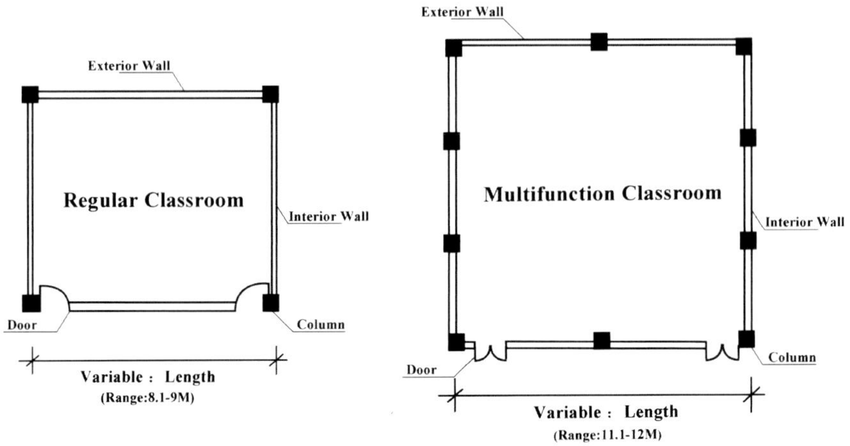

Fig. 7. Single classroom variable setting

2.4.3 Building Opening Angle

The building opening angle affects the shading between buildings and has an important impact on the study of solar heat gain of buildings. The building opening angle is set within a range of 20° for both the north and south sides of the building, taking into account the land use, utilization, and evacuation of the building.

2.5 Dynamic Information Modeling and Optimization Based on Solar Radiation Simulation

The article uses Rhino [11] and Grasshopper [6] platforms to construct dynamic information models, Ladybug plug-in which can be downloaded from the Food4Rhino [5] website to simulate the solar heat gain of buildings by inputting EnergyPlus Weather file (.epw) [2], and the highly visualized Wallacei plug-in [16] to perform multi-objective optimization (Fig. 8).

2.5.1 Reference Building

The reference building is located in a university in Jinan. The plan of the reference building is a U-shaped college building with 135° opening, and its dimensions are 75, 85 and 60 m. The building height is 21.6 m. The floor plan and the initial geometric model of the reference building are shown in Figs. 9 and 10.

2.5.2 Simulation Parameter Setting

In the setting of building parameters, the two wings of the U-shaped building can only be contracted inward in the east–west direction and cannot be extended outward because the whole building is limited by the site area in the east–west direction. As we can see from the plan, there are one multi-functional classroom and four general classrooms in the north block, considering the symmetrical arrangement of the classrooms, only one

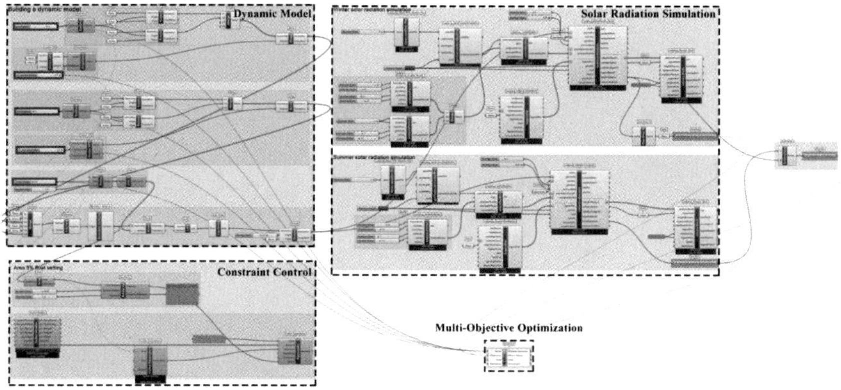

Fig. 8. Operation of optimization and simulation

multi-functional classroom and two general classrooms need to be optimized in terms of length. Therefore, the total optimization range R1 = R (multifunctional classroom) + R (ordinary classroom) * 2 = (12–11.1) + (9–8.1) * 2 = 2.7 m; the south side of the block has a multifunctional classroom and four ordinary classrooms, taking into account the symmetrical layout of the classroom, in the length of only one multifunctional classroom and three ordinary classrooms can be optimized, the optimization step length of 0.3 can be unified to set, multifunctional classroom length measured 12 m, the length of ordinary classrooms measured 9 m, so the total optimization range R2 = R (multifunctional classroom) + R (ordinary classroom) * 3 = (12–11.1) + (9–8.1) * 3 = 3.6 m. Considering the limitation of the site, the final optimized length range for the north side is 57.3–60 m, and the optimized length range for the south side is 71.4–75 m.

In terms of rotation angle setting, the north side building block is rotated 15° to the north and 5° to the south considering the site conditions, with the north side recorded as negative and the south side recorded as positive, while the south side block has a narrow initial state and can only be deflected to the south with a maximum offset angle of 20° and a rotation process step of 1° (Figs. 11 and 12). The height of each layer of the building is used as the optimization parameter, and the value is set with reference to the relevant national regulations for teaching buildings and the current function of the classroom, with 0.1 m as the step, the range is 3.6–4.5 m, and the number of layers is set as the initial state. The buildings in the site are arranged with openings to the west (Table 3). According to the above research results, the openings of buildings in cold regions of China can be to the west or to the east, and in order to increase the generality of the optimization experiment, it is necessary to discuss the typical building as an example in two cases of east and west openings respectively.

2.5.3 Solar Radiation Heat Gain Simulation

In the parameter setting of solar radiation simulation, the grid size of solar radiation calculation is chosen to be 2 m * 2 m with relatively high accuracy, and the grid offset distance of the test point is set to 0.01. By connecting to the CSWD climate database,

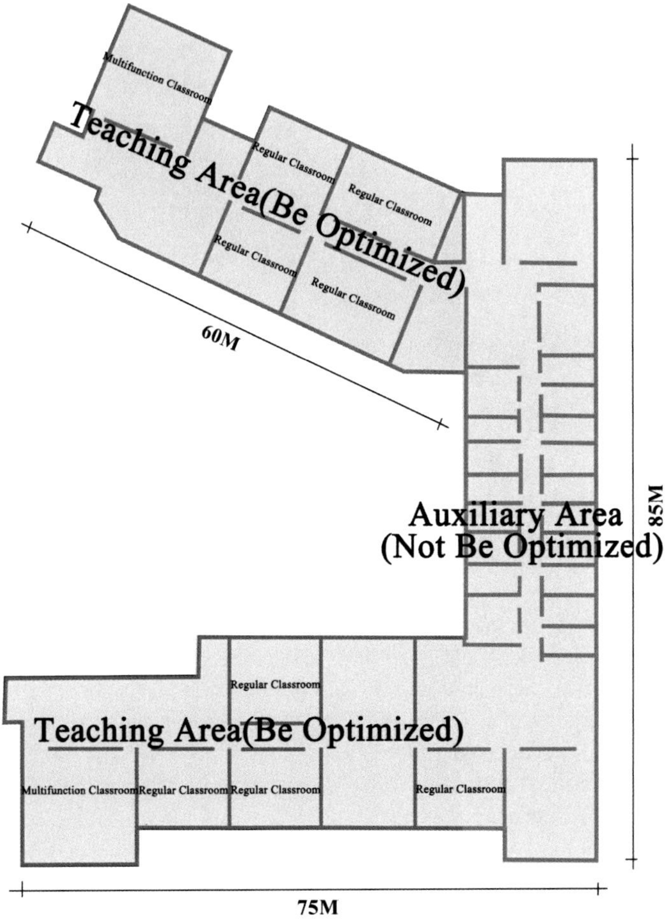

Fig. 9. Reference building plan

the solar heat gain of the building can be simulated for different time periods (Figs. 13, 14, 15, and 16). Considering that the subsequent optimization is based on the genetic algorithm and can only be optimized in the direction of the minimum value, the final solar radiation results in winter are taken as the reciprocal.

2.5.4 Multi-objective Optimization

The whole operation process is based on genetic algorithm (NSGA-II), and the size of each population generation is set to 20, the number of generations of simulated population is set to 100, the crossover rate is set to 0.9 in the genetic algorithm, and the variation rate keeps the default value (Table 4).

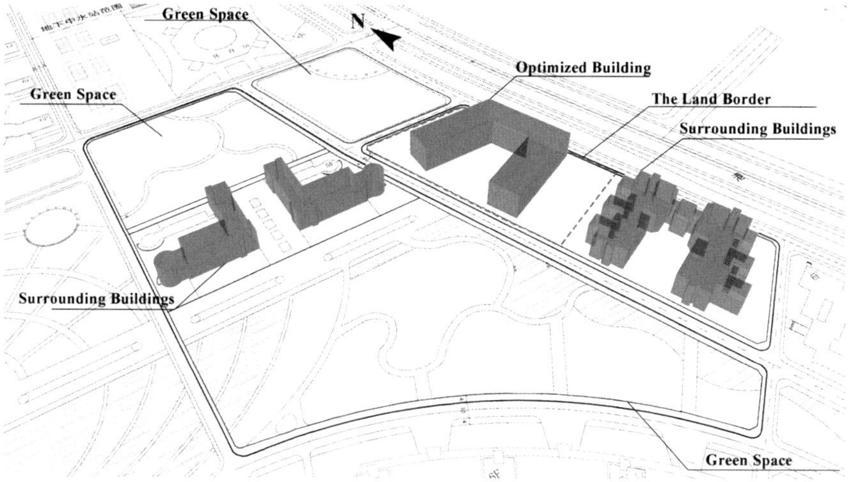

Fig. 10. Reference building location

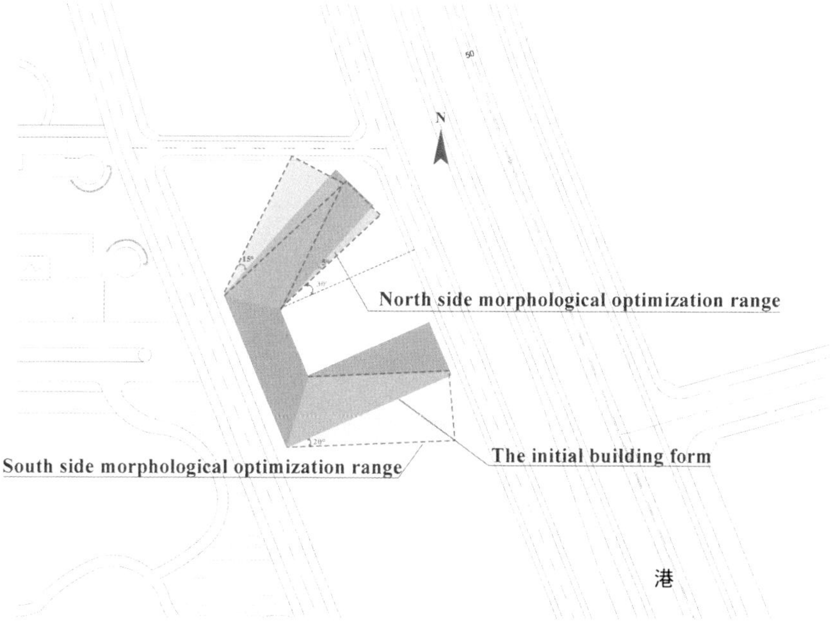

Fig. 11. Angle setting in case of eastward opening

3 Results and Discussion

After 50 generations, the optimization results in each direction tend to be smooth. A total of 2,000 optimization results were calculated, including a total of 205 Pareto front solutions (non-dominated solutions) for the case with the opening facing west and a total

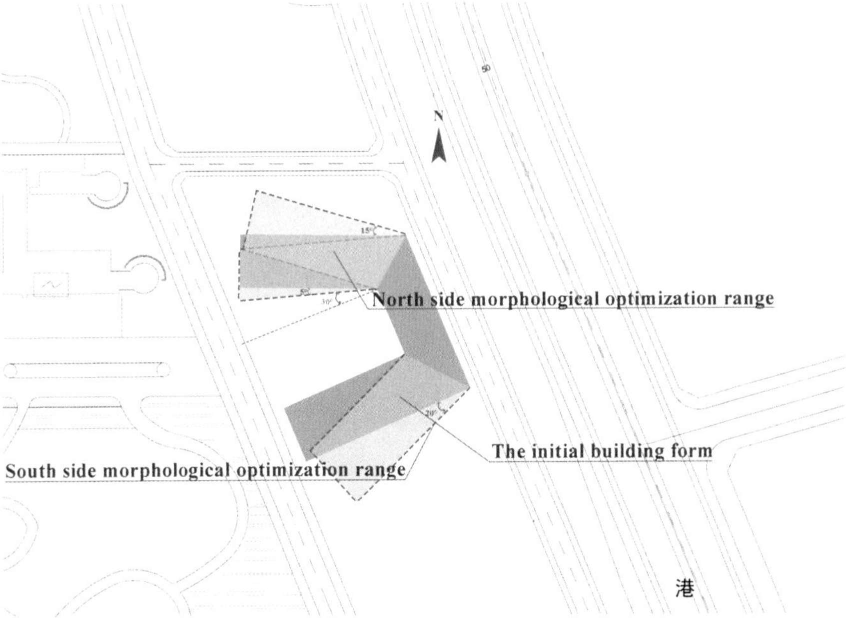

Fig. 12. Angle setting in case of westward opening

Table 3. The value ranges and the steps of variables setting of reference building

Variables	Unit	Value ranges	Steps
Optimized length of the north side	m	57.3–60.0	0.3
Optimized length of the south side	m	71.4–75.0	0.3
Height of each floor of the building	m	3.6–4.5	0.1
North side rotation angle	°	20	1
South side rotation angle	°	20	1

of 301 Pareto front solutions (non-dominated solutions) for the case with the opening facing east (Figs. 17 and 18). The actual running time of the whole process is about 6 h for a laptop computer with an i7-9750H CPU.

3.1 Evolution of Solutions

The gray area shown in the Figs. 19, 20 and 21 is the Pareto front surface of the feasible solution. As the optimization proceeds, the Pareto front surface shrinks continuously and moves closer to the optimization target, and the values of the three targets are significantly reduced in the case of the opening to the east, which means that the solar heat gain value of the building in summer decreases, the inverse of the solar heat gain

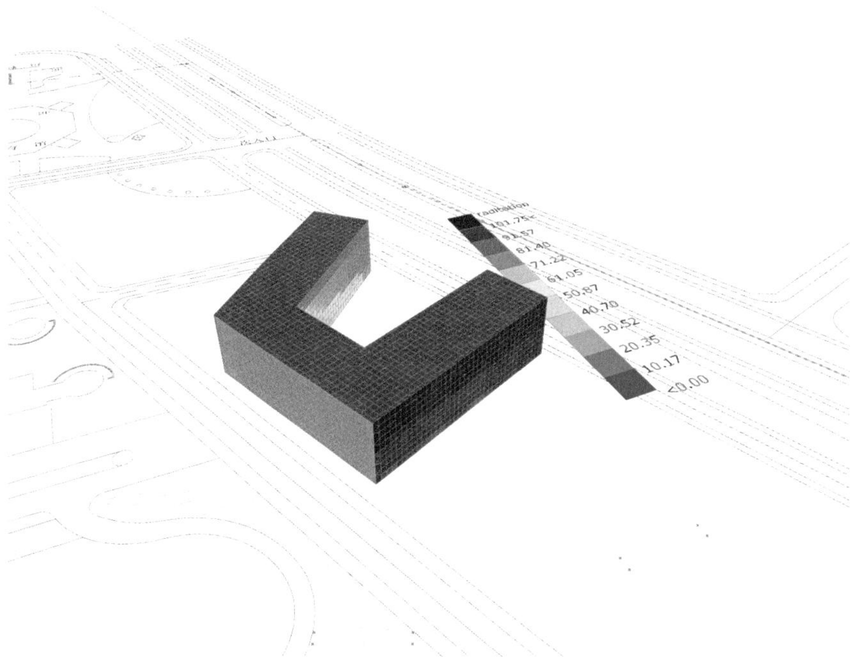

Fig. 13. Simulation of solar radiation heat gain in winter with an eastward opening

value of the building in winter decreases, and the difference between the solar heat gain value in summer and the solar heat gain value in winter decreases, which indicates that the performance of the feasible solution gradually improves.

The values of all three objectives decrease from the optimization to the 35th generation in the case of westward opening (Figs. 22, 23 and 24), which means that the solar heat gain value of the building in summer decreases, the inverse of the solar heat gain value of the building in winter decreases, and the difference between the solar heat gain value in summer and the solar heat gain value in winter decreases, which indicates a gradual improvement in the performance of the feasible solution.

3.2 Selection of the Optimal Solution

After finishing the evolution process of feasible solutions, it is necessary to select the optimal solution. The selection of the optimized solution is based on building performance indexes in the performance value range of more than 50%. The difference between the solar heat gain values in summer and winter represents the combined condition of the two seasons and is the basis for the experiment. By ranking the difference between the two according to the performance metric, the optimal solution can be filtered to ensure that no extreme cases occur. This process can be accomplished by a diamond diagram (Figs. 25 and 26), in which the values of each target are arranged on an axis, with the closer to the origin representing the better value of the target.

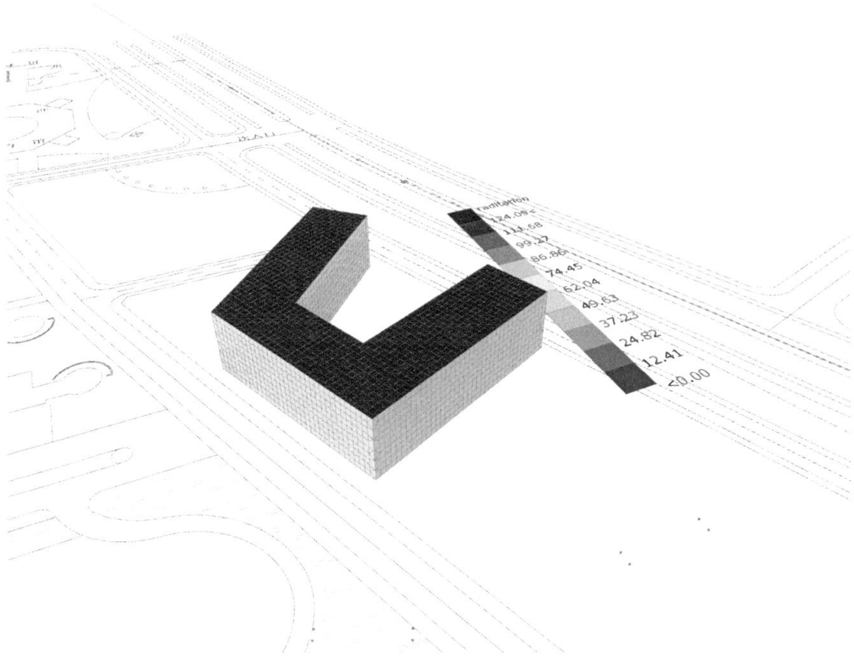

Fig. 14. Simulation of solar radiation heat gain in summer with an eastward opening

After filtering, a critical condition is found where the performance values of each objective are above 50%, beyond which there will be an objective that does not meet the 50% requirement. Within the critical condition, in the case of the east side opening, the length of the north side building should be reduced by 0–2.1 m, which means that the range of length change of the north side should be 57.9–60 m, and the length of the south side building should be reduced by 0–3.6 m, which means that the range of length change of the south side should be 71.4–75 m; the height of the floor should be 3.6–4.6 m; the north side building should not be turned more than 5° to the south and 14° to the north. In other words, the angle with the building opening direction is 25–44°, and the south side of the building should not be turned more than 16° to the south, which means the angle with the building opening direction is 0–16°.

In the case of west side opening, the length of north side building should be reduced by 0–2.7 m, that is to say, the range of length change of north side should be taken as 57.3–60 m, and the length of south side building should be reduced by 0–2.1 m, that is to say, the range of length change of south side should be taken as 72.9–75 m; the floor height should be chosen as 3.6–4.6 m; the north side building should not be turned to more than 5° to the south and 15° to the north, that is to say the angle with the building opening direction is 25–45°, and the south side of the building should not be turned more than 20° to the south, which means the angle with the building opening direction is 0–20°.

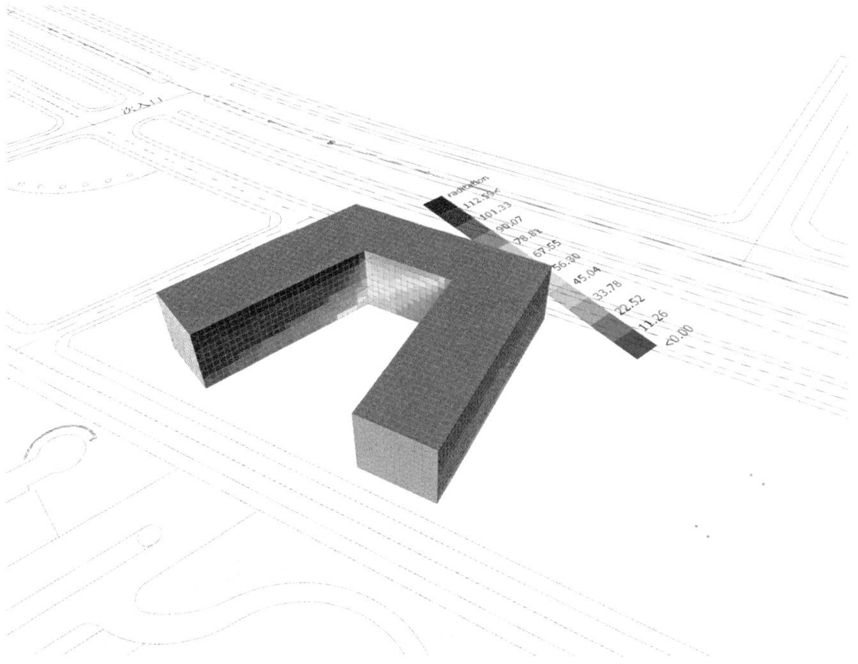

Fig. 15. Simulation of solar radiation heat gain in winter with an westward opening

To sum up, the parameters affecting the performance of the building in the case of the east side opening are the length of the north side building and the angle of rotation of the south side building, and the range of change of the length of the north side should be 57.9–60 m. The angle of the south side building and the direction of the building opening should be 0–16°, the parameters affecting the performance in the case of the west side opening are the length of the south side building, and the range of change of the length of the south side should be 72.9–75 m. As long as these parameters are ensured, the building can obtain high performance in all aspects within the set range.

4 Conclusions

In this study, a multi-objective optimization of U-shaped academic buildings in colleges and universities in cold regions of China was conducted to improve the winter solar heat gain and reduce the summer solar heat gain of the buildings. After 100 iterations, 205 non-dominated solutions were obtained for the westward opening case and 301 non-dominated solutions for the eastward opening case. The article discusses the value distribution of the optimization objectives. The parameter that affects the building performance in the case of east side opening is the length of the north side building and the rotation angle of the south side building, and the range of the variation of the length of the north side is 57.9–60 m, and the angle of the south side building and the direction of the building opening is 0–16°. The parameter that affects the performance in the case of the west side opening is the length of the south side of the building, and it is appropriate

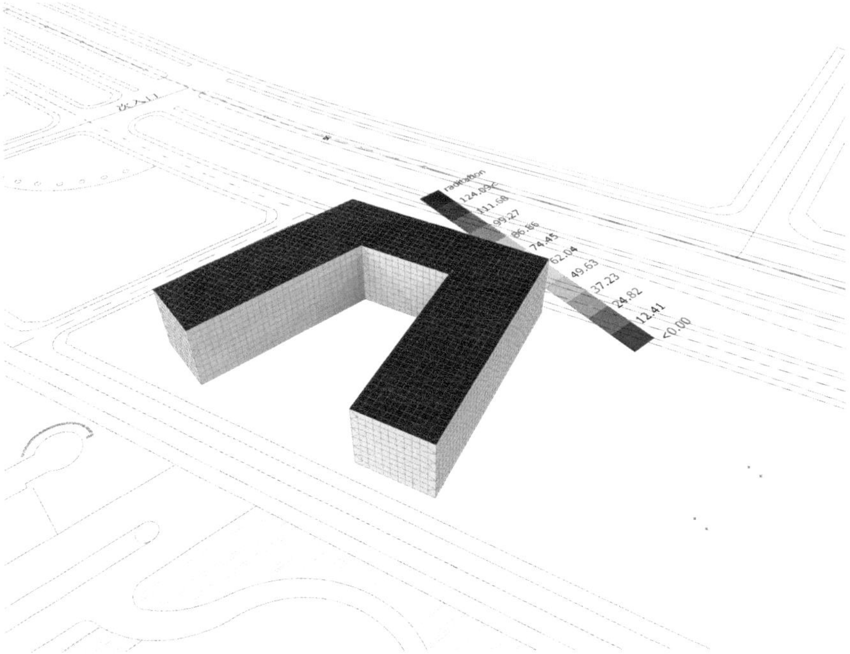

Fig. 16. Simulation of solar radiation heat gain in summer with an westward opening

Table 4. Settings of the optimization algorithm

Algorithm	Size of each population generation	The number of generations of simulated population	Crossover rate
NSGA-II	20	100	0.9

to take the range of 72.9–75 m. The performance of the building is improved through optimization, which proves that in the cold region of China, the reliance of the building on equipment can be reduced and the performance of the building can be improved through the optimization of the building form itself.

In order to improve the optimization efficiency and save the optimization cost, some auxiliary spaces are reduced and simplified, which makes the setting of building form-related parameters not comprehensive enough and difficult to take more factors into account. This is the insufficiency of this study, and the research on this aspect will be discussed in depth in the subsequent related work.

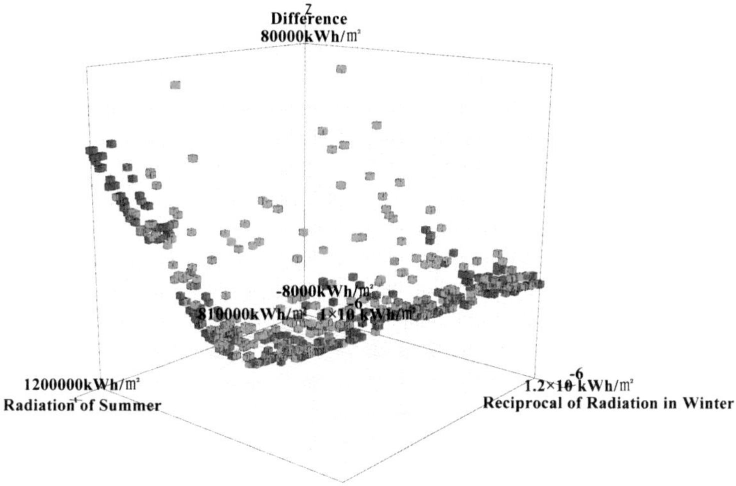

Fig. 17. Pareto front solution in the case of opening to the east

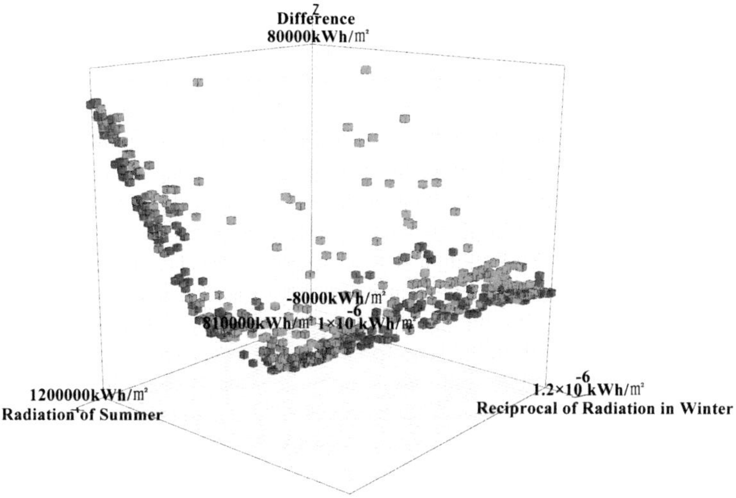

Fig. 18. Pareto front solution in the case of opening to the west

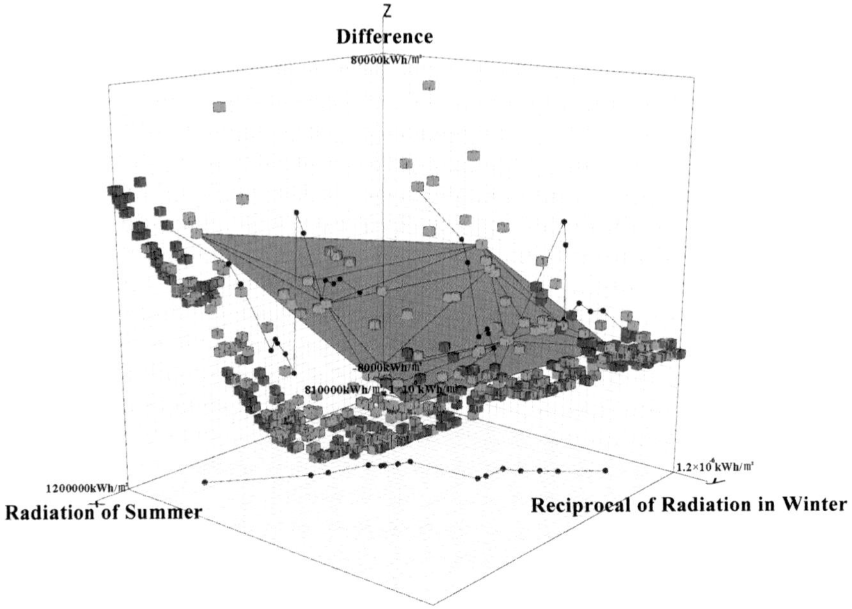

Fig. 19. Pareto front surface at the second generation in the east side case

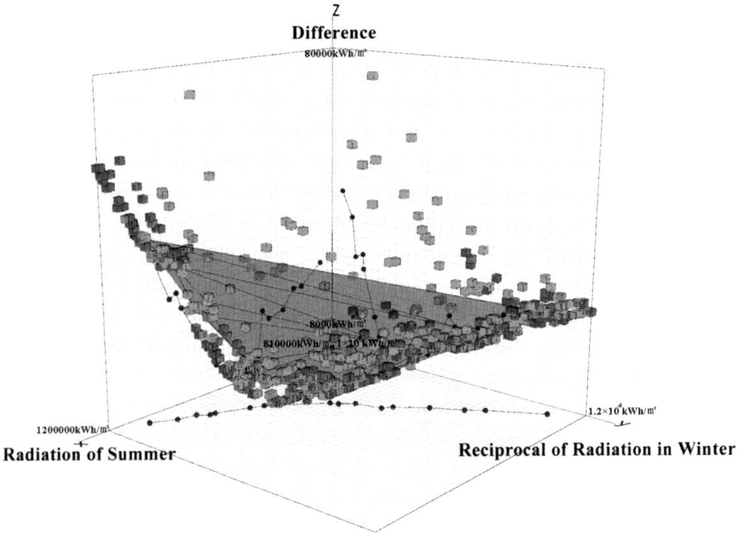

Fig. 20. Pareto front surface at the fifteenth generation in the east side case

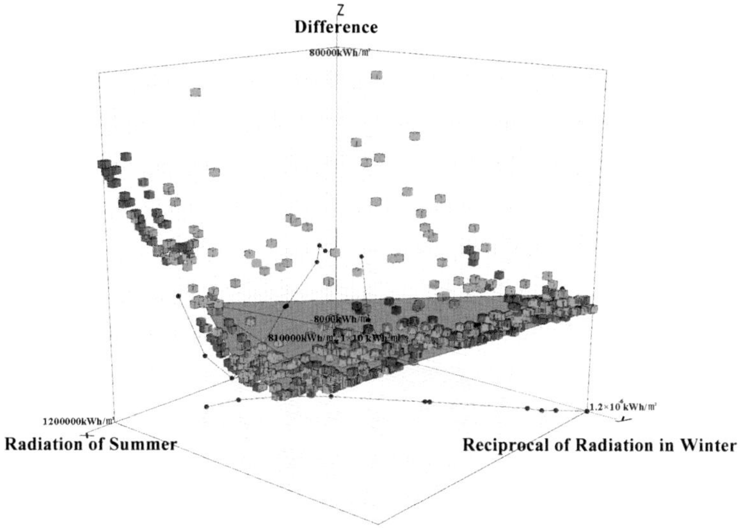

Fig. 21. Pareto front surface at the twenty-seventh generation in the east side case

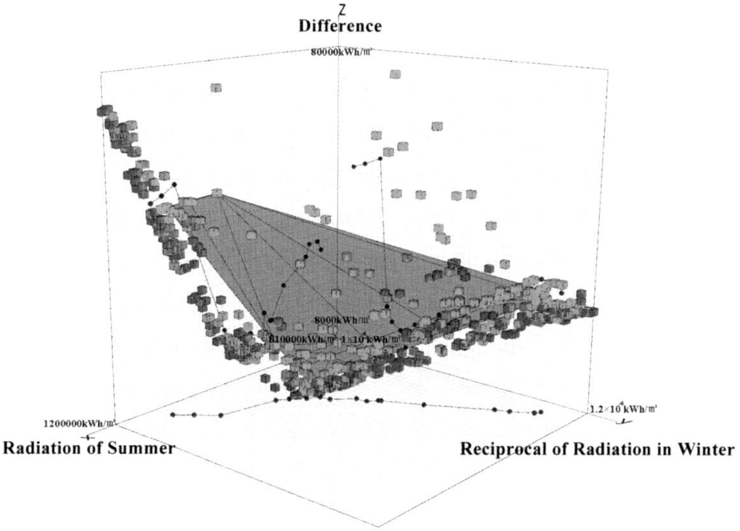

Fig. 22. Pareto front surface at the fifth generation in the west side case

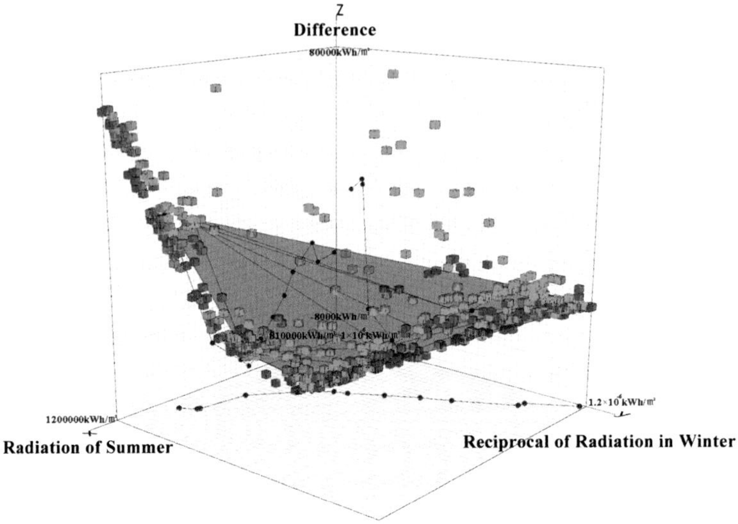

Fig. 23. Pareto front surface at the fourteenth generation in the west side case

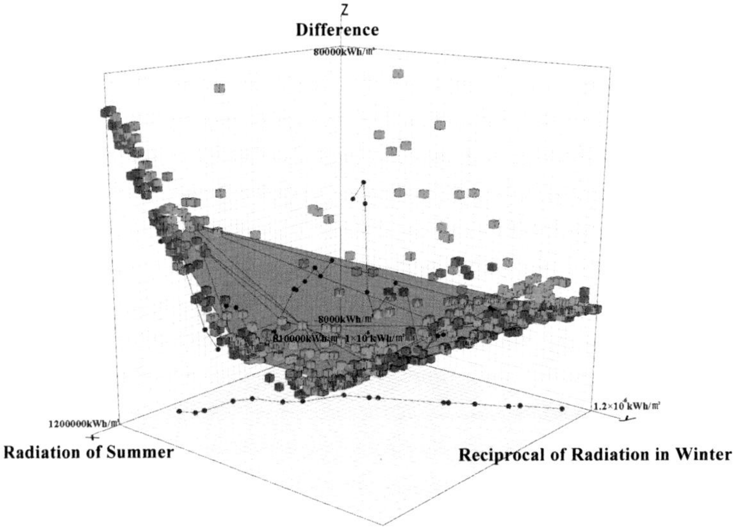

Fig. 24. Pareto front surface at the thirty-fifth generation in the west side case

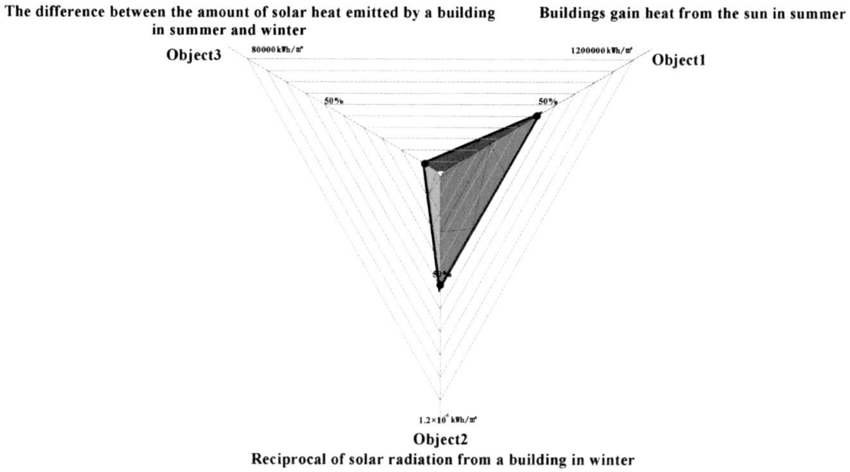

Fig. 25. Diamond diagram under critical conditions with the opening to the east

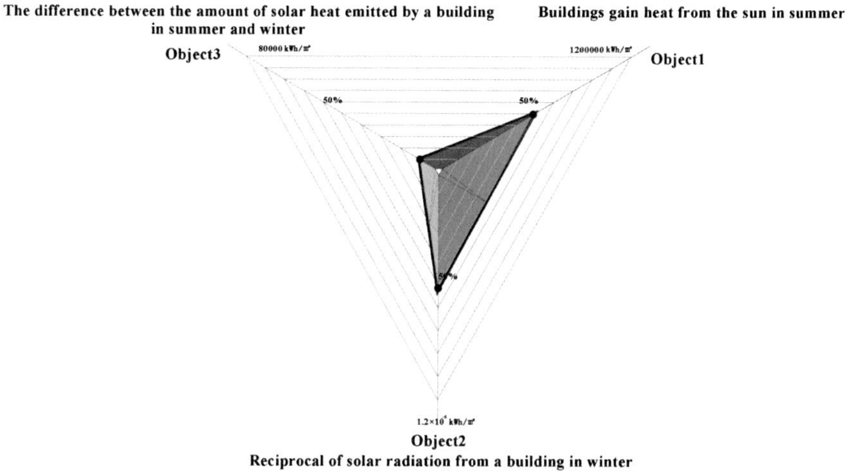

Fig. 26. Diamond diagram under critical conditions with the opening to the west

Funding. This research was funded by the Natural Science Foundation of Shandong Province of China (Project Number: ZR2021ME133) and the Opening Fund of Key Laboratory of Interactive Media Design and Equipment Service Innovation, Ministry of Culture and Tourism (Project Number: 20202).

References

1. Antón I, Tnase D (2016) Informed geometries. Parametric modelling and energy analysis in early stages of design. Energy Procedia 85:9–16
2. Auxiliary Programs. U.S. Department of Energy. DIALOG. https://energyplus.net/. Cited 3 Mar 2022

3. Diakaki C, Grigoroudis E, Dionyssia K (2008) Towards a multi-objective optimization approach for improving energy efficiency in buildings. Energy Build 40:1747–1754
4. Education Statistics. Ministry of Education of the People's Republic of China. DIALOG. http://www.moe.gov.cn/. Cited 3 Mar 2022
5. Food4Rhino Download. Available via DIALOG. https://www.food4rhino.com/en. Cited 3 Mar 2022
6. Grasshopper Software. DIALOG. https://www.grasshopper3d.com/. Cited 3 Mar 2022
7. Kaempf JH, Robinson D (2010) Optimisation of building form for solar energy utilisation using constrained evolutionary algorithms. Energy Build 42:807–814
8. Martins T, Adolphe L, Bastos L (2014) From solar constraints to urban design opportunities: optimization of built form typologies in a Brazilian tropical city. Energy Build 76:43–56
9. Meteorological Data. China Meteorological Administration. DIALOG. http://www.cma.gov.cn/. Cited 3 Mar 2022
10. Ministry of Housing and Urban-Rural Development of the People's Republic of China (2016) GB50176-2016.Thermal Design Code for Civil Building. Beijing, China
11. Rhino Software. DIALOG. https://www.rhino3d.com/. Cited 3 Mar 2022
12. Taleb S, Yeretzian A, Rabih AJ, Hajj H (2020) Optimization of building form to reduce incident solar radiation. J Build Eng 28:101025–101025
13. Tian Z, Zhang X, Jin X, Zhou X, Si B (2018) Towards adoption of building energy simulation and optimization for passive building design: a survey and a review. Energy Build 158:1306–1316
14. Teoman Aksoy U, Inalli M (2006) Impacts of some building passive design parameters on heating demand for a cold region. Build Environ 41:1742–1754
15. Vartholomaios A (2015) The Residential Solar Block envelope: a method for enabling the development of compact urban blocks with high passive solar potential. Energy Build 99:303–312
16. Wallacei Software. DIALOG. https://www.wallacei.com/
17. Wang W, Zmeureanu R, Rivard H (2005) Applying multi-objective genetic algorithms in green building design optimization. Build Environ 40:1512–1525
18. Zerefos SC, Tessas CA, Kotsiopoulos AM (2011) The role of building form in energy consumption: the case of a prismatic building in Athens. Energy Build 43:97–102
19. Longwei Z, Lingling Z, Yuetao W (2016) Shape optimization of free-form buildings based on solar radiation gain and space efficiency using a multi-objective genetic algorithm in the severe cold zones of China. Sol Energy 132:38–50

Using Pix2Pix to Achieve the Spatial Refinement and Transformation of Taihu Stone

Qiaoming Deng[1(✉)], Xiaofeng Li[1], and Yubo Liu[2]

[1] School of Architecture, South China University of Technology, Guangzhou, China
dengqm@scut.edu.cn

[2] State Key Laboratory of Subtropical Building Science, School of Architecture, South China University of Technology, Guangzhou, China

Abstract. Under the impact of globalization, the transformation of traditional architectural space is particularly important for the development of local architecture. As an important spatial component of traditional gardens, Taihu stone has the image characteristics of "thin, wrinkled, leaky and transparent". The "transparency" and " leaky" of Taihu stone reflect the connectivity and irregularity of the holes of Taihu stone, which are in line with the ideas of flowing space and transparency in contemporary architectural design. However, there are relatively few theoretical studies on the spatial analysis and design transformation of Taihu stone. The Pix2Pix model extracts the 3D spatial variation pattern by learning the variation pattern between two adjacent slices of Taihu stone. The trained Pix2Pix model can generate a series of continuous spatial sections with the spatial variation pattern of Taihu stone. Finally, the 2D sections are transformed into 3D building volumes to complete the spatial translation of Taihu stone in contemporary architectural design. In addition, this paper also provides a new idea for machine learning to master the continuous 3D spatial change pattern.

Keywords: Deep learning · Pix2Pix · Spatial transformation · Taihu stone

1 Introduction

After entering the twenty-first century, with the continuous development and construction of cities, Chinese urban architecture style is losing its own cultural personality. The reason is that the local architectural culture has lost its own individuality in the modern architectural design trend. Thus, the translation of local architectural culture in contemporary architectural design is particularly important for the inheritance and development of local architectural culture in contemporary times.

Classical gardens play an important role in traditional Chinese architectural culture. Taihu stone is an important spatial component in classical gardens, a sculptural language in Chinese gardening art, and an aesthetic expression of Eastern philosophical concepts. Traditional Chinese gardens are known for the subtlety of "though made by man, just like opening from heaven", and their space is therefore ambiguous and unqualified. The interior space of Taihu stone is a representative of this kind of space, and its unrepeatability and irreproducibility make it very precious. Taihu stone has the dual properties of

P. F. Yuan et al. (eds.), *Hybrid Intelligence*, Computational Design and Robotic Fabrication,
https://doi.org/10.1007/978-981-19-8637-6_31

building material and space, so this paper focuses on the transformation of Taihu stone into a design element of architectural space. However, the complex spatial relationship of Taihu stone makes the translation in contemporary architectural design a challenge.

The transformation of the space of Taihu stone in contemporary architectural design is of great importance. On a cultural level, the spatial translation of Taihu stone can advance the establishment of Chinese architectural systems in contemporary era. At the level of contemporary architectural design, the transformation of Taihu stone can bring more creative ideas to contemporary architectural design. At the same time the architecture transformed by the space of Taihu stone can be improved in performance. The reason is that the internal spaces of the building are connected to each other, which will bring better light and ventilation to the building. By this method, building energy consumption can be reduced.

With the rapid development of artificial intelligence technology, Generative adversarial network (GAN) has become a popular research direction in artificial intelligence, and the basic idea of GAN is derived from the two-person zero-sum game of game theory. The purpose is to estimate the potential distribution of complex data samples and generate new data samples. In this paper, we try to extract the logical relations of the complex space of Taihu stone with the help of GAN, so as to generate the architecture with the spatial change pattern of Taihu stone. The translation of the space of Taihu stone in contemporary times is accomplished through this way.

2 Background

In previous studies, research has focused on traditional gardening techniques and appreciation of Taihu stones. Ji [4] in "The Craft of Gardens" introduced the reasons for the formation of Taihu stones, materials and their role in the garden. Li [5] in "The idle feeling is sent occasionally" advocated the appreciation of Taihu stones in terms of "translucency", "thinness" and "leakiness". Feng et al. [2] in "Environmental Data-Driven Performance-Based Topological Optimisation for Morphology Evolution of Artificial Taihu Stone" presented a combination of CFD and BESO algorithm to topologically optimize the generation of Taihu stones. These studies have contributed to the development of gardening, especially in the selection of stone strategies for gardening and the creation of different spatial effects. However, these works rarely address the quantification of the internal space of Taihu stones and the transformation of spatial design.

For complex spatial logic relationships such as Taihu stone, its spatial distribution pattern can be extracted with the help of machine learning. Many recent studies have applied GAN models to layout generation and demonstrated that GAN can quickly grasp and generate complex spatial layouts. Huang and Zheng [3] proposed to use GAN to recognize and generate apartment floor plans. In the field of 3D machine learning, Zheng et al. [6] put forward to cut the 3D model into plans, then perform style transfer with the given style image and finally re-stack it into a 3D model. Del Campo et al. [1] proposed to express the 3D model as a 2D depth map, train it by CNN and style transfer, and then express the generated results back to the 3D state. However, the current research on 3D machine learning is mainly in the field of style transfer, and there is fewer research in 3D continuous space sequences.

From the above studies, it can be seen that currently in the field of machine learning mainly contains 2D machine learning and 3D machine learning. For the machine learning of such 3D spatial variation pattern of Taihu stone, this paper tries to propose a new idea to solve it. In this paper, the 3D Taihu stone model is extracted into multiple slices in sequence, and two adjacent slices are used as a set of original training samples. The Pix2Pix model obtains the overall 3D spatial change pattern by analyzing the variation trend of adjacent slices in each group of samples.

3 Methodology

The main process of the experiment to extract the internal spatial variation pattern of Taihu stone by Pix2Pix model is as follows (Fig. 1):

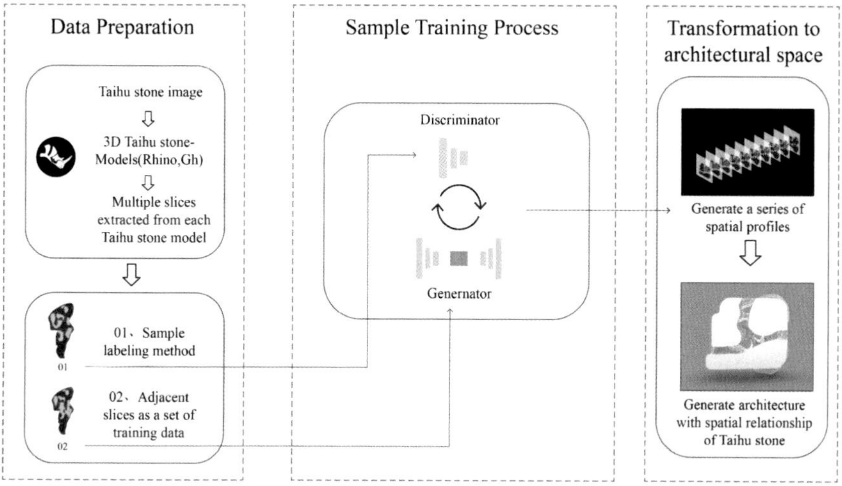

Fig. 1. Workflow of research.

(1) *Dataset establishment*. Firstly, we selected the images of Taihu stone with "transparent" and "leaky" characteristics from the web. Then, the 2D Taihu stone images are converted into 3D Taihu stone models by running Rhino and Grasshopper plug-ins. The profile slices extracted from the 3D Taihu stone model are served as the training dataset.
(2) *Sample processing and labeling*. Firstly, the spatially different elements in the samples are labeled with different colors, and then the two adjacent Taihu stone slices that have been labeled are used as a set of training samples.
(3) *Training and testing*. This experiment was conducted with a total of 520 sets of samples, including 500 sets for training and 20 sets for testing.
(4) *Generation of architecture*. By inputting one section, the trained Pix2Pix model is able to generate a series of consecutive sections with the spatial relationship of the Taihu stone. Finally, all the sections are combined to generate the architecture.

(5) *Experimental evaluation and analysis.* The experimental model is evaluated by the generation effect of the test set and the spatial effect of the generated 3D architecture.

3.1 Network Architecture

The traditional GAN consists of two parts, Generator and Discriminator. The Generator is designed to generate samples and the Discriminator is used to determine the authenticity of this generated sample.

During the training process, the goal of Generator is to generate as realistic images as possible to deceive Discriminator, whose goal is to try to distinguish the images generated by Generator from the real ones. As the two networks play against each other, both networks become more and more capable. The images generated by Generator become more and more like real images, and Discriminator becomes more and more capable of judging the authenticity of the images (Fig. 2). The Pix2Pix model adopted in this experiment is based on GAN to implement image-to-image translation. Therefore, we can generate very realistic images with generator once the Pix2Pix model is trained.

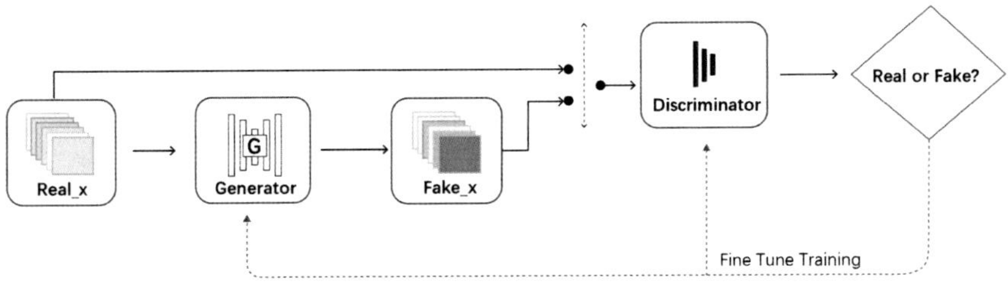

Fig. 2. The network of architecture of Pix2Pix.

3.2 Dataset

Taihu stone has the image characteristics of "thin, wrinkled, leaky and transparent", among which the "leaky and transparent" characteristics are more relevant to the contemporary architectural space. Therefore, in order to make the sections generated by the trained model have a better spatial effect, we selected the images of Taihu stones with obvious "leaky and transparent" features as the original samples.

3.2.1 Sample Processing

In this work, we have collected a total of 12 2D images of Taihu stone as the original samples. By running Rhino and Grasshopper plug-ins, the 2D original samples are converted into 3D Taihu stone models. Then, the 3D Taihu stone models are transformed into multiple sequential 2D profile slices (Fig. 3). The two adjacent slices are used as a set of original training samples.

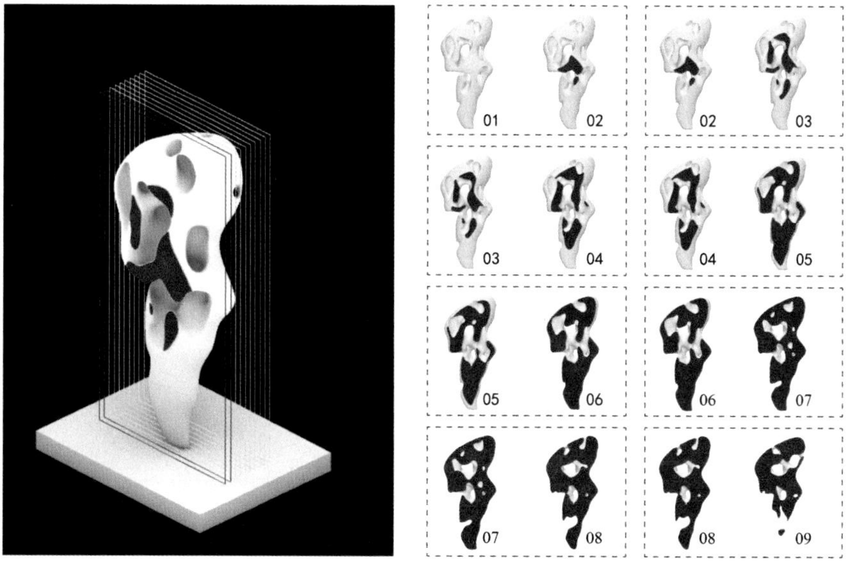

Fig. 3. Preparation of original samples of Taihu stone, left: 3D Taihu stone model, right: Grouped profile slices of the Taihu stone model.

3.2.2 Augmentation

For the purpose of better grasping the spatial distribution pattern of the Taihu stone by Pix2Pix model, 104 sets of original samples are rotated and mirrored, etc. Finally, 520 sets of samples are obtained, of which 500 sets of samples for training and 20 sets of samples for testing.

3.3 Labelling Based on Analysis

By analyzing the spatial elements of the Taihu stone, the Taihu stone profile is divided into three components: "transparent" space, "leaky" space and solid space. In this research, these three spatial elements are labeled with different colors in the sample processing (Fig. 4). The adjacent slices that are labeled are used as a set of training samples (Fig. 5).

4 Training and Analysis

4.1 Training Process

The data set of this experiment contains a total of 520 sets, 500 of which are applied for training and 20 for testing. The training process of the experiments is that the Pix2Pix model generates the Taihu stone profiles by learning the variation pattern of spatial elements between each group of samples. A total of 600 iterations are conducted during the experiment. From the training results (Fig. 6), it can be seen that the boundary of the image generated by the model is clear and the distribution pattern of the hole location

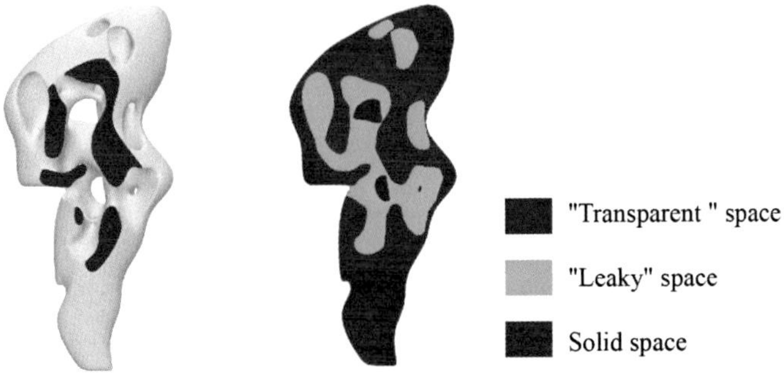

Fig. 4. Labeling based on spatial analysis, left: original slice, middle: labelled slice, right: labelling rule.

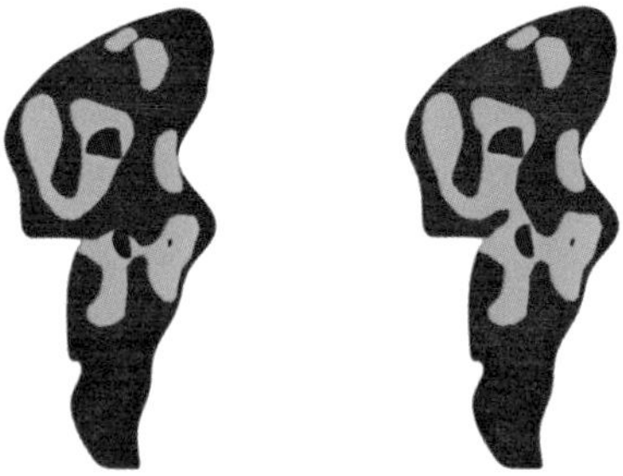

Fig. 5. Adjacent slices as a set of training samples.

is in line with the variation pattern between two adjacent slice samples. The "leakage" space is centered on the "permeability" space, and the generated profile boundary are consistent with the input samples.

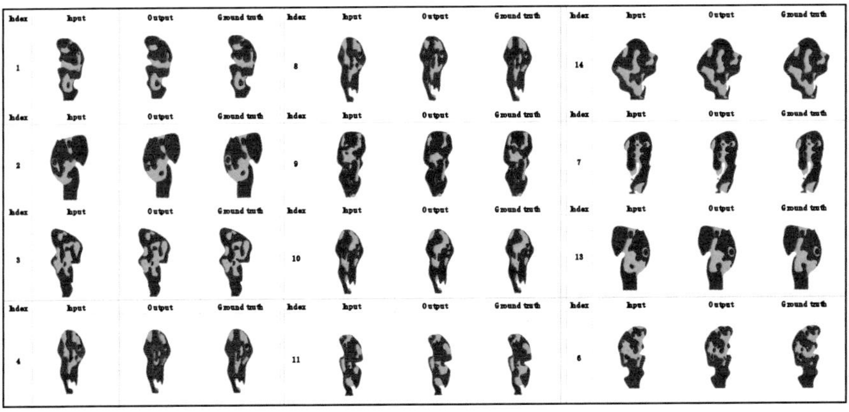

Fig. 6. The part process of the training.

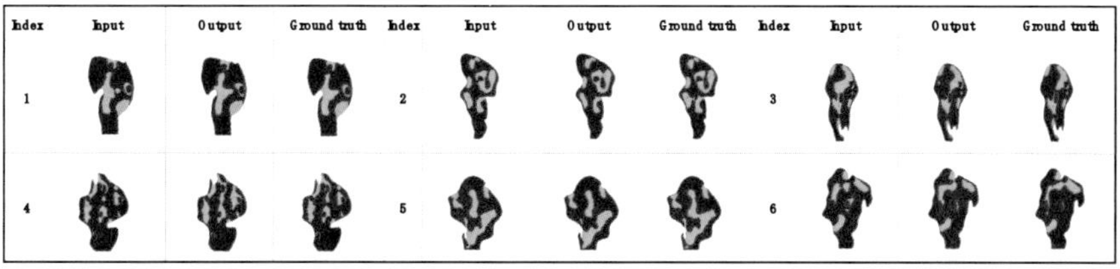

Fig. 7. The results of the testing training.

From the generated results of the test experiments (Fig. 7), it can be seen that most of the generated profile variation distributions follow the variation pattern of the hole distribution in adjacent slices. The difference between the test sample generation results lies in the magnitude of the profile hole variation. The generation results of the test experiments can prove that the Pix2Pix model has mastered the change pattern of adjacent slices in each group.

4.2 Generation of Continuous 2D Sections

In order to generate continuous architectural sections, the first section needs to be input and the trained Pix2Pix model can generate a second section with the spatial variation pattern of the Taihu stone. Then the second section is used as input to generate a third section. Through this method, the generation of all continuous architectural sections is completed (Fig. 8).

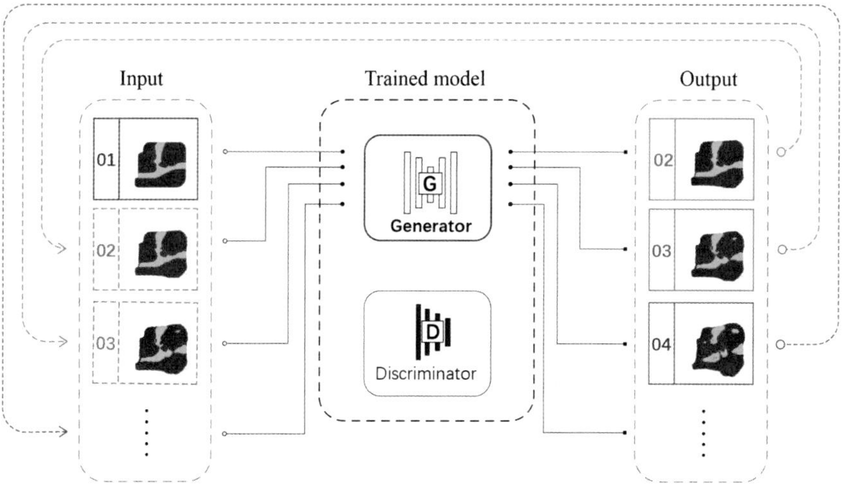

Fig. 8. The process of generating architectural sections.

The continuous 2D sectional images generated by Pix2Pix cannot be applied directly to architectural further design. Therefore, it is necessary to convert the 2D image into a modeling object in architectural language. The Rhino and Grasshopper plug-ins are applied to extract the boundaries of the different elements of the profile based on different colors, and then the boundaries are converted from pixels to curves (Fig. 9). The transformed sections are arranged equidistantly in sequential order (Fig. 10).

Fig. 9. Transformation of image to geometry.

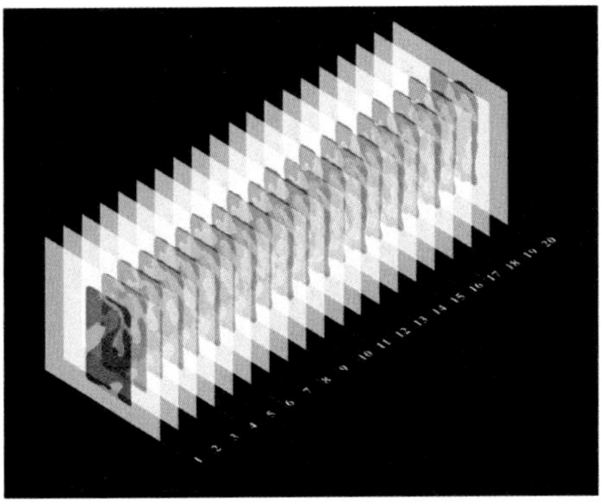

Fig. 10. Generation of multiple architectural sections.

4.3 Transformation of 2D Sections into 3D Model

By converting the building sectional curves into faces with Rhinoceros, and then extruding the faces in the same direction, the final 3D building volume is obtained. During the transformation process, the solid boundary curves are transformed into solid spaces, and the "permeable" and "leaky" spaces are transformed into the void spaces of the building. From the generated building (Fig. 11), it can be seen that the solid spaces in the building are interconnected and intricately changed, which is in accordance with the characteristic changes of "leakage" and "permeability" of Taihu Stone. At the same time, the solid space of the generated 3D architectural volume has similar qualities of flowing space and transparency in contemporary architectural design. Thus, the experimental results suggest that the transformation of complex spaces of Taihu stone in contemporary architectural design is achieved by this method.

Fig. 11. Architecture with the spatial pattern of Taihu stone.

4.4 Result Analysis

In terms of the generated results of individual 2D profiles, it can be found that the generated section variation pattern conforms to the spatial variation pattern of the Taihu stone slices. The boundary contours of the generated 2D sections are consistent with the first input section. The "leaky" space is centered on the "transparent" space. As the sections are generated iteratively, the "leaky" spaces are sometimes connected and then separated from each other.

Regarding the results of continuous 2D section generations, the similarity of adjacent sections is an important indicator of the generated results. By analyzing the continuous 2D sections, it can be found that the overlapping area of adjacent sections can reach 64.25% to 69.94%. This suggests that the variation between sections maintains a well continuity during the generation of continuous sections. At the same time, about 30% of the generated section can vary along a certain trend, which avoids the overfitting of the generated results (Figs. 12 and 13).

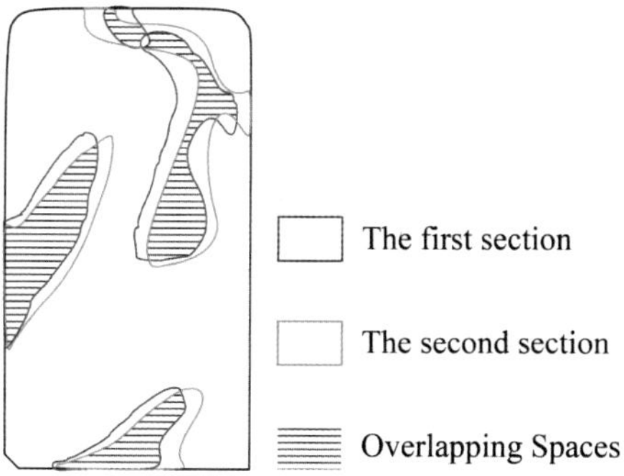

Fig. 12. Similarity analysis of adjacent sections

The results of the 3D model generation indicate that Pix2Pix is able to grasp the variation pattern of the "permeable" and "leaky" space of the 3D Taihu stone. The generated

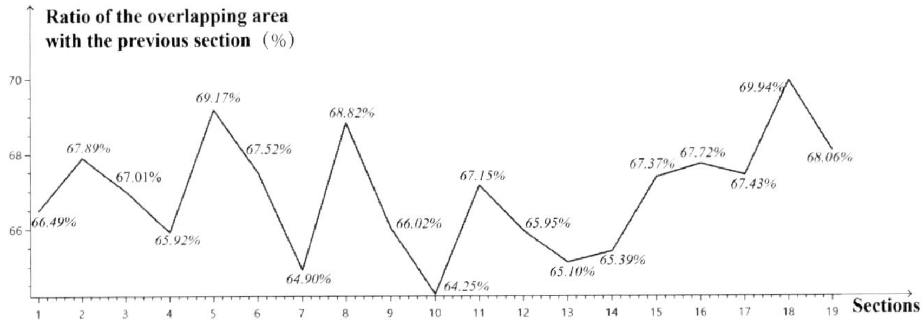

Fig. 13. Spatial continuity analysis of adjacent sections

3D void space model (Fig. 14) shows that the internal space of the building changes from the original separated state to the connected state, and then slowly separates. This is in line with Li Yu's statement in "The idle feeling is sent occasionally" that "the beauty of a mountain or a rock is in the three words: transparent, leaky and thin. This leads to the other, the other leads to this, if there is a road feasible, the so-called transparent; there are eyes on the stone, exquisite on all sides, the so-called leaky."[1] The effect of interconnection between building spaces maintains a high degree of consistency with his idea. Combined with the generated 3D building volumes, it can be illustrated that the Pix2Pix model can master the changing patterns of 3D complex spaces by learning 2D continuous profiles. In a word, the results of this experiment suggest that the transformation of the space of Taihu stone in contemporary architectural design has been realized.

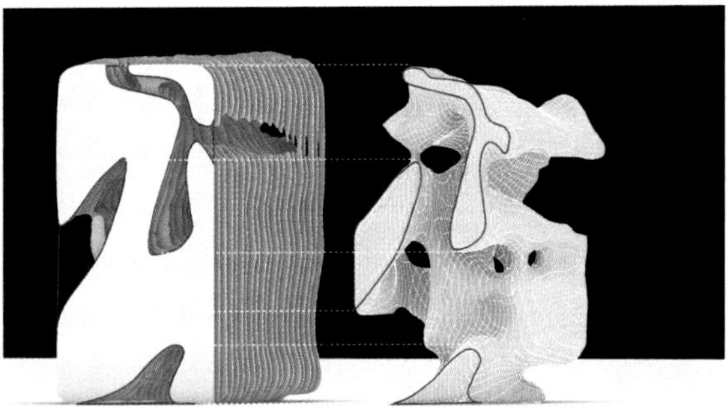

Fig. 14. Architectural solid space (left) and void space (right).

[1] Li, Y.: The idle feeling is sent occasionally. China Book Bureau, Beijing (2011).

5 Conclusion and Discussion

This paper is based on machine learning to extract and grasp the spatial variation patterns of 3D Taihu stone. This paper improves the effect of machine learning to master the change pattern of the 3D model by improving the labeling method and converting the 3D model into continuous 2D training samples. PIx2Pix achieves the extraction of complex spatial change pattern of Taihu stone by training 500 sets of samples and testing 20 sets of samples in this study. By analyzing the spatial effects of the generated 2D sections and 3D models, it can be proved that this experiment has accomplished the transformation of the complex space of 3D Taihu stone in contemporary architectural design. At the same time, this research provides new ideas for machine learning to master the 3D space variation law. More importantly, this research provides a new method for the translation of traditional Chinese architectural space in contemporary architectural design.

Of course, this study still has some limitations. First, after the Pix2Pix model is trained, the process of generating 2D sections is tedious and complicated. By inputting one section, only the corresponding next section can be generated. It needs to be iterated continuously by consuming more time to generate a series of consecutive sections. In the future, this problem can be solved by improving the neural network architecture as well as the structure of the training samples.

References

1. Del Campo, M., Carlson, A., & Manninger, S. (2019, October). Machine hallucinations: a comprehensive interrogation of neural networks as architecture design. In Proceedings of IASS Annual Symposia (Vol. 2019, No. 17, pp. 1-12). International Association for Shell and Spatial Structures (IASS)
2. Feng Z, Gu P, Zheng M, Yan X, Bao DW (2021, July) Environmental data-driven performance-based topological optimisation for morphology evolution of artificial Taihu stone. In: The international conference on computational design and robotic fabrication. Springer, Singapore, pp 117–128
3. Huang W, Zheng H (2018) Architectural drawings recognition and generation through machine learning. In: Proceedings of the 38th annual conference of the association for computer aided design in architecture, Mexico City, Mexico, pp 18–20
4. Ji C (1957) The craft of gardens. Urban Construction Press, Beijing
5. Li Y (2011) The idle feeling is sent occasionally. China Book Bureau, Beijing
6. Ren, Y., Zheng, H.: The Spire of AI - Voxel-based 3D Neural Style Transfer. In: Proceedings of the 25th International Conference on Computer-Aided Architectural Design Research in Asia (2020)

Collapsing Complexities: *Encoding Multidimensional Architecture Models into Images*

Viktória Sándor(✉), Mathias Bank(✉), Kristina Schinegger, and Stefan Rutzinger

Department of Design, University Innsbruck, i.sd, Technikerstr. 21, 6020 Innsbruck, Austria
{Viktoria.Sandor,Mathias.Bank-Stigsen,Kristina.Schinegger, Stefan.Rutzinger}@uibk.ac.at

Abstract. The paper details a 3D to 2D encoding method, which can store complex digital 3D models of architecture within a single image. The proposed encoding works in combination with a point cloud notation and a sequential slicing operation where each slice of points is stored as a single row of pixels in the UV space of a 1024 × 1024 image. The performance of the notation system is compared between a StyleGan2 and existing image editing methods and evaluated through the production of new 3D models of houses with material attributes. The uncovered findings maintain the relatively high level of detail stored through the encoding while allowing for innovative ways of form-finding—producing new and unseen 3d models of architectural houses.

Keywords: Form finding · 3D · Encoding · Point cloud · Machine learning · Architectural design

1 Introduction

Architectural design processes are increasingly situated in the digital space, resulting in a large amount of architectural 3D models. During all phases of an architectural design process, from concept to completion, digital 3D models increasingly act as the link between what we can think, and what we can build [4]. As a result, 3D models have become the core method for communicating and creating an architectural design. Compared with traditional 2D representations, the 3D model provides a more holistic representation of the spatial relationships that lay behind a given design [4], while offering an accurate and adequate representation of the architectural space and its proportions [7]. The transition from 2 to 3D has inevitably increased the complexity for architects to visualise and explore new ideas. This is particularly true early in the design process, where any promising design intent needs to be translated into 3D to be properly communicated. To overcome this new complexity, there is growing interest in working with artificial neural networks in digital design processes. Although artificial neural networks have the potential to address architectural complexities, training networks directly on 3D models remains a challenge.

P. F. Yuan et al. (eds.), *Hybrid Intelligence*, Computational Design and Robotic Fabrication,
https://doi.org/10.1007/978-981-19-8637-6_32

To establish usable artificial neural networks for design collaboration, we need to develop notations that can transmit the spatial qualities and complexities of architecture. Current research shows several attempts to train artificial neural networks on 3D models. Although some utilise a three-dimensional medium such as point clouds [1] or voxels, the development of 3D machine learning is still difficult for architectural purposes, due to availability, speed, and resolution. Therefore, our research focuses on exploring a 2D notation of digital 3D models, which can interact with a 2D machine learning framework to assist architects in the early stages of the design process.

1.1 Relative Work

Artificial neural networks trained on 2D data have shown a remarkable "talent" for grasping patterns and concepts within complex datasets; but how can we best encode 3D models into 2D mediums to benefit from this? Tomographic data is one such approach and is an inspiration for the work by Kench and Cooper. Their developed SliceGan can synthesise high fidelity 3D datasets using a single representative image. Their approach was successfully demonstrated on material microstructures [6]. Along a similar vein, with a more architectural agenda, Zhang and Huang presented their solution for a machine learning aided 2D–3D architectural form-finding, by introducing a sequential slicing of a given 3D model. In their method, a sequence of images, describing the 3D model, are stitched into a single image for compatibility with a 2D neural network [12]. With The Spire of AI project, Zheng and Ren introduced a method for voxel-based 3D neural style transfer using 2D slicing, in which pixel points of stylized 2D slice images are extracted and mapped into a 3D domain according to the slice order [9].

In contrast to the slicing approach Miguel et al., introduced a method for notating 3D models into a connectivity map utilising voxelated wireframes. Through variational autoencoders, this notation facilitated the generation, manipulation and form-finding of structural typologies [8]. Another example of an abstract notation takes place when storing data in pixels by plotting pixel plots. Although pixel plots are a popular format in data visualisation, they are also used for vertex animation textures. These textures are used to control morphing animation by mapping vectors to colours and storing them on the row corresponding to the animation frame [10]. This approach is often applied within the gaming industry since it is an efficient way to store a lot of data within an image while seamlessly connecting to the shader pipelines.

The solutions for compressing multidimensional data into 2D images are continuously evolving, nevertheless, it is still hard to properly store a complex architectural 3d model in a single 2D image.

1.2 Objectives

With these developments in mind, we want to present a new 3D to 2D notation system that through the manipulation of encoded 2D images can perform design operations on 3D models of architecture. The approach is combining a sequential slicing with a point cloud notation, where additional information such as the materiality is assigned to colours. The coloured point slices are then stored in a single image, encoding the points and their additional information to pixels. By testing the notation with artificial neural networks and traditional image editing techniques, we aim to explore new ways of form-finding in the early stages of a design process, while simultaneously evaluating the encoding methods' ability to transmit spatial qualities and complexities of 3D architecture models.

2 Methodology

The focal point of the research is the development of an encoding framework that allows the compression of a complex 3D model to a single, pixel image. In the proposed encoding method, point clouds play a significantly important role. While point clouds in architecture are mostly utilised in the digitization process of real-world objects, we chose them for their nature of describing digital surfaces and volumes in discrete format. Due to their multidimensional data-storage capacity, they are an ideal medium for representing a large number of (spatial) attributes. The proposed 3D to 2D encoding method in the research relies on the translation of 3D solids to 3D pixels and 3D pixels to 2D pixels that emerge into images. Since at the stage of 3D pixelization of models, it is advantageous for the resolution to remain adaptive and flexible, while irregularities do not interfere with the encoding method, 3D point clouds were chosen over voxels for the translation of 3D geometries.

Based on the desired image resolution of the encodings, the representative clouds are sampled. In this research we aimed for 1024×1024 pixel-sized image encodings, it being currently the most convenient image size for the training of StyleGans. The resolution of the encodings can be adapted for other purposes.

At this stage, the proposed method focuses only on the encoding of 3D solid compositions, where solids represent materials as spatially closed volumes to which colours are assigned. The resulting coloured volume compositions describe spatial concepts through labelled material distributions.

2.1 Encoding

The encoding consists of three main parts, that combine point cloud notation and sequential slicing operations on 3D models: Model Discretization, Colour Mapping and Legend construction.

2.1.1 Model Discretization

As the first step, we start with the discretization of the 3D model dataset. The discretization consists of four different steps. While each step differs in its resolution and sequence, they all preserve the spatial attributes of the original models (material, position, etc.) (Fig. 1).

Through slicing along a pre-defined axis, we divide the solid 3D model into groups of planar surfaces (sections). Each of the resulting surface sets, belonging to the same plane, is then split into 4 areas of similar size. The division can be defined by the dimensions of the whole model or by the local dimensions of each slice. In addition, the size and proportion of the segments can be uniform or customised based on the varying mass distribution of the model (Fig. 2). The four segments are further discretized by point cloud scattering (populating) on the corresponding surfaces.

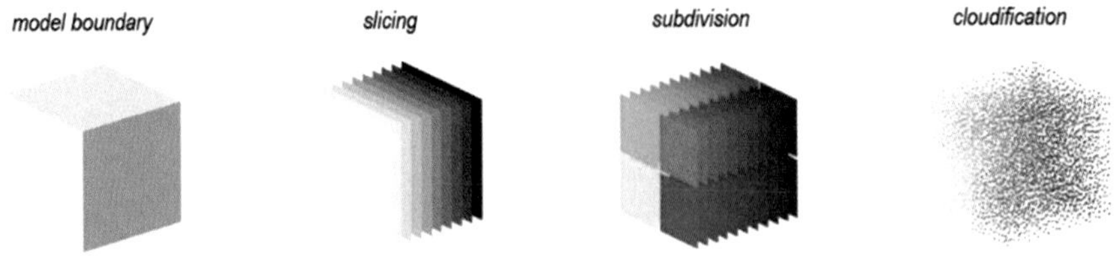

Fig. 1. Discretization Steps. Four model discretization steps of different resolutions and scales: model boundary, slicing, subdivision, cloudification.

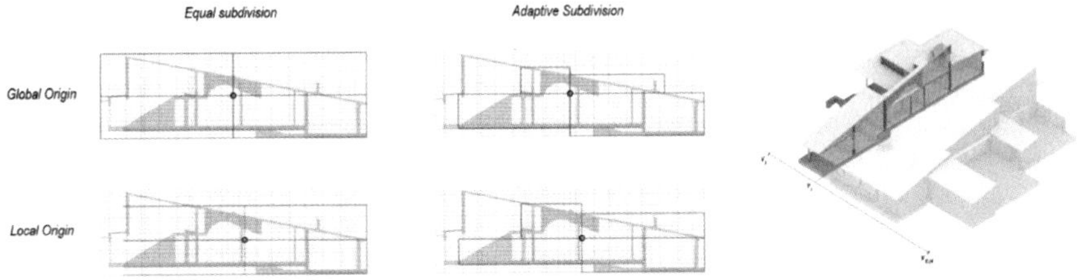

Fig. 2. Global versus local subdivision. Four distinct strategies for slice-subdivision. Top left: Equal slice-subdivision with global origin. Top right: Adaptive slice-subdivision with global origin. Bottom left: Equal slice-subdivision with Local Origin. Bottom-right: Adaptive slice-subdivision with local origin.

The discretization method, detailed above, provide us with precisely structured multi-dimensional 3D point clouds. Other than their densities, which represent the mass-void characteristics of the model, they also visualise the material attributes of space, based on the original materials, assigned to the 3D volumes.

2.1.2 Colour Mapping

The colour mapping of discrete 3D models (coloured point clouds) is based on 8-bit colour mapping of discrete positions in space. Using a similar technique to the previously mentioned vertex animation textures, the X and Y coordinates of each point in each slice segment are mapped to the green and blue channels of the RGB space. The original colour of the point, representing the material property in this dataset, is assigned to a predefined red channel value. As a result, we can express 255 × 255 positions in each slice segment with up to 255 different materials. Using this technique, we can represent 260,100 different positions on each slice of the model.

To bring the points of the 3D point cloud into a single 1024 × 1024 pixel image—pixel plot—we keep the discretization structure of the model and use it as the layout strategy for the image. The pixel plot of the 3D model is divided into 1024-pixel rows, four 254-pixel wide columns and an 8-pixel wide legend. The rows represent the slices, the columns their normalised segments, and the legend the position and scale of each segment (Fig. 3). Such structure provides us with the visualisation capacity of 1024 model slices each with a maximum of 1016 material labelled points. The resulting pixel plot stores a point cloud with the size of 1.040.384.

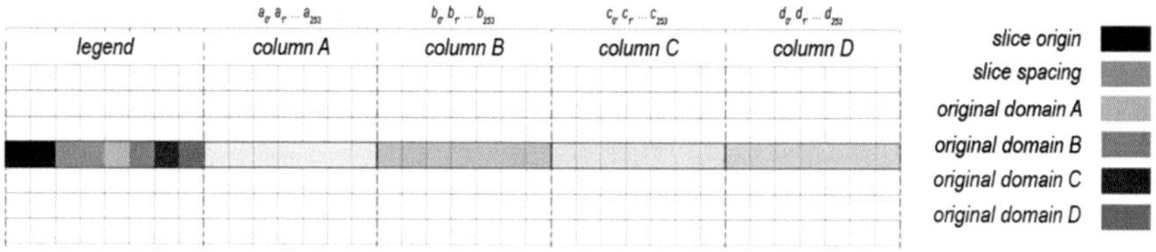

Fig. 3. Pixel plot structure, legend and columns. The four columns A, B, C and D, represent all slice divisions of the model. The legend on the far left of the pixel plot is responsible for controlling the model scale and overall geometry. The first two pixels contain the origins of each slice, 3–4 pixels store the slice spacing values, pixels 5–8 contain the original subdivision proportions of slices.

Since the number of expressible point positions in a slice segment is about 256 times larger than the number of storable positions in a 1024 × 1024 pixel plot, we had to sample the point cloud slices in many cases. After sampling, we tested four different strategies to sort the pixels representing points in the plots (Fig. 4).

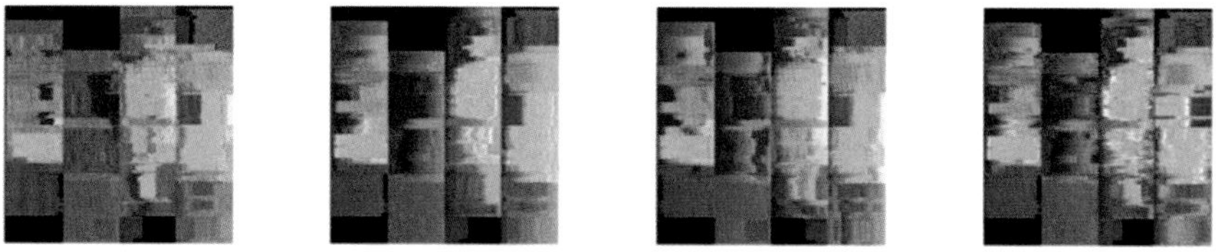

Fig. 4. Pixel sorting, material and position. 4 different strategies to sort the pixels of the plots. The examples show the 4 sorting results of Alvar Aalto's Louise Carre pixel plots. Left to Right: sorting based on materials (Red Channel); sorting based on position distance from Origin (0-GB); sorting based on position and material distance from Origin (0-RGB); sorting based on "travelling salesman problem" 2D

2.1.3 Legend

The first eight pixels of each row of the pixel plot is saved for the so-called legend. While the four 254-pixel-wide columns store the position of the model points mapped to a normalised domain, the legend stores the location of the origins and scales of the original cloud segments. To provide flexibility in the manipulation of 3D models, the legend is split into 6 sections. Pixels 1–2 store the local origin of each slice, pixels 3–4 store the spacing, and pixels 5–8 the original domain sizes of each slice segment (Fig. 3).

2.2 Decoding

The decoding of the pixel plot starts with the reading and structuring of the .raw image file. The next step is the extraction and decoding of the legend pixels to the slice origin, slice spacing and the four domain dimensions. The pixel channels of the remaining 4 columns, A, B, C and D are split and used to define the position and materiality of the model points. The green and blue channels are read as the x and y coordinates of the points while the red channel defines their materiality. To place the points of the columns to the right area of the slices, all green values of A and C and all blue values of C and D columns are inverted. At this stage, all decoded points lie in a single plane around the origin. To introduce the third dimension of the model, we ideally use the spacing values decoded from pixels 3–4 of the legend. To avoid segment distortion, the point coordinates are mapped to the original domain dimension, decoded from pixels 5–8 of the legend. In the case where individual origins are used for the segments, each segment is moved by the decoded vectors of legend pixels 1–2. If suitable, the legend can be replaced by other numeric manipulators.

3 Results

To evaluate the performance of the outlined encoding and decoding workflow, a dataset consisting of 50 labelled 3D models of famous architectural houses spanning 500 years of architectural history was used. All models were constructed from available published documentation and labelled according to the outlined method above. Each model furthermore has a similar level of detail, and a fully modelled interior (Fig. 5).

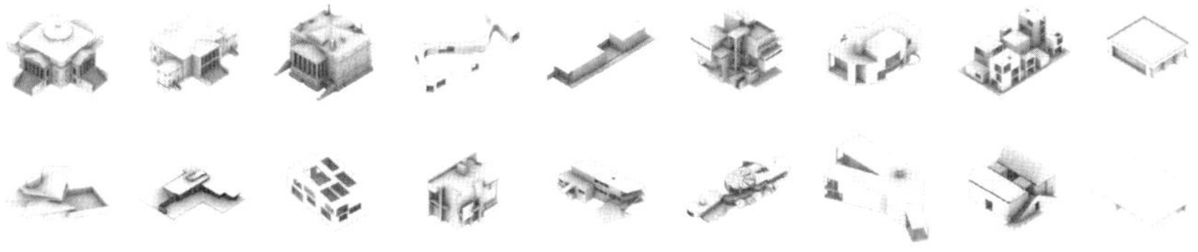

Fig. 5. Examples of 3D house models from the assembled dataset.

The form-finding performance of the proposed encoding–decoding workflow is compared in two different experiments. In the first experiment, we test conventional image editing methods, while in the second experiment, we use a StyleGan2 Ada [5] trained on pixel plots. In both experiments, the form-finding performance is evaluated through the production of new 3D models of houses, and their spatial segments with material attributes.

3.1 Image Editing

The encoding of a full 3D model into a single image offers a lot of new opportunities for manipulating the model through established image editing techniques. To properly test its potential, four housing models were encoded (Fig. 6), and image editing manipulations, such as Colour channel adjustments, Hue change, Saturation change, Blending, Collaging and Legend swapping were explored. Since the legend holds an enormous amount of power over the decoded results, it was excluded from the first five manipulations to better assess how the various editing operations perform.

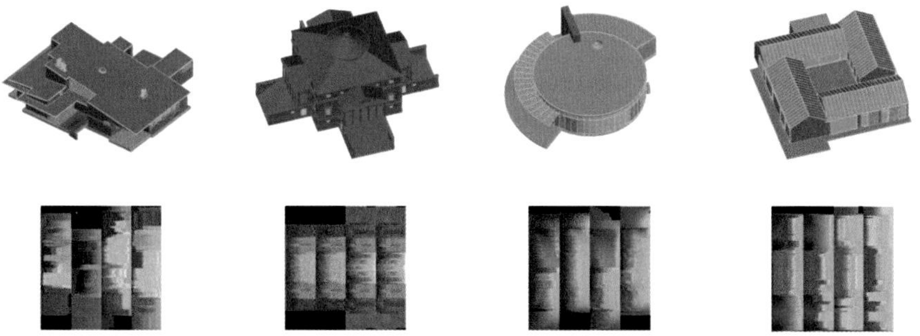

Fig. 6. The four 3D housing models and their corresponding pixel plots which is used for the six image editing manipulations. From left to right: Alvar Aalto, Maison Louise Carre. Andrea Palladio, La Rotonda. Arne Jacobsen, Leo Henriksen House. Lenshow Philmann, House at Mols Hills.

The results establish a small atlas of the potentialities in achieving new and varied 3D models through image editing techniques. Large colour adjustments through channels, hue or saturation easily end up distorting the initial 3D models beyond recognition, while smaller values, especially in the case with hue, show interesting deformations. The results (Fig. 7) from blending and collaging three different pixel plots produce exciting point cloud models, with architectural suggestions, highlighting the large and still mainly undiscovered repertoire in generating new shapes through these methods.

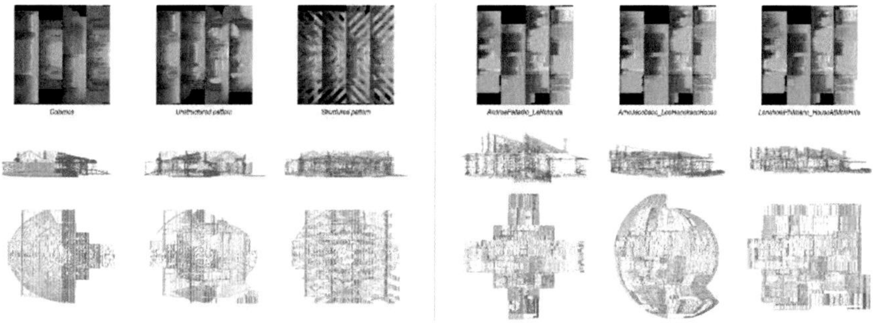

Fig. 7. Image editing results. Left: ***Collage***—Explores three different types of collages. The first exchanges full 256-pixel columns between three unique pixel plots. The second applies an unstructured circular brush pattern, while the third applies an organised repetitive pattern to collage three pixel plots together. The resulting models highlight the potential of using collaging techniques to create new models. Right: ***Legend***—The manipulation replaces the legend of a pixel plot with the legend from three other pixel plots (Fig. 6). As seen, this remaps the base model into the boundary shapes of the models, from which the replacement legends came. This is a powerful method to control and reconfigure complex models from one boundary representation into another

3.2 StyleGan

2D StyleGans have shown a remarkable ability for producing synthetic results that appear eerily similar to the data on which they were trained [11]. There are a plethora of different StyleGans and approaches for this purpose. For our experiments, we use the base repository for StylGan2 Ada from Nvidia [3] and test how well a network can generate new images based on our encoding structure. For the evaluation, we observed the extent to which architectural forms are reproduced in the decoded point cloud models.

3.2.1 Training on Segments

To generate sufficient data for training an artificial neural network, the full dataset of 50 architectural 3D models of houses were used. Following the outlined encoding method, 50 pixel plots were obtained from the 50 architectural 3D models for each slicing direction (Fig. 10). To increase the size of the training dataset, all pixel plots were divided along the columns into sixteen 256 × 256 pixel segments (Fig. 8).

Fig. 8. Segment training data versus StyleGan outcome. To the left is a subset of the segments used to train the network, while the right displays a selection of synthetic images produced by the trained StyleGan2 network.

The StyleGan2 Ada network was trained with a dataset of around 2000 images at 256 × 256 resolution. Furthermore, it was utilising transfer learning, training on top of the FFhQ-10 k dataset at 256 × 256 [2]. The network was trained for 3000 kimg. The produced synthetic images (Fig. 8) appear similar in layout but are limited in content. Although the resulting architectural properties of the synthetic pixel plots segments were vague, some of them showed recognizable spatial properties of the training data when decoded into point cloud models (Fig. 9).

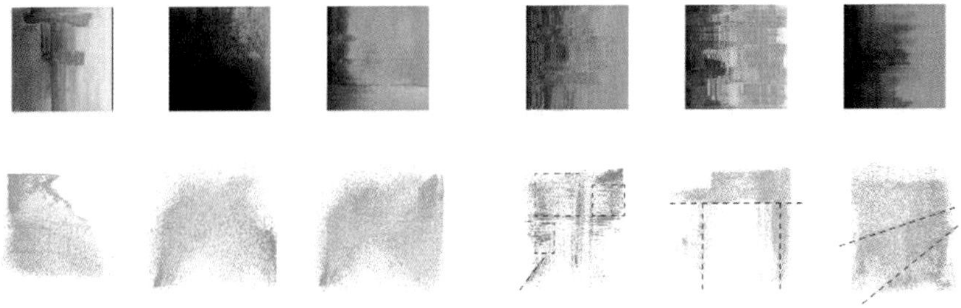

Fig. 9. Decoded results, synthetic segments. To the left are three common outcomes, while the right side displays three of the best, but also uncommon, outcomes

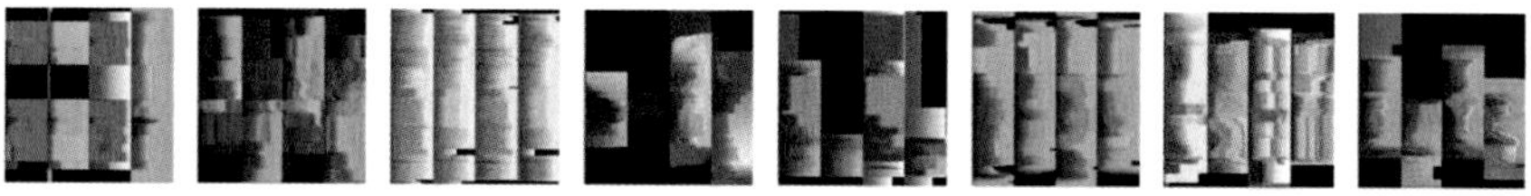

Fig. 10. Examples of pixel plots from 3D houses. A subset of the dataset used for training the StyleGan network on the full pixel plots.

3.2.2 Training on Entire Models

Similar to the segment training, a dataset consisting of the pixel plots depicting the 50 architectural 3D models was assembled (Fig. 10). This time, instead of splitting the pixel plots, the training dataset was multiplied by slicing each house in five directions, giving a total of 500 images after vertical mirroring. The legend stored the dimensions and the origin of the global domain in each pixel plot, which provided representative information about the dimensions of the model in the entire plot.

The StyleGan2 Ada network with the full pixel plots was trained with the same basic parameters as the segments but at a 1024×1024 resolution. Due to the relatively smaller dataset of only 500 images, the decoded results shown in Fig. 11, aren't very indicative of the feasibility of utilising a StyleGan to create new shapes. All the decoded images in Fig. 11 creates rather noisy and vague models without any particularly distinct spatial properties. In the most successful ones to the right, the legend and the noise appears less arbitrary. Here the distribution of materials in the clouds show some clarity that starts imitating the characteristics of the original 3D models.

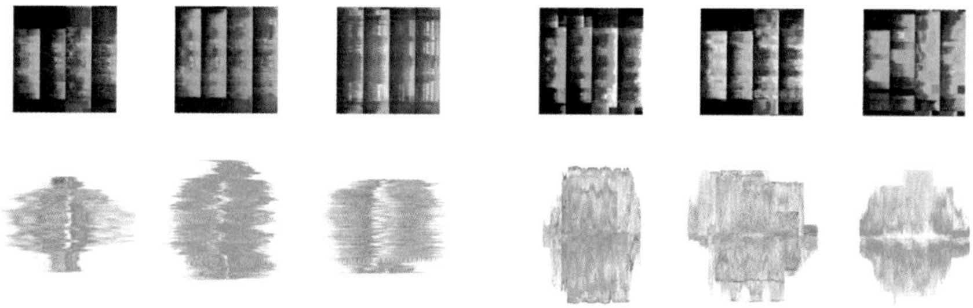

Fig. 11. Decoded results, full-size synthetic pixel plots. To the left are three common outcomes, while the right side displays three of the best, but also uncommon, outcomes.

4 Conclusion

In this paper, we presented a new method for architectural form-finding using existing concepts of houses. We showed that our method can be used to encode complex 3D models into a single image, decode a single image back to complex 3D models, and create architectural 3D models, using image editing techniques. Through the comparison of encoded images, one could recognize spatial patterns, representing the different material distributions in space. Thanks to the developed legend of the encodings, models could be merged, collaged and blended without losing the unity of decoded clouds. With our second method, we also showed how new, unseen 3D point clouds can be generated by a styleGAN, after being trained on the encoded images. While the "architectural" performance of the GAN generated clouds were not even close to the successful image edited concept-collages, -blends and -legend exchanges, in some cases we could still recognize the transferred spatial qualities of the original dataset. Since our current training data was rather small, the results are keeping us motivated.

However, during this research, we have encountered several limitations that need to be acknowledged. Our current encoding technique is working exclusively with solid models, making the preparation of the dataset time-consuming. Although the encoding and decoding of 3D models can be processed in the Rhino-Grasshopper environment, there is no direct connection yet between the image editing tools and the modelling software. This lack of immediacy makes it hard to properly explore and iterate within a design process.

Concerning the StyleGan our lack of training data generated results that often appeared spatially arbitrary. While the segmentation of pixel plots increased that dataset size drastically, successful training on entire pixel plots would also require more data. A larger training dataset would also allow us to train a StyleGan without utilising transfer learning, potentially improving the results. The supplementation of the authors' limited expertise relating to the customization of GAN setups would significantly push the outlined encoding implementation within artificial neural networks.

In this article, we have taken existing architecture as a basis for the discovery of new patterns. By using special encoding techniques and machine intelligence, we attempted to discover new ways of form-finding that transfer design knowledge to concept models at the early stages of architectural design. With this research, we introduced a new format for the representation of architectural concepts and aimed to highlight its potential as training data for artificial neural networks and also as new mediums for the creation of architectural concepts.

Acknowledgements. We would like to thank our students at the University of Innsbruck, Institute for Structure and Design, for their help in producing the dataset. The work was funded by the University of Innsbruck and the Austrian Science Fund (FWF) project F77 (SFB "Advanced Computational Design").

References

1. Achlioptas P, Diamanti O, Mitliagkas I, Guibas L (2018) Learning representations and generative models for 3D point clouds. arXiv:1707.02392 [cs]
2. Hellsten J, Karras T (2022) NVlabs/ffhq-dataset. NVIDIA Research Projects. https://github.com/NVlabs/ffhq-dataset. Accessed 16 Mar 2022
3. Hellsten J, Karras T (2022) NVlabs/stylegan2-ada-pytorch. NVIDIA Research Projects. https://github.com/NVlabs/stylegan2-ada-pytorch/blob/6f160b3d22b8b178ebe533a50d4d5e63aedba21d/README.md. Accessed 15 Mar 2022

4. Hirschberg U, Hovestadt L, Fritz O (eds) (2020) Atlas of digital architecture: terminology, concepts, methods, tools, examples, phenomena. Birkhauser, Boston
5. Karras T, Aittala M, Hellsten J, Laine S, Lehtinen J, Aila T (2020) Training generative adversarial networks with limited data. arXiv:2006.06676 [cs, stat]
6. Kench S, Cooper SJ (2021) Generating 3D structures from a 2D slice with GAN-based dimensionality expansion. arXiv:2102.07708 [cs]
7. Marinčić N (2019) Computational models in architecture: towards communication in CAAD. Spectral characterisation and modelling with conjugate symbolic domains. Birkhäuser
8. de Miguel J, Villafañe ME, Piškorec L, Sancho-Caparrini F (2019) Deep form finding using variational autoencoders for deep form-finding of structural typologies. In: Blucher design proceedings. Editora Blucher, Porto, Portugal, pp 71–80
9. Ren Y, Zheng H (2020) The spire of AI—voxel-based 3D neural style transfer. In: Anthropocene, design in the age of humans, vol 2. CAADRIA, Bangkok, Thailand, pp 619–628
10. Vasconcelos LO, Sato A (2020) Texture animation: applying morphing and vertex animation techniques. Wildlife Studios Tech Blog. https://medium.com/tech-at-wildlife-studios/texture-animation-techniques-1daecb316657. Accessed 15 Mar 2022
11. West J, Bergstrom J (2019) Which face is real? Which face is real? https://www.whichfaceisreal.com/methods.html. Accessed 15 Mar 2022
12. Zhang H, Huang Y (2021) Machine learning aided 2D–3D architectural form finding at high resolution, pp 159–168.https://doi.org/10.1007/978-981-33-4400-6_15

Material and Fabrication

Augmented Bricks: an Onsite AR Immersive Design to Fabrication Framework for Masonry Structures

Yang Song(✉), Asterios Agkathidis, and Richard Koeck

The Liverpool School of Architecture, University of Liverpool, Liverpool, UK
yang.song@liverpool.ac.uk

Abstract. The *Augmented Bricks* research project aims to develop an immersive design to fabrication framework for the assembly of masonry building components by incorporating robotic fabrication and augmented reality (AR) technologies. Our method incorporates two main phases: firstly, the design phase in which users' gestures and interactions are being identified in AR for the immersive design and simulation process; secondly, an innovative robotic assembly phase in which users can control a robotic arm for assembly by interacting with the AR user interface (UI). Our framework is validated by the design and assembly of four brick-based columns. Our findings highlight that the proposed design to fabrication framework offers a novel, intuitive design inspiration and experience beyond the traditional design methods. It returns the task of assembling parametric structures with high-tech equipment back to the designers, allowing them to master and participate in the entire design to the fabrication process. The impact of this practice-based research will allow architects and designers to modify and construct their designs more simply and intuitively through the AR environment.

Keywords: Augmented Reality (AR) · Immersive design · AR-assisted assembly · Robotic operation · Masonry structures

1 Introduction

The definition of AR appeared in 1997 and is described as a technological field that involves the seamless overlay of computer-generated virtual images aligned with the real world, and can be viewed and interacted with in real-time [1]. With the continuous development of technology and equipment, AR has gradually started to enter our daily lives. Particularly, in the past decade, with the invention of AR headsets and the popularisation of AR-ready smartphones and tablets, the research and applications of AR have grown explosively [5]. No matter how AR develops, it does not deviate from its purpose of bridging the gap between virtual data and the real world [2].

Furthermore, more and more architects are interested in AR because it arguably offers all kinds of new interaction scenarios in all the architectural fields [7]. Recently, AR technology has been applied and explored from finding and design, construction, visualisation to education, and more [11].

P. F. Yuan et al. (eds.), *Hybrid Intelligence*, Computational Design and Robotic Fabrication,
https://doi.org/10.1007/978-981-19-8637-6_33

Architectural design, being the quintessential 3D–4D design field, has throughout its history been limited by 2D or cumbersome 3D representation, such as sketching on the plane surface or building physical scale models [3]. Even though computer-aided architectural design and modelling software is widely used to produce digital 3D models, their preview is still limited to a 2D-based screen, which lacks an intuitive means of onsite visualisation and modification. Additionally, conventional screen-based visualisation methods for design and analysis are restrictive to how well the user understands the space on a computer, as the design is done outside the building site, hence there might be disparities between the design and final fabrication [9]. This limitation may be eliminated by AR technology, which has become readily available, together with tools facilitating the easy creation of 3D–4D models as holograms onsite. Furthermore, with its gesture and voice capture features, AR can increase the potential for interaction between humans and data [4].

Robotic fabrication, an emerging high-tech architectural digital fabrication method, has shown great potential for integrating architectural design and engineering practices, establishing a highly effective interplay between digital design and construction processes [8]. However, the robotic operation process requires complex knowledge and skilled programming code workers, which is an expertise that is traditionally not found in architectural practitioners [10]. Although there are already some robotic operation plugins within the *Rhinoceros/Grasshopper* platform, they require architects to visually programme the process, which is usually inefficient, complicated and accompanied by many debugging and instability errors. Robotic programming in *Grasshopper* also tests or challenges the traditional architects' logic. Even with these programme methods, there is a lack of security simulation and protection for inexperienced architects. Therefore, digital fabrication always needs the help of engineers. Due to the disconnection between architects and robotic engineers, uncertain situations often appear in the robotic fabrication process [6]. AR technology may avoid this limitation, which can capture interactive inputs through the UI and display onsite holographic simulation to provide an easy, safe, and low-threshold method for architects to control robots by themselves.

This paper proposes an onsite AR immersive design to fabrication framework by combining the above unique characteristics and functions of AR to find out how AR technology is changing and evolving the traditional design to assembly methods in architectural construction.

2 Research Methodology

Our *Augmented Bricks* research project proposes an onsite AR immersive design to fabrication framework for the assembly of masonry structures. The framework consists of two phases: (a) the algorithmic immersive design of the object and (b) the robotic fabrication of the object by a robotic arm (Fig. 1). To validate this AR-assisted framework, we conduct a design experiment, which includes the design and assembly of four parametric brick-based columns and evaluate its workflow as well as inspect the advantages and disadvantages of each step. The prototypes were designed and built with the styrofoam blocks (150 * 50 * 20 mm) as the prototype brick-based material for testing, which is suitable for parametric design, easy for AR devices to detect, and able to be picked and placed by the robotic gripper.

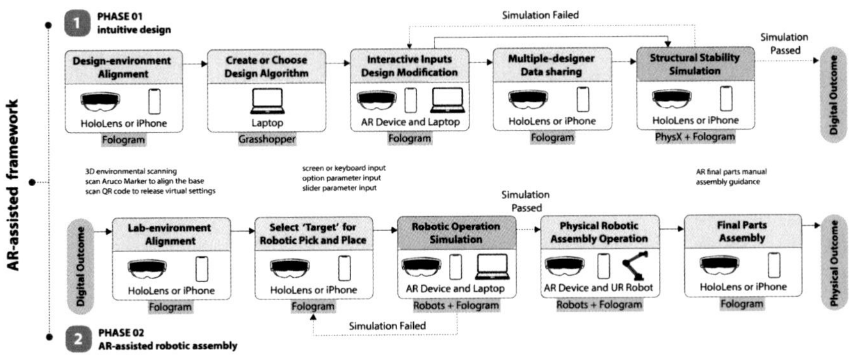

Fig. 1. The *Augmented Bricks* project AR-assisted framework flowchart. The framework is divided into two phases: immersive design and AR-assisted robotic assembly. The outcomes of each phase are a digital design and a physical structure.

Our software includes *Rhinoceros/Grasshopper,* which was applied for the development of the design algorithm, as well as the structural simulation plugin *PhysX*, the robotic fabrication firmware *Robots* and *Fologram* an AR plugin for *Grasshopper*. We use *Fologram* to identify interactions in AR from hand gestures or screen-based inputs; *PhysX* to give a real-time structural stability simulation and design modification feedback; as well as *Robots* to develop the robotic operation trajectory and gripper commands. Our original contribution is to integrate the advantage functions from various plugins and create an onsite AR immersive design to assembly framework for masonry structures.

Our hardware includes a handheld deviceis—*iPhone 11*, and a headset—*Microsoft HoloLens 1* for AR, as well as a *Universal Robots 10* robot arm with *Robotiq* 2F-140 grippers for the robotic equipment. We also use a laptop for back-end running and debugging. All of these devices are connected to a WIFI router in the same IP address network environment for transforming the data from different stages, and live streaming commends on design software and plugins to visualise and output response ports.

3 The Augmented Bricks Design Experiment and Outcomes

3.1 Phase 1: The AR Immersive Design

The AR immersive design process is the first phase of the *Augment Brick* experiment, which contains the 3D onsite environment scanning, gesture or screen-based interactive design input method, structural simulation feedback, and multiple-designer data sharing. The idea of proposing an immersive design method is to evolve the traditional design method by giving architects a 3D–4D modelling environment that could be shared and to provide them with an onsite virtual space experiment and structural rationale feedback before the structure is built.

The user requires the 3D scanning onsite design base before the AR immersive design. To achieve that, we provide two spatial environment scanning ports, an AR smart device (smartphone or tablet) and AR headsets. First, the user can activate the 'Track Scan' function in the *Fologram* plugin for real-time digitalised environment scanning. The scanning operation for users is to use a smart device or headsets by looking around with the camera smoothly in the onsite environment with a steady light source. Second, the physical environment will be transformed into a simple mesh in *Grasshopper* for architects to use as an onsite design base. Last, to adjust and align the digital environment or set the design boundaries, the user can use Aruco Markers to set the datum reference points physically upon the onsite base and convert them digitally by scanning the markers through AR devices. By doing that, the accuracy of the design plane is improved significantly (Fig. 2). The converted onsite base mesh is stored in a QR code for subsequent use. This method is only used for simple and basic onsite environments. For complex environments or uncertain onsite bases, we recommend the user to activate the *Capture App* for smart-device, or the spatial mapping function in *HoloLens*, to scan and import the corresponding highly accurate digital 3D meshes for further edition in software before using Aruco Markers to set the datum reference points.

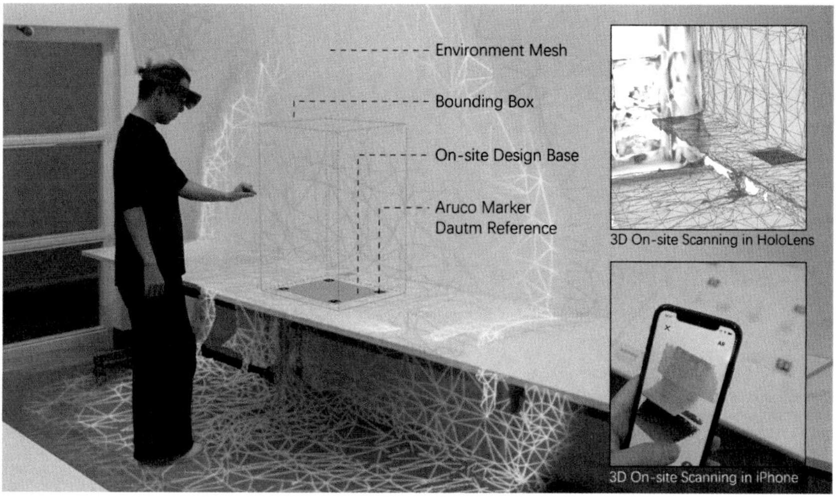

Fig. 2. The designer uses a 3D onsite environment scanning process, including the Aruco Markers datum reference points, to create the corresponding digitalised environment mesh for the immersive design base and bounding plane

For the immersive design process, we create an open design algorithm platform, followed by an AR immersive UI and a structural stability simulation feedback loop. First, the user can choose an algorithm from our design library, representing different brick-based structures shapes. These algorithms in our library follow the parametric design logic, which provides essential shape control and design constraints for AR interaction to reduce the impact of excessively active AR interactive inputs. The content of these algorithms will include the declared shape generation logic, interactive parameters, UI input factors, etc. Architects can customise the design algorithm according to their needs in our open platform. Second, scan the QR code in AR and release the onsite base data in the previous step. The virtual onsite hologram will be aligned immediately with the physical environment in the AR for the user to preview. Next, activate the AR immersive design UI, which is designed through the *Fologram* open-source function in AR devices. Users can use hand gestures or screen-based input methods to interact and adjust the parameter sliders on the AR UI in real-time. These design and modification inputs are connected to the design algorithm so that users can preview their design immediately as onsite 3D holograms, which can be previewed and experienced in real-time (Fig. 3). Besides that, our framework supports multi-participant for collaborating design on the same onsite base. Finally, the designed structure is simulated by *PhysX* for its stability. The user can preview the outcomes as holographic animations to find the fragile connection parts and modify them according to the framework feedback loop. After all the simulations and modifications are over, the design structure will be sent for robotic assembly.

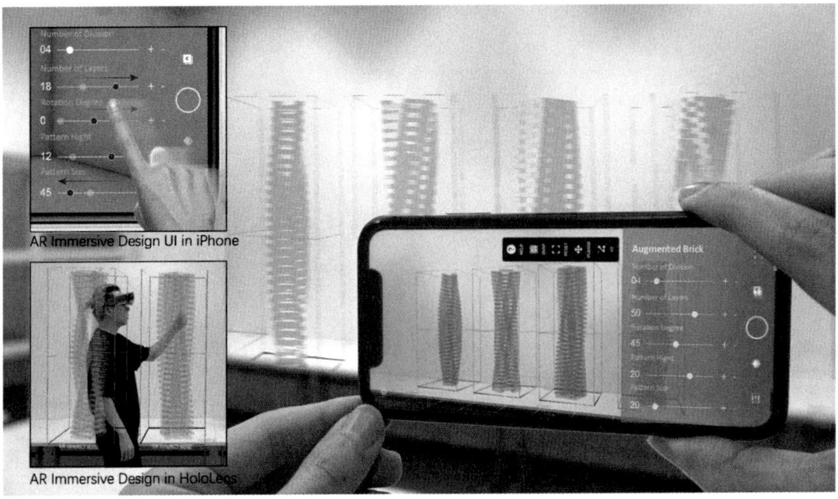

Fig. 3. The designer is using the AR immersive design UI to modify the structure and preview it in real-time onsite with an AR smart device (iPhone 11) and AR headset (HoloLens 1)

Phase 1 provides accessible QR codes, which contain the corresponding 3D onsite environment meshes, as well as the immersive design outcome models and data for users to access and align with the physical robotic operation base in phase 2.

Design Phase Findings

In summary, the AR immersive design process does fulfil our pre-determined assumptions. We successfully designed four brick-based columns in our AR immersive design framework. These four columns are applied to different shape generation algorithms to explore the impact of multiple parameter inputs, such as keyboard input, option input and slider inputs, on different design algorithms through the AR UI, as well as to explore the flexibility and friendliness of the customised design algorithm set up in the immersive design process. As a result, the immersive design phase is suitable for various interactive input modes and supports different customised algorithm settings. This onsite design and preview function break the conventional 2D-based design method, providing designers with a 3D–4D immersive perception in AR for more practical design. However, this process still has some limitations. For example, design algorithms have to be pre-set in the system. Since the current physical masonry structures are not made of interlocking units or are using adhesives, the structures rely on their own weight's structural stability, which significantly limits the diversification and complexity of design algorithms. Moreover, if the design algorithms can be set and realised in real-time by user interaction in AR, it will bring a qualitative leap to the user experience. However, it depends on software and equipment development capabilities. Finally, the natural onsite environment may not be as simple as the lab-based environment. Our system will cause tolerances in facing the complex onsite environment and unstable lighting. Therefore, extra sensors will be introduced into our system to improve the accuracy of dealing with complex environmental interference onsite.

3.2 Phase 2: The AR-Assisted Assembly

The AR-assisted assembly process is the second phase of the *Augment Brick* verification experiment, which contains the physical assembly segmentation and AR-assisted robotic operation. The idea of proposing an AR-assisted assembly method is to provide an easy, safe, and low-threshold method for architects to control industrial robots in the construction process by themselves without any computer science knowledge or coding skills. This unique assumption will reduce the design-build tolerances due to the architects' absence from operating and supervising the high-tech complex digital fabrication process.

Having completed the AR immersive design phase, the users need to upload their design output to our system with the help of the *Robots* plugin for robotic assembly. First, according to the operation radius of the robotic arm in our lab and the size of the structure, we set up an assembly segmentation range box (500 * 500 * 600 mm), which can be changed according to different brands of robots in different assembly situations. The designed structure will be divided into several parts according to this range box for the robotic operation because some structures will exceed the working radius of the robotic arm.

For the AR-assisted assembly process, our system will complete the design structure assembly of each part from the bottom up. First, the user needs to scan the QR code, which was generated from the phase 1, in AR devices to locate the virtual holographic world, including the virtual robotic arm, environment meshes, range box, and the part of the pre-designed structure that needs to be assembled, to the physical robotic operation site. Second, the structure will be divided into foam brick elements as targets in the robotic workflow. The user needs to manually point out the pre-designed structure hologram as the target, either by using hand gestures in headsets or by pointing at the screen through a smart device in AR. According to the user's interactive selection, the robotic operation trajectory will be shown as holographic lines immediately from the foam brick pick location to the target location. Then, the user can preview the robotic pick and place operation animation as holograms upon the entire construction set. We provide an AR-assisted robotic operation UI, in which the user can interact and adjust the robotic setting parameters, such as gripper open or closed commands, operation mode, operation speed, etc., during the holographic simulation process. After the simulation provides the expected results, the user can operate the robot by pressing the upload button through the AR UI for the automated robotic assembly process (Fig. 4). Moreover, the user could manually select each layer or even each brick as the target in AR UI, only when the special assembly sequence is required. The pick and place simulation and operation will be repeated on each brick or layer till the end. Finally, after the separate part constructions are complete, the user will manually assemble these parts in sequence according to the AR instruction (Fig. 5).

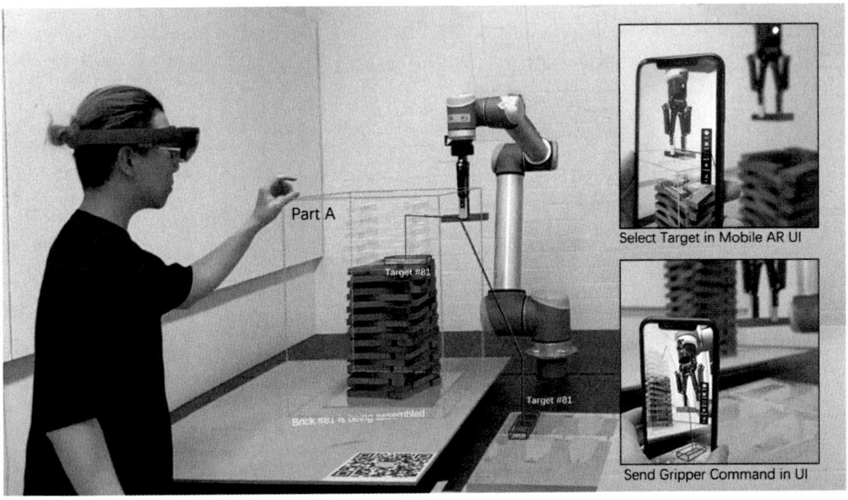

Fig. 4. The designer uses the AR-assisted robotic operation method to select the target hologram and send gripper commands (open or closed) through the AR environment and achieve the robotic pick and place operation to assemble the foam brick structure step by step

Fabrication Phase Findings

In summary, the AR-assisted assembly process indeed achieved a more accessible and intuitive robotic assembly operation for users based on our pre-determined assumptions. We finished assembling four brick-based columns efficiently and precisely. Even unskilled architectural students can easily manipulate this process. All the commands and processes have been pre-developed in our system, which means that the users do not need to be trained in how to use *Grasshopper* plugins or computer science language to control an industrial robot. They only need to manipulate the AR UI to preview the virtual simulation and realise the physical robotic operation, which is safer and more manageable for architects and designers to learn and use. In addition, this AR-assisted method can manually choose the robotic assembly order, pause and repeat at any time, which is more flexible than the traditional robotic operation method, and avoid the unstable connection due to the lengthy code generated by traditional methods. However, this process still has some limitations. We have currently only used a robotic gripper. Other robotic end-effectors such as hot wire cutting tools, and 3D printing tools, could be used in the future, allowing a much wider face of applications. Additionally, the current robotic pick

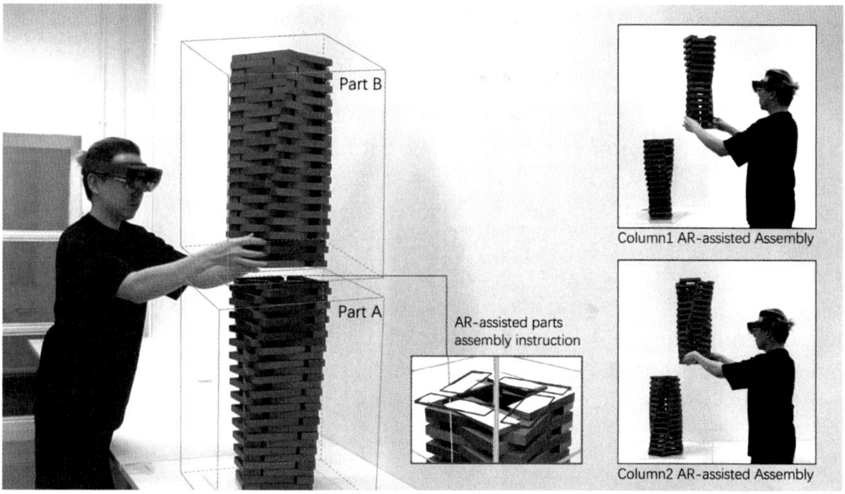

Fig. 5. The designer manually assembles these two parts in sequence according to the AR instruction onsite in the AR-assisted system. The user needs to align the bottom part of Part B with the red holographic instruction guideline to complete the assembly of the whole column

and place targets are based on the corresponding control points related to the AR design model. There are still tolerance issues if one is relying solely on gestures and interactions to command the robots via AR. More sensors will be applied to our AR-assisted system in order to improve the recognition ability and the physical and virtual alignment capabilities of the target location in the AR environment. Furthermore, these four masonry column outcomes are held together only by their own weight. Although they all passed the stability simulation before being built, the structures remain significantly unstable, especially with increasing height or environmental disturbances. Finally, the UI works well, but after completing several parts of the physical structure assembly, the shadow of the holograms and the physical bricks overlap, making it difficult for users to select visually. The visualisation of our UI should be further improved, for example, only the selected hologram target is displayed, and the rest are displayed or hidden in a wireframe, which is convenient for users' manipulation.

4 Conclusion and Discussion

The *Augmented Brick* research developed and verified an immersive design to fabrication framework, which operates successfully for the design and assembly of masonry structures with AR and robotic technologies. Our framework optimises the traditional architectural design to the fabrication process by providing users with the possibility of immersive spatial experience and design modification through AR immersive design methods and empowering them to control industrial robotic arms to achieve complex parametric shape construction through AR-assisted assembly methods (Fig. 6).

However, there are limitations and space for further improvement. Tolerance issues between the physical-virtual alignment and robotic grasping position are one of the most significant obstacles. The tolerance curing during the design process can be ignored because the slight hologram offset does not affect the immersive design, modification and preview. However, tolerances occurring during the assembly process need to be pre-calculated and incorporated into the assembly process as they affect the accuracy of the physical objects. We found out that feeding manually, brick after brick, to the robotic gripper reduces a certain amount of tolerances but also reduces the flexibility of the robotic automation. In further research, extra sensors, such as *Xbox Kinect* or *Azure Kinect,* are needed for teaching the robot to recognise and grab the bricks precisely to solve the feeding issue. Also, these sensors can help to scan the onsite design environment and the design base precisely, and to improve the recognition ability and the physical-virtual alignment capabilities of the target location in AR.

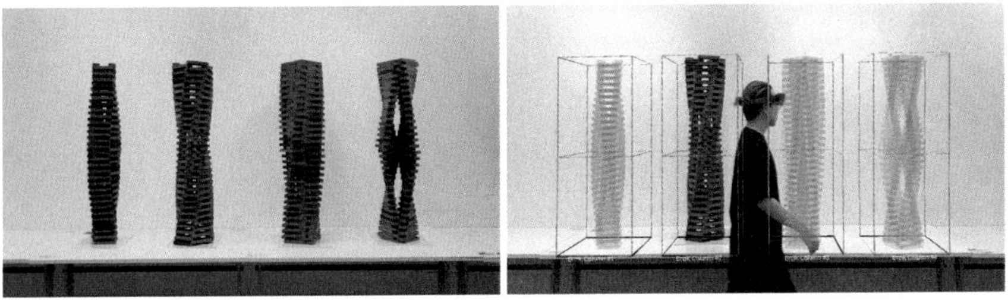

Fig. 6. Through our AR-assisted system, the entire process from design to assembly of four brick-based columns has been realised as preliminary physical tests

Additionally, the performance of our PhysX proved to be successful as it predicted the collapse of one of the masonry walls we tried to fabricate as shown in (Fig. 7). One can see that the collapsed structure is almost identical to the simulation model. Further research, could also investigate the development of interlocking brick joints as well as the use of adhesives, such as mortar and glue, to enhance the stability of structures to exploit the limitations. With the help of interlock joints or brick adhesive, the structure will no longer be constrained by gravity. More complex immersive design algorithms and more flexible interactive inputs will stimulate the creativity of architects. Moreover, the experiment needs to be repeated with real bricks in the future, as they may have different physical behaviours. We aim to optimise our ‘design to fabrication’ framework; thus, it can be applied to the design and construction of real architectural components.

Fig. 7. The simulation of the wall design (left) and the physical performance during the robotic assembly process (right). The structure does not contain any interlock joints or mortar between each brick to keep the structure stable.

Finally, we are also aiming to repeat the experiment by mounting the robotic arm on the MiA mobile robotic platform, which would liberate the fabrication process from spatial limitations appearing in the lab environment. The final goal is to achieve the onsite AR immersive design to fabrication framework in architectural scale applications. The Augmented Brick framework will bridge the gap between architectural design and high-tech construction techniques and place parametric design and high-tech manufacturing back into the hands of architects with the help of AR.

References

1. Azuma RT (1997) A survey of augmented reality. Presence 6:355–385
2. Azuma RT (2016) The most important challenge facing augmented reality. Teleoper Virt Environ 25(3):234–238. https://doi.org/10.1162/PRES_a_00264
3. Barczik G (2018) From body movement to sculpture to space, employing immersive technologies to design with the whole body. In: eCAADe 2018, vol 2, pp 781–788
4. Choo SY, Heo KS, Seo JH, Kang MS (2009) Augmented reality-effective assistance for interior design, focus on tangible AR study. In: eCAADe 2009, pp 649–656
5. Chu CH, Liao CJ, Lin SC (2020) Comparing augmented reality-assisted assembly functions, a case study on Dongong structure. Appl Sci 10:3383
6. Devadass P, Heimig T, Stumm S, Kerber E, Cokcan SB (2019) Robotic constraints informed design process. ACADIA 2019:130–139
7. Fazel A, Izadi A (2018) An interactive augmented reality tool for constructing free-form modular surface. Autom Constr 85:135–145
8. Mitterberger D, Dorfler K, Sandy T, Salveridou F, Hutter M, Gramazio F, Kohler M (2020) Augmented Bricklaying, human-machine interaction for in situ assembly of complex brickwork using object-aware augmented reality. Constr Robot 4:151–161
9. Nguyen DD, Haeusler MH (2014) Exploring immersive digital environments, developing alternative design tools for urban interaction designers. CAADRIA 2014:87–96
10. Schmidt B, Borrison R, Cohen A, Dix A, Gartler M, Hollender M, Klopper B, Maczey S, Siddharthan S (2018) Industrial virtual assistants, challenges and opportunities. In: ACM international joint conference, pp 794–801
11. Song Y (2021) Koeck R and Luo S (2021) Review and analysis of augmented reality (AR) literature for digital fabrication in architecture. Autom Constr 128:103762

Exploration and Design of the Contemporary Bracket Set Through Topology Optimization

Chengbi Duan[1], Suyi Shen[1], Dingwen Bao[2](✉), and Xin Yan[3]

[1] China University of Mining and Technology, Jiangsu 221116, China
[2] RMIT University, Melbourne 3000, Australia
nic.bao@rmit.edu.au
[3] Tsinghua University, Beijing 10084, China

Abstract. Dou Gong, pronounced in Chinese, and known as Bracket Set, is a vital support component in the ancient wooden tectonic systems. It is located between the column and the beam and connects the eave and pillar, making the heavy roof extend out of the eaves longer. The development of the bracket set is entirely a microcosm of the development of ancient Chinese architecture; the aesthetic structure and oriental artistic temperament behind the bracket make it gradually become the cultural and spiritual symbol of traditional Chinese architecture. In the contemporary era, inheriting and developing the bracket set has become an essential issue. This paper introduces the topological optimization method bi-directional evolutionary structural optimization (BESO) for form-finding. Through analyzing the development trend of bracket set and mechanical structure, the authors integrate 2D and 3D optimization methods and apply the hybrid methods to form-finding. This research aims to design a new bracket set corresponding to "structural performance-based aesthetics." The workflow proposed in this paper is valuable for architrave and other traditional building components.

Keywords: Bi-directional Evolutionary Structural Optimization (BESO) · Bracket set · Structural form-finding · Heritage building · Architectural component

1 Introduction

Chinese ancient architecture has a long history and a self-contained system based on a wooden framework. The development of the timber frame has also become the central vein of China's architectural development. The bracket set plays an essential role in the Chinese wood tectonic system. It symbolizes the feudal hierarchy, and an indispensable cultural character in traditional Chinese architecture. In an era where digital design is booming, how can the conventional bracket set be combined with advanced structural form-finding methods to create a new derivative of the bracket set that incorporates ancient and modern elements? The inheritance and innovation of the bracket set have become a significant concern.

P. F. Yuan et al. (eds.), *Hybrid Intelligence*, Computational Design and Robotic Fabrication,
https://doi.org/10.1007/978-981-19-8637-6_34

2 Translations and Derivation Practices of the Bracket Set

From a contemporary standpoint, a few architects have attempted to translate the traditional bracket set, employing the new material and techniques to integrate it into modern architectural systems. These can be broadly divided into three categories according to the structural role, form, and cultural symbolism.

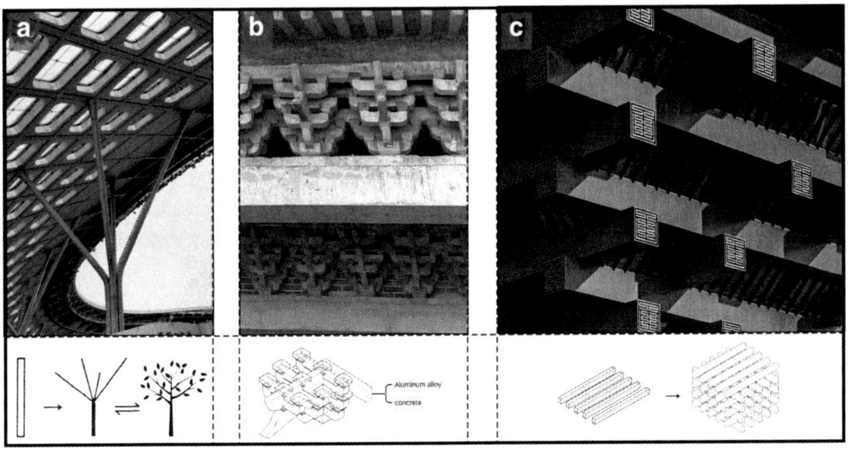

Fig. 1. **a** Tree-like column; **b** Archaize bracket set; **c** Bracket set on 2010 China Pavilion

2.1 The Dendritic Tree-Like Column with Structural Bionics Design

The application of dendritic columns in the spatial structure solves the problem that a single vertical column with an equal cross-section is challenging to set off the far-reaching eave, in accord with the function of the tiered projections of the arch. The dendritic column is upright and looks like a tree trunk. It has various branches at the top of the column, geometrically dispersed to form multiple points of support and radiating away from the centre of the column (Fig. 1a). Although the dendritic column satisfies the integration of "force" and "form", achieved straightforward transferred force and material efficiency, it differs from the bracket set in appearance.

2.2 Archaize Bracket Set Using Contemporary Tectonic Methods

Archaize bracket set is constructed from high-strength materials such as cement and aluminium alloy, imitating or copying the bracket set from various historical periods (Fig. 1b). The same are the beams, columns, arches and pillars in archaize buildings. Still, they are not optimized for the mechanical properties of the individual materials. As a result, the overall stiffness and the structural frequency of archaize buildings are excessive, which will lead to enormous material waste and the loss of characteristics of the traditional bracket set as simple, efficient, and exquisite.

2.3 Abstract Bracket Set Cultural Symbols

This type of building has a superficial structure that exhibits layer-by-layer stacking, achieving a cultural homage to the bracket set (Fig. 1c). In the case of the China Pavilion at the 2010 World Expo, the traditional bracket set is simplified through dislocation, orthogonality, and classification in three-dimensional construction. While the mechanical characteristics of force transfer from one layer to another are retained. However, this type of abstract translation detracts from the essential structural role of the arch and the corresponding scale of its components. It serves only as an abstract derivative and a metaphor for the bracket set culture.

3 Introduction and Analysis of Bracket Set

3.1 The Evolution of the Bracket Set

The evolution of the bracket set is a nearly comprehensive presentation of the development of architectural skills and the aesthetic trends of building components in each period of Chinese history. The bracket set is not only a single structural component, but also has multiple-layered significance in cultural inheritance, decoration and beautification, class identity, measurement, etc.

The development of the bracket set can be briefly divided into five stages. Firstly, from Western Zhou Dynasty to the Han Dynasty, the image of the bracket set was depicted in various historical relics (Fig. 2a). After the Han Dynasty, it appeared between the columns. Secondly, in the period of the Wei, Jin Dynasty, and Southern to Northern Dynasties, the form of bracket set gradually became standardized, and the classic forms of inverted-V brackets appeared (Fig. 2b). Thirdly, from the Sui and Tang Dynasties to the Five Dynasties, the bracket set was vigorously gorgeous and magnificent, characterized by the huge body (Fig. 2c), and played an indispensable role in the structure. Fourthly, during Song and Yuan Dynasties, the number of brackets set increased, and the volume decreased. A certain decorative effect was introduced. The size of components started to be unified, and the modular system appeared (Fig. 2d). The last was the Ming and Qing Dynasties. By the Qing Dynasty, the bracket set adopted the modular of the mortise of cap block, which was used to standardize the dimensions of each component of it (Fig. 2e). Decorative and colouredcolored paintings appeared on the surface of the bracket set, which further strengthened the decoration and weakened the structure roles.

3.2 The Structural Function and Mechanical Prototype of the Bracket Set

The bracket set, placed between beams and columns, plays the key role of a connecting link. And the load of the roof and the upper structure is transmitted to the spreading layer of brackets set, and then to the column and foundation. As a structural "transit hub", the arch body is paved up layer by layer, and the truss and purlin on the outermost layer are propped further for a longer distance, which supports the far-reaching eaves of the structure and makes it more imposing and magnificent.

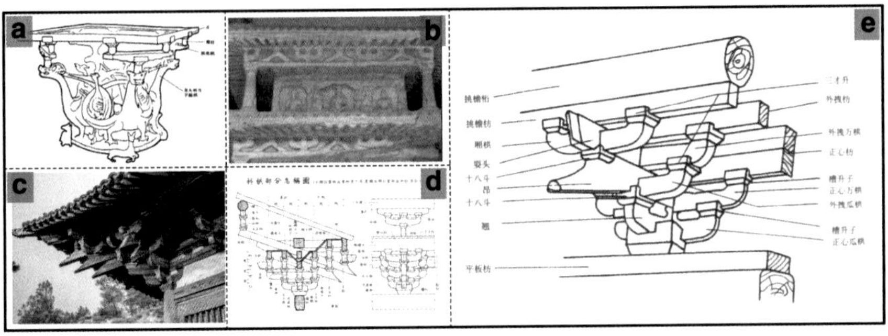

Fig. 2. The evolution of the bracket sets

The mechanical prototype of the bracket set is essentially a short cantilever beam, whose main structural function is to support the load of eaves. For the same amount of force on a cantilever beam, the larger the deformation of the cantilever end, the smaller the deformation of the support end. Therefore, it is necessary to increase the beam height from outside to inside to ensure the overall strength and stability. So, the bracket set adopts double eaves to hold the arch, and through the mortise-tenon joint, several beams become a whole, which increases the bending and shearing resistance of the bracket set. As supporting the eaves, the beam is too long, so the distance between the end of the bearing beam and the bearing support on the roof is too big, it causes the force arm becomes larger, and the bearing capacity of the beam becomes weaker, which is easier to break. Therefore, from the end to the starting point of the overhanging part, the bracket set adopts the way of layer-by-layer superposition and layered force transmission to ensure the overall stability.

The key mechanical components of bracket sets can be dissembled into buckets (Dou), arches (Gong) and levers (Ang). Specifically, the Dou is a square wooden block used to support the Gong and the Ang. The Gong is a bow-like part; the flower arm is vertical to the building facade, responsible for overhanging, while the horizontal arm is parallel to the building facade, which plays a balancing role. The Ang, an oblique cantilever beam structure and drooping frame, acts as a lever. The components of the bracket set are interlaced and stacked, full of rhythm changes.

4 Topology Optimization of the Bracket Set

In this experiment, the bi-directional evolutionary structural optimization (BESO) method has been applied to the design. The scale and position of the traditional bracket set are inherited. Also, the structural prototype of layer-by-layer force transmission is reserved. Through the digital form-finding and finite element analysis model, a new bracket set is designed on the premise of material efficiency and ensuring definite force transmission routes and a reasonable structure (Fig. 3).

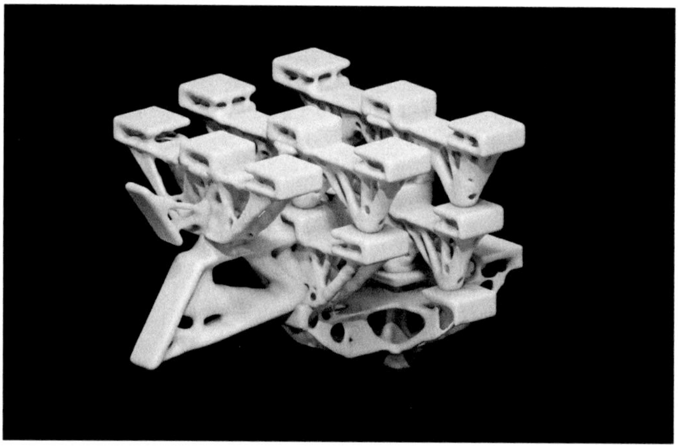

Fig. 3. Topological optimized contemporary bracket set

4.1 Design Methodology

The authors used the Ameba plug-in to carry out topology optimization and BESO algorithm based on the Rhino and Grasshopper platforms. In the application of topology optimization, the rational distribution of bracket set materials is realized by deciding the removal, reservation or supplement of materials through finite element analysis (Fig. 4).

The design transforms the bracket set in the Ming and Qing Dynasties. In that period, the Ang (lever) is the fake lever, which doesn't have the structural character of lower Ang in the Song Dynasty. The force transmission of this bracket set is neater and better aligned with the property of layer-by-layer force transmission from top to bottom. In topology optimization, it is also the most effective structural morphology to transmit the upper load with specific boundary support, material consumption, and material type. To keep to the layer-by-layer force bearing and transmission mode of each bracket set part, this study adopts the "zoning optimization" as the core logic of bracket set optimization. Specifically, the bracket set is divided into Dou, Dong, Ang, Qiao and other parts, and then they are respectively optimized by BESO /Ameba; later all optimized parts are reconstructed and assembled.

4.2 Topology Optimization Form-Finding of the 2D Bracket Set

The 2D topology optimization and planar force analysis are conducive to the subsequent 3D structure form-finding of the bracket set. In addition, the 2D optimized results can be compared with the 3D optimized ones to decide the rationality of the results and provide a reference for adding supplements, resetting boundary conditions, and optimizing parameters. In the zoning optimization of the 2D bracket set, the primary parts, including cap block, small block, oval arch, regular arch and flower arch, are extracted. The axial bucket oval arch is selected as representative of component composition simulation analysis.

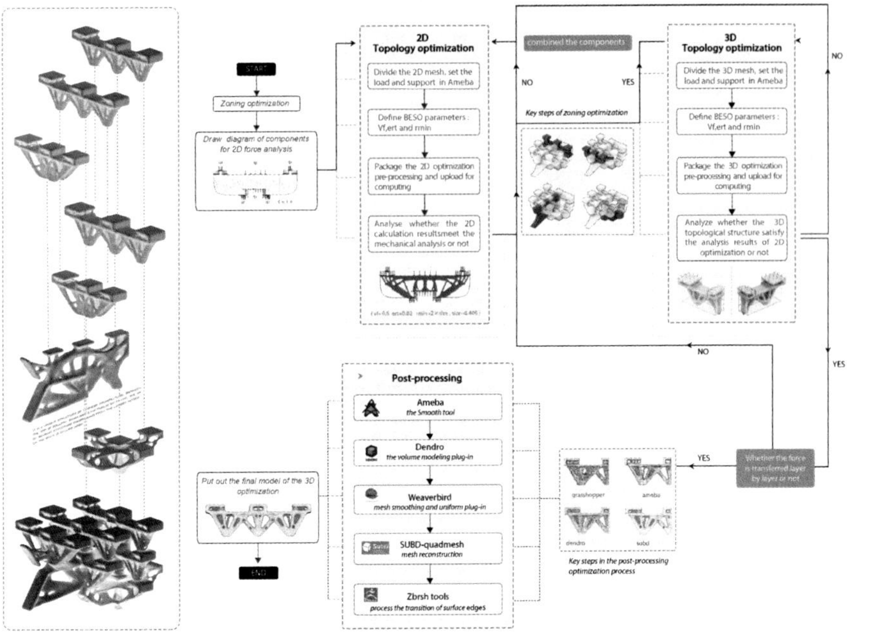

Fig. 4. Workflow of the bracket set optimization & exploded diagram of the optimized bracket set

First, the planar angle of each component is selected, and the ratio scale of each element is determined according to the Dou-Kou module system in the Qing Dynasty. Taking "Dou-Kou" as the modulus unit, an accurate 2D model of each component is constructed. Then according to the descriptions of force bearing of each part of the structure and the connection mode between Dou and Gong in Ying Zao Fa Shi and Engineering Practice Rules and Examples from Qing Dynasty, the force diagram of components is drawn for 2D force analysis. Next, the load and support are set as boundary conditions after generating the mesh element in Rhino/Ameba, and parameters for preprocessing are set to run the BESO algorithm for seeking optimal 2D topology structural forms (Fig. 5).

The 2D topology optimization results and force analysis obviously show that the entasis part's material deletions outnumber that of other parts. The corner in the traditional bracket set is a non-force bearing part or a part with less bearing load. The ancients artificially omitted the corner material based on their construction experience, thereby developing the practice of entasis. This is an experimental result of artificially saving materials and efficiently expressing structural force transmission. In the 3D topology optimization, the bracket set prototype is simplified into a more fundamental geometrical morphology without entasis, aiming to observe whether the original model without entasis will produce similar morphology such as entasis and chamfer after topology optimization. Finally, the critical sections are determined; in the subsequent zoning optimization process of 3D components, some parts are further resolved as a whole.

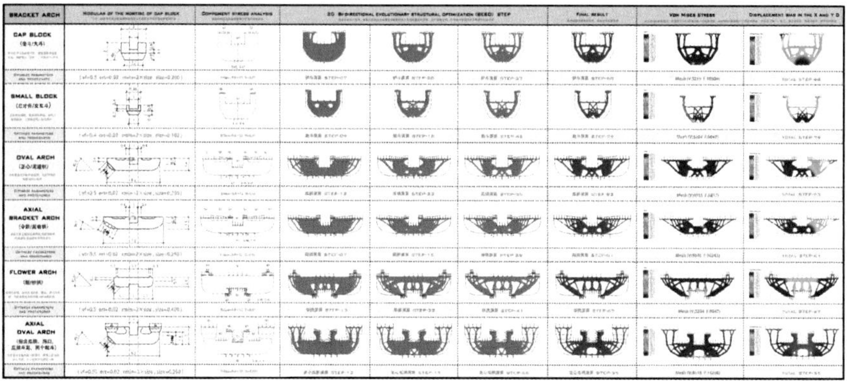

Fig. 5. 2D topological optimised bracket set components

4.3 Topology Optimization Form-Finding of the 3D Bracket Set

After previous 2D component decomposition optimization and considering the different shapes and low strength of the bracket wall after optimization, the small blocks were combined with the arch body in the 3D, prototype and a typical combination of overlapping arched and bucket-shaped blocks (one bracket set with three small blocks) were formed (Fig. 6). During the 2D optimization, the dangerous critical cross-section. The rabbets of the cap block, the first layer oval arch and the first layer flower arch complement each other with overlapped dangerous sections with hidden safety hazards. Therefore, we combined the three components with the central part as the integrated component to further supplement the strength of the critical section.

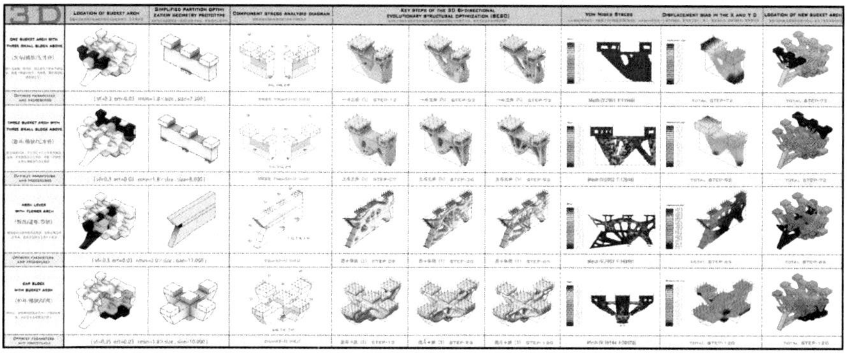

Fig. 6. 3D topological optimized bracket set components

During the optimization, the geometric prototype of the bracket set was simplified, and the entasis chamfer was cancelled to form a simplified, integrated cube. The purpose is to determine whether material deletion and chamfer would reappear at the original entasis position in the optimized structure. At last, four combinations were formed: one bucket arch with three small blocks, three buckets with three small blocks, the combination of Ang and lower arch, and the combination of cap and block flower arch. Then 3D optimization was performed and comparatively analyzed with the 2D optimization results. Parameters were adjusted to achieve the optimal solution for bracket set design (Fig. 7).

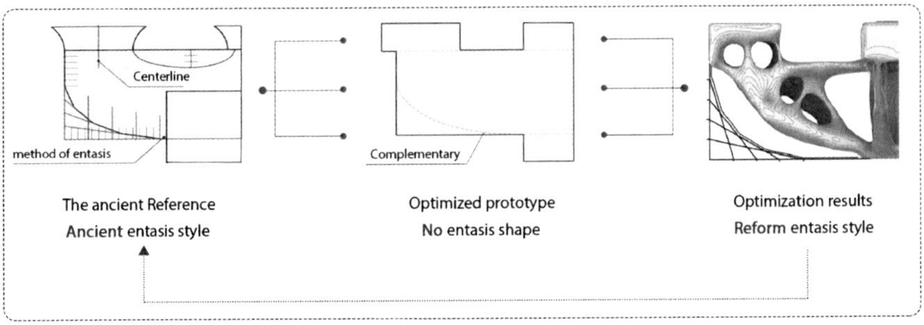

Fig. 7. 3D topological optimized bracket set components

According to the results of the 3D optimization, several bracket set combinations have presented a structural shape of oblique fractal bifurcation from the support to the load end. This mechanical structure is similar to the structural prototype of the tree-like column and the fractal structure of tree branches in nature, proving the high correlation and similarity between bracket sets and tree branches in the mechanical prototype. The upper and lower surfaces have complete materials for the structure of small blocks, while the middle part is a porous structure. Massive materials were cut for the original entasis position or even "entasis style" was formed. The chamfering amplitude of these designs was more obvious than in traditional bucket arches (Fig. 8). The topology optimization results also reflected the wisdom of ancient craftsmen and the excellent craftsmanship of the bucket arch, which is absolutely an architectural gem that integrated form and force (Fig. 9).

4.4 A Case Study and Topological Optimization Experiment of the Hall of Prayer for Good Harvests

Architrave is also an important part of the Chinese timber frame. Usually, an architrave is installed on the top of a column to link the bracket sets between columns with the load-bearing horizontal member. In some architectures, large and small architraves were stacked and juxtaposed with the middle connected by a clamp pad. These architraves were simplified as a complete geometric graph and were used as the optimized prototype together with the eave column. During the architrave topology optimization of the Hall of Prayer for Good Harvests, the combined member of a ring of architrave and eave column can be disassembled into six same units according to the symmetry. Then the 1/12 minimum optimization unit can be obtained based on the mirror-symmetric properties of the units (Fig. 10).

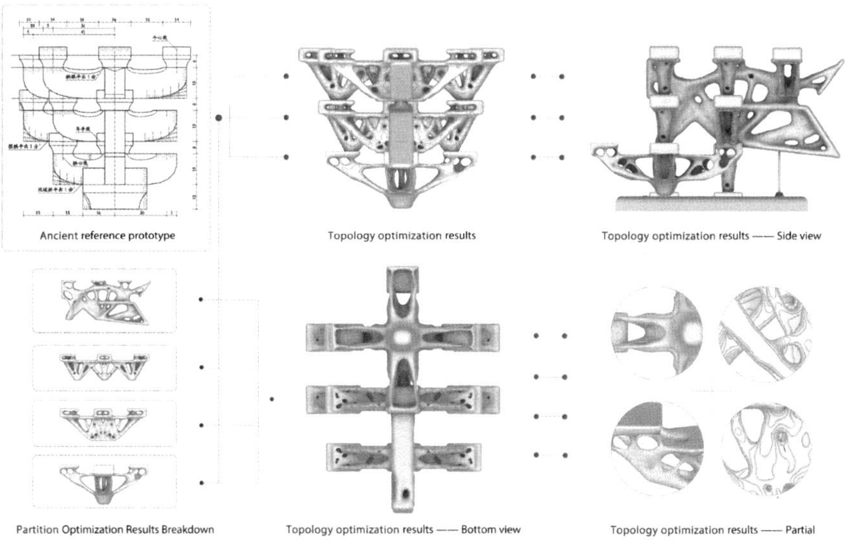

Fig. 8. Topology optimization results of the entasis part of the bracket set

Fig. 9. 3D topological optimization

For the load arrangement, the downward uniform load can be set on the upper critical plane connecting the architrave and bracket set, and a side thrust can be imposed on the sparrow brace, which ensures that all the units are closely connected in the partition optimization. The base was installed at the position of the symmetrically optimized vertical cross-section and column bottom. The overall effect after optimization was shown in Fig. 11.

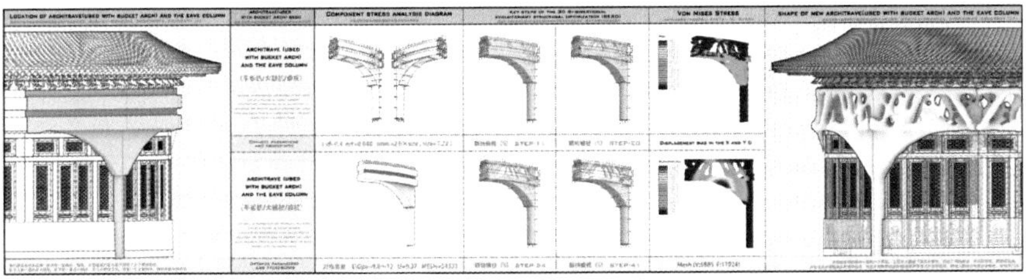

Fig. 10. 3D topological optimised architrave bracket set components

Fig. 11. 3D topological optimized architrave bracket set components

After the optimization of the architrave, we combined the results with the previous optimization results of the bucket arch. We replaced the corresponding original structure of the Hall of Prayer for Good Harvests while retaining the traditional paintings, doors, windows, tiles and other decorative components. The overlap of new and old elements formed a stark contrast in appearance (Fig. 12) while maintaining the same mechanical essence. This kind of hidden link ensures a more rigorous connection between culture and structure on the basis of the unity of aesthetic form and force in contemporary structure.

5 Conclusion and Future Work

Bi-directional Evolutionary Structural Optimization (BESO) was performed on the bracket set in this study. 2D and 3D comparison and partitioned optimization were applied to the design. The new bracket set was designed based on high and reasonable structural performance, material efficiency, traditional tectonic rule, with the features of aesthetic form and force-form united.

Fig. 12. Topological optimised Hall of Prayer for Good Harvests

The attempt of the new bracket set design can trigger the thinking about the new value of bracket set as a symbol of architectural culture in contemporary times. Meanwhile, we also applied the complete workflow to the architrave-eave column combination. The results also turned out to be enlightening. Through the attempts, we aim to encourage more scholars to explore the possibilities of inheritance and derivation of traditional architectural components from the structural perspectives.

References

1. Liang SC (2003) Diagram for Gong Cheng Zuo Fa Ze Li. II. Science Press, Beijing
2. Li J (1983) Ying Zao Fa Shi. China Architecture & Building Press
3. Wei GA (2007) Mechanical behavior and ANSYS analysis of Dougong in Chinese ancient timber building. Xi'an University of Architecture and Technology
4. Lv X (2010) Mechanical behavior of bracket set in Chinese ancient timber buildings. Beijing Jiaotong University
5. Lin KQ (2020) Research on structure morphology collaborative design for digital architecture. South China University of Technology. https://doi.org/10.27151/d.cnki.ghnlu
6. Xie YM, Zuo ZH, Lv JC (2014) Architectural design through bi-directional evolutionary structural optimization. Time + Architecture (5):20–25. https://doi.org/10.13717/j.cnki.ta
7. Cao P, Wang QH (2010) Ilustration for the construction history of building group of the hall of prayer for a good harvest of the temple of heaven in Beijing. New Archit (2):116–121
8. Wang PY (2011) Research on the Qi Niandian of temple of heaven of reinforced concrete structures. North China University of Technology
9. Yu MH, Oda Y, Fang DP, Zhao JH, Zhang DL, Zhu RX,Che AL (2006) Advances in structural mechanics of Chinese ancient buildings. Adv Mech (1):43–64
10. Zhao B, Chen TY (2016) Application of topology optimization to architectural design. Archit Cult (11):104–105
11. Zhou ZY (2021) Research on topology optimization design of leisure chair structure form. Central South University of Forestry & Technology. https://doi.org/10.27662/d.cnki.gznlc

12. Qian WL, Huang Y, Li H (2020) A review of topology optimization methods applied in building intelligent design. Intell Build Smart City (7):6–10. https://doi.org/10.13655/j.cnki.ibci
13. Yuan F, Chai H, Xie YM (2017) Special issue towards an integration of architecture and structure performance design. Archit J (11):1–8
14. Xu X, Liu JH (2018) Inheritance and innovation of traditional ancient architectural culture—a modern interpretation of the bucket arch. Nei Jiang Ke Ji 39(11):112–113

Bio-digital Sand Logics: Dune Sand Material and Computational Design

Marcus Farr(✉)

AUS / Tongji, Shanghai, China
marcusfarr@mac.com

Abstract. This paper discusses the creation of a new sand-based material, performative testing, and the computational logic involved in the design of a prototypical architectural system. Dune sand is known to be an unstable material compared to river or marine sand and as a result it is not normally used for construction. Because of this, desert regions have grown a reliance upon imported materials creating massive sustainability issues due to large scale global shipping, importation and resource extraction. This research indicates there is a viable opportunity to leverage dune sand as an ongoing line of inquiry for material science and design in local desert regions. It establishes that there is very little architectural research being done in this particular area. The methodology begins with experiments in bio-material using dune sand as a compound, and then establishes a construction system based upon a manifold of experiments. Along with material investigations, the process uses a Scientific Testing Method (STM) and Hypothesis in Action (HIA) as part of the testing methodology.

Keywords: Technology · Desert environments · Bio-materials · Bio-digital design

1 Introduction

The research project asks how a regionally appropriate architectural system might integrate with computational process to allow for the use of a new material agenda using dune sand from local deserts. Sand is a global necessity. The UN 2060 projections indicate that sand is the most widely consumed construction resource [5], and past UN studies in both 2011 and 2017 have accurately predicted the future need of sand and small aggregate to advance beyond other materials and will remain massively mined and utilized well into the future creating environmental problems due to mining, habitat loss and global importation. To facilitate this line of inquiry, the research followed a hypothesis using sodium and sodium mixtures to create bio-synthetic material from local dune sand. Cities in the Middle East and North Africa (MENA) import massive quantities of river sand from Southeast Asia and Australia to fuel the need for concrete and masonry products [13]. This thesis tested sodium and sand together as a superheated mixture and concluded that in certain situations it has characteristics that rivals concrete or traditional masonry.

P. F. Yuan et al. (eds.), *Hybrid Intelligence*, Computational Design and Robotic Fabrication,
https://doi.org/10.1007/978-981-19-8637-6_35

The research points to the fact that local sand material can be used to create typological prototypes such as *mashrabiya* and *jalis* or other facade units for spaces requiring breezes, which can create better situations for humans in harsh desert environments. With this in mind, an algorithmic design workflow was generated using computation to articulate specific architectural outcomes that have regional relevance. Units that emulate breeze blocks are employed to create a variety of performance-based compositions that can be highly programmed to accommodate site specific conditions. Using Grasshopper Ladybug we can determine additional performance-based logic for the compositions on a given site based upon sun and wind data.

The project responds to pressing global issues as identified by the UN Sustainable Development Goal SDG11, "Sustainable Cities and Communities" [16] by helping to design resilient, and sustainable cities and human settlements, and designing material that aids in the reduction of environmental impacts of cities and city building by fostering resource efficiency in sustainable and resilient building processes. It also responds to UN Sustainable Development SDG13, "Climate Action" by reducing greenhouse gases due to transportation and importation on a global scale. Because dune sand is known to be an unstable material, desert regions have grown a reliance upon imported materials. This work indicates there is a viable opportunity to leverage dune sand as an ongoing line of material inquiry, establishing that there is very little being done in architectural research that responds to the populations living in harsh desert environments partnered with bio-sand material and computational design. The methodology begins with experiments in bio-material with sand as a compound, and then, through empirical testing, establishes an ongoing construction sequence selected from a manifold of recreations based on successful experiments. This process uses the Scientific Testing Method and Hypothesis in Action, allowing the results to inform "design by research", followed by application.

Computational parameters are used to determine a series of performative results for regionally appropriate construction based upon the characteristics of the material strength and dimension. The material demonstrates it can be used in a variety of capable configurations that make unit based, porous constructions such as those used for thick stereotomic wall types, and ventilated *mashrabiya* and *jali* walls. A jail, meaning "net", is the term for a perforated stone or latticed screen, usually with an ornamental articulation based upon geometry or natural patterns. This form of architectural decoration is common in Indo-Islamic architecture [14]. These are computationally designed to be culturally appropriate to this region. Grasshopper is used to arrange the units into compositions that are also environmentally responsive and controllable. The computational patterns can be opened or closed in varying degrees to adjust for orientation and site conditions. The work illuminates possible solutions for the regional problem of building in the Sahara and Arabian desert utilizing a surplus of dune sand to construct locally appropriate wall types that build upon vernacular traditions but offer a more performative result. Compared with other methods such as concrete or masonry, which use high degrees of imported material, this solution can be more sustainable. To meet the UN sustainability goals, we must research and discover more applications for available resources in extreme environments. This paper documents a method, a material, and an outcome with a variety of potential possibilities (Fig. 1).

Fig. 1. Nachna Parvati, Temple Jali, India, Gupta Period, and Jali at Bibi Ka Maqbara, Aurangabad, India.

2 Location, Properties and Need

To place the project in a regional environment, the Arabian desert was used as a context. The vernacular traditions in this region are limited by their material resources. Mud, arish, coral stone and rock were used traditionally when available, especially along the coasts [8]. Fabric was used by nomadic tribes who circulated through the deserts seeking favorable weather patterns as it was lightweight and offered architectural flexibility and could be transported on camel back [9]. The question arises as to what local materials are widely available now, and what can be leveraged for research as a further line of inquiry relative to architecture. These regions have a massive surplus of sand, and yet very little of it is utilized for construction. Modern cities such as Dubai, Sharjah, Abu Dhabi, and Jeddah are literally surrounded by sand yet it is mostly unused. Instead, other types of sand are imported from other countries for construction because river sand is deemed to be of higher quality for architecture and manufacturing. In these regions, imported sand is brought in from Southeast Asia and Australia [3] by boat to manufacture concrete, glass, mortar and block which increases the carbon footprint and overall sustainability factor for every architectural project in the region. So, what is the role of the desert in modern construction? What opportunities arise from this situation?

Sand is a resource heavily in demand. It is the most widely used resource in the world besides air and water [2]. It is in most everything we use in modern society, cell phones, glass, concrete, roads, computers. Yet the largest concentration of sand is not used. The importance of understanding how desert sand can be used is a critical environmental concern especially for the regions living closest to it (Fig. 2).

“Natural sands are eroded or weathered particles of rocks. Sand is made by simply grinding up rocks into increasingly smaller pieces. Sand can also be made out of living creatures, from shells and other organisms of the living world, and many beaches are composed of pulverized animal shells. Sand grains can originate from catastrophic geologic phenomena, as when molten lava erupts from volcanoes and shatters in the air, scattering particles across the oceans to land as tiny grains” [7]. Sand is different depending on its origin and the potential uses are different as well compared to origin. Desert sand has a different set of physical characteristics compared to river sand. River sand is sharp and angular as the grains were conditioned by water rather than air, it

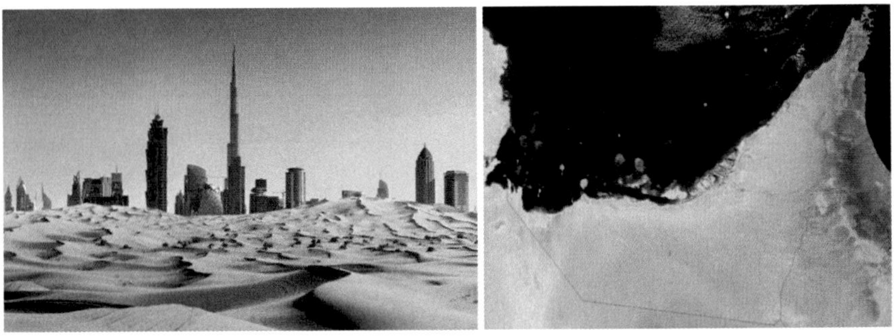

Fig. 2. Dubai with UAE desert in foreground, and satellite image of Rub' al Khali desert (the Empty Quarter), Getty Images.

contains higher concentrations of quartz, and as a result it compacts successfully as a mixture for concrete. Desert dune sand is round and does not self-organize in the same way river sand does. Because it was conditioned by air, the grains are round in shape, so it organizes like a bag of marbles. The grains have a smooth surface finish and the particle size of desert sand is very fine, it is slightly alkaline in nature, and it is very dense, similar to dirt, all of which make it less useful for modern construction. Sand is a self-organizing material, as are all aggregates, and adheres to a consistent behavior when poured. As the architect Frei Otto noted with extensive studies of spoil piles and sand, these materials have "a funnel that is formed within the granule mass with a natural angle of repose" [11]. Measuring a material angle of repose produces slightly different results [10], however in sand it is normally a 34° angle, which limits it [7], and asks for additional binders to be involved in this material research.

Sand is a global necessity and UN reports suggest his will continue, for example the UN 2060 projections indicate that sand is most widely consumed construction resource, and past UN studies in both 2011 and 2017 have accurately predicted the future of sand and aggregate to advance beyond other materials [15], and even though research through sustainability is allowing some material mining to slow down, sand and aggregates will remain massively mined and used well into the future. It is estimated to be a widely used material even in 2060 projections which also illuminate the further environmental problems that will follow with mining, habitat loss and global importation.

Because this is a problem that has both regional and global implications, it was deemed appropriate to study as a further line of inquiry. In the UAE, almost all materials are imported or made from imported resources so the importance of employing regional resources and finding new ways of building and designing with local sand as an option for regional materiality can lead to productive solutions for architecture and the UN Sustainable Development Goal for life on land. In order to move forward with this research, the project re-considered what regional architecture is in this location, how computational approaches can intervene with new material options, and how an advent of bio-technology/bio-engineering can alter our current understanding.

3 Methodologies

For this project, the methodology manifests in two parts. The first being a material study for the creation of a bio-sand unit that can be used architecturally. The second being an overlay of computational technology to begin understanding what can be constructed from this bio-material performatively. Experiments using bio-material with sand are very complicated and often unsuccessful due to the nature of the material. Previous precedents have been successful in making structurally stable units using resins and binders that are unsustainable and toxic. This study called for an approach that was not using resins or toxic binders. It began with recreations of experiments using scientific testing method (STM) that implemented a workflow for a hypothesis in action, design by research and application. The project followed this method using desert dune sand sourced from the wild in the desert of the UAE paired with three biological substances that could act as a sustainable binder, Sporosarcina Pasteurii, sodium thiosulphate, and urea.

Sodium thiosulphate was the material that worked most consistently with the project trials and subsequently became the dominant material additive with the sand for the project. This material was used to create a series of additive molding trials based on thick casts and thin casts. Sodium thiosulfate is a colorless crystal of sodium, sulfur, hydrogen, and oxygen. Both the Environmental Protection Agency (EPA) and Federal Food and Drug Administration (FDA) in the United States consider it a safe substance and permit its inclusion in human foods such as table salt [17]. It is often present at therapeutic bath spas or thermal spas in contact with the human body, and other uses include waste water treatment plants where it is used to clean water before releasing it back to a river. In principal it is more un-toxic than table salt. It is on the World Health Organization's List of Essential Medicines as one of the safest and most effective medicines needed in a health system [18] (Fig. 3).

Fig. 3. Sodium Thiosulphate, Dune sand brick "growing", final sodium sand brick.

This additive is effective because it changes states with temperature variation allowing the sand and salt to bind in a superheated mixture and upon cooling this process forms a composite unit. The strength varies based upon the amount of sodium and sand added together. It works by creating a monolayer of sodium around the sand, binding it at the same time, and results in the production of multiple layers of hardened sand material solidified by the sodium as it dries into a hardened state.

This material combination ultimately was used to create a series of bio-synthetic units made from sand that simulated standard masonry sizes. Rhino and Grasshopper were then used to demonstrate the material in a series of compositional arrangements essentially by creating a surface in the X–Z directions, adding an ability to move freel and create twisted openings while stacking the units, and orienting the bricks in variou positions so as to become performative in a variety of different situations.

4 Results

During the course of research for this project, four distinct and separate processes wer tested all using varying concentrations of dune sand and sodium as a mixture. The natura self-organizing behaviors of sand were observed and also researched. One can observ the behaviour of sand as a self-organizing material especially with pourings and pilings Although pourings and pilings were studied initially, molds and casting became th primary vehicle for making the architectural units in this study. Different thicknesse were tested with these processes to understand the material behavior in a solid stat ranging from thin (6.35 mm) to thick (76.2 mm).

Out of the tests conducted, Experiment 3 entitled "Solid State Sand as Thick Mate rial" performed with the most consistency and the most strength. This material wa generated by combining one part dune sand to one part Sodium Thiosulphate (sand & sodium equal by percentage), which created a unit of 76 mm in thickness. This wa achieved by heating solid state sodium to a melting point, superheating to reach boil ing temperature, removing from burner, creating a "Solid State" to "Liquid State". Th super-heated bath of liquid state sodium thiosulphate was poured into a sand-filled mol 3″/76 mm deep. Solidification was achieved through stirring the mixture as the sand pa ticles were allowed to bind with sodium and the mixture cooled. The solid-state sodiur thiosulphate is now effectively combined with sand as a hardened material, proving th hypothesis that when melted sodium thiosulphate comes into contact with sand, the form a bond (a biological cementation), creating a sandstone-like biomaterial.

The creation of the bio-material was a significant component to this research projec however the second phase was to test how this material can be used in an architectur scenario. In the Middle East, *mashrabiya* and *jalis* are commonly used architectur features. Their existence dates back hundreds of years and were originally used as wate storage areas in houses so that the ventilated openings could cool the water [1]. Th continued use of the *mashrabiya* and *jalis* allowed for its evolution from these wate storage areas to devices that protruded or cantilevered over irregular plots in dens urban areas to correct the language of the architecture and increase the sizes of space on the upper floors of stacked housing without changing the size of the ground floor c moving past the plot limit.

They were also used for privacy which is an important concern in Middle Easter architecture. They are found in both historical and contemporary architecture as a com monly used feature for regional architecture. For this reason, it was used as an initi typological reference. In grasshopper, a definition was made that arranges stacks c sandstone units. These stacks were given inputs the same size as the units made fror the final sandstone test mold. With grasshopper, the units were then allowed to rotat from one unit to the next so as to create a ventilated façade. The control of the opening can be calibrated to be fully open or fully closed, with a vast array of potentials betwee (Fig. 4).

Fig. 4. Salt brick, compression test & resultant brick.

The strength of the sand brick was tested for compression using a calibrated Form + Test M1 3000 kN machine. The material proved to be very durable and held up well under continuous compression with 1750 kg/m^3 being used for "Bulk Density", and 1511.6 kN being applied for "Normalized Compression", and 51.4 MPa for "Strength", where one MPa is equal to one million pascals (Pa); a pascal is one newton of force per square meter, a megapascal is one million newtons per square meter. As a rule, the higher the MPa of a material, the stronger the material will be, and the less likely it is to fail. For example a 32 MPa (at 4,600 psi) for concrete is often used in the region [3]. However, it is important to note that "Load" was not directly calculated as it was not an ASTM test.

The use of Grasshopper is critical as it allows for the remapping of geometry from one axis-system to another, and allows the units to be transformed more systematically. With this process, the Grasshopper definition allows for a flexible *mashrabiya* and *jalis* to be modeled along with the synthetic bio-material. This acts as the primary design tool that facilitates a workflow for a bio-synthetic *mashrabiya* and *jalis*. The system can move from a closed wall surface to a system with various openings and closing in a very systematic way. This design workflow can benefit regional architecture and can be custom controlled to allow for different lighting situations, privacy requirements, ventilation needs and can accommodate the breeze patterns for a particular area in a city (Figs. 5 and 6).

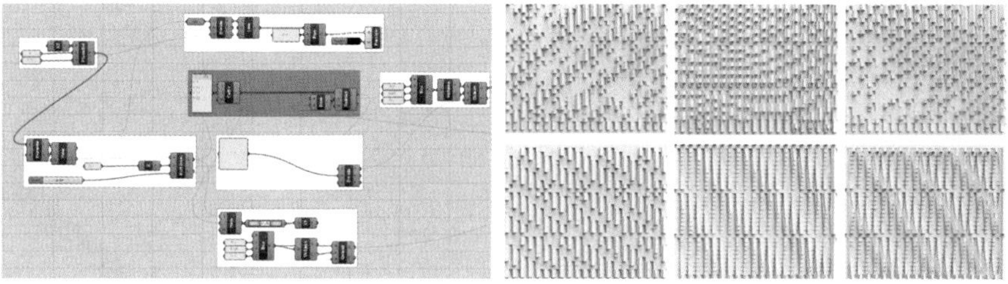

Fig. 5. Grasshopper jail/mashribiya arrangements

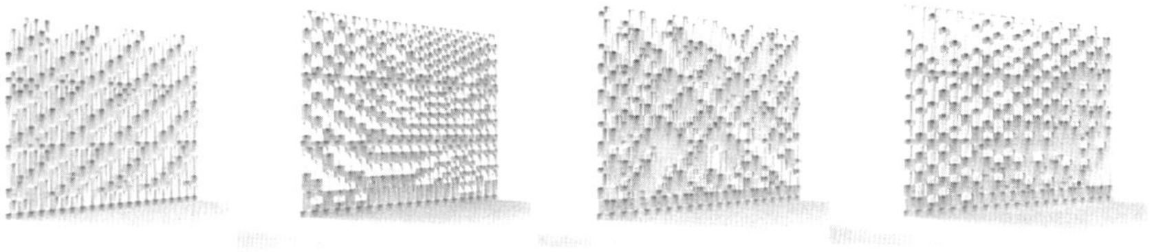

Fig. 6. Jali/mashrabiya compositional arrangements

5 Conclusion

The research project asks how a regionally appropriate architectural system might integrate with a computational process to allow for the use of a new material agenda using dune sand from local deserts. To facilitate this line of inquiry, the research followed a hypothesis using salt and salt mixtures such as Sodium Thiosulphate to create bio-synthetic material from local dune sand. This was important due to the fact that desert dune sand is rarely used in modern construction. Cities in the Middle East and North Africa import massive quantities of river sand from Southeast Asia and Australia to fuel the need for concrete and masonry products. Based upon previous scientific evidence, the project research found that sodium thiosulphate could successfully help to create sand-based material systems. This thesis tested Sodium Thiosulphate and concluded that in certain situations it is a viable material outcome that can have the strength needed for material assemblages. It is relatively safe and easy to mold. Concerns however are the amount of Sodium required to create the material and the durability of the material if exposed to weather and water. The referenced test in this paper (Experiment Three) concludes that with a dimension of 76 mm or more, it is a stable unit capable of being stacked and arranged in compositions similar to masonry construction, lending additional viability due to the skillsets of local craftsmen and their familiarity with brick.

A design workflow incorporating this bio-material can be combined with computation to articulate specific architectural outcomes that have regional relevance. The research points to the fact that local sand material can create typological prototypes such as mashrabiya and jalis or façade units requiring breezes, offering a method for viable human scale architectural prototypes to be further explored and designed. Further research can be done on the material responses to weather and the structure limitations of the material (Figs. 7 and 8).

This is needed to continue conversations regarding how designers can replace current proprietary media such as mass-produced masonry and concrete created from unsustainable methods using imported sand. The project responds to pressing global issues as identified by the following UN Sustainable Development Goals: SDG11, "Sustainable Cities and Communities" by helping to design resilient, and sustainable cities and human settlements, and designing material that aids in the reduction of environmental impacts of cities and city building by fostering resource efficiency in sustainable and resilient building processes; SDG12 by fostering responsible consumption and production in order to ensure sustainable consumption and production patterns, and to ensure good use of resources; SDG13, "Climate Action" by reducing greenhouse gases due to transportation and importation on a global scale [2].

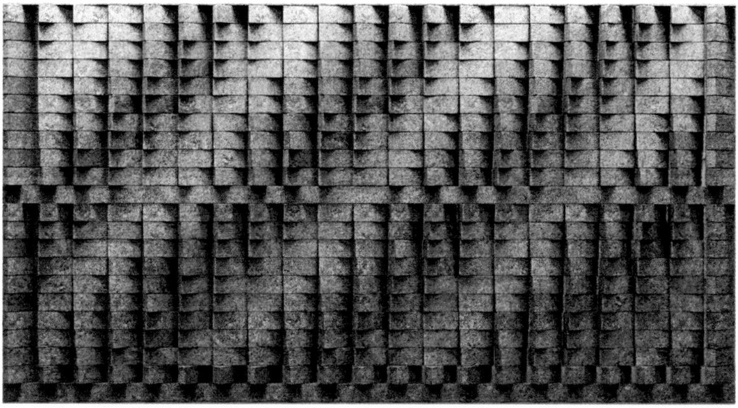

Fig. 7. The result of the grasshopper defined bio-synthetic masonry material composition

Fig. 8. Bio-masonry mashrabiy/jali prototype

References

1. Ashour, A. (2018). Islamic architectural heritage: Mashrabiya/jalis, islamic heritage, architecture and art II. WIT Press
2. Beiser V (2016) BBC, BBC future. https://www.bbc.com/future/article/20191108-why-the-world-is-running-out-of-sand#:~:text=Sand%2C%20however%2C%20is%20the%20most,be%20found%20together%20%E2%80%93%20every%20year. Accessed 18 Mar 2022
3. Burke G (2016) North stradbroke Island sand mining to end early. http://www.abc.net.au/news/2016-05-26/north-stradbroke-island-sand-mining-to-end-2019-early/7446860. Accessed 1 Mar 2022
4. Fagan J (2020, November 1) UN United Nations Environment Programme. UNEP Commons. https://www.unep.org/. Accessed 12 Dec 2021
5. Fritts R (2019) Science direct, UN Report. https://www.science.org/content/article/world-needs-get-serious-about-managing-sand-says-un-report. Accessed 19 Mar 2022
6. Giancoli D (2000) Physics for scientists & engineers, 3rd edn. Prentice Hall
7. Glover TJ (1995) Pocket reference (for mechanics). Sequoia Publishing
8. Hattstein M, Delius P (2000) Islam: art and architecture. H.F.Ullmann Publishing

9. Ibid
10. Nichols EL, Franklin WS (1898) The elements of physics, vol 1. Macmillion
11. Otto F, Rasch B (1992) Finding form. Deutscher Werkbund.
12. Rael R, San Fratello V (2018) Printing architecture. Princeton Architectural Press
13. Rayasam R (2016) BBC, UN Report. https://www.bbc.com/worklife/article/20160502-even-desert-city-dubai-imports-its-sand-this-is-why. Accessed 18 Mar 22
14. Thapar B (2004) Introduction to Indian architecture. Tuttle Publishing, Singapore
15. Torres A, Simoni M (2021) Sustainability of the global sand system in the anthropocene, one earth, vol 4, no 5. ScienceDirect
16. United Nations (2020, October 12) UN sustainable development goals. https://sdgs.un.org/goals. Accessed 14 Jan 2021
17. WHO (2019, October 12). World Health Organization. Model list of essential medicines. https://apps.who.int/iris/bitstream/handle/10665/325771. Accessed 10 Dec 2021
18. ibid

Hexagonal Responsive Facade Prototype in Responding Sunlight

Tria Amalia Ningsih(✉), Abraham Chintianto, Cahyo Pratomo, Muhammad Haikal Milleza, Muhammad Arif Rahman, and Intan Chairunnisa

Faculty of Engineering, Department of Architecture, Universitas Indonesia, Depok City, Indonesia
tria.amalianingsih@gmail.com, intanisac@ui.ac.id

Abstract. This paper discusses an architectural responsive façade system using hexagonal parametric forms and kinetic mechanism which responds sunlight. Its purposes are to buffer excessive sun exposure that goes through interior space and maximize the covering area with an incremental rotational joint system. The study aims to explore responsive façade system as second skin for architectural building, focusing on design, mechanism, and fabrication processes. The prototype consists of three parts: the hexagonal modules where the membranes and its frames are compacted; a series of levers to synchronize the movement of opening the membranes from each module; and a structural framework to hold each module as united kinetic façade system. As a preliminary prototype, the system can potentially be applied to several types of existing buildings and easily installed in various sizes and configuration. This kinetic mechanism can decrease sun radiation up to 50% than unprotected window façade.

Keywords: Responsive façade · Parametric modelling · Tessellation · Kinetics mechanism

1 Introduction

Responsive architecture underpins the physical objective of buildings. It reconfigures the needs of variable mobility, location, or geometry [1]. Responsive facade refers to a façade—main or additional ones—reconfigured in response to certain variables based on the building's context. The purpose of using this responsive facade system is to answer specific geographic or climate issues, one of which is climates with a prominent level of solar exposure.

Currently, many buildings implement a responsive facade. In Dubai, the *Al-Bahar* Tower is knowingly famous for its responsive facade that takes *Mashrabiya* cultural elements [2]. This facade membrane can open and close automatically when the sun passes through one side of the building. A facade system like this has also been implemented at the *Syddansk Universitet* in Denmark with a triangular shape made of perforated metal that regulates the amount of light entering [3]. Simple buildings, such as houses, also use

P. F. Yuan et al. (eds.), *Hybrid Intelligence*, Computational Design and Robotic Fabrication,
https://doi.org/10.1007/978-981-19-8637-6_36

a responsive facade. An example is the tropical cave house in Vietnam, whose facade can rotate simultaneously with a manual system [4].

Most responsive facades are designed in a modular form arranged in a grid (vertical or horizontal multiplied). This modular is due to facilitate the development of the module and its operation when applied to the building [5]. In the development of the module, the exploration process was greatly assisted by Parametric Design and Digital Fabrication. The parametric approach makes the design exploration process easier by providing specific indicators such as shape, dimension, and thickness, then modifying these parameters according to context needs [6]. This parameter can create a module for a responsive facade based on the given performance, thus creating a design that responds to need and adapt other contexts only by adjusting to specific parameters. Using the parametric design also describes the opening and closing process on the responsive facade, which becomes more manageable and well described. Modelling that uses precise measurement gives an idea of how the mechanism will work in the real world. However, minor adjustments are applied at the realization stage to respond to material characteristics that are not described in the parametric design.

To realize the design into a physical form, Digital Fabrication provides various conveniences and adds value in the development of a responsive facade. One of the advantages is to ease of production in a short or predictable time [7]—cutting various patterns with laser cutting machine or making objects with 3d printing machine. At the same time, precise product results can also be generated through data input, adapted to digital model data [8]. In addition, digital fabrication also provides options with various several types of materials by adjusting the fabrication machine used.

It is established that a responsive facade can create dynamic protection in the building against the surrounding environment. Then, in the design development process, several architects have applied a parametric approach to get data from the context such as sun exposure, wind, other values that affect design results. However, few have explored the relationship between design and prototyping processes in developing responsive facades. Furthermore, further processes are needed, namely re-checking the results of the design and prototyping, whether they are following the needs of the building.

Therefore, the study aims to explore the hexagonal responsive façade system that responds to sunlight and works as a second skin for architectural buildings; it focuses on design, mechanism, and fabrication processes. Few objectives of this study: (1) we identify the excessive sun exposure that goes through interior space, then analyze the requirement for covering the window area. (2) We investigate responsive façade systems in design, kinetic mechanism, and its construction details using the digital modelling and parametric approach. (3) After finalizing each component's design of the responsive façade system, we carry out the prototyping process using digital fabrication techniques to test the kinetic mechanism in modular scale. (4) based on digital modelling and fabrication processes, we evaluate the limitations and potential of the hexagonal responsive façade for the sun shading.

2 Literature Studies

2.1 Responsive Façade in Tropical Architecture

A responsive façade is a skin of a building that actively reacts or change toward specific climate and environmental aspects (light, wind, and temperature) to suit the daily user needs inside a building [9]. Its commonly regulated by mechanical, passive, or electro-mechanical system that consist of sensing (perceive environmental data), control (translate), actuating (movement), and structural components. Mechanical system generates motions in a simple machine like a kinetic façade and can be operated by hand hence easily adapt to the user's preferences and needs. This system offers a relatively long lifespan but tends to subside more shortly than those made of non-mechanical parts [10].

Warm climates-such as tropical climates-have a primary demand for cooling in a building's thermal comfort, and passive cooling strategies are the most effective way to respond to it. There are 4 types of passive cooling: shading, window-to-wall ratio, glazing type, and ventilation strategy. Shading and ventilation have the highest cooling savings potential in a controlled simulation out of the four. Both shown coherence on average savings on warm-dry and warm-humid climate groups from the review, but climate conditions are still crucial in their effectiveness [11].

2.2 Kinetic Mechanisms Using Aperture Morphology/Analogy

A kinetic mechanism is a synergistic kinetic movement of each part in a system that produces an effect or a response. This understanding become the basis for the development of a kinetic facade system, a façade that–for the example-can switch from an open to a closed condition [12]. Kinetic facades are innately complex systems that consist of corresponding components that work across various material domains [13].

The evolution of kinetic façade comes along with the development of its pattern and movement. Pattern development focuses on three functional aspects: aesthetic element, environmental controller, and alternative producer of renewable energy. Movement development consists of rotation, elastic, retractable, sliding, and self-adjusting [13]. Inspired by a camera's lens movement, Aperture can be defined as the opening in a lens through which light passes to enter the camera [14]. Through its iris blade components, it can regulate the amount of light exposure and alter the condition of the system from one situation (closed) to another (opened).

2.3 Digital Fabrication: From Parametric Modelling to Fabricating Process

Digital fabrication lies under the scope of CAD (Computer-Aided Design) and CAM (Computer-Aided Manufacturing) as it utilizes computer-driven tools to build or cut parts [15]. It can be classified into five techniques namely sectioning, folding, contouring, forming, tessellation [15]. The latter are commonly used for responsive façade since it has characteristic in terms of generating form as a collection of pieces that fill the gaps in a plane or surfaces. Tessellation can be made through parametric approach, process that enables designers to define the relationship between elements or groups and to assign values or expressions to organize and control those definitions [5]. It offers opportunity to explore unexpected solutions and minimalized complicated reworking.

Ladybug tools enable simulate sun exposure and façade's efficiency rate. It is open-source computer software that connects CAD interfaces to a host of validated simulation engines to generate an interactive 2D and 3D graphical diagram [16]. This software acts as a plugin for Grasshopper3D and can simulate both visual and energy consumption of the parametric modelling generated earlier [17]. High validity and integration with modelling software results in its wide uses ranging from architectural research to large-scale projects.

Manufacturing techniques can be classified into formative, additive, and subtractive [18]. While the formative technique shape materials and additive techniques add material, subtractive techniques cut and/or etch flat materials into custom shape patterns using laser cutting machine –thermal-based fabrication process [19]. It works well with various sheet-shaped materials (paper, wood, plastic, acrylic, and metal), able to create high precision model, and has adjustable laser modes allowing various artistic and functional possibilities. This series of benefits eventually attract its utilization in fabricating high-quality architecture models such as those for façade [5].

3 Methodologies

3.1 Parametric Design Approach

The design of a responsive facade is developed using 3D modelling software Rhinoceros and Grasshopper [20] to identify the design requirements and qualifications. We alter Eqs. (5) through parametric design such as grids, geometry, orientation, and dimension in different values as performance considerations. This visualization interrogates the feasibility of a responsive façade and calculates the proportion of design required for responding to sunlight (see Fig. 1).

Fig. 1. (left to right): Al-Bahar Tower, Syddansk Universitet, Tropical Cave House

3.2 Digital Fabrication for Prototyping Process

There are five different modules in the prototyping process; each module is made based on purposes and modifications from previous modules. Thus, this research utilizes digital fabrication to generate the prototype of a responsive façade due to its affordability and efficiency. We explore both subtractive and additive manufacturing techniques to fabricate parts, including hexagonal modules, the mechanism, and the frame. We attempt to use FDM 3D printing [7] to create small parts using PLA filament materials with additive manufacturing. Meanwhile, we exercise subtractive manufacturing with laser cutting machines to cut several materials (paper, cardboard, and plywood 3 and 6 mm) and produce modules and frames (Fig. 2).

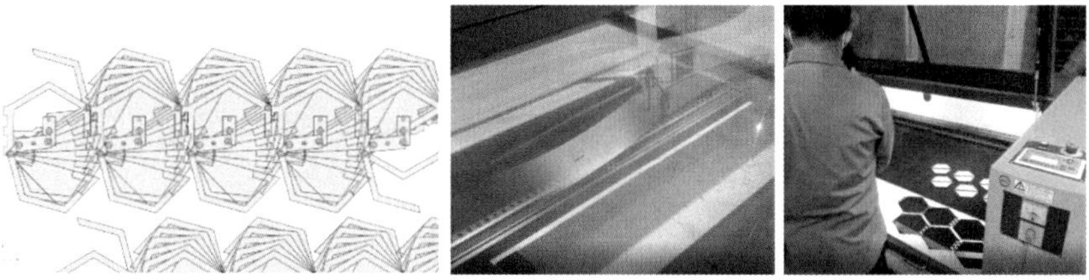

Fig. 2. (Left) Parametric design for generating responsive façade (middle and right) Laser cutting process to create parts of modules

3.3 Sun Radiation Simulation with Ladybug Tools

The purpose of this responsive façade is to tackle sun exposure pass through the openings. We perform the sun radiation simulation to verify the feasibility of responsive façade in responding sunlight. Using Ladybug Tools, an interactive visualization of climate data [16], we evaluate the responsive facade modules that block glass window from sunlight-hours (Fig. 3).

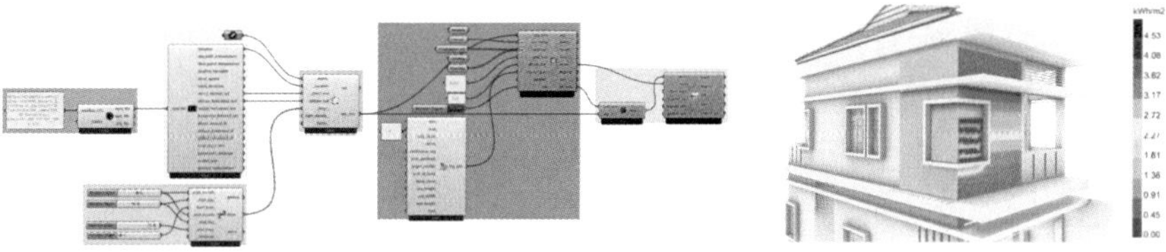

Fig. 3. Sun Radiation using Ladybug Tools

We picked a two-storied conventional house situated in West Java, Indonesia. The house is located at −6.43° north (latitude) 106.78° west (longitude). It sits at the intersection of *Kemang* Street and *Alifah* Village in *Sawangan* District, Depok. The responsive façade is placed outside the building mass facing northeast, where it will cover a single glass window measuring 1000 × 1200 mm. The variable on this simulation is only focused on the solid area (surfaces with solid materials) and void area (the surface with glass) since Ladybug tools will not be able to calculate the material aspect (U value) (Fig. 4).

Fig. 4. (left) Site location ; (right) Single Glass Window as the context for the façade

4 Result

4.1 Parametric Modelling for Creating Responsive Façade

This project is developed upon two primary concepts, a tessellation pattern known for its coverage and adjustability combined with incremental and space-saving ability of an aperture mechanism. Serves as main organizer, tessellation is firstly explored through comparison of various basic geometries. Hexagon shape is then discovered as the suitable option for several reasons including circular shape tendency, no residual space, and the dividable into smaller groups or rows. Moreover, hexagon's 60° sides offer a fresh diagonal arrangement compared to the conventional x and y-axis (Figs. 5 and 6).

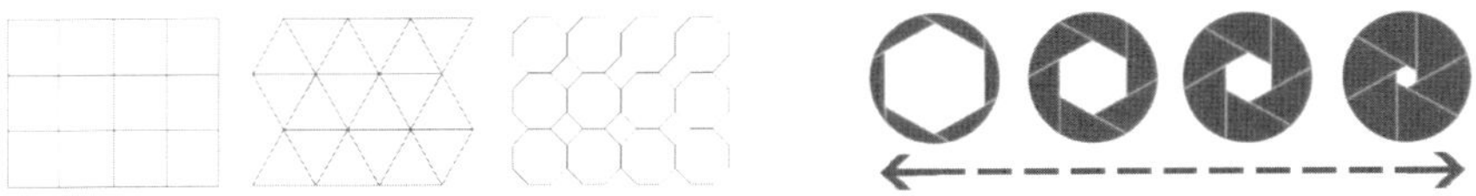

Fig. 5. Example of: (left) Tessellation Pattern; (right) Aperture Movement Principle

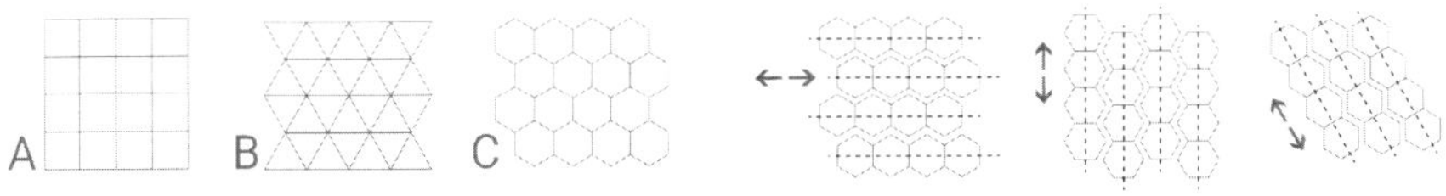

Fig. 6. (left) Tessellation Grid Exploration; (right) Arrangement Possibilities

Preliminary context analysis shows a gradient of sun exposure zones in the shape of three 30 × 100 cm long rectangles. Hexagonal pattern is then grouped into 3 rows of 5 identical modules positioned above each zone that span 30 cm wide to shade each corresponding zone individually. A single hexagonal module consists of a body, pair of arms, pair of forearms, handles, and two sets of membranes. Body serves as a platform to attach other components via a rotatable and stopper joinery. Modules are statically connected to each other and to its frame using finger joinery while a lever will collect each module's handle to arrange simultaneous movements. Lastly, semi-transparent blades are added between forearm and body to become a radiation barrier upon activation and stay transparent when not used. Array polar calculation help set 5 pieces of membrane that rotates at a 9° angle to achieve accumulative opaqueness when fully deployed (Fig. 7).

Fig. 7. Sun exposure zone and hexagonal module dimension and position adjustment

4.2 Digital Fabrication for Responsive Facade Prototype

Preliminary manual modelmaking followed by digital fabrication techniques are used in the creation of the façade that is divided into 5 stages. It starts with the creation of a paper-based prototype-1 aimed to materialize the digital concept and primary mechanisms. Suitable joinery is then explored in the second prototype that have a tiny strand of thread tied at both ends that creates a smooth rotational joint. The cardboard-made model offers proportional geometry but is too-easily bent and has loose joinery (Fig. 8).

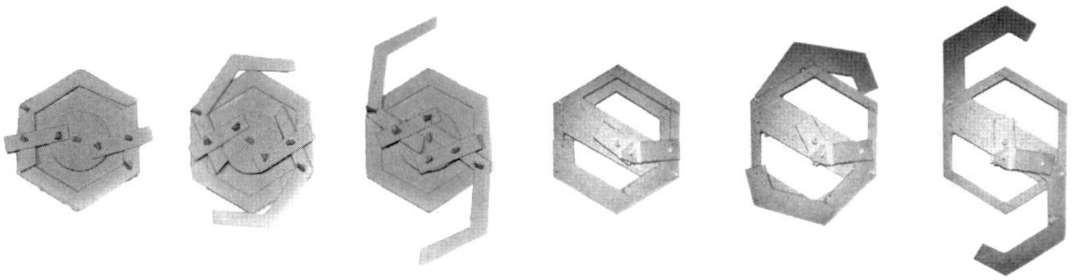

Fig. 8. Manual modelmaking: (left) first prototype to materialize primary component; (right) Second prototype to find suitable joinery with proper geometry and dimension

Digital fabrication implementation becomes the highlight of the 3rd and 4th stage that focuses on rationalizing the details. Plywood-based prototype-3 features a 3D-printed ring plate that creates consistent gap between pivoted components for a sleek and sturdy movements but still lacks tightness and has collision risk between contiguous modules. The 4th prototype sees wooden dowels replaces 3D-printed joineries-for a tighter and durable pivot-and the addition of finger joints to the sides for ease of assembly. It also features a longer lever integrated with five handle points and opposing placement of the two sets of forearms to avoid neighbouring collisions (Fig. 9).

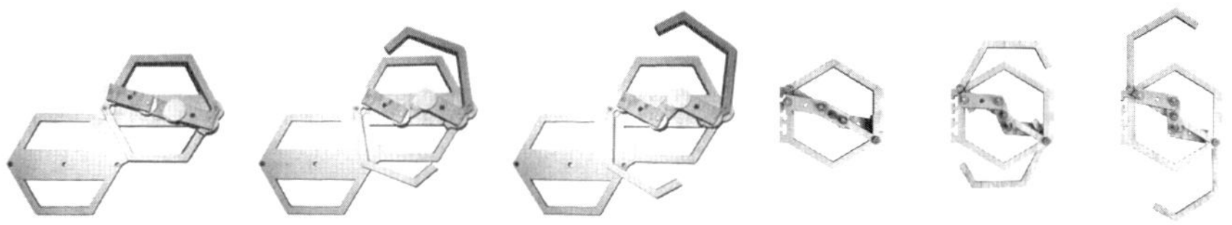

Fig. 9. Digital Fabrication: (left) Third prototype explores digital fabrication in creating parts, (right) Fourth prototype features adjusted joinery and parts reconfiguration

Membrane and frame integration into the modules mark the creation of final prototype. Membranes are fastened to the module via strands of yarn running through holes in its edges with a larger central hole to prevent jam. Finally, Plywood-based frame measuring 1100 × 950 × 70 mm are integrated. It can be disassembled into three main parts that each one can holds a row of modules (Fig. 10 and Table 1).

In this development, sensing and operating aspect of the facade depend on manual user intervention. The mechanism is deployed (closed) and reversed (open) by hand-rotating a handle located in the middle of a module row. Different rows will have their own handle to achieve independent levels of shading control. Rotation of the handle

Table 1. Prototype development

Protoype	Purposes	Result	Feedback
Module 1 (Paper)	Finding a suitable **mechanism**	Can rotate the whole part with a single handle	Needs to find the proper dimension and consider the position of parts
Module 2 (Cardboard & thread)	Find the mechanism that has proper **geometry, dimension, and material**	The whole part can rotate smoothly without colliding between the membrane and wings; the material is sturdier; can use one lever to rotate 2 modules	Needs sturdy joinery between parts (only use thread); the frame is too thick
Module 3 (Plywood 6 mm and PLA plastic)	Find the mechanism by the plywood material and the **accuracy of digital fabrication** in creating parts	The whole part has more durability; it can rotate smoother since there is a gap between parts	There is a limit of rotation (appx. 45 degrees), otherwise the membrane cannot return; joinery is still loose
Module 4 (Plywood 6 mm and wooden dowel)	**Modify module** prototype; configurate all modules with lever; modify joinery with dowel	Joinery with dowel can rotate the module perfectly, can rotate all 6 modules in single leverage	The position, the amount, and dimension of membrane in module
Module 5 (Plywood 6 and 3 mm, Duplex board, and translucent mica paper)	**Finalize modules** with membrane; create frames to simulate the primary facade	The membrane can work perfectly to cover sunlight; the structure can withstand the whole modules	The material of membrane and its joinery can be explored; the lever can be improved with single leverage; the gap between modules can be closer

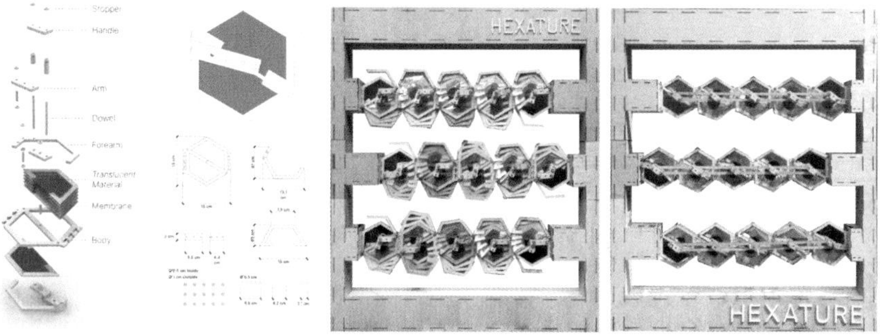

Fig. 10. Fifth Prototype; (top) membrane and frame integration with the module; (bottom) final assembly of the prototype

will trigger the movement of the forearm and membranes through its connections with the arm and forearm. When the forearm rotates, it will pull five layers of membrane with it as they are connected by a thread and share the same pivot point. These sets of reactions will result in the facade being deployed (closed). Levers that connect all centre handles inside a single row will transfer the user's hand rotation to all adjacent modules simultaneously.

Automated sensing and rotating mechanism are one of the possible future developments of this system. A sensor will attach at the exterior side of the window and detect incoming light intensity which will be translated as an input to generate a specific row configuration (opened or closed). This scheme allows the façade to work autonomously based on weather condition and only low degree of user intervention is needed (Fig. 11).

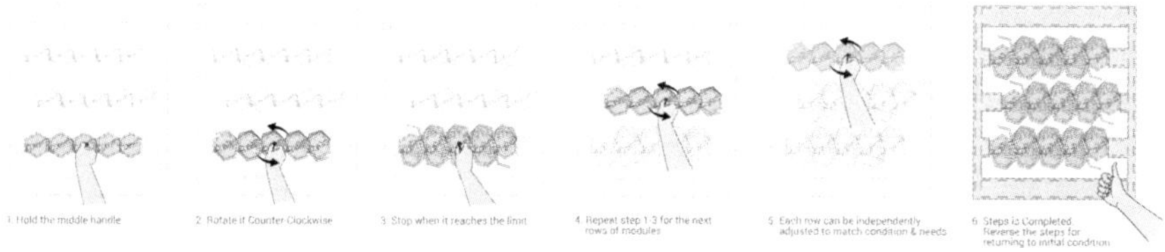

Fig. 11. Opening and closing mechanism of the façade

4.3 Shading and Sun Radiation Simulation with Hexagonal Responsive Façade

The shading performance of the facade is simulated digitally in a 3d model. Based on Fig. 12, we can see that the facade could affect the daylight that enters the interior through the translucent membranes. The intensity of daylight in every degree of openness is diverse and adjustable to suit the needs of the interior. The membrane is the main element of the diverse shading and makes it different than the facade without membranes.

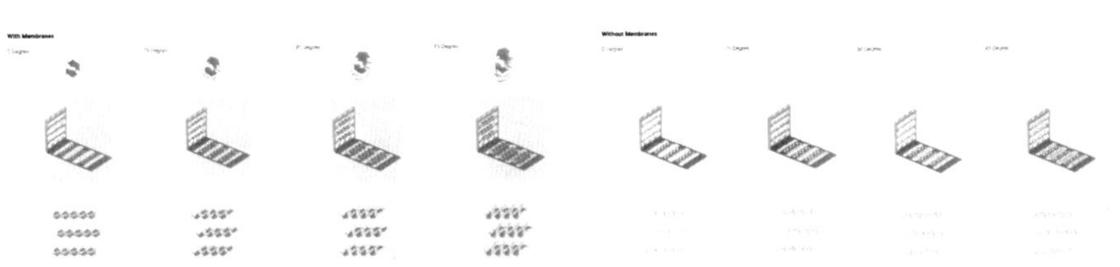

Fig. 12. Shading diagram with and without membranes

The façade's performance is also simulated digitally with the help ladybug software to analyse the ability to filter sunlight by measuring sun radiation parameters in the interior. Tropical climate data of Depok, West Java, as the location is inserted into the software as a contextual situation. Preliminary simulation for full year shows a period between June and July has the highest sun radiation level in the interior, with 2–2, 4 kWh/m^2. This timespan becomes a fixed variable for comparing various module phases in the most extreme radiation condition possible.

The facade has two primary phases, a fully closed module with 0 degrees rotation and a fully opened module with 45° rotation. Upon its installation into the window, a fully closed module (0° rotation) can already reduce radiation level down to 2–1.6 kWh/m^2 and portrays 60% received radiation compared to the condition without the facade. It corresponds to a reduction of yellow colour shades in the sun radiation diagram that indicates a 2–2, 4 kWh/m^2 radiation level.

An incrementally opened module (45° rotation) can limit incoming sun radiation decline to 1, 2–0, 4 kWh/m^2 or 50% of condition without the facade. This figure can be explained visually in the radiation diagram below depicting more distribution of dark blue shades compared to the two previous tests. The dark blue colour represents sun radiation levels of 0–0, 4 kWh/m^2 hence more shades of blue indicate higher façade performance (Fig. 13).

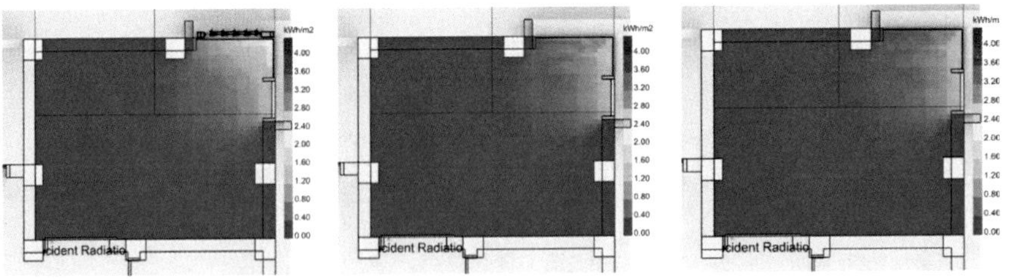

Fig. 13. Sun radiation diagram (left to right): without prototype, with prototype on 0°, and with prototype on 45°

5 Discussion

5.1 Problems that Occurred in the Production Stage of the Hexagonal Module

In the process of making the final prototype, there were several obstacles that hinder the production of the hexagonal module. The first problem is the difficulty of measuring the length of the rope that connects one membrane to another. The reason is the knots that hold the ropes together on the membrane are conventionally made by hand. This can be solved by measuring and cutting the required number of strings before attaching them to the membrane assembly.

Then, the joint becomes stiff after gluing the stopper and dowel which results lack of smooth rotation. The adhesive liquid blocked the stopper and the dowel hit the surface. This issue is resolved by sticking the stopper first into the join. After that, then the series of joints are installed on the body and forearm without applying any adhesive substance.

Lastly, the constraint that seriously hinders the production process of the hexagonal module is the friction between one component and another. an example of this problem occurs on the surface of the body with the forearm which makes the process of opening and closing the hexagonal module hampered. The solution to this problem lies in providing the distance between one component layer and another using the thickness of the joint to avoid friction between the forearm surface and the body and so on.

5.2 The Potential of Hexagonal Responsive Facade for Sun Shading

Environmental simulation in the previous chapter shows a 40–50% efficiency rate that is met under several situation to be considered as follow. Firstly, ladybug tools have a limitation in calculating the effects of semitransparent materials to the shading quality, leaving a slight possible shift in the overall efficiency figure. Secondly, weather data used in this simulation comes from an authorized yet open-source database (ladybug.tools/epwmap) that have a limited available weather station point, with the closest point to the site being 30 km away causing further figure shift possibility.

Regarding its application toward the multitude of architecture, hexagonal module has the potential to be used not only in residential buildings but also in the openings of a high-rise building. Constructed upon a modular system, each module can be adjusted to match existing opening properties through the parameters of dimension, number of modules, and control row or group division. Exposure simulation run through the various sizes of modules and windows shows a relatively consistent shading quality with the optimum shading performance comes from 15 to 120 cm module (Table 2).

6 Conclusion

This study develops the hexagonal responsive façade system that responds to sunlight and works as a second skin for architectural buildings. The remaining stages include design, mechanism and fabrication processes. As a tropical country, a house in Indonesia can receive sun radiation around 2–2, 4 kWh/ m2 during the day. This excessive could generate enormous amounts of energy to set up thermal comfort in interior space. While

Table 2. The potential of hexagonal responsive façade

Exisitng Window 60 cm × 45 cm	Apply at 30 cm × 30 cm Façade	Apply at 90 cm × 90 cm Façade	Apply at 120 cm × 120 cm Façade
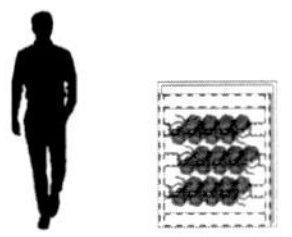	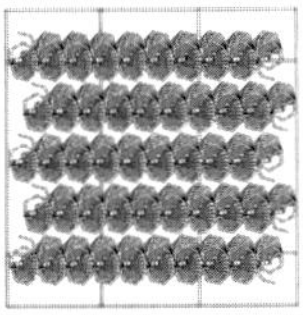	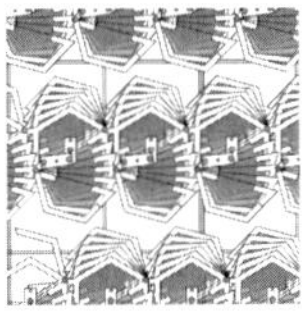	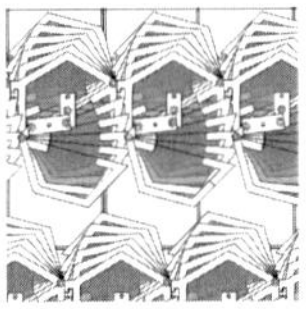
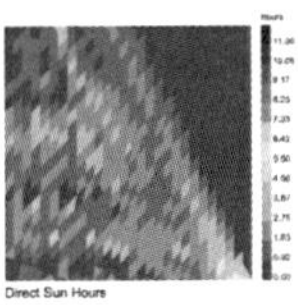	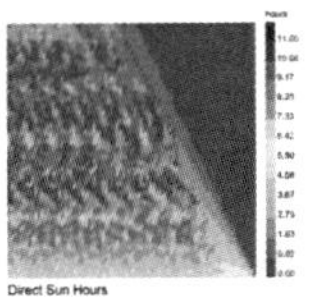	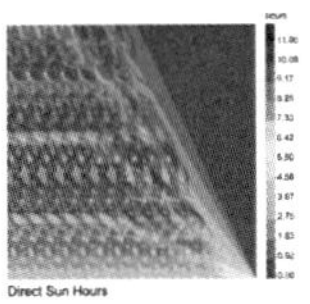	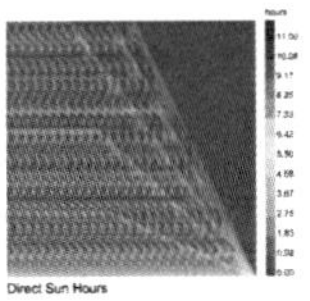

the glass window's purpose is for the openings and accessibility of the atmosphere to come through interior space, it is necessary to add more layers on the top of the window to block unnecessary light and heat that interferes with human comfort. Thus, a secondary façade requires incremental coverage and adjustability aspects to respond to the changing sun exposure throughout the day.

Hexagonal responsive façade has capabilities in reducing sun exposure with mechanical technology–operates by hand–which is suitable for a tropical house due to its efficiency and affordability. The tessellation configuration of the modules solves the covering areas where a dimension of 150 × 150 mm modules (in fully open condition) can cover 150–300 mm width of the sun radiation zones. The module consists of 4 distinct parts, and each part has a specific purpose: body, arm, forearm, and membrane. The levers support these modules to generate the responsive façade movement and the framework to hold modules and attach them into the primary façade. The final prototype is fabricated using laser cutting machines and plywood materials. However, it is possible to substitute the materials with the more durable ones with tropical weather (heavy rains and extreme hot season). For 1100 mm height, 950 mm width, and 70 mm thickness, the responsive façade can cover 1000 mm height and 1200 mm width windows.

A fully close hexagonal responsive façade system with 0 degrees rotation can reduce sun radiation level down to 2–1.6 kWh/m^2 or 40% radiation reduction. Ones the façade is open to 45°, sun radiation decreases to 1, 2–0, 4 kWh/m^2 or up to 50% radiation reduction; some areas have possibilities to completely unattached to the suns –the areas that further away from the window. In conclusion, this responsive façade could be beneficial in decreasing excessive sun radiation coming to interior space and can be implemented in several typology of buildings in tropical countries. The evaluation remains in digital simulations, in future research, it is recommended to test the prototype in practical conditions using sun radiation sensor to see the effectiveness of the responsive façade.

References

1. Pesenti M, Masera G, Fiorito F, Sauchelli M (2015) Kinetic solar skin: a responsive folding technique. Energy Procedia 70:661–672
2. Cilento K (2012) Al Bahar towers responsive facade/Aedas. Archdaily. Archdaily.com
3. Brake AG (2015) Henning Larsen's university building has a facade that moves in response to changing heat and light. Dezeen. dezeen.com
4. Archdaily (2020) Tropical cave house/H&P architects. Archdaily
5. Dunn N (2012) Digital fabrication in architecture. Laurence King Publishing
6. Hernandez CRB (2006) Thinking parametric design: introducing parametric Gaudi. Des Stud 27(3):309–324
7. Weiner H (2019) Additive versus subtractive manufacturing–simply explained. https://all3dp.com/2/additive-vs-subtractive-manufacturing-simply-explained/
8. Maing M, Vargas R (2013) Digital fabrication processes of mass customized building facades in global practice.
9. Romano R, Aelenei L, Aelenei D, Mazzucchelli ES (2018) What is an adaptive façade? Analysis of recent terms and definitions from an international perspective. J Facade Design Eng 6(3):65–76
10. Heidari Matin N, Eydgahi A (2019) Technologies used in responsive facade systems: a comparative study. Intell Build Int 1–20
11. Prieto A, Knaack U, Auer T, Klein T (2018) Passive cooling & climate responsive façade design: exploring the limits of passive cooling strategies to improve the performance of commercial buildings in warm climates. Energy Build 175:30–47
12. Fox MA, Yeh BP, (eds) (2000) Intelligent kinetic systems in architecture. In: Managing interactions in smart environments. London: Springer, London
13. Sharaidin K, Burry J, Salim F (2012) Integration of digital simulation tools with parametric designs to evaluate kinetic façades for daylight performance
14. Mansurov N (2020) Understanding aperture in photography. https://photographylife.com/what-is-aperture-in-photography
15. Iwamoto L (2013) Digital fabrications: architectural and material techniques. Princeton Architectural Press
16. Roudsari MS, Pak M, Smith A, (eds) (2013) Ladybug: a parametric environmental plugin for grasshopper to help designers create an environmentally-conscious design. In: Proceedings of the 13th international IBPSA conference held in Lyon, France
17. Valitabar M, Moghimi M, Mahdavinejad M, Pilechiha P (2018) Design optimum responsive façade based on visual comfort and energy performance
18. Redwood B, Schöffer F, Garret B (2017) The 3D printing handbook: technologies, design and applications: 3D Hubs
19. Obudho B (2019) What is a laser cutter?–simply explained. https://all3dp.com/2/what-is-a-laser-cutter-simply-explained/
20. Yoon J (2019) SMP prototype design and fabrication for thermo-responsive façade elements. J Facade Design Eng 7(1):41–62

Large-Scale 3D Printing Using Recycled PET. The Case of Upcycle Lab @ DB Schenker Singapore

Associate Professor Felix Raspall[1(✉)] and Associate Professor Carlos Bañón[2(✉)]

[1] School of Design, Adolfo Ibáñez University, Diagonal Las Torres 2700, Peñalolén, RM, Chile
felix.raspall@uai.cl

[2] Architecture and Sustainable Design Pillar , Singapore University of Technology and Design, 8 Somapah Road, Singapore, Singapore
carlos_banon@sutd.edu.sg

Abstract. Large-scale additive manufacturing for architectural applications is a growing research field. In the recent years, several real-scale projects demonstrated a preliminary viability of this technology for practical applications in architecture. Concurrently, the use of recycled polymers in 3d printing has progressed as a more sustainable feed for small-scale applications. However, there are limited empirical examples on the use of additive manufacturing using recycled polymers in large-scale and real-life architectural applications. This project develops two design and fabrication approaches to large-scale manufacturing using recycled Polyethylene Terephthalate (PET) from single-use bottles into large design elements and tests them in a real-life project. The two designs are discussed in detail: a 4 m diameter dome-like chandelier printed with a robotic extruder using recycled PET pellets, and a 3.5 m diameter chandelier using a Fused Deposition Modeling (FDM) printing farm. The paper covers the state of the art of related printing technologies and their gaps, describes the printing process developed in this research, details the design of the domes, and discusses the empirical evidence on the benefits and drawbacks of large-scale additive manufacturing using recycled polymers. Overall, the research demonstrates the possibilities of large-scale additive manufacturing using recycled polymers, adding findings form a real-life project to the growing body of research on additive manufacturing in architecture.

Keywords: Additive manufacturing · Digital design · Interior design · Polymer recycling

1 Introduction

Large-scale additive manufacturing using polymer extrusion is gaining traction among digital design and manufacturing research. In recent years, several real-scale projects started to demonstrate preliminary viability of this technology for practical applications in architecture. Concurrently, the use of recycled polymers in 3d printing has progressed, delivering commercial filaments and pellets which serve as a more sustainable

P. F. Yuan et al. (eds.), *Hybrid Intelligence*, Computational Design and Robotic Fabrication,
https://doi.org/10.1007/978-981-19-8637-6_37

feed. However, limited empirical examples on the use of additive manufacturing using recycled polymers at full-scale architectural applications exist. This project develops two approaches for large-scale 3D printing system using recycled Polyethylene Terephthalate (PET) from single-use bottles into large design elements and tests the process within a real-life architectural commission. The research is the result of a collaboration between the authors' laboratory AIRLab, the multinational company DB Schenker and the research agency NAMIC, during 2020–21. The project produced a calibrated bespoke printing process (including transforming bottles into pellets and filaments and the actual large-scale printing), the digital design of large-scale lighting fixtures and the fabrication and assembly of the printed elements. Two designs are discussed in detail: a 4 m diameter dome-like chandelier printed with a robotic extruder using recycled PET pellets (Figs. 1 and 2), and a 3.5 m diameter chandelier using Fused Deposition Modeling (FDM) printing farm (Fig. 3). In total, the project salvages more than 400 kg of plastic.

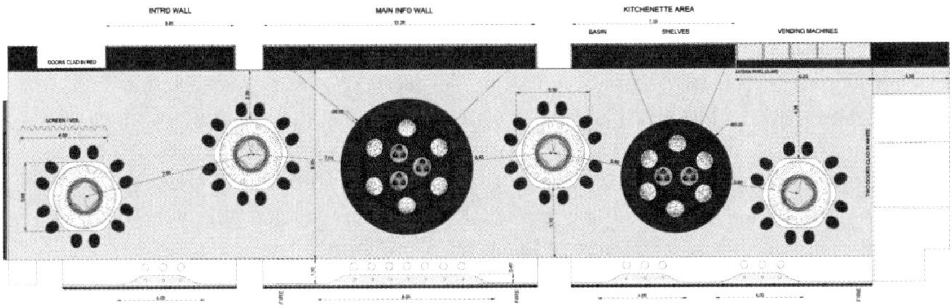

Fig. 1. Plan of the project. The circles are the seating spaces, covered by the printed chandeliers.

The paper covers the state of the art of related printing technologies and their gaps, describes the printing process developed in this research, details the design of the domes, and discusses the empirical evidence on the benefits and drawbacks of large-scale additive manufacturing using recycled polymers. The discussion includes the digital design process, the technical detailing, the production of polymer feed from bottles, the printing time and costs using an FDM printing farm and a robotic arm, the assembly strategies, process parameters, and bottlenecks and solutions, among others. Overall, the research demonstrates the possibilities of large-scale additive manufacturing using recycled polymers, adding findings from a real-life project to the growing body of research on additive manufacturing in architecture.

2 State of the Art

The research covers two main areas of research: large scale additive manufacturing at an architectural scale, and 3d printing using recycled polymers. The state of the art in these areas is presented.

2.1 3D Printing at Architectural Scales

Additive manufacturing technologies have been originally developed for applications smaller than those required for architectural applications. Architectural research has

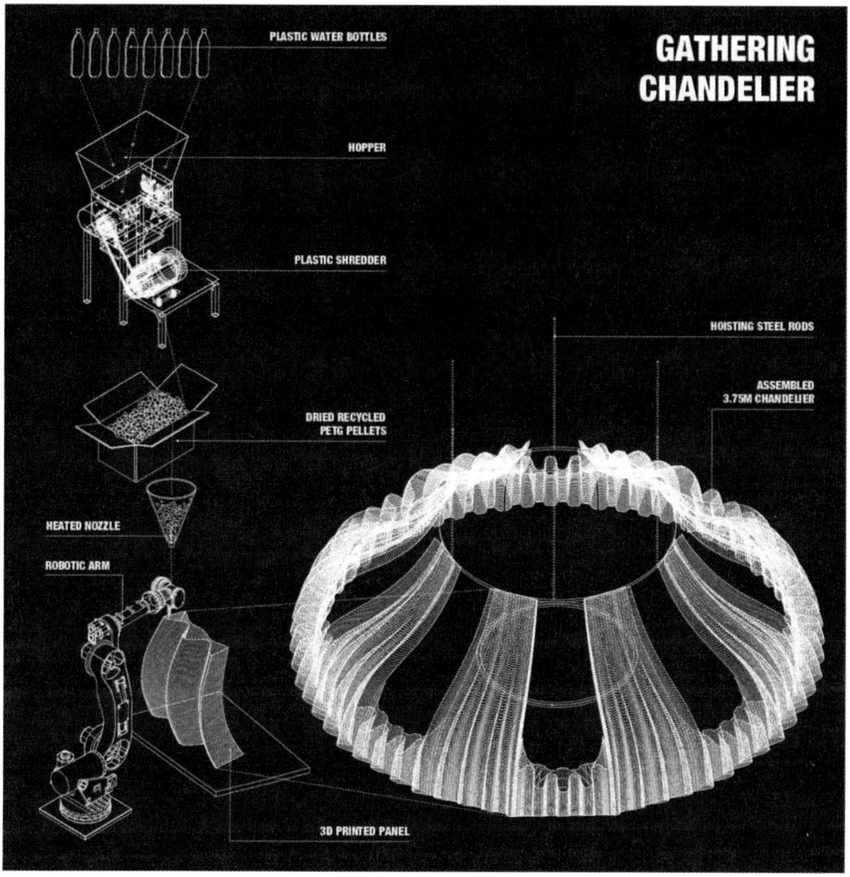

Fig. 2. On the left, a diagram of the material lifecycle from single use bottles, which are shredded into flakes that is used as feed for a robotic printing system. On the right, an axonometric view of the chandelier.

advanced technologies to bring 3d printing into a functional construction method. General literature on 3D Printing in Architecture can be found on books by Rael and San Fratello [1], Bañón and Raspall [2]. Both books argue for and demonstrate with built examples the possibilities of a different 3d printing technologies for architecture and construction.

3D Printing with polymers in architecture in a particularly active field of research. The key challenge is that commercial printing technologies, due to the relatively small print volume, is not immediately applicable for most architectural uses. Two main approaches to overcome this issue stand out.

First, research on large assemblies using relatively small components printed with commercial printers farms. An example on this approach is the work by Taseva et al. [3], which proposes a facades system using multiple components and an intricate system of connection between them.

Fig. 3. View of the chandelier from below. The ring contains the lighting, which is projected to the translucent PET material.

The second approach involves the development of large-scale printers for polymer printing. Typically, this is achieved by installing a polymer extruder to an industrial robot arm. An example of this system was used by Yuan et al. [4] for a pedestrian bridge design.

2.2 Printing with Recycled Polymers

Substantial research from material science and manufacturing engineering has demonstrated how 3D printing can operate with recycled polymer. Zander et al. [5] reviews the advances in the area. These advances led to numerous commercial filaments and pellet suppliers. An example of a recycled PET filament is commercialized by B-Pet (https://bpetfilament.com/).

From the field of digital manufacturing in architecture, several research projects have focused on the use of recycled polymers in architecture. For example, Bruce et al. [6] demonstrates the reusability of plastics for filament production of a screen system. Large scale robotic printing with reclaimed PLA has been tested by Shiordia [7], and applications in large scale fabrication has been tested by Shiordia [8]. These academic research projects produce valuable small-scale prototypes. However, full scale architectural projects are still missing.

3 Large Scale 3D Printing with Recycled PET

To breach the gap between academic research and real-life application of recycled polymer 3D printing in architecture, this research proposes, develops, and documents a real-life architectural commission that implements this technology to solve a specific design brief.

In 2020, the authors were approached by international logistics company DB Schenker to design the interior architecture of the lounge space for their new headquarter building. The design concept proposed a space that repurposes several materials that the company often discards, transforming waste into objects of design value. Packing cardboard and foams, and pallets were transformed into various furniture pieces. In addition, PET bottles discarded by the company were repurposed into large dome-like chandeliers of intricate geometry, to be manufactured using 3D printing. This aspect of the project is discussed in this paper.

The function of the PET chandeliers was to break down the large lounge space where employees have lunch breaks and gather into more contained, intimate spaces. Figure 1 shows the plan of the project. The position of the chandeliers coincides with the hexagonal tables circular lounge spaces.

Two different types of chandeliers were developed, aiming to advance research on the two most distinct polymer printing processes in architecture reviewed in the literature: Large assemblies of components using FDM desktop printers and a custom built FDM robotic system. In both cases, recycled PET was tested as main feed, in filament and pellet form, respectively.

3.1 Gathering Chandelier

The *Gathering Chandelier* is a 4 m diameter dome-like lighting fixture, made of 16 large, robotically printed pieces. Its design was parametrically produced to be manufacturable with a single, continuous extrusion. Each component's bounding box size is around 1.50 × 0.60, 0.40 m. Figure 2 shows the fabrication process and the geometry of the chandelier. Figures 3 and 4 show the end-product. The chandelier was suspended from the ceiling using a top ring. On the bottom part, a compression ring contains the lighting elements.

Fig. 4. View of the Gathering Chandelier in the interior space, with the lounge chairs made from reclaimed packing foam and the low table made of reused cardboard.

The fabrication process was developed and calibrated by the research team. It consists of a large ABB robotic arm, to which a 1.00 × 0.60 m heated build plate was mounted. The extruder, mounted stationary on a frame, was fed with recycled PET pellets. For this project, two extruders were tested. On a first stage, a DIY filament maker was repurposed as an extruder. This DIY extruder successfully produced the first two pieces, taking 40 h to print.

At a later stage, a professional grade extruder model MDPH2 by company Massive Dimension replaced the original one for faster material output. With this extruder, the manufacturing time for each component was 20 h. Figure 5 shows the printing process. The total amount of recovered plastic for the whole dome was 160 kg. The plastic pellets were procured from a PET recycling company in China.

3.2 Dining Chandelier

The Dining Chandelier tests the creation of a large-scale architectural fixture using conventional FDM 3D printers and recycled PET filament. Four identical 3.5 m diameter chandeliers were created for this project. The domes consist of 10 rows of 45 pieces each, accounting for a total of 450 pieces per dome. Figure 6 shows the fabrication process and the geometry of the chandelier. Figures 7 and 8 shows the end-product.

Fig. 5. Printing process for a large piece of the dome. The 1.5 m tall piece is almost complete. The Robot positions a heated built plate (bottom of the image), while the extruder is stationary on a frame (top of the image).

Each of the 450 components was designed to be printable with a standard FDM printer, with built volumes smaller than 200 × 200 × 100 mm. The geometry of each tile was based on a Schwarz D ("Diamond") minimal surface, and it was designed to be printable without supports, taking around 6 h to print and enabling 2–3 prints per machine per day. The uppermost element in the dome was specifically designed to interface with aluminum ring that suspends the dome from the ceiling. The lighting was solved with an LED ring suspended in the center and projecting light upwards towards the chandelier. Careful attention was given to the connection between components, which was solved

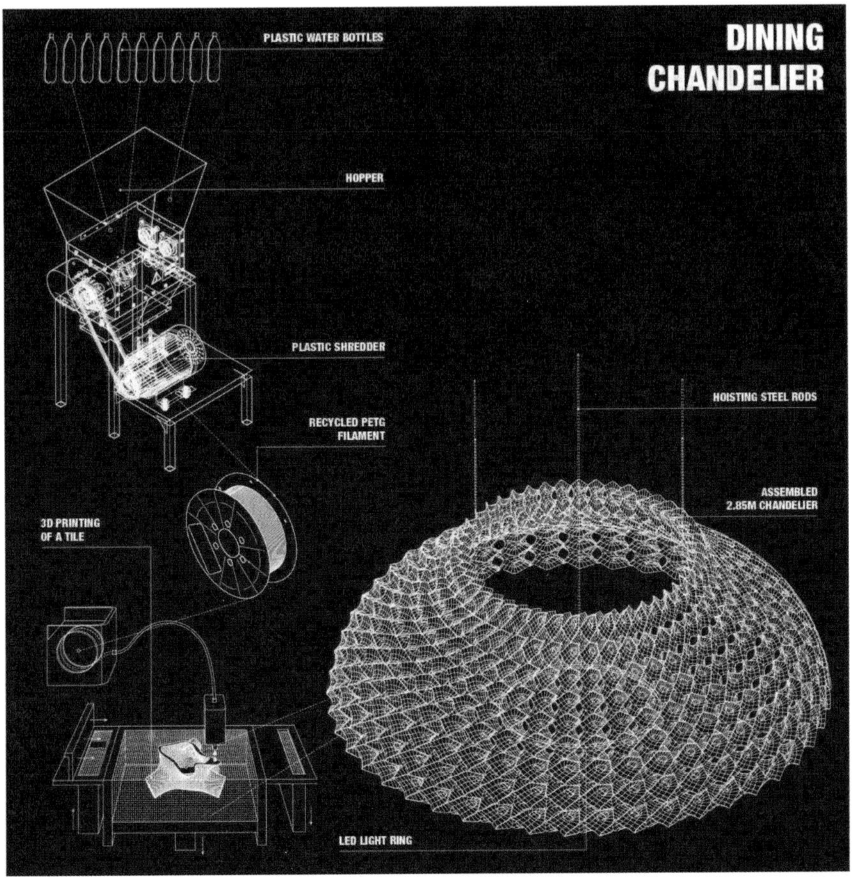

Fig. 6. On the left, a diagram of the material lifecycle from single use bottles, which are shredded into flakes and extruded into 1.8 mm filament, which are then used in desktop FDM printers. On the right, an axonometric view of the chandelier.

with features that concealed nylon nuts and bolts. The total plastic recovered for each of the four domes is 60 kg.

3.3 Processing of Discarded Bottles into Pellets and Filaments

As part of the investigation, the research developed a production line to transform single use bottles into the raw materials needed to produce the chandeliers. The system includes a plastic shredder, a plastic dryer, and a plastic filament extruder. Figure 9 shows some of the components of the system. The process of creating printable flakes for the robotic system was successful, but the feed of flakes (rather than conventional nurdle shape pellets) in the extruder led to occasional clogging. This compromised the printing pace

Fig. 7. Left: view of the chandelier from below. Right: detail of the chandelier articulated texture, which is the result of the repetition of a minimal surface geometry.

Fig. 8. View of the Dining Chandelier in the interior space, with the large tables made recycled pallets.

and reliability of the printing process. Several modifications to the extruder were tested, such as adding a small vibrator, which improve the performance but did not completely solve the clogging issues. Further research on the flow of plastic flakes is needed.

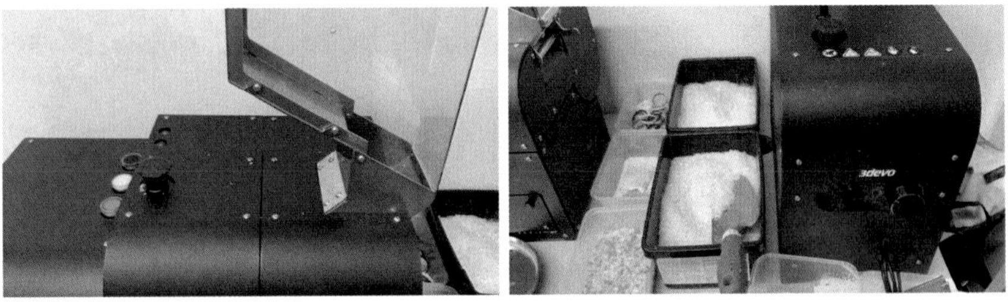

Fig. 9. Left: Plastic shredder. Right: Plastic flake dryer.

4 Results and Discussion

The research output includes the design and production of the five domes (1 × Gathering Chandelier and 4 × Dining Chandelier). These items were successfully manufactured and assembled; and the project was completed with very positive appraisal from the client. Through this built example, the project demonstrates that fabrication of large, complex geometry and functional architectural components using recycled PET is commercially viable using 3D printing technologies.

Further research on the processing of discarded material is needed to reduce issues with the extrusion of plastic flakes or their processing into pellets. Despite the two presented prototypes are interior chandeliers, these applications can be considered applicable for bespoke façade elements. Relevant aspects, such as insulation, wind loads, transparency, airtightness, and aging or deformation should be further evaluated.

Acknowledgements. This research was supported by NAMIC Singapore. The research team included Sourabh Maheshwary, Wan Mengcheng, Simon Rocknathan, Tay Boon Kiat, Kwang Kai Jie, Alba Lombardía Alonso and Nahaad Vahid.

References

1. Rael R, San Fratello V (2018) Printing architecture: Innovative recipes for 3D printing. Chronicle Books
2. Bañón C, Raspall F (2021) 3D printing architecture: workflows, applications, and trends. Springer, Berlin/Heidelberg, Germany
3. Taseva Y, Eftekhar N, Kwon H, Leschok M, Dillenburger B (2020) Large-scale 3D printing for functionally-graded facade
4. Yuan PF, Chen Z, Zhang L (2018) Form finding for 3d printed pedestrian bridges
5. Zander NE, Gillan M, Burckhard Z, Gardea F (2019) Recycled polypropylene blends as novel 3D printing materials. Addit Manuf 25:122–130
6. Bruce C, Sweet K, Ok J (2020) Closing the loop-recycling waste plastic
7. Shiordia Lopez R (2020) Large format FDM printing of recycled plastic pellets: closing consumer cycles in the fabrication of meso scale objects
8. Nováková K, Prokop Š, Vele J, Achten H (2018) PET (s) culpt-crowd-printing recycled polyethylene terefphthalate
9. Strauss H, Envelope AM (2013) The potential of additive manufacturing for facade constructions. TU Delft

Towards a Digital Repertoire: Design and Fabrication of a Robotically-Milled Brass Chandelier

Professor Paul Loh[1](✉) and David Leggett[2]

[1] Bond University, 14 University Dr, Robina, QLD 4226, Australia
ploh@bond.edu.au

[2] Architectural Research Lab, LLDS, Preston, Australia

Abstract. The paper described the design and fabrication of a robotically-milled brass chandelier using a bespoke vertical axial revolving material holder as a robotic fixture. While the technique described is for a chandelier design, it has potential architectural applications, as demonstrated by architects such as Barkow Leibinger. The significance of this research lies in the increased flexibility of the technique performed using a robotic arm compared to the current industrial method using tubematic laser cutter. In addition, the paper outlined the design of the robotic fixture and the computational workflow to create an integrated design-to-fabrication workflow. The research highlighted robotic systems as a potential design environment through reflection on Material Engagement Theory (MET) framework. Critically, the workflow constructed design feedback as robotic agencies that provide affordances through the fabrication setup. Such affordances contribute to the designing process and refine craftsmanship by creating transactional relationships between tools and material as a digital repertoire. This emerging design environment extends robotic research into design practice.

Keywords: Robotic fabrication · CNC milling · Agency · Robotic workflow · Material engagement

1 Introduction

Computer Numeric Controlled (CNC) metal tube cutting techniques used in the manufacturing industry predominantly utilised an axial revolving material holder coupled with a solid-state laser or plasma cutter, sometimes known as tubematic cutter with a range of up to 1500 mm in tube diameter [1]. Alternatively, CNC tube cutting can be performed using an industrial robotic arm with a plasma or oxyfuel cutter end-effector on a horizontal axial revolving holder [2]. While an industrial tubematic cutter is typically designed to produce effective tube fabrication, it also means that it is a piece of single task machinery and is often limited to specialist engineering and infrastructure projects.

The research project explores a novel setup using a robotic arm with a milling router and a vertical axial revolving holder to fabricate a brass chandelier. The advantage of the setup using a robotic arm with a more common end-effector, such as a milling spindle,

P. F. Yuan et al. (eds.), *Hybrid Intelligence*, Computational Design and Robotic Fabrication,
https://doi.org/10.1007/978-981-19-8637-6_38

allows designers and architects to better access the technique for flexible and small-batch manufacturing [3] without compromising the fabrication quality.

The technique and outcome of the research, while limited to the design and fabrication of a chandelier, have a more comprehensive architectural application, as demonstrated by Barkow Leibinger [4, 5]. The practice has explored using such techniques to develop architectural applications such as lighting fixtures, spatial installation, sunscreens and façade rainscreen. The primary interest is in the ornamental quality that such technique provides as a serial effect as an outcome. However, their material research in ornamentation is merely the aftermath of the method without intervening in the fabrication process.

This research extends Barkow Leibinger's material research to understand better how the affordances provided by the technique (the ornamental quality) can be productive in the design and fabrication process. The key contribution is exploring how feedback from the robotic setup provides design agency–moving away from using the CNC technique as a means to an end. The research questions: How can aesthetic quality through digital fabrication bring about a transactional relationship with robotic agency towards a productive digital repertoire? A form of hybrid intelligence.

The following section outlines the design intent, research objectives and critical theoretical framework of the research. In Sect. 3, the paper outlines the technical setup of the robotic system–particularly the robotic fixture for the material and the design to fabrication workflow. Section 4 identifies the robotic agency and discusses the transactional relationship between tools and material as affordances for design.

2 Background

The brass chandelier consists of 37 brass tubes individually trimmed using a 7-axis robotic arm at the Architectural Research Lab (Fig. 1) in Melbourne, Australia. It is a bespoke commission installed in a double-height space for a private dwelling. The design intent that emerged through dialogue with the client was to create a sculptural object that would act as a focal point to the space and provide background lighting. Brass is chosen to complement the interior finishes of the room, but more importantly, it is soft and can be milled, unlike steel. The research takes place over 12 months, from initial material exploration to the design and fabrication of the brass chandelier.

2.1 Design of the Brass Chandelier

The primary structure of the chandelier consists of a hanging frame–composed of a steel rod hanger with top and bottom plates, which the pipes hook onto (1e). A set of LED lights is positioned in the centre of the cluster (1f). The tubes are arranged so there are 20 mm gaps between them to allow light to spill out. There are four sets of cutting operations performed on each pipe as illustrated in Fig. 1: the top (1a) and the bottom (1b) cut, three openings facing the LED light on the inside (1c) and up to 3 openings facing towards the outside of the chandelier (1d). The pipes' top and bottom (1a and 1b) were trimmed diagonally to reveal fluid sections that bundled together to create an illusion of tubes suspended in space. Three openings face the hanging structure (1c).

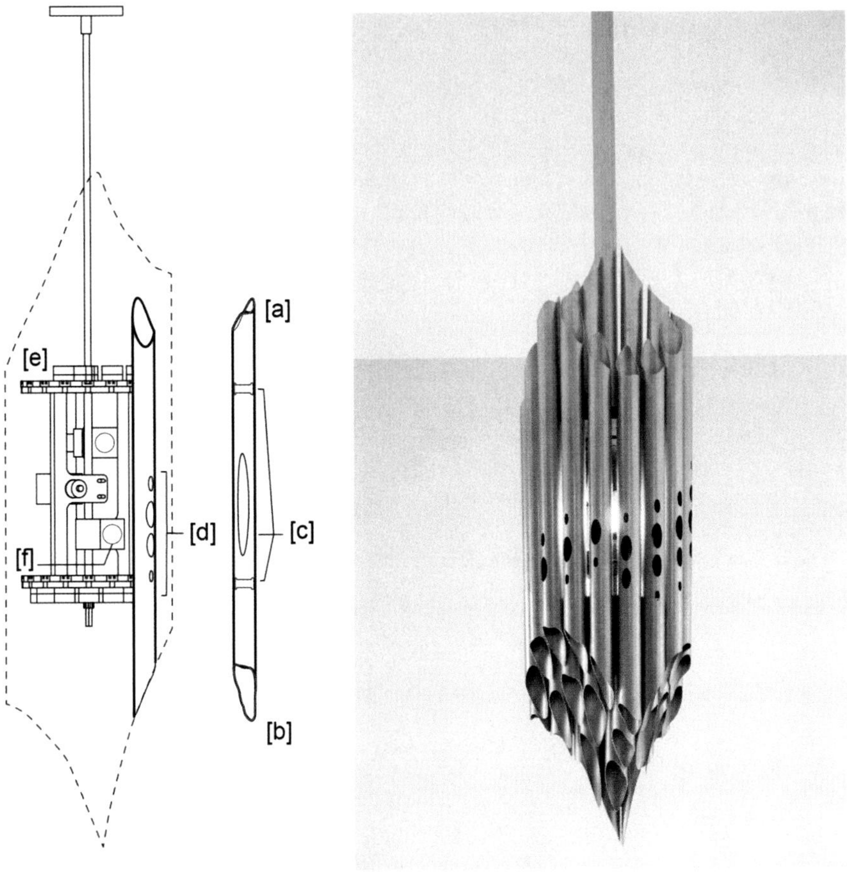

Fig. 1. Robotically fabricated brass chandelier. Left, Diagram illustrating the overall design with the hanging frame structure and a typical pipe with the various openings. Right, the installed final outcome. Photography by Ben Hosking.

Two of them are for hanging the pipe onto the frame, and a larger hole allows the LED light to illuminate the inside of the tube. The openings in the middle sections of the tube (1d) create visual interest and will enable the light to spill out sideways.

2.2 Research Objective

The objectives of the research are:

- To explore the feasibility of an alternative robotic revolve cutting using the milling technique.
- To design and prototype a robotic fixture to enable flexible manufacturing using a robotic arm.

- To examine the agency of robotics in designing. In other words, to use the robotic setup as both a fabrication and deigning tool through an integrated design and fabrication workflow.

2.3 Theoretical Framework: Material Agency as Craftsmanship

The research is informed by Material Engagement Theory (MET). Malafouris [6–8], hypothesis a design environment where the designer enters a level playing field with the tools, material, and techniques deployed during the fabrication process. In this environment, they are all actors or agents of the system whose agencies are relative to the fabrication tasks.

Centred in the discussion of MET is the agency of material. Malafouris argued from an archaeological perspective that it is a quality that is not limited to humans and can be satisfied by an object in-so-far as the object (tools and technology included) can become an extension of the person. Malafouris [6], (pp. 169–177) highlighted material agencies by examining the cognitive intent of knapping hand axe. He demonstrated that the act of knapping flint is an exercise of multiple agents at work, including the hand of the knapper, the knapping stone, and the stone being knapped. Each subsequent strike of the flint determines the angle of the next strike.

This idea of iterative negotiation of agencies shifted our understanding from the traditional concept of a preconceived image of an axe head within the flint toward one where design is a negotiation between the human action, material and tool. He states, 'There are no fixed agentive roles in this process; instead, there is a constant struggle towards a "maximum grip" ([6], p. 176)'. As the maker actively engages with the material to produce the form, the process of Making is an enacted embodied engagement. Malafouris suggests the form of an intended object is not external but learned and sustained as an idea and developed through the Making process as an explicit 'sense of agency' ([6], p. 176). Similarly, David Pye stipulated that the Making process contains the intentionality of the maker expressed through the agency of tools and material [9, 10]. Malafouris ([6], p. 140) posits, 'intention no longer comes before action, but it is in the action'.

MET hypothesised that the formal outcome (as an artefact) does not result from preconceived ideas. Instead, the design intention is developed through the experience of Making or 'directly embodied and realised in the hybrid space of situated action' [11]. In other words, through the act of designing, prototyping and Making. It is not a total surrender of the designer's role to material effects but rather a careful and measured sense of agency. Critically, this is not just an interaction of agents but a transactional relationship, where the outcome is not merely a by-product of the exchanges between agents but transformed by their agency; Ihde and Malafouris [11] rightly termed this "becoming".

The following section will examine the robotic fabrication before returning to discuss the robotic agency and the transactional relationship evident in the fabrication process.

3 Robotic Fabrication

The brass tubes that make up the chandelier are trimmed using HSD 18 kW milling electrospindles attached to a Kuka KR120 robotic arm coupled with a turn-table. A bespoke robotic fixture design to hold the tube for milling is anchored to the 7th axis turn-table. Geometric design and robotic programming are performed through McNeel Rhino with Grasshopper 3D and Kuka PRC.

3.1 Robotic Fixtures Design

There are several challenges in designing the robotic fixture. First, unlike most tubematic cutter setups where the revolving axial is horizontal and typically pinned on both ends, the 7th axis turn-table conditions the axial revolving holder to be vertical and anchored at the base only. This produces a situation where the vibration from the milling process would cause the tube to oscillate, resulting in inaccurate trimming.

Second, the brass tube must be easily replaced for subsequent milling to ensure an efficient fabrication procedure. As described in Sect. 2.1, each tube is trimmed on both ends with a series of hole cut-outs. Through prototyping process, we observed that the further we cut the tube from the base of the fixture, the greater the oscillation. For the final version of the fixture design, the maximum acceptable distance from the base for any trimming operation must be within the 900 mm range. To avoid accessive oscillation for longer tubes (with a length greater than 60% of the acceptable distance), the researcher decided to flip the tube at mid-point to reduce the distance between the milling and the base of the fixture. Therefore, each tube must be removed from the fixture and manually flipped after the initial cutting procedure.

Figure 2 illustrates the design of the vertical axial revolving holder with a wide base (2b) anchored to the turn-table (2a). The tube holder (2c) is designed to be extended to accommodate a longer tube (2d) which could increase the range of the trimming operation. A steel tube (2e) is bolted to the holder (2c) to hold the brass tube (2f). In addition, a set of solid MDF blocks is required to infill the core of the brass tube to reduce vibration caused by the tool during the milling process (2g). The solid MDF blocks (2h) are inserted into the steel tube (2f), and they can be added or removed depending on the length of the tube to be cut.

These challenges inform the fixture's design, but more critically, the outcome as a series of test prototypes (successes and failures) inform the development of the repertoire. Three fixtures were developed, each incrementally refining the workmanship of the technique by reducing the vibration caused by the milling process. A smaller version of the chandelier with eight tubes (Fig. 2) is developed as a proof-of-concept for the hanging structure and testing of the robotic workflow in generating the design constraints. These developments allowed the research team to design the automated workflow using affordance produced from the technique.

3.2 Design Workflow

Figure 3 outlined the chandelier's computational and robotic workflow from design to fabrication within a single visual script using Grasshopper 3D. The workflow integrates a

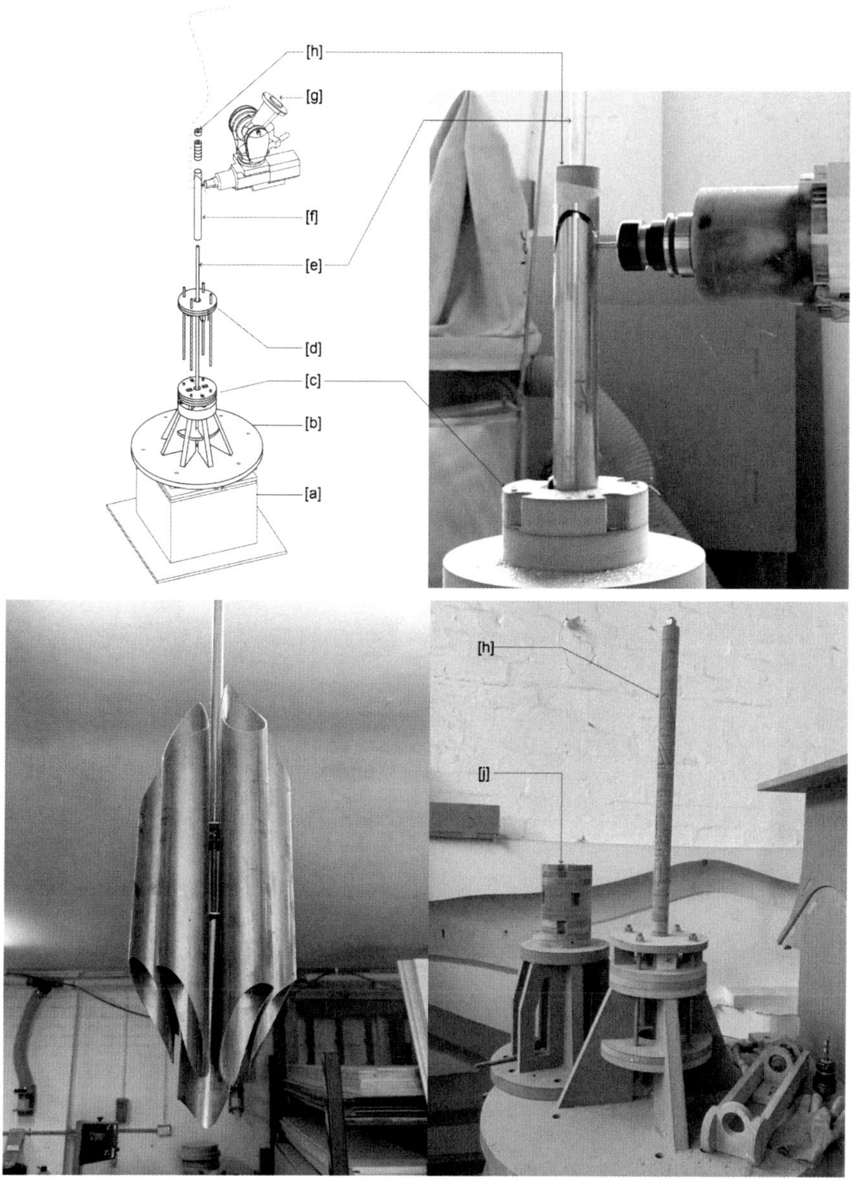

Fig. 2. Top left: Design of the fixture holding the metal tube. Top right: Milling of the brass tube using the fixture. Bottom left: 8-tubes chandelier prototype. Bottom right: Robot fixtures indicating the solid MDF block [h] infill to the brass tube and early fixture prototype [j].

seamless data flow scripted from the basic design parameters (3A) consisting of the plan layout, where the length of each pipe (L_T) considers the structural frame height (L_F), a constant gap around the frame (2x) and the trimming curve (y sin θ). The optimisation process (3C) considers the range of the robotic trimming operation discussed in Sect. 3.1.

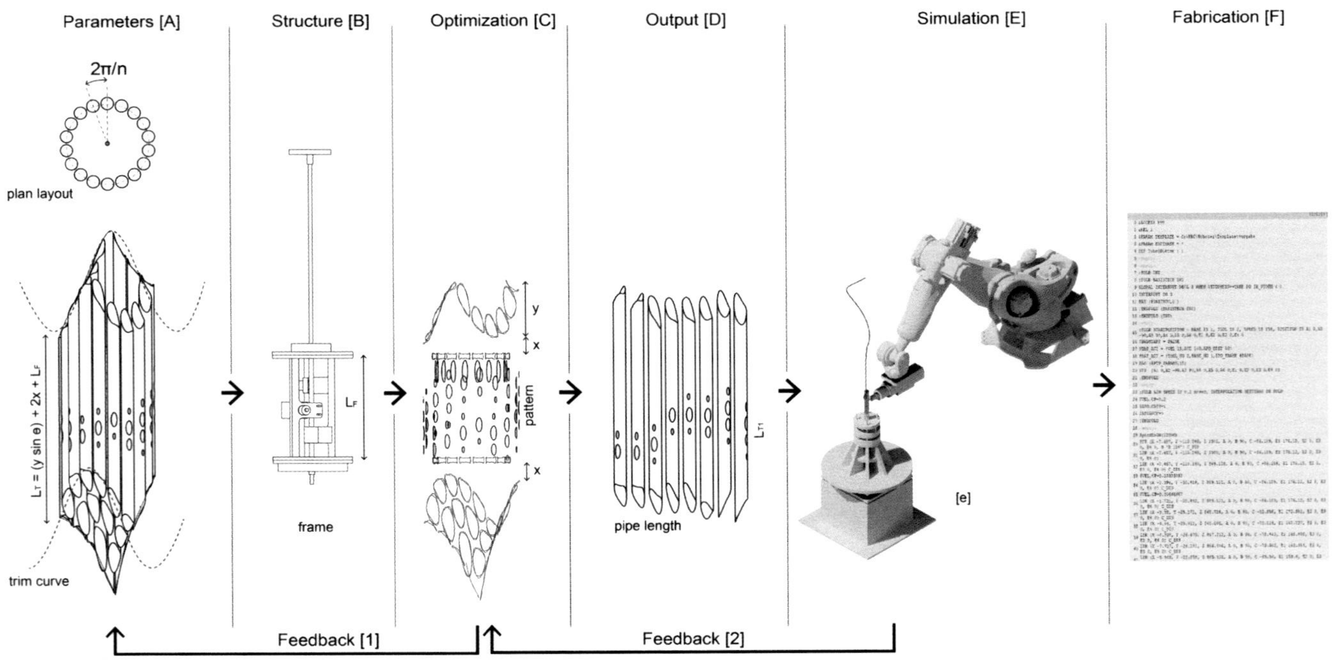

Fig. 3. Computational workflow for the brass chandelier from design to fabrication. The workflow integrates design parameters, including the structural framing within the workflow. These parameters interact with the robotic process to provide feedback in the designing process.

Again, this is constrained by the length of the structure frame as it contains the main LED fixtures, which condition where the openings in the middle sections of the tube could be placed (see Fig. 1d).

4 Towards a Digital Repertoire with Robotic Agency

The research begins to identify the transactional relationships between tool and material as affordances in the designing process. This section discusses the contribution of robotic agencies toward the design process and the concurrent refinement of craftsmanship.

4.1 Feedback as Robotic Agency

The workflow diagram reveals at least two moments of interaction between the scripting process and the tooling. The first feedback (Fig. 1) process occurs when designing the pattern (Fig. 3c) to provide side illumination to the chandelier. This is related to and conditioned by the second feedback (Fig. 2), which occurs when the robotic trimming is simulated using Kuka PRC. The simulation allowed us to check for clashes with the robotic arm and conditioned a zone where the pattern could be placed. This is, in relation to both ends of the pipe, as each tube varies in length. Subsequently, the simulation informed the amplitude of the sine curve that is used to trim the top and bottom edges.

4.2 Designing with Robotic Agency

The robotic fixture, technique and workflow provide affordances for designing on three levels.

First, affordances are generated by producing aesthetic quality as values for the design and evaluation of craftsmanship. The development of the robotic technique and physical prototypes allowed the designer to evaluate craftsmanship through the tooling process. The objective is to align the design intention with the physical outcome [10] using the constraints provided by the techniques to condition the design, including the form and pattern distribution. The desire to bring design intention closer to the outcome as a form of refinement of workmanship through robotics demands a transactional relationship between the various agents. Here, traditional static design is replaced with designing through agencies–a negotiation between the tools, material, and technique. The robotic agency replaces preconceived forms through the making activity to create an emerging effect. This is evident in the 8-pipe chandelier and the fixture prototypes. Within this exchange of agencies, nothing is fixed but in a constant state of negotiation [6]. In doing so, the robotic agency begins to shape the design.

Second, the robotic fixture and workflow function as an extension of the designer toolset. The technique becomes part of the digital repertoire where its application can extend beyond the design and fabrication of a chandelier. This research setup a concurrent design and robotic parameters to create a dialogue between design and fabrication. This is different from Barkow Leibinger's exploration of the tubematic technique, where design is an after-effect of the system.

Third, creating design feedback within the robotic system develops a transactional relationship between the tooling and the material, transforming the robotic agency into an emerging design environment. The design of the robot fixture is evidence of such a transactional becoming. It acts as a jig for the robotic process to counteract forces enacted by the robotic arm to hold the material in position for shaping. Thereby transforming the material (a brass tube) into an artefact guided by the design intention and sustained through a cumulative set of aesthetic values. The emerging aesthetic values generated through the trimming procedure maintained the design intention and allowed the designer to act and modify the outcome by incrementally refining the technique.

4.3 Future Exploration

The robotic agency described above is embedded within the workflow and fabrication process. Here, we have only explored the automated agency through constraints and in-build parameters within a design environment. The next steps will be to develop a learning environment within the ecology to allow self-population of the patternation at the individual tube and global level using Artificial Intelligence.

5 Conclusion

The paper outlined a design to fabrication workflow and setup to robotically milled tubular brass resulting in a chandelier design. While the technique described is used for prototyping a chandelier, it has potential architectural applications. In this research, we further the design research by Barkow Leibinger using the tubematic technique by examining the relationship between fabrication and its affordances. The study explored the technique's feasibility and outlined the design of a robotic fixture to enable flexible manufacturing using a robotic arm. Feedback is constructed within the computation workflow to condition the final design outcome in form and pattern distribution. The paper reflects on Material Engagement Theory (MET) to explore how robotic systems can create agency for designing through affordances generated through the transactional relationship between robotic fixture, technique and workflow. The emerging aesthetic quality provides design value for evaluating workmanship by negotiating between the tools, material, and technique to create ornamental effects. The robotic fixture and workflow form a digital repertoire that extends the designer's toolset. Doing so extends the potential of the method beyond the current application. Last, in an emerging design environment, the feedback within the robotic system develops a transactional relationship between tooling and material. This transformational process impacts the designing process, which incrementally refines the technique. As an emerging design environment, the robotic system extends robotic research into architecture practice.

Acknowledgements. The project acknowledges the technical contributions of Ryan Huang and Yuhan Hou (LLDS). We also like to extend our thanks to the client whose patience allowed the research to take its course.

References

1. Trumpf (2022) Laser tube cutting machines. https://www.trumpf.com/en_INT/products/machines-systems/laser-tube-cutting-machines/ Accessed 15 Mar 2022
2. Microstep (2021) Microstep. https://www.microstep.eu/machines/pipe-profile-cutting-machines/. Accessed 23 Nov 2021
3. Pichler A, Wogerer C (2011) Towards robot systems for small-batch manufacturing. In: 2011 IEEE international symposium on assembly and manufacturing (ISAM). IEEE, pp 1–6. https://doi.org/10.1109/ISAM.2011.5942336
4. Barkow F, Leibinger R (2009) An atlas of fabrication. AA Publications, London, England
5. Barkow F, Leibinger R (2014) *Spielraum.* Hatje Cantz: Ostfildern
6. Malafouris L (2016) How things shape the mind. The MIT Press
7. Malafouris L (2020) Thinking as "Thinging": psychology with things. Curr Dir Psychol Sci 29(1):3–8. https://doi.org/10.1177/0963721419873349
8. Malafouris L (2021) How does thinking relate to tool making? Adapt Behav 29(2):107–121. https://doi.org/10.1177/1059712320950539
9. Loh P, Burry J, Wagenfeld M (2016) Reconsidering Pye's theory of making through digital craft practice: a theoretical framework towards continuous designing. Craft Research 7(2). https://doi.org/10.1386/crre.7.2.187_1
10. Pye D (1968) The nature & art of workmanship. A & C Black, Chippenham, England
11. Ihde D, Malafouris L (2019) Homo faber revisited: Postphenomenology and material engagement theory. Philos Technol 32(2):195–214. https://doi.org/10.1007/s13347-018-0321-7

Soft Pneumatic Robotic Architectural System: Prefabricated Inflatable Module-Based Cybernetic Adaptive Space Model Manipulated Through Human-System Interaction

Si-Yuan Rylan Wang(✉)

School of Architecture and Engineering, Nanchang University, Nanchang, China
siyuanwang1321@gmail.com

Abstract. In this paper, a cybernetic adaptive space model based on prefabricated inflatable modules and physical interaction manipulation is introduced. The research aimed to redefine an intelligent and organic trend of residing and working by providing an adjustable and performative space system. The conjunction of human-space interaction, as well as the soft and hard architectural elements adaptive to dynamic living modalities and environmental conditions, are included in the methodology. The datasets based on the human body posture are collected through IMU sensors to provide coding inputs for defining modular inflatable structures, which anticipate generating heterogeneous morphological variations apt for flexible scenarios. The elaborated pre-fabricated samples successfully conform to the expected inflating behavior through silicone patterns. The results demonstrated the possibility of future architecture as an unrestrained configuration. Integrating the shape-shifting space within modular manufacturing and interactive technology can deprive the performance of many constraints. It can render a responsive ecosystem through a behavioral transformation of the in-habitants.

Keywords: Tangible interaction · Material morphology · Pneumatic robotics · Responsive system

1 Introduction to the Robotic Architectural System

Nowadays, modern cities around the world have been clustering into large-scale living compounds in both real and virtual ways rapidly, people are more mobile and networked than ever before. Consequently. It leads to high demand for personalization and flexibility, and people can no longer be satisfied with rigid and standardized housing units. A clash between the mass production of architecture and the growing need for customization by the individual is assessed. However, the current building industry has not caught up with this trend. People are still dwelling with solidified facilities the same as in the last centuries. With new technologies and the digital revolution, the needs of the public, as well as the desires of an individual, can be accomplished. In this direction, architectural works show needed functions, provide the framework for the place, and express its function at a particular occasion. Meanwhile, with the accelerated pace

P. F. Yuan et al. (eds.), *Hybrid Intelligence*, Computational Design and Robotic Fabrication,
https://doi.org/10.1007/978-981-19-8637-6_39

of modern life and the rapid development of new technologies, architecture faces the challenge of rapidly changing scenarios over time. In the new era, architecture should be created as a transformable and vivid structure enriched with complete resources and more possibilities for civic lives.

In this situation, deformed architecture should be refined through a technological blueprint. Simultaneously, the emergence of interactive technology and new material prompted the traditional architecture via intelligent transformation in terms of morphological property. Of note, the invention of pneumatic soft robotics has made the morphological shift and its cybernetic formula possible. Numerous research cases proved that the outstanding paradigm as a soft responsive architecture via technical, functional, and actuated directions had been endowed. In this work, a set of criteria for creating relevant components is discussed. Based on these criteria, a scheme of the cybernetic architectural systems is developed, which utilizes the space interaction with people through morphology-variating modules and body tracking signals. With flexibility increased significantly by the incorporation of deformed components. The combination of two mechanistically synergistic circuits formed an interactive architecture system.

2 Literature Review: From Theory to Practice

The previous architectural research, technological practices, and social proposals mentioned the initial architectural transformation by shifting shapes only. Essentially we focus on the structures that live according to the changes in their environment. They breathe differently in different conditions and make changes in accord with the environment [1]. Thus, they may have a little bit of their own life. The original idea considers something close to an animal, namely an idea about re-establishing a relationship with architecture. Among the explorations of the forerunners of this kind of soft architecture, like Archigram, Buckminster Fuller, and Yona Friedman [8], the Fun Palace conceived by Cedric Price in 1964 was almost taken into practice. While developing his design for the Fun Palace, he described his visions for such a place Stanley Mathews [6]: "Old systems of learning are now decayed. The new universities will be of the world and in each man… The variety of activities cannot be completely forecasted; as new techniques and ideas arise they will be tried. The structures themselves will be capable of changes, renewal, and destruction. If any activity defeats its purpose it will be changed (Fig. 1)." Price pointed out that in allowing for change and flexibility, it is essential that the variation provided does not impose a discipline which may only be valid at the time of design; It is easier to allow for individual flexibility than organizational change–the expandable house; the multi-use of fixed volumes; the transportable controlled environment Charles Jencks and Karl Kropf [2]. In this way, the concept of improvisational architecture was introduced through, an entity, whose essence was in a continuous process of construction, dismantling, and reassembling by permitting multiple and indeterminate uses. Gorden Pask, the technical consultant of the Fun Palace committee, emphasized the functions, which are performed by human beings or human societies, and claimed that a building cannot be viewed simply in isolation, but meant as a human environment. It perpetually interacts with its inhabitants by serving them and controlling their behaviors [11].

New correlations in terms of the crucial relation of structure with its target were developed with another theory by Herman Hertzberger. He introduced the linguistic concept

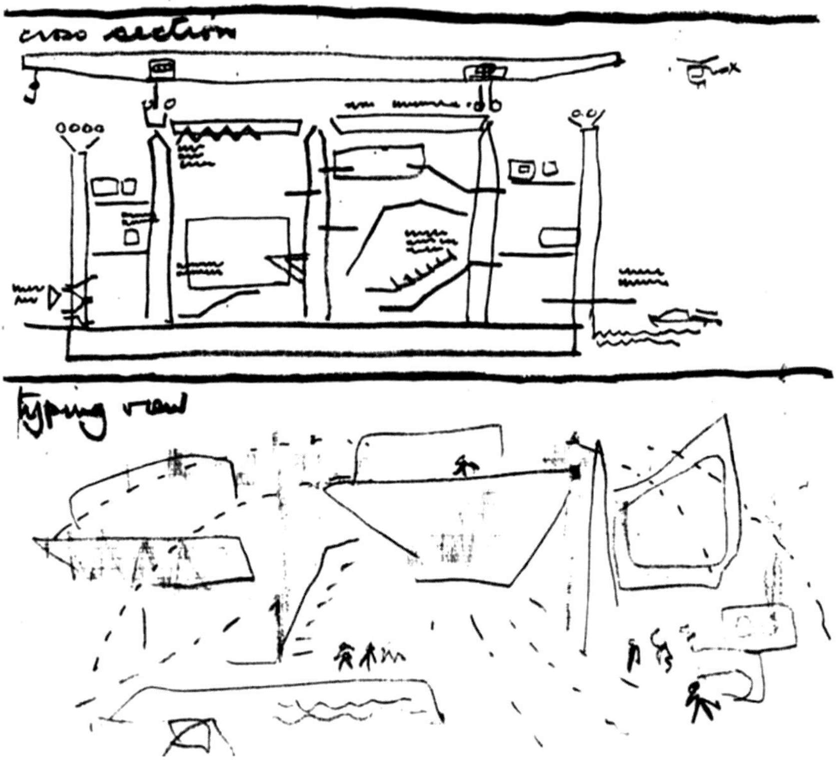

Fig. 1. Cedric Price, fun palace, sketches and notes, 1964, Cedric Price Archives, Canadian Centre for Architecture, Montreal

of ‘competence’ and ‘performance’ addressed by Chomsky to illustrate the essence of architecture [2]: Competence is the knowledge that a person has of his or her language, while performance refers to the use he makes of that knowledge in concrete situations; It can indeed be established with architecture, In architectural terms, competence defines a form’s capacity to be interpreted, and performance, stated by the form, interprets a specific situation.

All previous relevant research advocates a new kind of active and dynamic architecture to permit multiple uses and constantly adapt to change. It can be a network of diverse events and a space of oscillation between incongruous activities simultaneously. To redefine the users, the architectural space, and scenes adapting to different functions and interactions of humans with space, a creative and flexible control system is needed by the users in a dynamic space pattern for various daily activities. New technology like artificial intelligence and cybernetics is expected to be followed. Besides, the research of material systems and soft robotics in search of an alternative to rigidity can be accepted. As Nicholas Negroponte proposed that responsive architecture during the spatial de-sign problems can be explored by applying cybernetics to architecture [9]. Meanwhile, by forming a new frontier of kinetic designing, the domain of soft robotics has created an exciting and highly interdisciplinary paradigm in engineering, which provides a method for revolutionizing the role of robotics in architecture.

Soft robots are primarily composed of easily deformable matter such as fluids, gels, and elastomers with characteristics matching biological tissue and organs' elastic and rheological properties [4]. Till now, understanding and controlling the shape of thin, soft objects has been the focus of significant research efforts among physicists, biologists, and engineers for decades. For Example, so far, the researchers of the Interactive Architecture Lab designed a pavilion with cybernetic pneumatic silicone-made components [5]. Challenging conventional design thinking about adaptive architecture, the experiments outlined in this research suggest approaches to building soft responsive architecture. Soft responsive architecture acts as a proxy for the improvisational performance of architecture, enabling the theoretical visions of deformed architecture through technology. Although the soft robotics system is technologically viable, the inconvenience in manufacturing and excessive variates need further development to accommodate the popularization and diversification of space use. Thus, the proceeding development of multifunction should be advanced in the shape-shifting approach. One way of empowering diversity is to standardize the intelligent components with varying soft elements in specific-sized modular and freely combine them for diverse spaces.

3 Methodology Based on Structure and Performance

In order to provide a shared space of hybrid functions satisfying the compound responsive structure, this research is mainly focused on the adjustability, behavioral orientation, and performative aspect of the architectural system. Spaces have been considered introverted or extroverted atmospheres by categorizing them depending on the diversity of activities such as connected vs closed, private vs open, and stable vs dynamic. To figure out one's expectations for improvisational space, some inter-views were performed to summarize several types of scenes in different modes. However, the list was not comprehensive, as the variety of activities could never be precisely forecast (Fig. 2).

3.1 The Exploration of Material Agent

While selecting the material as the systemic agent, the hardness and deformability were the first two prime factors. Stiffness is related to spatial properties, while deformability affects the flexibility of spatial transformations. On the balance of the architectural hardness and soft deformability, in this study, the inflatable silicone as the material agent is chosen, and it is combined with the hard plate as the architectural component (Fig. 3).

3.2 Inputs, Outputs, and Interaction

The responsive approach to any architecture design affects certain environmental conditions or users' needs in terms of simulative responses to them. This kind of system works with two components, namely the sensors, which are the input, and the actuators, which are the output signal. The sensors measure real-time data such as light,

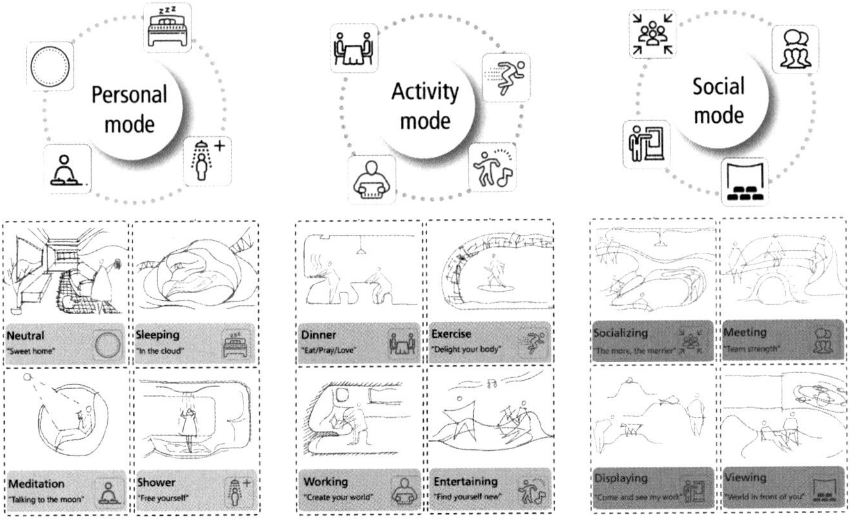

Fig. 2. Space scenarios adaptive to multiple activities

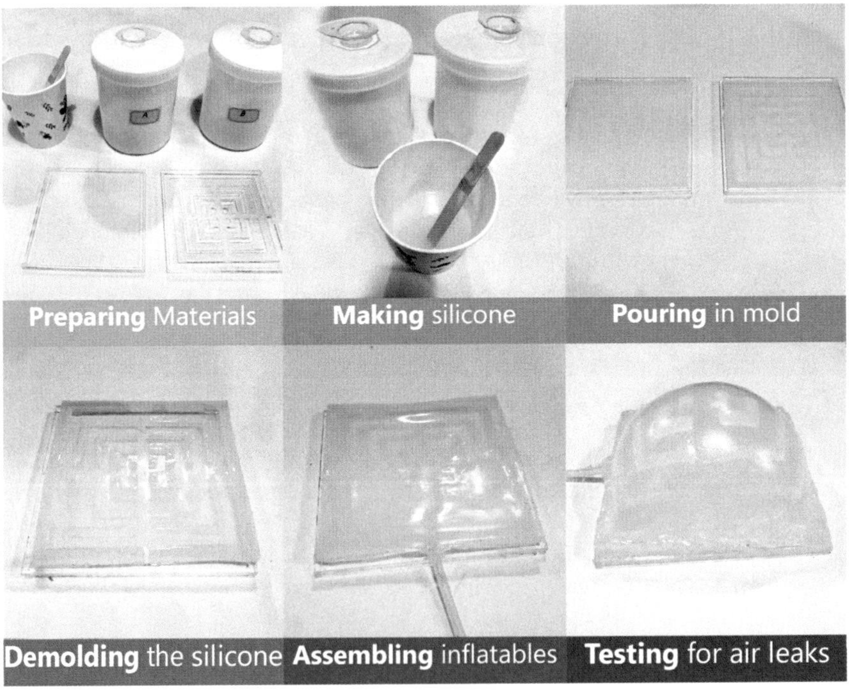

Fig. 3. Material fabrication

temperature, humidity, movement, position, speed, etc. This data is fed to the system to trigger the actuator, which performs in terms of change in its shape, color, size, position, and geometry. The main medium of architecture in this cybernetic system was set as the basic elements such as the smart floor, ceiling, and walls. As the proxy of the morphological changes of the architecture, the soft modules generate outputs to render the performative outcomes of the different activities, a unique atmosphere, and separate privacy. The kinetic actuation can be collected from the scene switching (e.g., Reshaped lounge/Partition rise up/Seating zones of different undulating surfaces), the output devices include an actuation motor, kinetic driver, VR glasses, odor transmitter, temperature regulator, glass transparency regulator, tilt brush, music player, visual projections, and view wander. The interaction input can be set by the fingerprints, VR waves, body temperature, sunlight, voice, panel setting, press projected spot, facial capturing, and body tracking. An interactive performance can be better actuated and set into different levels by synergizing all these mentioned points. It is expressed as follows: Level 1—Interaction in the unconscious state: Pressure controlling (spontaneous deformation): Sofas and beds for sitting and lying; Level 2—Interaction in transforming mode: Body-pose controlling via skeleton tracking (smart deformation): Space transform into different scenes.

In this experiment, Level 2, the human-system interaction, was further investigated. The IMU sensor was used to record the tilt angle of the user's body as input. The human body posture captured by the IMU sensor is defined through coding to the modular inflatable structure. The combination and changes of different modules were proceeded to create a lively scene suitable for different functions and a flexible change of life mode (Fig. 4).

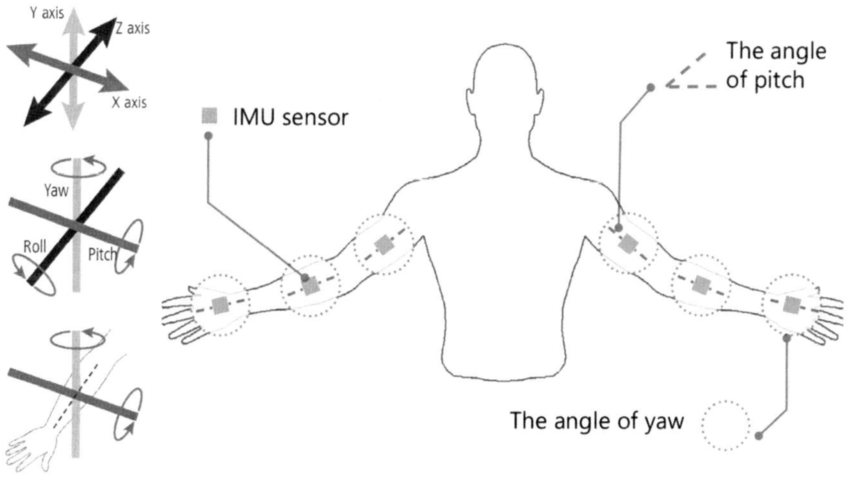

Fig. 4. Sensing mechanism and inputs

The current investigations involved the construction of an interactive system and its sampling by analyzing combined morphological models in the diverse predesigned patterns. In order to realize the space-change of a scene transformation, component modules were elaborately designed in specific patterns consisting of different calculated material attributes. The components inflate into conditional shape to meet the requirement of various scenes. The tilt angles of the body along 2 axes of the IMU sensor using 3 left and right arms positions were used as the input data, and the inflatable structure modulating in 3 * 3 m as the material agent was measured for the space performance output. The silicone material was selected for its elasticity to conform to the cambered surface, durability, and adaptability for flexible arrangement, and it can also be forged tightly with other materials. The assembled architectural system is displayed as a Soft Pneumatic Robotic Structure (SPRS) with silicone inflatables on the 3D printed acrylic plates controlled by mechanical-inflatable algorithms. In addition, a set of the tracheas, valves, manifolds, adapters, connectors, circuits, Arduino boards, wires, etc. is formed in the system (Fig. 5).

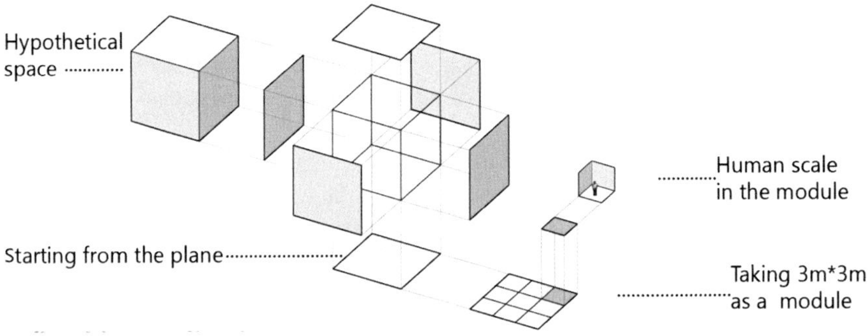

Fig. 5. Module configuration

3.3 Fabrication and Systemization

So far, the fabrication research of soft modules includes several aspects to be focused on. The manufacturing procedure faces its most duration for the casting complement of the rubber after blending two raw materials, since the inadequate time may not allow the downright weld of the weak joints, which leak the gas and result in the inaccuracy of the experiments. The ratio of the raw materials also cast a sequential effect on the toughness of the finished rubber, which directly influences the swelling capability of the soft samples. The silicone modules were connected to the inflator bump to generate the expected inflation, controlling the air pressure to the modules in graded amounts. This method obtains the diverse surface variations of the materials. Although two samples were noticed at risk of breaking the elastic extent, leading to abnormal morphology at the highest level of air compression, the variations with respect to others are negligible. In addition, the results confirm that the different morphologies are the consequence of the inconsistencies in the pattern, their curvature rates and inflation sizes are obtained under the elaborated manipulation. The air pressure in the module decides the volume of the morphology of a specific shape and the bending orientation depends on the main direction of the actuating material. Meanwhile, the combination of multiple samples created flexible abundant shapes of space, and it is also testimony to the necessity of the initial design of variable module patterns. As one part of the pneumatic interaction system, SPRS combines human-capturing censoring and developed through space monitoring manifestation (Fig. 6).

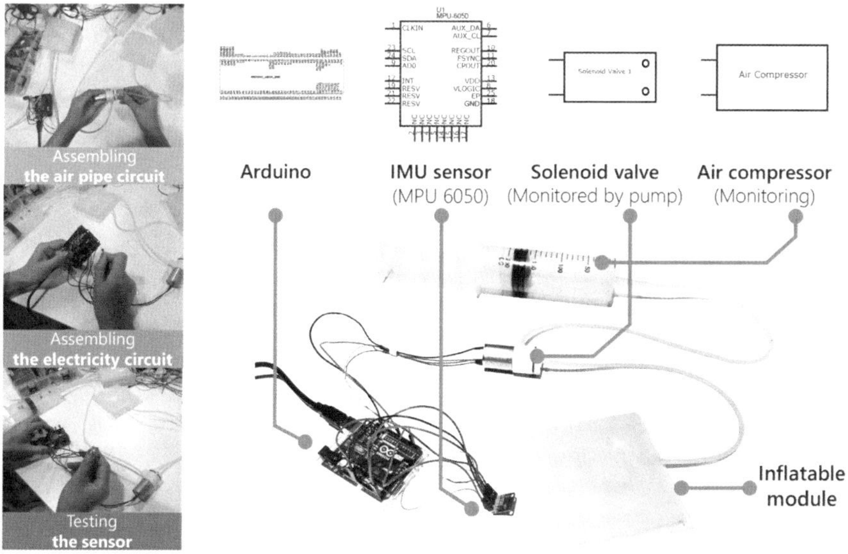

Fig. 6. System assemblage

4 Results as a Synergistic System

All the material, interaction, and fabrication studies are integrated into an interactive synergistic control system, as can be seen in the following figure. The control system SPRS includes two parts: The electrical circuit composed of Arduino, IMU sensor, solenoid valve, and relay, and the air pipeline consisting of the solenoid valve, air compressor, inflation module, etc. Fig. 7 shows the structural results obtained from it. It performs a robust actuation in desired features by the digital control of human interaction and the SPRS (Fig. 7).

The samples are dimension-adaptive for the common functions and scenarios, providing the precise exploration of the pattern design of the modules (Fig. 8). A quantitative analysis was applied to determine the swelling behavior based on the maximum curvature of the calculated levels within the considered position and scale (Table 1). Based on this approach, the prediction for the average uncertainty of the model in this study slightly exceeds the acceptability limit defined by the previous research. Nevertheless, these results suggest that data obtained using SPRS to simulate material inflation and space construction can provide more information for assessing the impact of performative strategies than that of the traditional setting of collaging configuration (Fig.8).

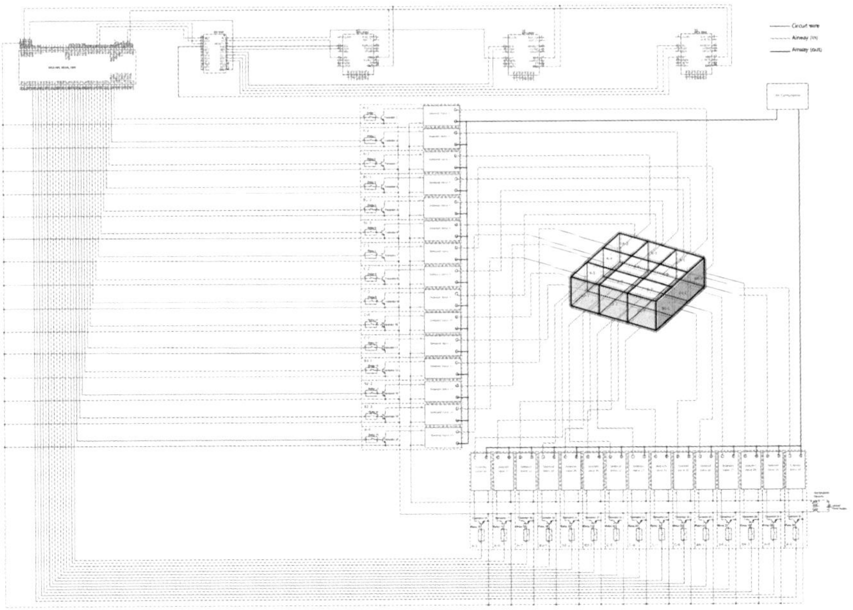

Fig. 7. The control system (SPRS)

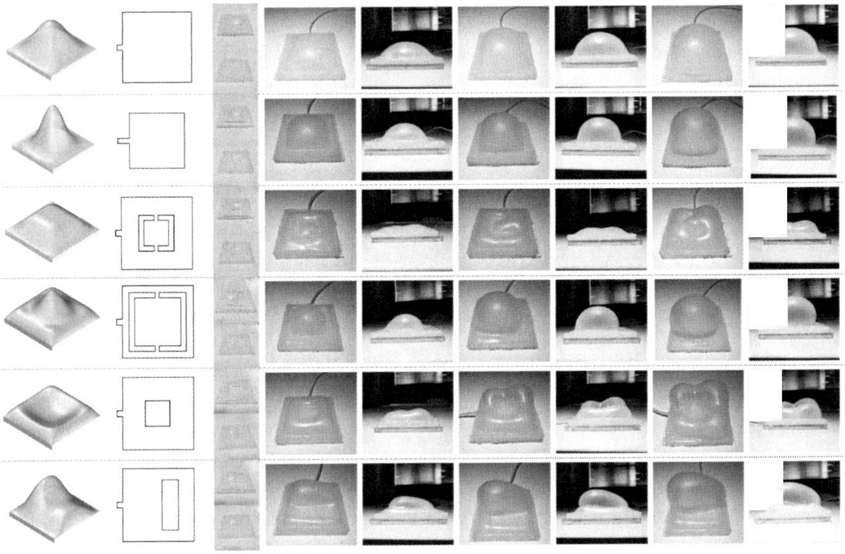

Fig. 8. Sample swelling testing and morphology detection

Table 1. Module information analysis

Module	Pattern	Pressure I		Pressure II		Pressure III	
		Altitude	Area	Altitude	Area	Altitude	Area
No. 1	None	0.3	0.8	0.4	0.9	0.5	1
No. 2	Central	0.3	0.6	0.45	0.7	0.6	0.8
No. 3	Average	0.1	0.8	0.2	0.9	0.3	1
No. 4	Average central	0.3	0.7	0.45	0.8	0.6	0.9
No. 5	Periphery	0.2	0.9	0.3	0.95	0.4	1
No. 6	One-side	0.3/0.1	0.5	0.4/0.2	0.55	0.5/0.2	0.6

Of note, it can be observed that the intermediate zone created by the arrangement and combination of the diverse samples is partly out of expectation. There are striking richnesses noted when each of the modules forms a synergic entity of the morphological space and structural system in the adaptive intelligent SPRS for physical interaction (Fig. 9).

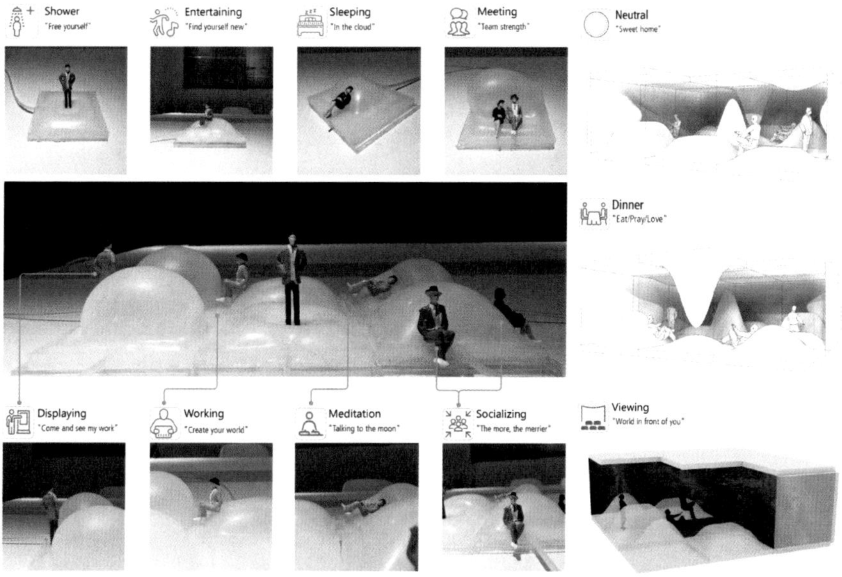

Fig. 9. Space generation and scenario presentation

5 Discussion and Future Work

In terms of constant change, impermanence, process-oriented, and interchangeability in improvisational and responsive architecture, the research work presented a scheme of technological realization and a systematic approach for the deformed architecture using cybernetic robotics. In this study, the testing was extended to the diverse surface variations among the materials. In order to create free-style and abundant shapes of space by developing the combination of multiple samples, the open-ended possibilities for a series of unexpected scenarios were included. Based on this approach, the prediction of the performance of the model allows the indeterminacy, and the dynamic equilibrium of the improvisation in the activity and scene adaptability through the synergistic behaviors of these samples is obtained. Thus these findings extend the pneumatic robotics architecture, confirming a more flexible, circumstantial, and biotic facet of architecture as an interactive environment. Most notably, the study investigates the soft robotics architectural system by integrating multi-faceted impact upon intelligent architecture in material, space, interaction, fabrication, and cybernetics. The structure with its inhabitants and the interplay between them form the entity of an evolving ecosystem mediated by each of them.

However, although the hypotheses were supported statistically, indeterminacy, the crux collectively pointed out by this work and the relevant research seem contradicted to the cybernetics mechanism. As Van Oyen [10] mentioned, while material agency denotes the possibility that things can act, the material objects have an effect on the course of action that is irreducible to direct human intervention Astrid Van Oyen [10]. Thus the counterbalance between the morphological manipulation of the cybernetic system and the autonomy and randomness of the space usage calls for discreet consideration in the human-system interaction.

Future work should include the following focus points. The research on the fabrication and performance of the soft modules on the architectural scale is to be furthered. And the specific performance strategy of the space interaction, which relied on cybernetics as a dynamic system including behavioral goals out of the realistic flow standards, should be explored. Furthermore, oriented on the equilibrium and inclusivity, the resilient control system based on feedback investigation, game theory, behavioral neurology, and brain-computer interaction can be introduced to study the long-term performative strategies. What needs to be retrospect is that, as a foothold for this study, even though Fun Palace represented an unprecedented architectural synthesis of technology, its birth was motivated socially, by the emancipation and empowerment of the individual. As Price had been quite explicit: "Fun Palace wasn't about technology. It was about people" Stanley Mathews [7].

References

1. Eero L, Juulia K (2018) Another generosity. Dezeen. https://www.dezeen.com/2018/05/25/nordic-pavilion-venice-architecture-biennale-inflatable/. Accessed 20 Mar 2022
2. Hertzberger H, Ghäit L, Rike I (2005) Lessons for students in architecture. 010 Publishers, Rotterdam, pp.92–121.

3. Jencks, C., & Kropf, K. (2006). Theories and manifestoes of contemporary architecture. Chichester, England: Wiley-Academy.
4. Majidi C (2013) Soft robotics: a perspective—current trends and prospects for the future. Soft Rob 1(1):5–11. https://doi.org/10.1089/soro.2013.000
5. Mangion F, Zhang B (2014) Furl: soft pneumatic Pavilion. Interactive Architecture Lab. http://www.interactivearchitecture.org/lab-projects/furl-soft-pneumatic-pavilion. Accessed 10 Mar 2022
6. Mathews, S. (2006). The Fun Palace as Virtual Architecture. Journal of Architectural Education, 59, 39–48
7. Mathews, S. (2005). The Fun Palace: Cedric Price's experiment in architecture and technology. Technoetic Arts, 3(2):73–92. https://doi.org/https://doi.org/10.1386/tear.3.2.73/1
8. Neeraj B (2013) Crazy-radical soft architecture, from the 1950s to today. Architizer. http://blog.fabric.ch/index.php?/archives/2013/12/16/C8.html. Accessed 20 Mar 2022
9. Negroponte N (1975) Soft architecture machines. MIT Press, Cambridge, MA. https://doi.org/10.7551/mitpress/6317.001.0001
10. Oyen, A. V. (2018). Material Agency. The Encyclopedia of Archaeological Sciences, 1–5. https://doi.org/10.1002/9781119188230.SASEAS0363
11. Pask G (1969) The architectural relevance of cybernetics. Archit Des 7(6):68–77

Organized references

12. Bhatia, N. (2013). Crazy-Radical Soft Architecture, From The 1950s To Today | #architecture #soft. fabric | ch. http://blog.fabric.ch/index.php?/archives/2013/12/16/C8.html. Accessed 20 March 2022
13. Hertzberger, H. (2005). Lessons for students in architecture (Vol. 1). 010 Publishers.
14. Jencks, C., & Karl, K. (2006). Theories and manifestoes of contemporary architecture (2nd ed.). Wiley-Academy.
15. Lundén, E., & Kauste, J. (2018). Another Generosity. 2018 International Architecture Exhibition, Venice. https://www.lunden.co/research/another-generosity/. Accessed 20 March 2022
16. Majidi, C. (2013). Soft Robotics: A Perspective—Current Trends and Prospects for the Future. *Soft Robotics, 1*(1):5–11. https://doi.org/10.1089/soro.2013.0001
17. Mangion, F., & Zhang, B. (2014). Furl: Soft Pneumatic Pavilion. Interactive Architecture Lab, London. http://www.interactivearchitecture.org/lab-projects/furl-soft-pneumatic-pavilion. Accessed 10 March 2022
18. Mathews, S. (2005). The Fun Palace: Cedric Price's experiment in architecture and technology. *Technoetic Arts, 3*(2), 73–92. https://doi.org/10.1386/tear.3.2.73/1
19. Mathews, S. (2006). The Fun Palace as Virtual Architecture. *Journal of Architectural Education, 59*(3), 39–48. https://doi.org/10.1111/j.1531-314X.2006.00032.x
20. Negroponte, N. (1976). Soft Architecture Machines. The MIT Press. https://doi.org/10.7551/mitpress/6317.001.0001
21. Pask, G. (1969). The Architectural Relevance of Cybernetics. *Architectural Design, Computational Design Thinking 7*(6):68–77
22. Van Oyen, A. (2018). Material Agency. In The Encyclopedia of Archaeological Sciences (pp. 1–5). https://doi.org/https://doi.org/10.1002/9781119188230.saseas0363

Morphology of Free-Form Timber Structure Determination by LSTM Oriented by Robotic Fabrication

Yiping Meng[1(✉)], Yiming Sun[2], and Wen-Shao Chang[1]

[1] School of Architecture, University of Sheffield, Sheffield, UK
ymeng16@sheffield.ac.uk

[2] Department of Automatic Control and System Engineering, University of Sheffield, Sheffield, UK

Abstract. Robotic arms are increasingly being used as an automation tool in non-standardized fabrication and construction, while the mechanical characteristics can also impact the accomplishment or the accuracy of the components. Timber is regularly used in different scales of a non-standard free-form structure fabricated by the robotic arm. The anisotropic mechanical characteristics of timber constrain the structural morphology. Developing a method of determining the morphology that meets the technical restrictions of the robotic arm and the material properties of timber is the aim of this research. In this paper, taking Centre Pompidou-Metz as a geometric case, glue-laminated timber as the main construction material, LSTM is applied for predicting the shape of the element. The geometric data is transformed into the fabrication data to testify to the kinematic singularities. The limitation of the workspace is derived from the Monte-Carlo method based on the DH model of the robotic arm. The experimental results show that the proposed method is effective in predicting the curves that match the characteristics of timber materials and robotic fabrication constraints.

Keywords: Free-form structure · Timber · Morphology · Robotic fabrication

1 Introduction

The design for free-form timber structures fabricated/constructed by robotic automation is a complex and multi-disciplinary system. To clarify the research questions, different factors related to the research field need comprehensive considerations.

Robotic fabrication: Since the 1990s, industrial robots dominated robotics research, and the technical necessities determined areas of investigation for robotics [1]. Efficiency is highly valued to achieve production sustainability and economic growth in the manufacturing industry. Under the notion of "Industry 4.0" calling for a highly automated, autonomous, flexible system, robotic automation as a new technique has been gradually applied to construction in academics and industries.

Free-form timber structure: The current standard form of architecture could not meet the variety of demands of human Aesthetic needs. Non-standard and free-form

P. F. Yuan et al. (eds.), *Hybrid Intelligence*, Computational Design and Robotic Fabrication,
https://doi.org/10.1007/978-981-19-8637-6_40

architecture becomes more and more acceptable. Based on the attention paid to sustainable environmental design, timber is a perfect building material to meet the measurements for environmental construction efficiency. Freeform structures using timber as the main materials are driven by digital technologies in design methods and product fabrication for irregular geometries [2].

One question for robotic timber fabrication for free-form structure is how to consider the technical aspect and material properties throughout the whole design process. The conventional way of taking a robotic arm as a fabrication method is shown in Fig. 1a–c. The research aims to extract the appropriate features of geometry for the LSTM method and to develop a method to transform the predicted geometry data into robotic information for fabrication.

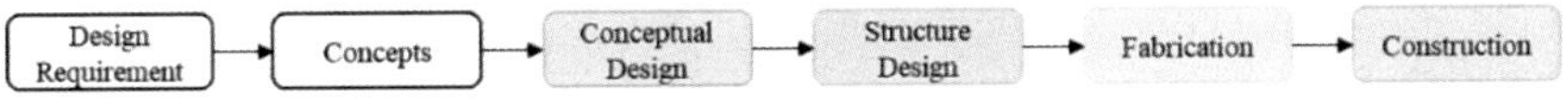

(a) Workflow for 'design-fabrication'

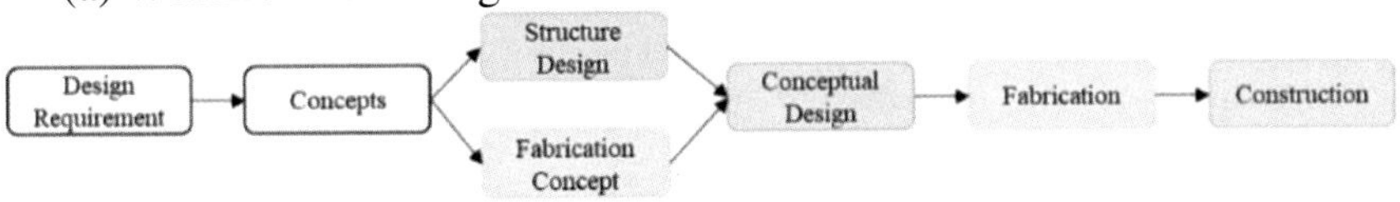

(b) 'Structure & Fabrication' oriented design after concepts

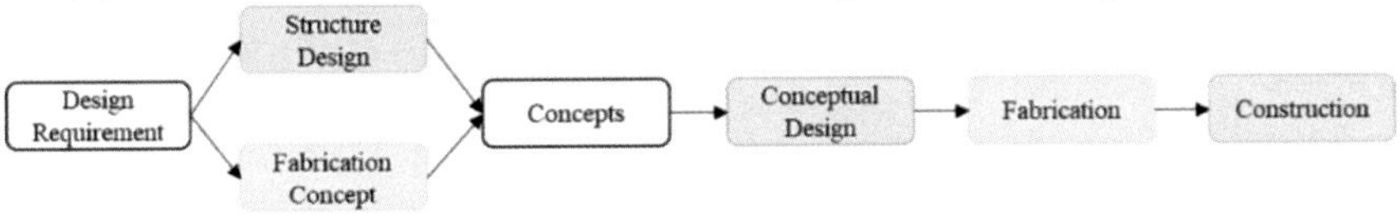

(c) 'Structure & Fabrication' oriented design before concepts

Fig. 1. Fabrication consideration in a different stage of the design process

2 Related Work

Digital fabrication technologies have enabled timber structures becoto become irregular and complex. Compared with Computational Numerical Control (CNC) technique, the mobility, and not high requirements for the working condition of robotic fabrication system are more flexible [3]. And this advantage conforms to the development trend–that is to take the design information as the input to produce construction automatedly [4]. Now, the robotic timber fabrication technique has not only been researched in laboratories but also applied in some large-scale construction applications [3, 5–7]. According to different forms of the components designed by different based-factors, different robotic fabrication types are needed such as cutting, milling, sewing, driand lling on different timer products from natural wood to engineering timber products [8]. More and more cases of using robotic fabrication automation have been demonstrated [9–12]. In summary, robotic fabrication can be applied throughout the whole design process in different phases from preparation to customised fabrication.

The rationality of free-form morphology means the design can be fabricated and constructed which is a complex work for architects and engineers [13]. Machine learning (ML) is derived from statistics, and the quality of an ML system is determined by the low error rate of prediction or classification [14]. ML can generate data which can be applied in generative design work using technique like deep neural networks (DNNs) [15]. One of the DNN model that has demonstrated the ability to be applied in geometry generation is the generative adversarial network (GAN) [16]. The model built through GAN can learn from the existing 2D images and transfer the empirical data into the generative design in 2D form [17]. 2D application of GAN is only limited to the 2D plan or façade generations. In ML area, many attempts have been made to generate 3D objects.

Other machine learning networks has potential to deal with 3D geometry. The data type of the current LSTM applications is in time-sequence, and the results prove the effectiveness and accuracy in predicting time sequential data like wind or air quality [18, 19]. As for the image or 2D data type which is not in sequence technically, one step to transform image into sequence data is needed additionally [20]. As for 3D geometric data, LSTM or other machine learning methods have not been widely used, especially in architectural design field.

In this paper, LSTM method is applied to predict the free-form surface morphology to improve the rationality of free-form structure considering material properties of timber and robotic fabrication. The first step is to transform the geometric data into sequential data types, and the simulation environment for the robotic is set up. The experiment of applying LSTM model to predict the morphology of free-form curves would be taken in to testify the feasibility of the transformed data to be applied in LSTM. After the prediction experiment, the methods of testifying for the singularities and the limitation of the robotic fabrication would be operated. The results of the experiments and the method would be discussed in the discussion and conclusion part.

3 Methodology

3.1 Workflow

To improve the rationality of the free-form morphology design, the impact of characteristics (6 Degree-of-freedom) and constraints (the type of fabrication, the dimensions of working space) of robotics as a technique on architectural geometry design are considered. To generate the initial geometry of free-form timber morphology, this research would discuss using LSTM machine learning to fulfil the constraints of timber material and robotic fabrication in conceptual design.

Based on the morphological design requirements of free-form timber structures, the pathway of machine learning for predicting the curve is developed as follows, shown in Fig. 2:

1. Choose the appropriate input and output.
2. Transform the input and output into a training set (in numbers or figures).
3. Select the training method.
4. Test the training accuracy.

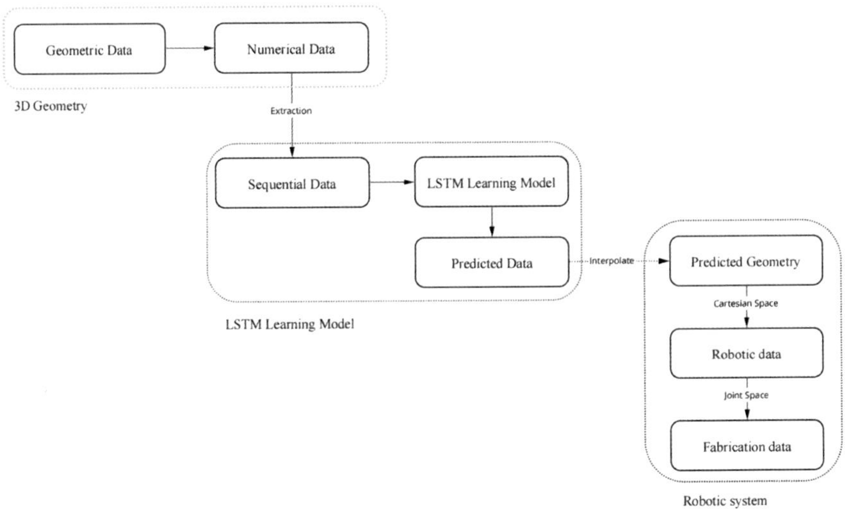

Fig. 2. Workflow of the LSTM prediction and robotic testing

3.2 Data Transformation

To complete the prediction learning task, the appropriate free-form model mattes in the learning task. This research takes the Center Pompidou-Metz Model as a case to extract the data for LSTM (Fig. 3).

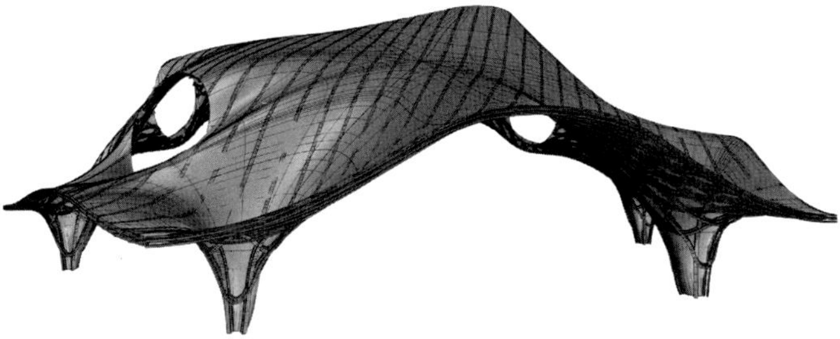

Fig. 3. 3D model of the case

As LSTM works well for the sequential data, the geometric data of the model would be transformed. The main difficulty is the geometric design is stored in a three-dimension form (like 3 dm, obj) while machine learning deals with numbers. If the model is presented in the figures from perspective views, there would be a loss of geometric information. The idea of data transformation is to find the proper way to store geometric information that fits the LSTM method and could be exported to the robot arm to generate fabrication commands (Fig. 4).

For the data transformation in this condition, assuming the number of curves to be analysed in N, every curve has been divided into $(M - 1)$ parts evenly. There are eight

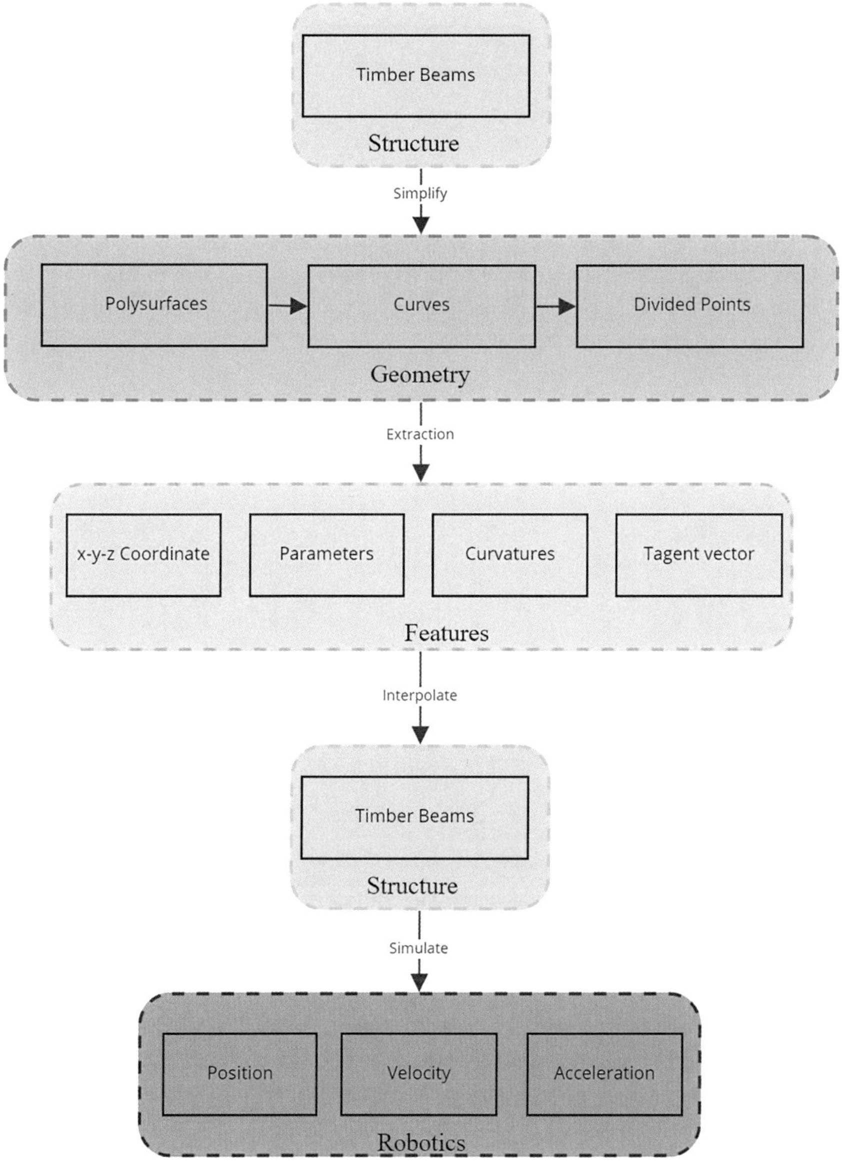

Fig. 4. Data transformation

parameters to describe this curve, the position of the division-point P(x, y, z), curvature K, the position of the point on the curve t_1, the tangent of the points $T(a, b, c)$. Every curve can be described by a matrix, which is $Q_{m\times 8}$.

The detailed of matrix Q is shown as:

$$Q_{M \times 8} = \begin{bmatrix} x_1 & y_1 & z_1 & t_1 & K_1 & a_1 & b_1 & c_1 \\ x_2 & y_2 & z_2 & t_2 & K_2 & a_2 & b_2 & c_2 \\ \cdots & \cdots & \cdots & \cdots & \cdots & \cdots & \cdots & \cdots \\ x_M & y_M & z_M & t_M & K_M & a_M & b_M & c_M \end{bmatrix} \tag{1}$$

According to the features of the discrete numbers extracted from the curves of the timber columns and beams, the LSTM training model is selected to predict the six variables of every curve for the best result. The workflow is shown in Fig. 5. After the LSTM learning network, the predicted parameters can interpolate the predicted curve which can be compared with the test one.

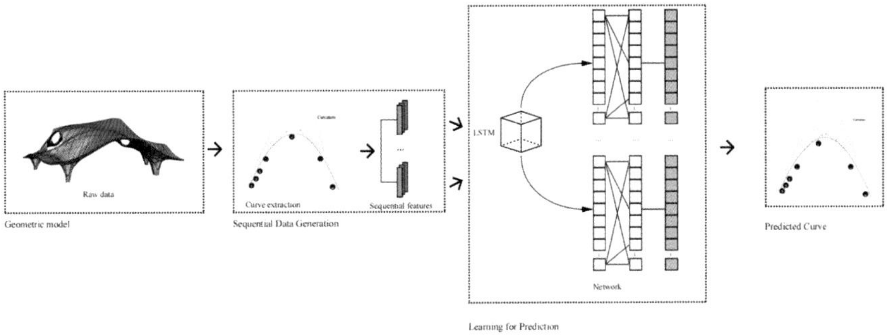

Fig. 5. LSTM learning process

3.3 Robotic Setup

To operate the technical analysis of robotic in both Cartesian space and joint space which is unique for robotic system, DH method is applied to build the model of robotic arm. The x–y-z coordinate of the components can connect the geometry model of the component with robotic arm.

In robot motion control, there is a corresponding matrix mapping between the joint velocity and the corresponding end-effector velocity and angular velocity as in the correspondence of the previous section, and this mapping reflecting the interrelationship between joint velocity and end velocity is known in robotics as the "Jacobi matrix ". It is expressed as follows,

$$V_e = J[q] \cdot \dot{q} \tag{2}$$

Whether $|J[q]| \neq 0$ is the way to testify the singularities.

4 Experiment

4.1 Training

Four timber beams from the geometric model are selected, and 16 corresponding boundary curves of the 4 beams are extracted to generate the division points. In the prediction for 21 divided-points, the geometric information of 15 curves are selected as training data and one curve is set as the test data and the training process is shown in Fig. 6. Figures 7 and 8 present the prediction error compared with the test data. To further test the prediction accuracy, the predicted tangent vector and the corresponding curvatures of the divided points are applied to fit the curve. The angles between the predicted tangent vector and the original one are shown in Fig. 9.

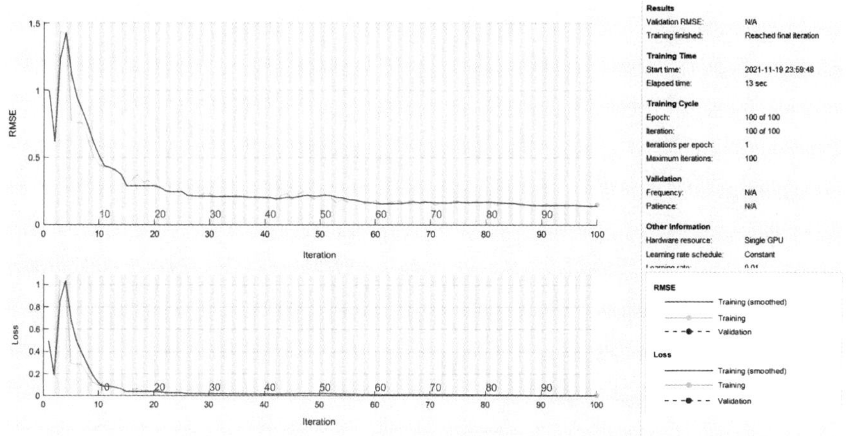

Fig. 6. Training results with 21 points

4.2 Robotic Simulation

Based on the robotic working space cell, the limitation including the obstacles and the range or the movement can be visualised as shown in Fig. 10.

The determination of whether the toolpath of the fabrication of the free-form components satisfies the constraints of the robotic arm is shown in Fig. 11.

5 Discussion and Conclusion

The results of the prediction experiment shows that the transformed 3D geometric data in sequential form fits for the LSTM to operate the prediction learning task. The converged training results illustrates the feasibility of LSTM in predicting the morphology of free-form. By comparing the test and predict datasets, the predicted vectors match the test data sets while the predicted curvature is more oscillatory and deviates to some extent from the test ones. By transforming the predicted curve into 3D model, the predicted geometry can be compared with the test model directly in 3D environment. Figure 9 shows the deviation between the predicted vectors and original vectors where color blue stands for the predicted curve. Based on the predicted curve model, the curve is extended into timber beam which would be fabricated by robotic arm. The robotic work cell case

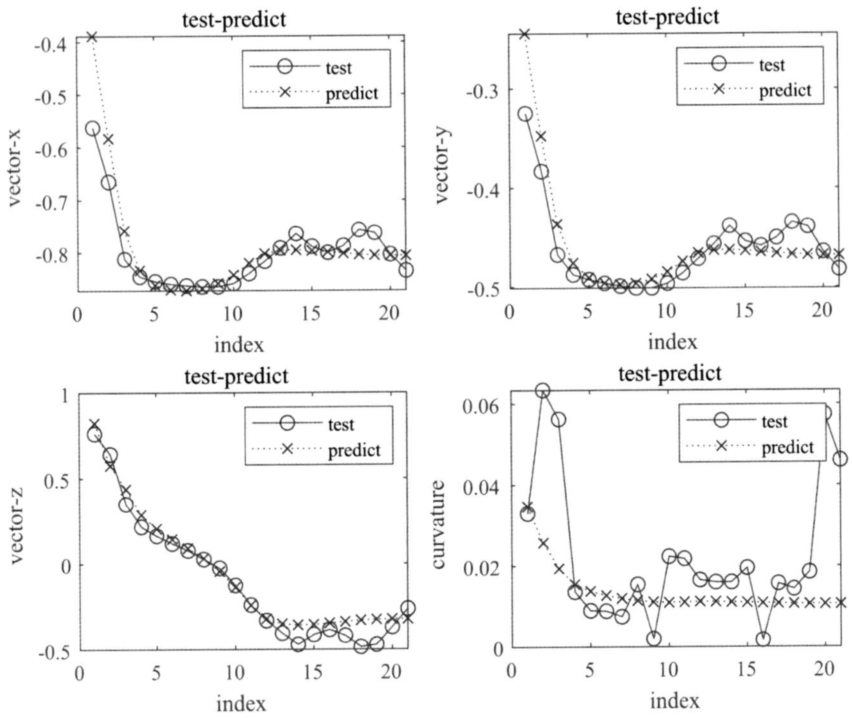

Fig. 7. Test results with 21 points

presents the workflow of transforming the predicted timber component information into the fabrication data which can be turned into robotic commands.

In conclusion, this research first proposes one method to extract the geometric features to describe the free-form curve which could be transformed into the sequential data for LSTM prediction learning. The experiments demonstrate the workflow of taking LSTM to predict the curve with the curvatures that meet the restrictions of timber properties and the results proves the effectiveness of LSTM taking $\{x, y, z, t, K, a, b, c\}$ as sequential features. When applying the robotic arm to fabricate the structure component which is transformed from the predicted free-form curve, DH method is applied to build the model of robotic arm in Matlab which a process to connect the geometric information in Rhino to the robotic simulation and analysis in Matlab. The working space limit can be computed by Monte Carlo method and the singularities of the robotic arm are derived based on the Jacob matrix to testify the tool path of the predicted free-form structure components.

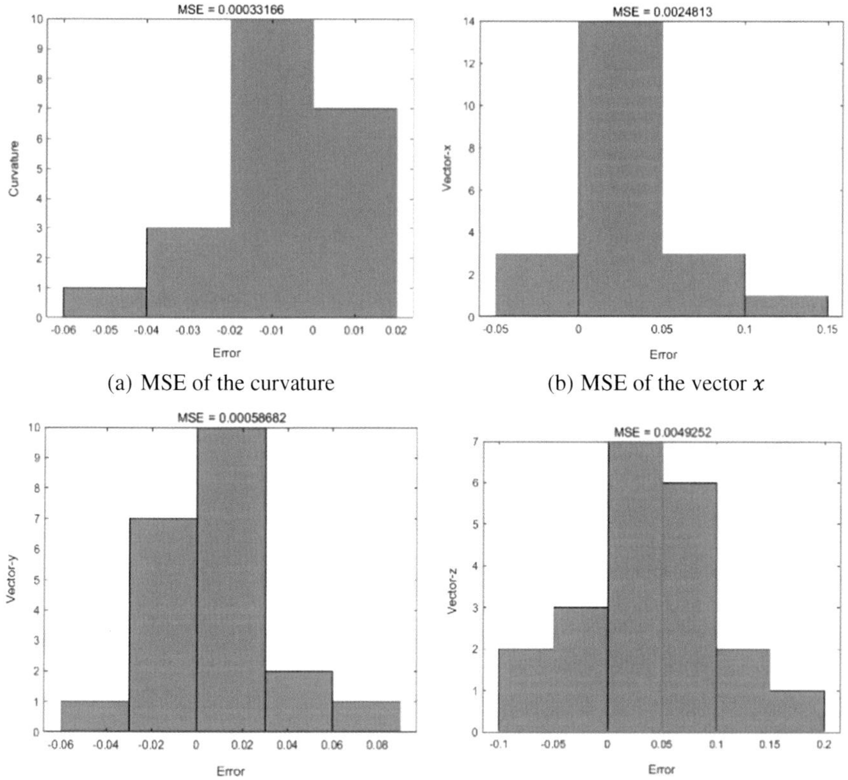

(a) MSE of the curvature

(b) MSE of the vector x

(c) MSE of the vector y

(d) MSE of the vector z

Fig. 8. Histogram of the MSE of the prediction

Fig. 9. Comparison of the generated curve and the tested curve

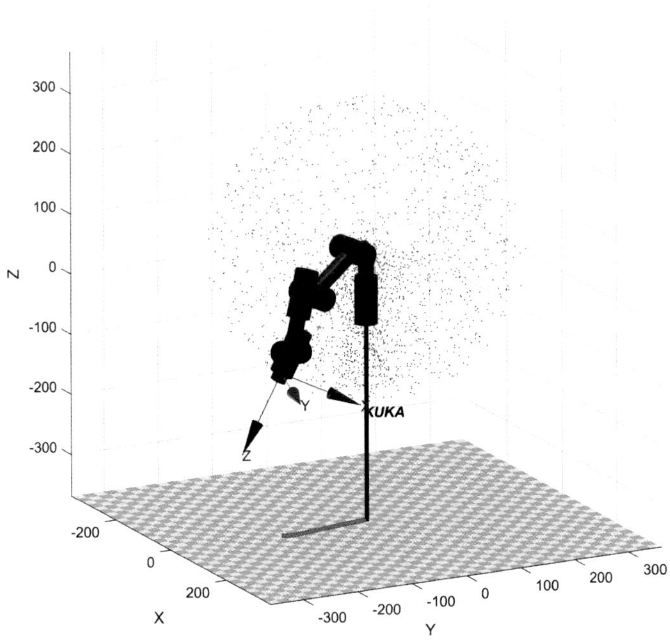

Fig. 10. Robotic working limits

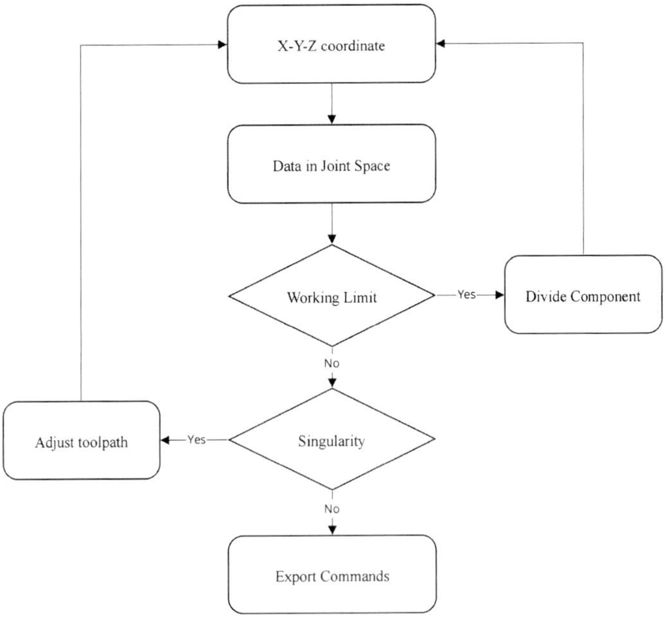

Fig. 11. Workflow of the robotic testification

References

1. Garcia E, Jimenez MA, De Santos PG, Armada M (2007) The evolution of robotics research. IEEE Robot Automat Magaz 14(1):90–103
2. Monier V, Bignon JC, Duchanois G (eds) (2013) Use of irregular wood components to design non-standard structures. Trans Tech Publication
3. Willmann J, Knauss M, Bonwetsch T, Apolinarska AA, Gramazio F, Kohler M (2016) Robotic timber construction—expanding additive fabrication to new dimensions. Autom Constr 61:16–23
4. Bock T (2015) The future of construction automation: technological disruption and the upcoming ubiquity of robotics. Automat Constr 59:113–121
5. Menges A, Schwinn T, Krieg OD (2016) Advancing wood architecture. Taylor & Francis
6. Vercruysse E, Mollica Z, Devadass P (eds) (2018) Altered behaviour: the performative nature of manufacture Chainsaw Choreographies+Bandsaw Manoeuvres. Springer
7. Williams N, Cherrey J (2016) Crafting robustness: rapidly fabricating ruled surface acoustic panels. In: Robotic fabrication in architecture, art and design 2016. Springer, pp 294–303
8. Menges A, Schwinn T, Krieg OD (2016) Advancing wood architecture: a computational approach. Routledge
9. Brell-Cokcan S, Braumann J (2013) Robl Arch 2012: robotic fabrication in architecture, art and design. Springer Science & Business Media
10. Reinhardt D, Saunders R, Burry J (2016) Robotic fabrication in architecture, art and design. Springer
11. Willette A, Brell-Cokcan S, Braumann J (2014) Robotic fabrication in architecture, art and design 2014. Springer
12. Willmann J, Block P, Hutter M, Byrne K, Schork T (2018) Robotic fabrication in architecture, art and design 2018: foreword by Sigrid Brell-Cokcan and Johannes Braumann, association for robots in architecture. Springer
13. Menna C, Mata-Falcón J, Bos FP, Vantyghem G, Ferrara L, Asprone D et al (2020) Opportunities and challenges for structural engineering of digitally fabricated concrete. Cem Concr Res 133:106079
14. Hastie T, Tibshirani R, Friedman J (2009) The elements of statistical learning: data mining, inference, and prediction. Springer Science & Business Media
15. Larochelle H, Bengio Y, Louradour J, Lamblin P (2009) Exploring strategies for training deep neural networks. J Mach Learn Res 10(1):1532–4435
16. Goodfellow IJ, Pouget-Abadie J, Mirza M, Xu B, Warde-Farley D, Ozair S, et al (2014) Generative adversarial networks. arXiv preprint arXiv:14062661
17. Oh S, Jung Y, Kim S, Lee I, Kang N (2019) Deep generative design: Integration of topology optimization and generative models. J Mech Design 141(11):1050–0472.
18. Han S, Qiao Y-H, Yan J, Liu Y-Q, Li L, Wang Z (2019) Mid-to-long term wind and photovoltaic power generation prediction based on copula function and long short term memory network. Appl Energy 239:181–191
19. Qin Y, Li K, Liang Z, Lee B, Zhang F, Gu Y et al (2019) Hybrid forecasting model based on long short term memory network and deep learning neural network for wind signal. Appl Energy 236:262–272
20. Xie K, Wen Y (eds) (2010) LSTM-MA: A LSTM method with multi-modality and adjacency constraint for brain image segmentation. IEEE

Performative Ornament: Enhancing Humidity and Light Levels for Plants in Multispecies Design

Andrea Macruz[1](✉), Mirko Daneluzzo[2], and Hind Tawakul[2]

[1] Tongji University, Shanghai, China
andrea.macruz@uol.com.br
[2] Dubai Institute of Design and Innovation, Dubai, UAE

Abstract. The paper shifts the design conversation from a human-centered design methodology to a posthuman design, considering human and nonhuman actors. It asks how designers can incorporate a multispecies approach to creating greater intelligence and performance projects. To illustrate this, we describe a project of "ornaments" for plants, culminating from a course in an academic setting. The project methodology starts with "Thing Ethnography" analyzing the movement of a water bottle inside a house and its interaction with different objects. The relationship between water and plant was chosen to be further developed, considering water as a material to increase environmental humidity for the plant and brightness through light reflectance and refraction. 3D printed biomimetic structures as supports for water droplets were designed according to their performance and placed in different arrangements around the plant itself. Humidity levels and illuminance of the structures were measured. Ultimately, this created a new approach for working with plants and mass customization. The paper discusses the resultant evidence-based design and environmental values.

Keywords: Posthumanism · Nonhuman · 3D printing · Thing ethnography · Multispecies design

1 Introduction

The human-centered paradigm has dominated design for over three decades. Designers are currently being challenged to work on complex socio-technical systems due to the ramifications of technology and environmental transitions [1]. The paper shifts the design conversation from a human-centered design methodology [2] to a posthuman design [1], considering human and nonhuman actors. It asks how designers can incorporate multispecies [3, 4] and "design for all-life" [5] approaches to create projects with greater intelligence and performance. The study includes research on biomimicry strategies to investigate how ergonomics operate when designing for nonhuman bodies and better design performance.

The paper provides a posthuman reading of the notion of "user", moving beyond anthropocentrism [6] towards biocentrism [7]. It examines examples of emerging design

P. F. Yuan et al. (eds.), *Hybrid Intelligence*, Computational Design and Robotic Fabrication,
https://doi.org/10.1007/978-981-19-8637-6_41

techniques that stress the rising discourse of posthumanism [8] and suggests that non-human lifeforms should be included as partners in design research, either as informants and co-designers or as users. To illustrate this, it describes a project that designed a set of "ornaments" for plants, which culminated from a course in an academic setting. As a novel contribution to the field, the work extends traditional understandings of the "user" to nonhumans and questions the field of action for the contemporary design practitioner.

The project methodology starts with "Thing Ethnography" [9] analyzing the movement of a water bottle inside a house and its interaction with different objects. The relationship between the water and plant was chosen to be further developed, considering that the lack of humidity and light in internal environments affects plants and that water can be used as a material to create a wetter microclimate and enhance brightness through reflection and refraction of light. The plant could be relocated to another space under different conditions, but sometimes, a plant is wanted in a specific place, regarding the functionality of an internal environment. So, water was introduced into the project to amplify light using 3D printed biomimetic structures as supports for water droplets. These design decisions became beneficial and evolved according to their performance. The structures can be combined and arranged in different ways around the plant as architectural elements staged around the individual plant specimen, as long as they don't block the sunlight of the leaves. These arrangements were important because they allowed for an evidence-based design tuned for maximum performance and plant health. Humidity levels and illuminance of the structures were measured using an in-door hygrometer, Inkbird ITH-20, and a digital lux meter HoldPeak 881D.

Ultimately, this created a new and novel approach for working with plants as nonhuman actors and recognizing the potential of "other-than-human" perceptual capabilities and mass customization. The results pointed us to a deeper understanding relative to nonhuman design, ergonomics, and design performance to gain insight into desirable nonhuman situations.

2 Posthuman Design and Multispecies Theory

The traditional dualistic systems of natural and artificial, human and animal, human and machine, are blurring because of technological advancements. They emphasize the new sorts of agency that nonhumans, whether environmental or technological, have in the world. A rising body of social theory has arisen around concepts that aim to make sense of this blurring of boundaries and introduce hybrid, non-dual, relational modes of thinking [3]. This paper focuses on one of these hybrid modes of thinking, particularly the posthuman, for a design practice that considerably increases our understanding of the many agencies, dependencies, entanglements, and relationships following current problems and questions facing the design field [1].

The human-centered design methodology is a practice where designers focus on human needs [2]. However, humankind is just one of many variables that shape our environments, and its agency is constituted through relational processes rather than predetermined. As design moves into the social sector and deals with issues in complex socio-technical systems, it's critical to change this approach to a broader one, such as a posthuman design, considering human and nonhuman actors. The work being done in post-humanist and post-anthropocentric disciplines argue that we should start thinking of manufactured systems as networks that encompass a variety of living creatures and the agency structures that act in and around them to understand our planet's transitions better.

Also, design can function as a process-oriented critical instrument rather than being interpreted as an affirmative discipline. One area open to exploration concerns the design of re-interpreting man's relationships with other species and the environment, proposing future multispecies cohabitation scenarios. Exploring multispecies worlds to learn about environmental challenges requires designers to become familiar with alternative methodologies, intended to be more than just collections of procedures but also distinct ways of behaving, thinking, and experiencing [10].

The word "multispecies" first appeared in the fields of biology and ecological sciences more than a decade ago to describe patterns of co-construction of environmental niches and wildlife management [3]. Its recent arrival into the discipline of anthropology has aided in the development of new interpretations of the notion. The concept of multispecies provides a starting point for reconsidering the role of nonhuman participants in the design and related processes. Also, it extends the concept of sustainability because examining environmental challenges from nonhuman perspectives could lead to different outcomes rather than those envisioned through technocentric methods [10].

This paper describes a project that proposes a speculative design that engages human and vegetal perspectives to create a multispecies reading of the concept of light quality, reflectance, refraction, and humidity in interior spaces.

3 Theoretical Underpinnings for Methodology

3.1 "Thing Ethnography"—Water

Ethnographic research is a design approach that uses observation and interview techniques to learn about the people who will use the designed objects. The fundamental concept is that a detailed understanding of people's lives, habits, motivations, and challenges can lead to better products or, at the very least, more relevant design proposals. Ethnographers and anthropologists spend extended amounts of time engaged in the people they study, enjoying daily life with them. They observe, photograph, record, take notes, interview, etc. and generate hypotheses, texts, and publications based on these activities to establish their academic credibility [11]. Designers are exploring this knowledge and ethnographic tools to come up with new ideas, develop concepts, and put them into action [12].

However, beyond standard ethnographic research, this project was based on "Thing Ethnography" based on Elisa Giaccardi and other authors' study with nonhumans [9]. They state that we shape objects as humans, and objects also shape us. Recognizing this continual interplay between people and objects necessitates anthropological and design methods that allow both parties an equal voice. They also affirm that things that inhabit our houses demonstrate different degrees of dynamism and emplacement. In our project, "Thing Ethnography" was developed by analyzing the movement of a water bottle inside a house and its interaction with different objects (Fig. 1). Water was chosen because it is necessary for humans' and nonhumans' survival. Also, a water bottle is one of the most dynamic objects in our homes as it moves through space, occupying multiple ecosystems.

According to "Thing Ethnography", paying attention to movement reveals how objects interact: they live in communities of things and behaviors that are sometimes distinct and sometimes overlapping. This knowledge raises concerns about the habitats in which they operate. Our research wasn't equipped with software or sensors differently from Giaccardi's one; ours was a simple analysis of the movement of a water bottle throughout the day. However, it was interesting to understand which objects a water bottle interacts with day-to-day and could potentially develop a better relationship. In this case, the one between a water bottle and the plant. So, the project's starting point was the following question: how could the relationship between water and plant be enhanced?

This research also tackles the concept of co-performance. According to Giaccardi and Kuijer [13], co-performance views artifacts as capable of learning and performing alongside humans. Co-performance recognizes the dynamic contrasts in capacities between humans and objects from this perspective and emphasizes the inherently cyclical relationship between professional design and use. In our proposed scenario, we decided to focus on the capacity of artifacts to perform alongside humans. Both plants and humans establish a mutualistic relationship within a win–win scenario. Nonhumans and humans need water and light to survive, so when designing to enhance one's performance, the other will automatically increase. The idea is that when there is co-performance, both species can thrive and generate greater cooperation between humans, nonhumans, and the earth. Lovelock extends this concept by posing the earth as a self-regulating entity [14].

Fig. 1. "Thing Ethnography": the study of the movement of a water bottle inside a house and its interaction with different objects.

3.2 Biomimicry Strategies

With the analysis of the plant and water relationship, the "Thing Ethnography" led to the initial idea of designing 3D printed structures embedded with water that would create a wetter microclimate for the plant and enhance the light in internal environments through reflection and refraction. This design necessity was observed because of discoloration on certain leaves due to a lack of humidity and light in specific situations. Two biomimicry strategies were used to support the design of the structures in this project.

The first one was related to how nature optimizes sunlight. There is a silvery-whitish-gray shrub native to deserts of the southwestern United States called The Desert Holly, scientific name *Atriplex hymenelytra;* its strategy to deal with sunlight in the desert is very effective. It needs sunlight to do photosynthesis, and mainly it gathers different sunlight rays, such as the morning and the evening sun, due to its fragmented and tilted in high angles leaves. So, the plant takes advantage of the sunlight that is not so strong during these periods, but at the same time, it has a silver-colored feature to help reflect sunlight when it is too strong [15].

The second one was related to how nature collects and retains water. The Namib desert is home to Darkling beetles, and to survive, some species of these beetles include unique points and bumps in their wing scales that aid in water collection. Water droplets occur as air condenses on the tips, which run off the bumps and into the beetle's mouth [16]. Warka Tower, a biomimetic project inspired by the beetle, similarly gathers atmospheric water vapor from rain, fog, or dew which condenses on the polyester mesh's cool surface, generating droplets of liquid water that flow down into a reservoir at the structure's bottom [17].

These two survival strategies of the Desert Holly and Darkling beetles were applied to the shape and geometry of the structures for better performance (Fig. 2).

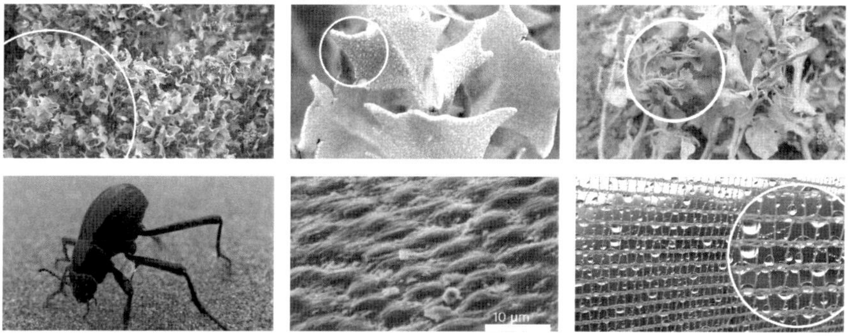

Fig. 2. Desert Holly, angulation of leaves; Darkling Beetle; scanning electron micrograph of the textured surface of the depressed areas on Stenocara surface; atmospheric water vapor condensed on the Warka Tower mesh's surface.

4 Methodology

The design of the structures started based on the different angles of the fragmented leaves of the Desert Holly as a strategy to capture more sunlight. One structure was composed of a hexagon with six triangles bent in opposite directions due to a "v" shape join design on the geometry. The connections between them were thought to be on the extensions of the edges of two triangles, implying different design typologies. The structures were 3D printed using Polylactic Acid (PLA). An object that would interact with the plant should be composed of natural and compostable materials just like the plant, in this case, PLA and water. Also, they should be lightweight to better adjust to the plant's ergonomics. They were printed as flat structures and then folded for efficient storage and transport, like origami (Fig. 3).

Fig. 3. First 3D printed structures: flat, bent, and a different typology showing the connection.

In parallel, textures were studied based on the Darkling beetle and Warka Tower to understand how water droplets could be stored within a mesh. However, the 3D printed structure would not harvest water from the air, but just store sprayed water differently from the examples above. The structures were used as supports to accommodate the water droplets to create a wetter microclimate for the plant and increase the light reflection and refraction. Since water is denser than air, light is refracted as it enters the drop, which can illuminate a larger space [18].

A parametric definition was generated using Rhinoceros and Grasshopper. Different meshes with 1, 2, and 3 layers of 3D printed PLA were tested, and the one that performed better was the two-layered one. The one-layered mesh was not enough to store a good amount of water droplets, and the three-layered one, although it could hold more water droplets, was not malleable and quite heavy (Fig. 4).

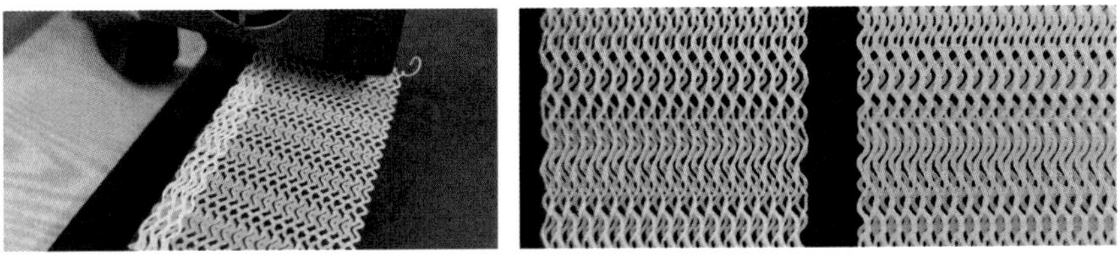

Fig. 4. Meshes with one, two, and three layers of 3D printed PLA.

Observing that the mesh had undulations on the outer part and that this could be used to join the structures together, it was decided that there weren't going to be different typologies but rather one structure that would allow multiple connections. An open-ended possibility of combinations was created using an interlocking system of a standard module.

After the tests above, the structure's design was simplified: the hexagon in the middle was substituted by triangles to increase the number of distinct angles and better capture different stages of sunlight. Other joints were created on the structure as part of the weaving strategy for the triangles to bend in different directions. Also, the undulations on its outer part were exaggerated to make the interlocking system more stable and expand the plant's grip-ability (Fig. 5).

Next, 25 structures were arranged around a plant in an indoor environment that didn't receive constant and direct sunlight to prevent the evaporation of the water embedded in the structures. They were combined and arranged in different ways around the plant itself, as an architectural element to be staged around the individual plant specimen and according to how much weight the plant could support in each part and culminating in different performative qualities. These arrangements were important because they allowed for an evidence-based design tuned for maximum performance and plant health.

It was measured the humidity level with an in-door hygrometer, Inkbird ITH-20, and the illuminance using a digital lux meter HoldPeak 881D.

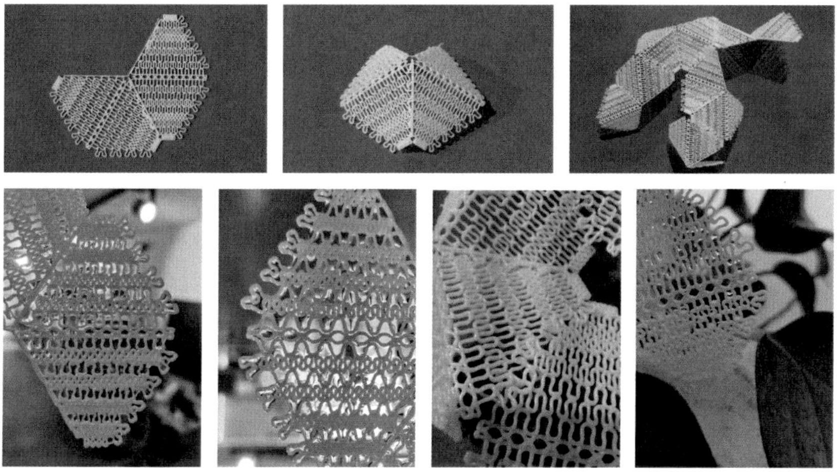

Fig. 5. Final structures: flat, bent, connected with others; structure with water; a close-up of the structures' interlocking system; and the structures arranged on a plant.

5 Results and Conclusions

The structures were tested by measuring the humidity level with a standard in-door hygrometer, Inkbird ITH-20. The initial humidity level was 39%, and after water was sprayed on them, the humidity level increased to 53%. When this last step was repeated with the structures arranged around a plant, the humidity level changed to 62% (Fig. 6). As indoor plants need humidity levels between 40 and 60% [19], the structures provided quite an effective change in humidity.

The illuminance of the structures was also measured using a digital lux meter HoldPeak 881D with a rotating head sensor head. The illuminance measured on the plant was 44.5 lx, and subsequently, on the structures without water arranged around the plant, it increased to 78.5 lx. So, it is possible to notice an increase of 34 lx just with the addition of the structures. However, this tool was not accurate enough to measure the difference with the water in the structures because water reflection (from 5 to 80%) depends on the angle of incident radiation. The reflection coefficient of water is higher when the radiation angle is low: during the sun rising or setting. At 90°, the water body absorbs a great deal of sunlight, and only a small portion of the sunlight is reflected [20]. Also, the refractive index of water is 1.3, while air is 1.0, and it is important to take into consideration [21]. Therefore, it was noticed that the light quality increased, but further investigations must be conducted to gather quantitative data to measure the enhancement of light reflection and refraction of water in the structures.

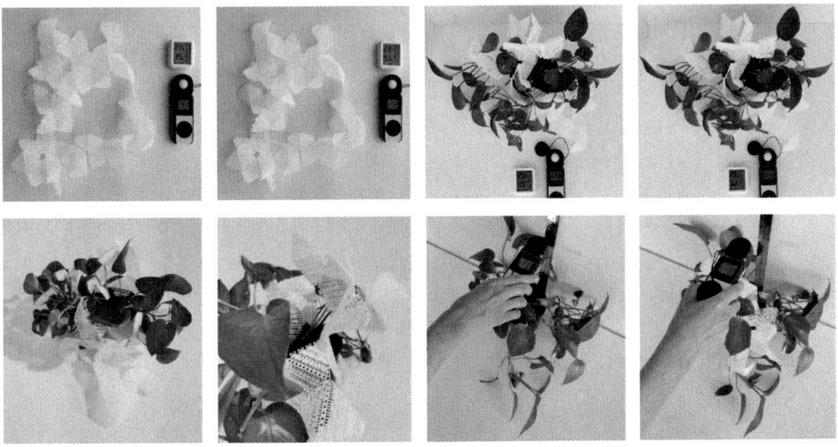

Fig. 6. Photos of the structures without water (39% of humidity) and with spayed water (53% of humidity); structures arranged around a plant without water (39% of humidity) and with sprayed water (62% of humidity); photos of the structures arranged around a plant; the amount of lux was measured just on a plant (44.5 lux) and subsequently on the structures arranged around the plant without water (78.5 lux).

This paper moves the design conversation away from human-centered design and toward posthumanism, which considers both people and nonhuman actors, usually overlooked in design processes. It begins with an interdisciplinary conversation that connects new research in nonhuman design, ergonomics, and design performance to acquire insight into desirable nonhuman settings.

This study details the development of a project of "ornaments" for plants as part of an academic course. It begins with a "Thing Ethnography" and further research on the lack of humidity and light in internal settings and water employment to increase humidity levels and enhance brightness through reflection and refraction of light. 3D printed biomimetic structures were designed and tested to support water droplets and then placed in various patterns around a plant. The findings reveal that the structures increase humidity and light levels, but more research is needed to evaluate the augmentation of light reflection and refraction of water. They also imply that considering nonhuman actors can lead to different design proposals and a mutualistic relationship scenario based on the concept of co-performance.

This paper expands our sense of inter-dependence with other species. It creates a new approach for working with plants as nonhuman actors and recognizing the potential of "other-than-human" perceptual capabilities and mass customization. The study explores the evidence-based design that resulted and the environmental ideals we want our society to embrace, with important implications for future design rules and practice.

References

1. Forlano L (2017) Posthumanism and design. She Ji: J Design Econ Innov https://doi.org/10.1016/j.sheji.2017.08.001
2. Norman D (2013) The design of everyday things: revised and expanded edition, 2nd edn. Basic Books, New York
3. De Ruiter P, Wolters V, Moore J (2005) Multispecies assemblages, ecosystem development and environmental change. In Dynamic food webs: multispecies assemblages, ecosystem development and environmental change, vol 3. Academic Press, Burlington
4. Kirksey E (2014) The multispecies salon, 1st edn. Duke University Press Books, USA
5. Boradkar P (2015) Design for all life. Is it time to re-examine human-centered design? Core77. https://www.core77.com/posts/31264. Accessed 13 Mar 2022
6. Crutzen PJ, Stoermer EF (2021) The 'Anthropocene' (2000). In: Benner S, Lax G, Crutzen PJ, Pöschl U (eds) Paul J. Crutzen and the anthropocene: a new epoch in earth's history. The Anthropocene: Politik—economics—society—science, vol 1. Springer. https://doi.org/10.1007/978-3-030-82202-6_2
7. Lanza R, Berman B (2010) Biocentrism, 1st edn. BenBella Books, Texas
8. Braidotti R (2013) The posthuman, 1st edn. Polity Press, Cambridge
9. Giaccardi E, Cila N, Speed C, Caldwell M (2016) "Thing ethnography": doing design research with non-humans. In: Proceedings of the 2016 ACM conference on designing interactive systems (DIS '16). pp 377–387. https://doi.org/10.1145/2901790.2901905
10. Gatto G, Mccardle J (2019) Multispecies design and ethnographic practice: following other-than-humans as a mode of exploring environmental issues. Sustainability 11(18):1–18. https://doi.org/10.3390/su11185032
11. Nova N, Léchot-Hirt L (2016) Beyond design ethnography: how designers practice ethnographic research, 1st edn. Provinces Press, Vancouver
12. Van Dijk G (2012) Design ethnography: taking inspiration from everyday life. In: Stickdorn M (ed) This is service design thinking: basics, tools, cases, 1st edn. Wiley, New Jersey
13. Kuijer L, Giaccardi E (2018) Co-performance: conceptualizing the role of artificial agency in the design of everyday life. In: CHI '18: Proceedings of the 2018 CHI conference on human factors in computing systems. https://doi.org/10.1145/3173574.3173699
14. Lovelock J (2016) Gaia: a new look at life on earth, Illustrated. Oxford University Press, United Kingdom
15. AskNature Nugget Ep. 3: Desert Holly (n.d) Available via BIOMIMICRY 3.8. https://www.youtube.com/watch?v=YVq9GXJzrrE. Accessed 13 Mar 2022
16. The Beetles That Drink Water from Air: Darkling beetles (n.d) Available via ASKNATURE. https://asknature.org/strategy/water-vapor-harvesting/. Accessed 13 Mar 2022
17. Lightweight Water Collection System Inspired by Darkling Beetles: Warka Water (n.d) ASKNATURE. https://asknature.org/innovation/lightweight-water-collection-system-inspired-by-darkling-beetles/. Accessed 13 Mar 2022
18. Bansod VR, Wandile AA (2015) Study on solar water bulb-a liter of light. IJIRST–Int J Innov Res Sci Technol 1(10):256–259. Available via ACADEMIA https://www.academia.edu/15897424. Accessed 13 Mar 2022

19. Jackie Carroll (n.d) Raising humidity. In: Gardening know how. https://www.gardeningknowhow.com/houseplants. Accessed 08 May 2022
20. Absorption/reflection of sunlight (n.d) Understanding global change. https://ugc.berkeley.edu/background-content/. Accessed 08 May 2022
21. Light and sound–reflection and refraction (n.d) BBC. https://www.bbc.co.uk/bitesize/guides/zdwnb9q/revision/4. Accessed 08 May 2022

Regression-Based Inductive Reconstruction of Shell Auxetic Structures

Tomás Vivanco[1](✉), Juan Eduardo Ojeda[2], and Philip F. Yuan[3]

[1] Pontifical Catholic University of Chile, Tongji University, Shanghai, China
tvivancol@uc.cl
[2] Pontifical Catholic University of Chile, Technical University Darmstadt, Darmstadt, Germany
jeojeda@uc.cl
[3] Tongji University, Shanghai, China
philipyuan007@tongji.edu.cn

Abstract. This article presents the design process for generating a shell-like structure from an activated bent auxetic surface through an inductive process based on applying deep learning algorithms to predict a numeric value of geometrical features. The process developed under the Material Intelligence Workflow applied to the development of (1) a computational simulation of the mechanical and physical behaviour of an activated auxetic surface, (2) the generation of a geometrical dataset composed of six geometric features with 3,000 values each, (3) the construction and training of a regression Deep Neuronal Network (DNN) model, (4) the prediction of the geometric feature of the auxetic surface's pattern distance, and (5) the reconstruction of a new shell based on the predicted value. This process consistently reduces the computational power and simulation time to produce digital prototypes by integrating AI-based algorithms into material computation design processes.

Keywords: Artificial intelligence · Material computation · Auxetic structures · Computational design

1 Introduction

The emergent applications of Artificial Intelligence algorithms in architectural and design practices opened a wide range of novel methods and processes to envision, create and optimize the design process, varying from the scale of urban design to the exploration of synthetic spaces. Naturally, data and the way how are organized plays a crucial role in the entire process, which can vary from datasets of images to numeric values. Nevertheless, applying AI-based algorithms to material computation requires a linear workflow to generate a specific dataset of values that can precisely represent a geometry. Requires that the design process be based on geometrical features that represent its physical characteristics.

P. F. Yuan et al. (eds.), *Hybrid Intelligence*, Computational Design and Robotic Fabrication,
https://doi.org/10.1007/978-981-19-8637-6_42

This research project utilizes a regression Deep Neuronal Network to predict the distances of each cell's patterns from an auxetic surface that was previously actively bent, based on its Gaussian curvature, osculating point, pattern distance and the applied force parameters.

Auxetics are metamaterial structures with a negative Poisson's ratio, in which the mechanical performance relies on the geometry rather than on the material itself. When stretched, they become thicker in the perpendicular direction to the applied force. This phenomenon occurs due to their internal structure and how this deforms when the sample is uniaxially loaded.

This makes them a material system with a wide range of applications on the architectural scale, for example, by reducing the amount of energy needed to create a three-dimensional shape [1], by distributing the internal mechanical forces of the system, achieving a relaxed and stable form.

Computing bent activated auxetic surfaces requires simulating each cell's behaviour that composes the surface and its global deformation, demanding a significant computational power. By creating and training a tailored regression DNN model, the simulation time and power can be reduced by just a fraction, offering the users an interactive tool to input the desired performance and receive a precise predicted pattern.

1.1 Auxetic Structures

Auxetic structures present the unique capacity of becoming wider when stretched and narrower when compressed [2]. The word auxetic comes from the Greek word αὐξητικός (auxetikos) which means "tends to increase". Some of its edges and vertices work under compression to give a material the capacity to extend, reducing one axis its length, thus giving space for the other edges and vertices to elongate and, consequently, make the system extend. This relationship between compression and traction forces is defined as the Poisson ratio (v), which is the ratio of the transverse contraction of a material to the strain in the direction of the stretching force [3]. A negative Poisson ratio occurs when compression forces are applied, and, in contrast, the Poisson ratio is positive when there is tensile deformation (Fig. 1). The Poisson ratio values could have a wider range of values for anisotropic materials than isotropic materials.

Bi-dimensional auxetic structures are results from the tessellation of a given plane with periodic regular polygons [4] working as an individual but concatenated cell, where its deformation magnitude and direction are directly related to the Poisson ratio. Because of this, auxetic structures materials should have a low density or be flexible. From this basic definition of the deformation ratio, auxetic surfaces have specific characteristics that contribute to a more refined understanding of their variabilities and geometric properties [5] as a material system:

- Shear properties: shear modulus can be similar to the bulk module, meaning that the structure becomes hard to break but easy to deform.
- Indentation resistance: Hardness can increase in an auxetic material due to its negative Poisson's ratio.
- Fracture toughness: because of the displacement for geometry, it possesses more crack resistance to fracture.

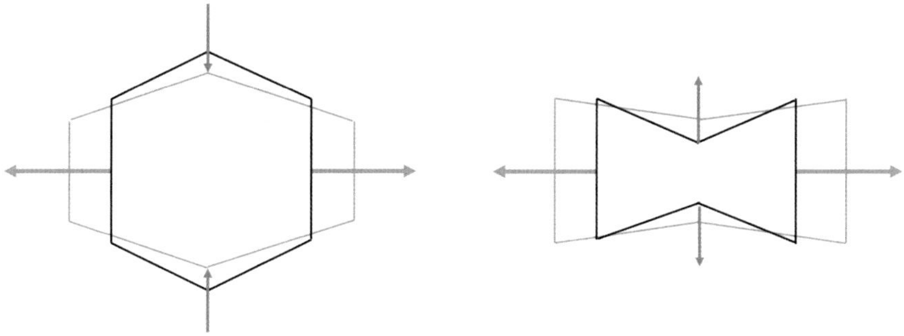

Fig. 1. Left: non-auxetic material with a positive Poisson's ratio under tensile stress. Right: auxetic material with a negative Poisson ratio under tensile stress. *Source* From the authors.

- Synclastic curvature: will form a synclastic curvature geometry by a natural distribution of its internal forces.
- Energy absorption: auxetic structures have a high capacity to absorb constant deformation loads in low frequencies.
- Variable permeability: The pore-opening properties offer a high filtering capacity on the micro and macro scale.

1.2 Generative Particle-Spring System for Material Behaviour Simulation

A particle-spring system consists of particles given mass and position, which are connected by springs with stiffness and rest length [6]. By applying an external force over the network of particles and springs, each particle moves, affecting the others producing a concatenated deformation because of the springs, distributing the embedded energy to find its equilibrium state.

Physics simulators engines for generative design software such as Kangaroo Physics are powerful tools to simulate and fast preview the physical deformation of geometries submitted to external forces over a predefined geometrical system. Offering the capacity to compute the synclastic geometries generated by the application of tension in opposite axis directions on auxetic structures. Particle-Spring system simulations have been successfully applied to test different auxetic properties and the influence on the Poisson ratio when the internal angle of the hexagons of an auxetic cell is changed [7].

Despite the good performance of particle-spring physics simulators to quickly simulate geometric deformations, this process is based on iterative methods. It requires computing several variations of the same model demanding high computational power to simulate complex models.

1.3 Artificial Intelligence for Material Prediction

The application of machine learning algorithms to find a specific solution to a given problem is not new. In the early 90's, shallow learning algorithms were used in the process of using inductive systems in knowledge acquisition [8] for the application of different civil engineering purposes, improving the understanding of a given domain through the systematic development of a system of decision rules governing that domain [9]. Problems that share the same domain are among the most common potential applications of trained Artificial Neuronal Network (ANN) models [10]. Offering one substantial difference from conventional iterative processes for searching for potential solutions. Because solutions emerge from local rules, exploring new outcomes relies on the generation of new global results [11].

These processes follow a common strategy: (1) the generation of a geometry-based dataset after optimization or physical simulation process; (2) architecture definition and training of an Artificial Neural Network (ANN); (3) prediction of the output value; (4) reconstruction of the global geometry.

Because Linear Regression (LR) models can only predict a specific value with a linear relationship between the features and the target, the ANN models require the target values to be continuous from an interval. Because Deep Learning algorithms work like the human brain neurons, it consists of an Input layer, Hidden layers and an Output layer, which can learn the complex relationship between the features and target due to the presence of activation function in each layer [12]. For this, a Forward propagation process for multiplying weights of each feature by adding them and a Backward propagation for updating the weights–requiring optimization and loss functions in the model- are enabled.

2 Research Objectives, Methods, and Processes

This research aims to build a clear workflow to predict an auxetic structure three-dimensional deformation pattern from a series of given properties as inputs to reconstruct the structure computationally. For this, a dataset generation is required from a base geometry with a defined set of rules after being subjected to a simulation of physical deformation activated with a vertical force in an active bending process. A Particle-Spring (PS) physics simulator is used to study the morphological modification of the deformed structure by measuring its geometrical features and then exported as a value dataset. Due to the high computational power required to simulate complex or large geometries, a trained ANN model with the dataset is used to predict, bypassing the physics simulator, and reducing computational power and time considerably. To achieve the research follows a workflow composed of five steps:

- Computational simulation of the mechanical and physical behavior of an activated auxetic surface.
- Parametric workflow for the generation of datasets of material features.
- The construction and training of a regression Deep Neuronal Network (DNN).
- Prediction of a specific feature of the geometry (output_Y) from given features values (input_X).
- Reconstruction of the material system geometry of the predicted structure based on feature inputs.

2.1 Generative Physics Simulation

A ten-by-ten cell of 1 by 1 unit auxetic structure geometry was used as basic geometry. The edges of each cell work like pivots that moves towards its centre point at a relative distance between its centre point and the global force point. This movement gives the auxetic properties to the structure. If a cell centre point is closer to the vector force, the displacement is bigger, increasing its flexibility; on the contrary, the greater the distance, the displacement value is lower increasing its stiffness (Fig. 2).

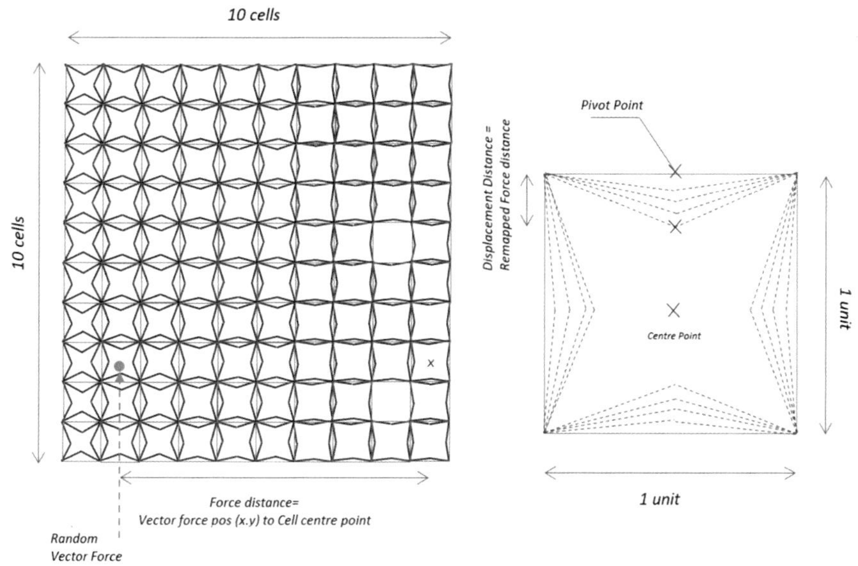

Fig. 2. Auxetic structure, cell, force, and pivot displacement. *Source* By the authors.

The auxetic structure was input as a rigid body in a PS physics simulator. Each of the four vertices of the structure served as anchor points. The vector force was the force that randomly changed its position and amplitude, generating different outcomes from the simulator.

After applying a horizontal force to the global surface, the pivot point of each cell was deformed and displaced, generating a vector pointing to the interior of each cell. Finally, vector normals were reoriented to each cell, and its length was used to rebuild the surface in two dimensions. That length becomes extremely important in this process; the surface can be rebuilt in two dimensions to generate a specific three-dimensional deformation and simplify the potential manufacturing process.

2.2 Dataset Generation

The simulated geometry in the PS simulator was analyzed to extract the values of each geometric feature (Fig. 4), the six global features [13]: the osculating point and the Gaussian curvature of the global deformation, the start and endpoints from the U and V coordinates, and the displacement distance of the pivot point (Fig. 3). Also, the local cell data was extracted: the centroid position in x, y and z, U and V nodes coordinates, and each pivot point displacement distance.

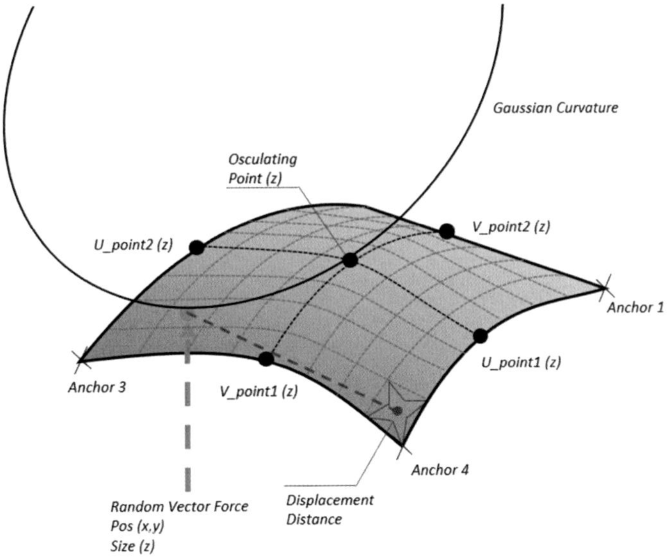

Fig. 3. Auxetic deformation features analysis for the generation of the dataset: osculating Point U and V coordinate start and endpoints and displacement distance. *Source* From the authors.

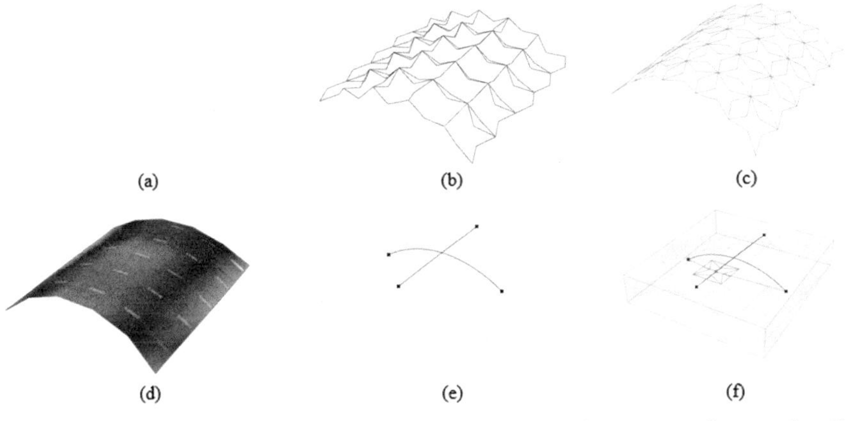

Fig. 4. **a** A deformed shell structure with 25 cells. **b** deformation patterns from each cell. **c** Reconstruction of the 3D pattern cell of the original deformed mesh. **d, e** Global Features extraction: gaussian_curvature, osculating_point, u_node1, u_node2, v_node1, v_node2, extract global x bounding box dimension, extract global y bounding box dimension, extract global z bounding box dimension. **f** Local Features extraction of each cell: centroid x position, centroid y position, centroid z position, distance u_node1 to reference cell, distance u_node2 to reference cell, distance v_node1 to reference cell, distance v_node2 to reference cell, pattern distance 1, pattern distance 2, pattern distance 3, pattern distance 4. *Source* From the authors.

With this data a dataset of 3.000 values per feature was generated and to define the *input_X*–features to feed the network–, and *output _Y*–feature to predict-, which is associated with the *pattern_distance*.

2.3 Deep Neuronal Network Architecture Model

A Principal Components Analysis (PCA) was initiated to understand the correlation between the global features of the system and the local features of each cell, in order to detect and select the input features for training the model and to generate the output feature to predict (Table 1), the value from which the structure will be reconstructed.

After several iterations, an ANN Model was composed of six Dense Layers with a rectified linear activation function ReLU, and one Dense layer with a sigmoid activation to predict the value. The model was trained with the dataset produced under different geometrical deformations (Fig. 5), with 100 epochs and a validation split of 20%.

2.4 Geometry Reconstruction

The predicted values of the displacement were associated to each of the four pivots of each cell in x, y and z dimensions (Fig. 6). This allowed the rebuilding of the shell structure by reversing the parametric design process and bypassing the PS simulator (Fig. 7), at the same time, the final pattern was redrawn in two dimensions for a potential 3d printing manufacturing process.

3 Conclusions

This workflow offers a novel way to optimize and reduce the computational power needed to compute the three-dimensional physical deformation of structures and to invert the design process allowing the designer to input the desired parameter retrieving the optimal solution within the AI-based design model.

The material strength can be considerably optimized through geometry by applying an integrated material intelligence workflow to develop digital prototypes, decreasing the amount of embedded required energy during the design and manufacturing processes.

The next steps could be based on the comparison and analysis of the predicted values between this research workflow with Regression models plugins and Particle-Systems simulators in a generative software. Along with testing the process with other shell structures by increasing the heights and the tridimensionality of the structures.

Table 1. Data sample of each geometrical feature. The first ten features were used as inputs and the last one 'pattern_distance´ as output to predict.

Iiteration	X_movement	Angle_1	Angle_2	Gaussian_curvature	Osculating_point	u_node 1	u_node2	V_node1	V_node2	Vector_angle	Pattern_disance
0	5.083703	13.574596	13.574596	0.001217	10.81259	3.337135	3.337135	2.305343	15.106804	1.062343	2.469323
1	1.74068	−15.079885	−18.616449	−0.000417	4.188744	1.431219	1.438955	12.831266	0.078778	1.956758	4.395734
2	5.397657	16.265634	9.192507	0.000811	11.233674	3.432897	3.444401	3.134234	15.11984	1.858934	2.377169
3	2.054634	−12.388847	−22.998538	−0.000411	4.590853	1.555104	1.57966	13.207677	0.081383	1.497747	2.425326
4	5.711612	18.956672	4.810417	0.000571	11.735675	3.537905	3.57563	4.159424	15.207417	1.599719	2.359344
5	2.368589	−9.69781	−27.380627	−0.000377	5.084956	1.693404	1.724839	13.51206	0.085329	1.278189	2.470359
6	6.025566	21.647709	0.428328	0.000529	12.405408	3.681105	3.716511	6.020141	15.525343	1.360134	2.361207
7	2.682543	−7.006772	28.237283	0.000578	6.506339	2.10046	2.014346	11.238372	0.274817	2.215165	2.077381
8	6.33952	24.338747	−3.953762	0.000576	12.81505	3.766593	3.802761	7.572598	15.565353	1.214877	2.353249

Source From the authors

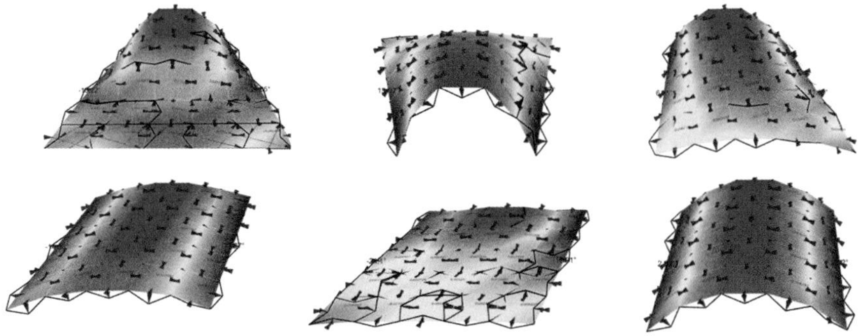

Fig. 5. Variations of the geometrical deformation of the system were used to train the regression Model. *Source* From the authors.

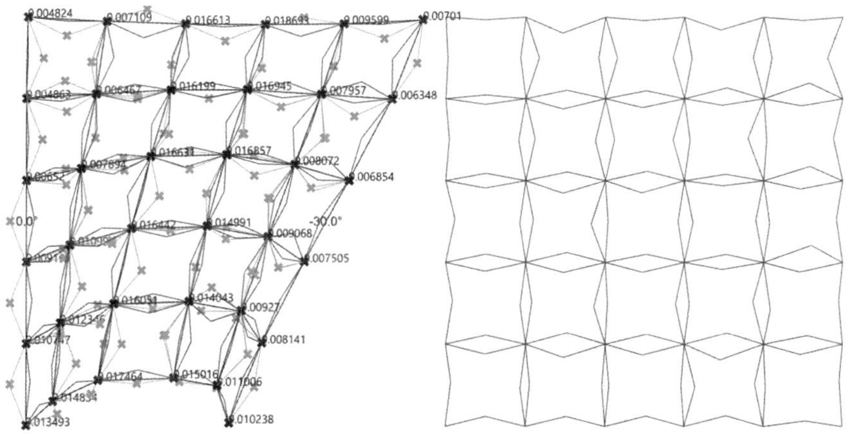

Fig. 6. In blue is the reconstructed two-dimensional pattern using the predicted value of pattern_distance. In red is the original surface. *Source* from the authors.

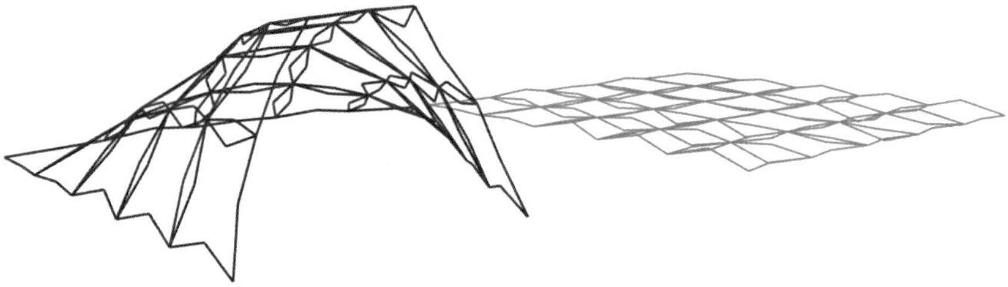

Fig. 7. In blue is the reconstructed three-dimensional shell-like structure from the predicted value of pattern_distance. In green is the bi-dimensional pattern for 3d printing.

Acknowledgements. Antonia Valencia, Tomas Sanchez for the initial studies, Faculty of Architecture, Design and Urban Studies and the Digital Fabrication Laboratory of the Pontifical Catholic University of Chile. Fab Lab Barcelona and the Institute for Advanced Architecture of Catalonia.

References

1. Vivanco T, Valencia A, Yuan PF (2020) 4D printing: computational mechanical design of bi-dimensional 3D printed patterns over tensioned textiles for low-energy three-dimensional volumes. In: Proceedings of the 25th CAADRIA conference
2. Evans KE, Nkansah MA, Hutchinson IJ, Rogers SC (1991) Molecular network design. Nature 353:124
3. Evans KE (1991) Design of doubly-curved sandwich panels with honeycomb cores. Compos Struct 17:95–111
4. Daekwon PJ, Romo A (2015) Poisson's ratio material distributions. Emerging experience in past, present and future of digital architecture. In: Proceedings of the 20th International conference of the association for computer-aided architectural design research in Asia (CAADRIA 2015)/Daegu 20–22 May 2015. pp 735–744
5. Liu Y, Hu H (2010) A review on auxetic structures and polymeric materials. Sci Res Essays 5(10):1052–1063
6. Bertin T (2001) Evaluating the use of particle-spring systems in the conceptual design of grid shell structures. Thesis Master of Engineering in Civil and Environental Engineering, Massachusetts Institute of Technology
7. Naboni R, Mirante L (2015) Metamaterial computation and fabrication of auxetic patterns for architecture. 129–136. https://doi.org/10.5151/despro-sigradi2015-30268
8. Hanna S (2007) Inductive machine learning of optimal modular structures: estimating solutions using support vector machines. Artif Intell Eng Des Anal Manuf 21(4):351–366. https://doi.org/10.1017/S0890060407000327
9. Arciszewski T, Ziarko W (1990) Inductive learning in civil engineering: rough sets approach. Comp. Aided Civil Infrastr Eng 5:19–28. https://doi.org/10.1111/j.1467-8667.1990.tb00038.x
10. Reich Y, Barai SV (1999) Evaluating machine learning models for engineering problems. Artif Intell Eng 13(1999):257–272
11. Aksöz Z, Preisinger C (2020) An interactive structural optimization of space frame structures using machine learning. In: Gengnagel C, Baverel O, Burry J, Ramsgaard Thomsen M, Weinzierl S (eds) Impact: design with all senses. DMSB 2019. Springer, Cham. https://doi.org/10.1007/978-3-030-29829-6_2
12. Razi M, Athappilly K (2005) A comparative predictive analysis of neural networks (NNs), nonlinear regression and classification and regression tree (CART) models. Expert Syst Appl 29(1):65–74
13. La Magna R, Knippers J (2018) Tailoring the bending behaviour of material patterns for the induction of double curvature. https://doi.org/10.1007/978-981-10-6611-5_38

Ceramic Incremental Forming–A Rapid Mold-Less Forming Method of Variable Surfaces

Yuxuan Wang, Yuran Liu, Riley Studebaker, Billie Faircloth, and Robert Stuart-Smith(✉)

University of Pennsylvania, 210S 34th St, Philadelphia, PA, US
yuxuan_1909@163.com, rssmith@design.upenn.edu

Abstract. Following architectural practice's widespread adoption of 3D modelling software, the digital design of free-form surfaces has enabled more heterogeneously organized architectural assemblies. However, fabricating envelope components with double-curved surface geometry have remained a challenge, involving significant machine time and material waste, and great expense to produce. This proof-of-concept project proposes a rapid, low-cost, and minimal-waste approach to forming double curved ceramic components through a novel approach to Ceramic Incremental Forming (CIF), using a 6-axis industrial robot, a passive flexible mold, and a custom ball-rolling tool. The approach is comparable to Single Point Incremental Forming (SPIF) that is used for forming complex shapes with metal sheets. This method promises to achieve high-quality, ceramic building envelope components, while eliminating the need to build proprietary molds for each shape and reducing the waste in the forming process. Compared with other architectural mold-less forming methods such as clay 3D printing, the approach is more time and material efficient, while being able to achieve similar levels of complexity. Thus, CIF may offer potential for further development and industrial applications.

Keywords: Ceramic · Digital fabrication · Architectural robotics · Mass customization · Incremental sheet forming · Flexible mold

1 Introduction

Novel manufacturing techniques for architectural ceramic components are increasingly of interest to designers who see value in leveraging clay's geometric variability, durability, and climatic properties [1, 3, 10]. The most widely used industrial manufacturing technique is extrusion, which is used to mass-produce architectural ceramic components through customized extrusion dies [5]. However, this method cannot efficiently support components with geometric variation or double curvature. The traditional way of forming architectural components with double curvature is to make dedicated, custom molds for each shape, leading to substantial material waste and increased fabrication costs [5]. Due to this, architects must align architectural ceramic assembly designs with manufacturing constraints and typically reduce geometric varation to an arrangement of a selected

P. F. Yuan et al. (eds.), *Hybrid Intelligence*, Computational Design and Robotic Fabrication,
https://doi.org/10.1007/978-981-19-8637-6_43

number of repeatable shapes which limits variability within an overall facade assemblage. In contrast, a universal mold-less forming method for double-curved geometries that is sufficiently rapid to support industrialization would enable architectural designs to embody greater degrees of variability.

This research proposes a novel approach to Ceramic Incremental Forming (CIF), a rapid, moldless method for forming bespoke double-curved ceramic components. This method involves the development of hardware and software including: a passive flexible mold and ball-rolling tool (Fig. 1), and a toolpath generation method related to material characteristics that incorporates feedback from RGBD camera (RGB + Depth) data. The approach is demonstrated through the prototyping of highly accurate double-curved surfaces (Fig. 2), fabricated with minimal waste and without the need for custom molds. Further directions for development are also proposed in this paper.

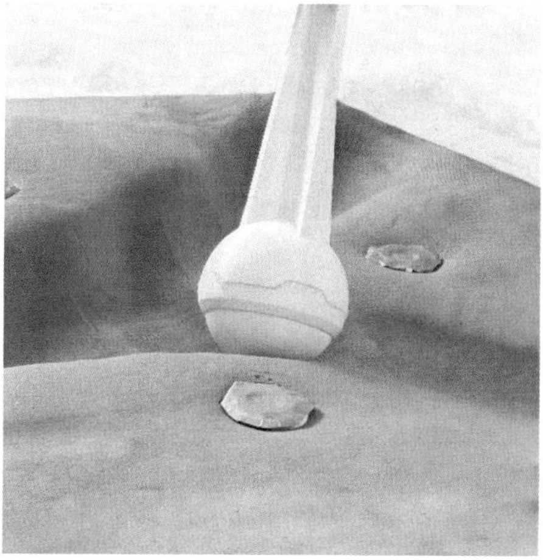

Fig. 1. Photo of test

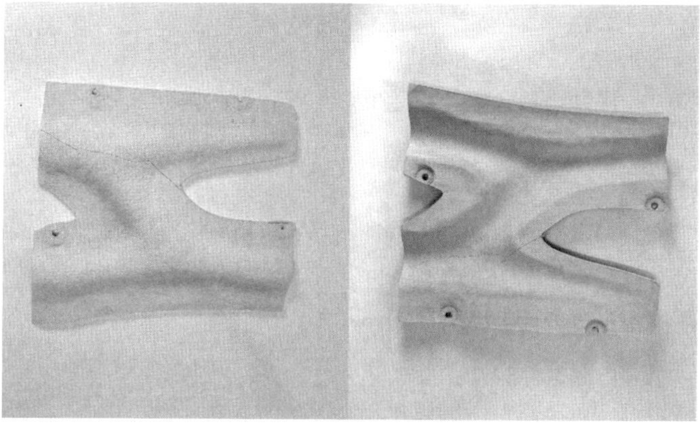

Fig. 2. Both sides of one test panel

2 State of the Art

Several recent developments in forming methods for architectural ceramic panels have utilized ceramic 3D printing or Modular Mold Forming (MMF) [2]. The ceramic 3D printing method reduces waste by alleviating the need for custom molds. However, while suitable for small complex volumetric parts, for double-curved surface geometries, 3D printing requires a mold or scaffolding approach which adds complexity, while the method remains limited in manufacturing speed and cost, making it difficult to meet industrial demands of large-scale architectural applications. MMF, like other adaptive mold forming methods, uses an adjustable modular mold to form clay components by using a series of individually adjustable pins. After the mold is adjusted and covered with a layer of interpolator membrane, an extruder 3D prints clay on the modular mold. Robot milling is used as a post-processing operation to enhance the surface quality, enabling the high-quality production of variable, double-curved ceramic panels. This method is an ingenious way to create an adaptable mold for variable shapes, minimizing the waste problem of custom molds for each shape. However, adjusting the mold for each shape, combined with 3D printing and milling, still takes a relatively long time for mass customization [2]. In addition, because the MMF is formed by slumping, the shapes it can produce are limited in draft angle. For example, sudden steps, overhangs, and undercuts are hard to achieve by this method.

Alternative manufacturing approaches used with other materials can offer greater possibilities for ceramic forming. A state-of-the-art method for mold-less metal sheet forming is the Single Point Incremental Forming (SPIF) [6], which is made possible by the development of Computer Numerical Control (CNC) [9]. SPIF uses the force and motion of a CNC-controlled stylus and leverages the ductility of sheet metal to gradually transform the metal sheet into a shape. SPIF requires no die or mold but does require clamping to hold the position of the sheet in either a vertical or horizontal orientation along its edge.

3 Methods

This research primarily includes the design and testing of a 6-axis robotic workflow for CIF including the creation of a passive flexible mold, the design of ball-tip end effectors, and tool path planning including the role of feedback during the fabrication process. Ceramic, double curved building envelope components are prototyped in a highly efficient way with minimal waste to study the opportunities and constraints of the workflow and to identify technical challenges.

3.1 Passive Flexible Mold

While SPIF is an effective rapid mold-less forming method, challenges exist for its application to clay materials. A clay slab is too soft to be fixed by clamping. In the absence of support, the clay slab will fall downward due to its weight and tear adjacent to the edge. To address this issue, a tooling strategy was developed that provides support to the soft clay slabs while allowing them to be freely formed.

Whereas MMF uses an array of adjustable pins and a membrane overlay to support the forming process. CIF proposes a passive flexible mold that can, like sand in a box, be shaped and reshaped under pressure (Figs. 3 and 4). Contained in a box with an open top face, a bed of polystyrene balls (2–4 mm in diameter) supports a continuous membrane on top of which a clay slab can be gradually transformed. The polystyrene balls are locally compressible and can generally provide the necessary support for the clay slab while they also act like a fluid and can balance pressure through. A continuous neoprene fabric membrane (consisting of 90% polyester and 10% spandex and 1.5 mm thickness) is laid over the top of the balls containing them while providing a continuous, smooth and supple support surface for the clay slab to rest on. The neoprene's four-way stretch capacity of 15%, makes it well suited to supporting clay slabs during double-curved surface forming. The fabric selection was based on available products in the market, other membrane materials (more elastic and breathable) could be tested in further research. Several adjustable supports are introduced in the flexible mold near the bolt-through connections. The number of supports is determined by the desired shape and the number of connections that fix the finished ceramic panel to other building components. There is a fixing knot placed on the clay slab at the top of each support to preserve the accuracy of each connection. It is fixed to the support by rare earth magnets in both horizontal and vertical directions and can hold the clay slab to prevent unexpected movement by its bumpy texture.

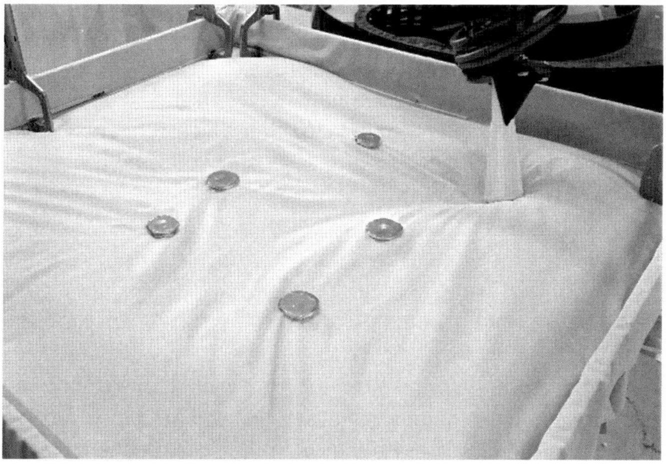

Fig. 3. Photo of passive flexible mold

3.2 External Forces–Shear and Friction

In incremental forming processes, Emmens et al. teach that materials are subject to two types of external forces, shear, and friction [7]. Studying those forces on clay materials is important for toolpath planning because clay materials, unlike sheet metal, lack rigidity and are highly plastic.

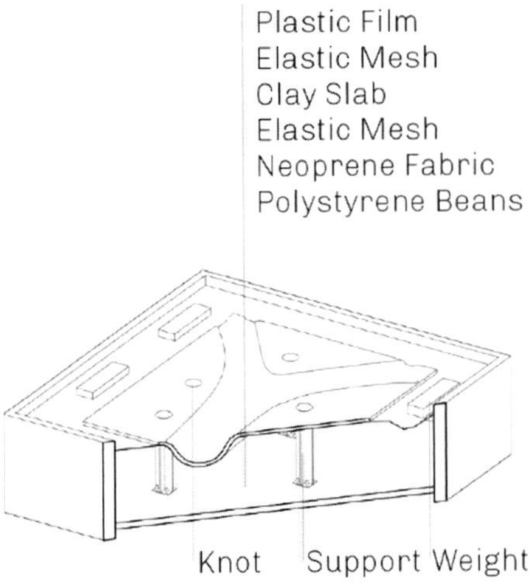

Fig. 4. Section of passive flexible mold

In the CIF method, the clay is transformed from a slab into a shell using shear forces. By adding a force that is normal to the target surface, the clay slab will bend, allowing the shear to work on the clay slab. Most SPIF methods use 3-axis CNC machines, which can only provide shear force in a fixed direction [7]. In contrast, the application of more-than-5-axis machines can provide multi-degree freedom to the tool to control the direction of shear force [6]. As shown in Fig. 5, forming with a fixed direction has a higher probability of tearing and cracking the clay when positioned at a steep angle, compared to forming with variable forming direction which only provides shear force. In theory, more-than-5-axis machines can even form undercuts and overhangs. Therefore, we used an ABB IRB-4600 robot with 6° of freedom (DOF) to develop a prototype surface panel (5 DOF were used). To avoid collision between the tool or robot and the clay slab, the midline of the angle between the normal direction of the surface and the vertical direction was used as the normal of the target plane.

Generally, SPIF researchers would like to avoid friction which can lead to cracking, low accuracy, and other failures [8, 9]. When using SPIF on sheet metal, a lubricant is applied to the sheet metal to minimize friction between the metal sheet and the stylus. However, in CIF, friction does not necessarily need to be avoided. Because clay is inherently plastic, friction can be used to achieve some desired effects. One tested method is to use a low amount of friction and push the clay in a specific direction. This method redistributes the clay slab's density, much like a craftsperson's hands might distribute clay when forming a clay vessel. For example, it is possible to push clay from a flat place toward a sloped edge to make the edge steeper or, to push a maximum amount of clay into an area for substantial deformation.

3.3 End Effectors

In the tests, two kinds of end effectors have been tested. One adapts to a reciprocal pressing action, one adapts to moving forward in a rolling manner (Fig. 6).

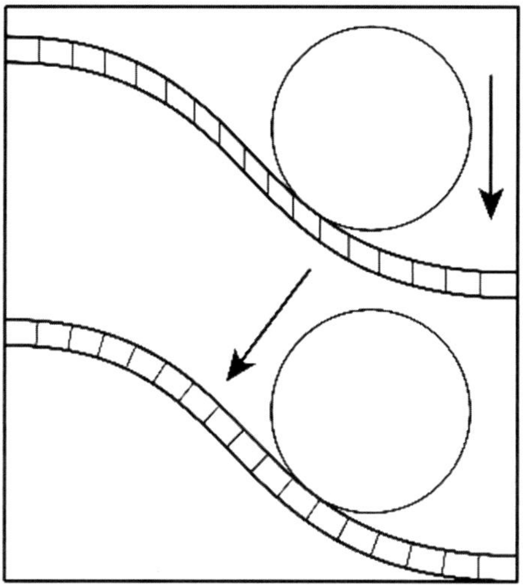

Fig. 5. Forming directions' effect on clay slab

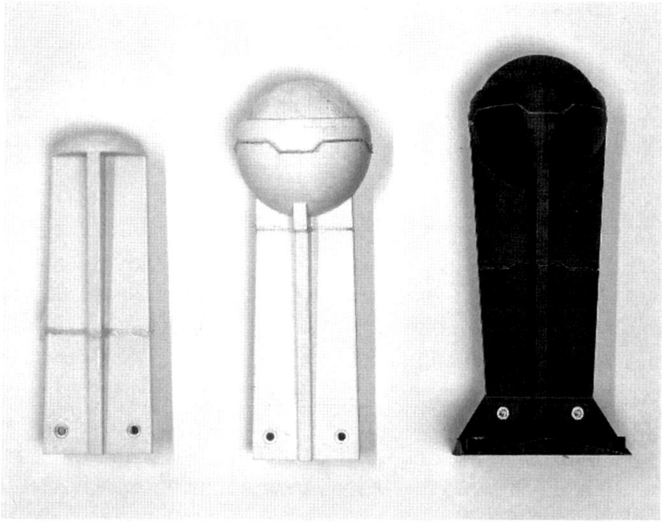

Fig. 6. End effectors

The first strategy employs a pressing tool, adding a target plane to the normal of existing target planes, thus placing pressure on the target surface. This approach avoids friction and has a lower requirement for stepping down, which is good for avoiding cracking in the clay slab. The absence of friction reduces the overall impact radius of the tool and improves accuracy during the forming process. However, this approach results in a long toolpath and a relatively long fabrication time. In addition, frequent acceleration and deceleration places wear and tear on the robot. Thus, this strategy was deemed less desirable and was discontinued.

The second strategy uses a rolling tool to connect each target plane and place pressure on the clay surface by rolling. Friction is unavoidable in this strategy. Even if a ball is used to avoid dynamic friction during the rolling process, static friction and forces perpendicular to the normal direction during forming are unavoidable. In order to avoid failure, it is necessary to reduce their effects, and the step down needs to be controlled, ideally within 1/10 of the ball's diameter. This strategy is faster, but it is slightly less accurate when compared to the pressing tool, and has a slightly higher risk of failure. The size of the ball is influenced by several factors [4, 6], which needed to be further studied.

In the last prototype, an 80 mm diameter ball was used to form a shape with a maximum depth of 40 mm on a 12 mm thick, 500 mm wide clay slab as shown in Fig. 2. In earlier prototypes, both an 80 mm diameter pressuring tool and a ball-rolling tool were tested on a shape with a maximum depth of 60 mm. The results showed that, for complicated shapes, using both tools would be helpful.

3.4 Toolpath Planning

In order to test this method's ability in a venation pattern self-supporting wall, tests were based on two identical shapes, a "Y" shape and an "H" shape. Both have a double-curved area in the center and a flat area along the boundary. The flat area was also used to allow for several connections to fix the finished panel with other building components which require higher accuracy. The following toolpath strategies are based on these shapes.

3.4.1 Overall Toolpath Strategy

In order to control the accuracy of the final outcome, the clay slab was placed a bit higher than the highest elevation of the target shape before forming, so that all of the areas would be formed from above in the forming process to avoid unadjustable deviations which are lower than the target surface. As the area around the formed region is slightly deformed during the forming process by internal forces in the clay slab and neoprene fabric, and due to the polystyrene balls' fluid-like displacement, this process should be carried out as early as possible so that it can be adjusted in the following forming process. Therefore, the curved part should be formed first in the panel forming process, followed by the flat part, and finally by adjusting all the parts by 3D scanning (Fig. 7).

3.4.2 Toolpath Strategy for the Curved Part

In the forming of the curved part, because of clay's high plasticity, a layer-by-layer toolpath planning logic similar to that in 3D printing is applied. First, the input shape is sliced into multiple layers by step down. Similar to the out-to-in and in-to-out spiral path-planning logic for metal sheets [6], there are two logics for CIF. The out-to-in logic in each layer was applied in the prototypes: tracing the boundary first, followed by the infill part. This logic can define the boundary of each layer and avoid the forming processes' influence on other areas to achieve higher accuracy. The in-to-out logic in each layer may cause deviation from its surroundings, and it is hard to control the accuracy (Fig. 8).

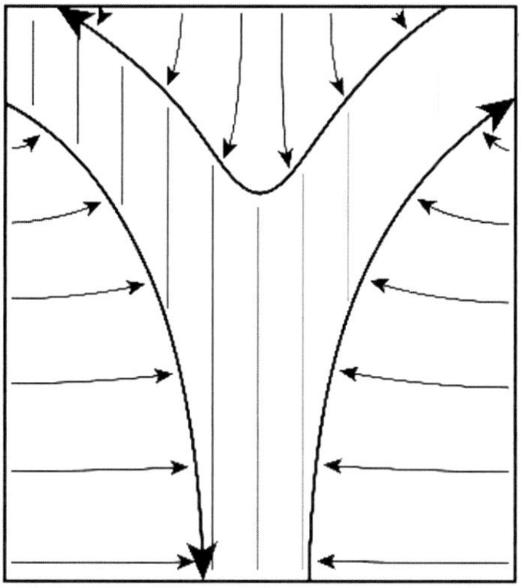

Fig. 7. Overall toolpath strategy

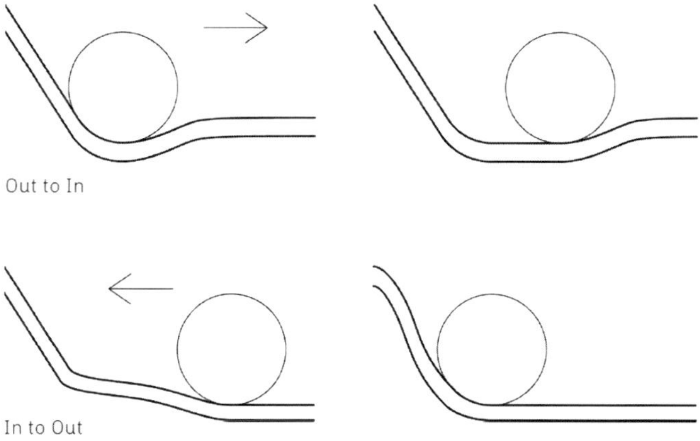

Fig. 8. Effects between two toolpath strategies

3.4.3 Toolpath Strategy in the Flat Part

For the toolpath in the flat area, friction should be considered. The steep edge is difficult in most incremental forming, but by planning toolpath toward the steep edge, with the help of friction, a small part of the clay body can be redistributed to the edge, thus improving the molding accuracy [6]. This has been proved by tests, in which an edge of 60° was formed with relatively-high accuracy.

3.5 Ball Filling's Pressure Control

In the forming process, as the total volume of polystyrene balls is continuously compressed, the support force to the clay slab will gradually approach or even exceed the slab's bending resistance, which maintains the slab's form, and further leads to deviation in the form. Therefore, it is necessary to release the polystyrene balls' pressure during the forming. In the prototype, several weights were placed on the neoprene fabric around the clay slab. The pressure was controlled by adjusting the weights to allow the polystyrene balls to flow out from the forming area. In future experiments, several pressure sensors and pumps could be applied to control the total amount of polystyrene balls, to achieve automated adjustment.

3.6 Surface-Protective Materials

In order to guarantee surface quality and protect the end effector, protective materials are needed. In the prototype, a layer of elastic mesh fabric (90% Nylon and 10% Spandex, which has 0.35 mm thickness, and mesh openings with a diameter of about 1 mm) was placed on both sides of the clay slab (Fig. 4). This kind of fabric is very ductile, more than neoprene fabric's, and will impose almost no additional restrictions in the forming process, making it a good medium to avoid adhesion between clay slab and membrane. It can also prevent cracking while transferring dried clay plates. During the forming process, it is recommended to place a rough plastic film between the mesh fabric and end effector to further reduce friction and prevent clay particles from getting into the ball roller and clogging the tool. Both the fabric and plastic film can be reused to minimize waste.

3.7 Feedback Forming

Because the passive flexible mold has a spring back effect, it is difficult for most of the shapes to be formed accurately at one time. To address this, following the forming process, a feedback loop is introduced (Fig. 9). First, an RGBD camera (RealSense D435i) scanned the surface and generated a 3D model to compare with the target model. Figure 10 shows a sample of the comparison data. Then an additional forming process was performed on mesh scan data vertex positions that remained higher than the desired geometric outcome. In the prototypes, relatively high accuracy was usually achieved after one feedback loop. Since clay would have a larger deformation in the drying process, it was not necessary to perform more feedback loops before drying. To mitigate deformation during drying, another feedback loop was employed after the clay slabs were leather dry.

4 Results and Discussion

4.1 Surface's Accuracy and Quality

Clay is subject to a significant shrinkage rate during its drying prior to bisque firing, which can result in non-uniform shrinkage. For plates with a certain thickness, different drying rates on the two surfaces can lead to different shrinkage rates and the occurrence of bending. This is the main source of errors for this method.

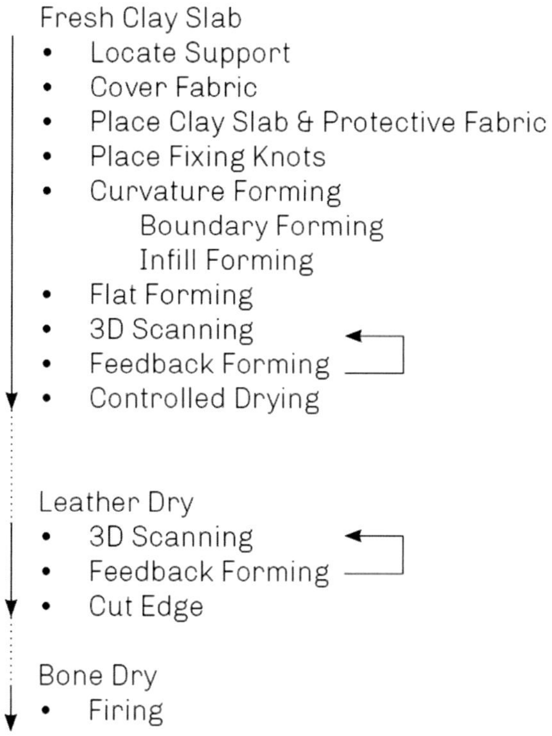

Fig. 9. Full process work flow

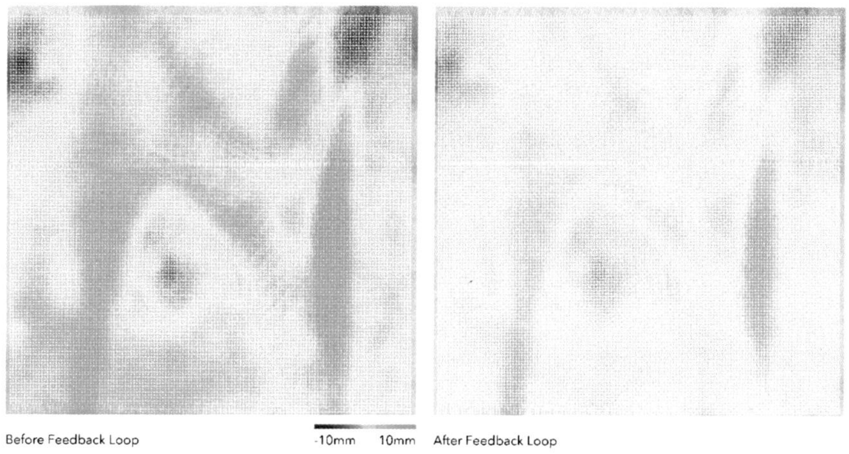

Fig. 10. Scan results before and after feedback forming

In the prototype, Laguna #10G EM101 sculptural clay was used, which has a relatively low shrinkage rate and can mitigate this effect. In the experiment, most of the shrinkage occurred during the process of drying from wet to leather dry, and only limited shrinkage occurred during the process of drying from leather dry to bone dry. The clay can still perform a small amount of forming work while it is leather-dried. So, performing another feedback forming in this state would help to drastically improve the final accuracy.

4.2 Cracks and Textures

In the prototype, most of the failures did not happen during the forming process, but during the transferring process. Therefore, surface protection methods are necessary. In the transferring, protective fabrics were used to avoid deforming. Inappropriate cooling speed after bisque firing would also lead to cracks.

Different protective materials left textured imprints on the clay. The mesh fabric left a mesh texture on the surface, which would provide interesting pixel-like results when glazed (Fig. 11). The plastic film leaves a leather-like texture caused by the clay's shrinking. More surface effects could be tested in further research.

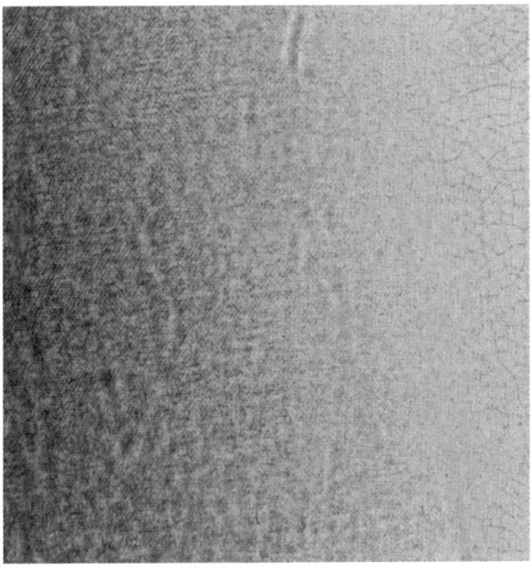

Fig. 11. Pixel-like texture after glazed

5 Conclusion

A CIF method successfully produced accurate double-curved ceramic panels with no formwork or material waste within a short operating time, demonstrating great possibilities for industrial applications requiring mass customization. As an emerging field of research, further studies are needed to address some of the qualitative and quantitative issues that were observed but were not engaged with in the research due to the time and resource limitations. In Fig. 12, a self-support venation structure design was proposed to demonstrate potential application scenarios. Due to restrictions in academic timetable, a "Y" shape and an "H" shape were made with this method, but could not be included in the assembled structure.

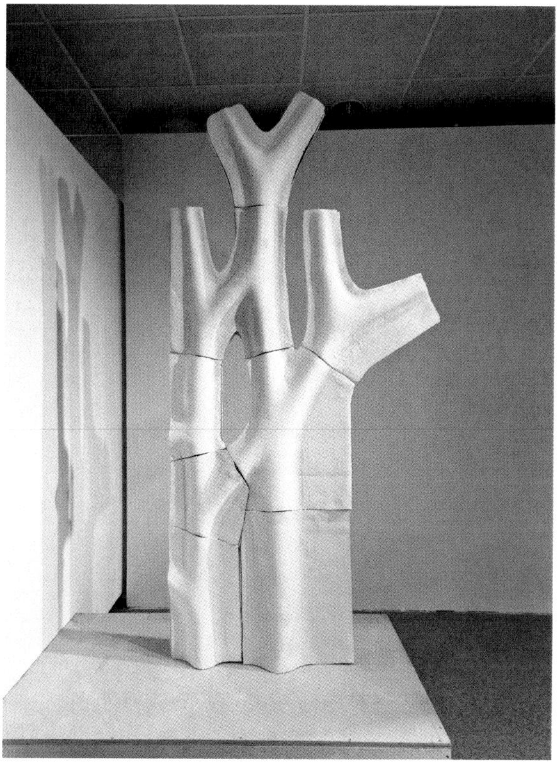

Fig. 12. Self-support structure design proposal

5.1 Cracking Study

Cracks, as the most common failure in incremental forming, need to be further investigated as to why they are generated and how they can be avoided in this method. Parameters such as the threshold of shear force, the slab's minimum thickness, and the end effector's size can have an impact on the generation of cracks [7]. In addition, the texture on the slab's surface may also be a cause of cracking. It needs to be studied whether the mesh fabric will bring a greater risk of cracking while making beautiful surface textures.

5.2 Fluidized Bed Technology

The fluidized bed is a common technology in the chemical industry. By feeding compressed air through the particulate medium, an air film is created between the particles, which eliminates the friction and turns a tight particulate medium into a fluidized one [11]. Therefore, a possible method is proposed here: using inelastic and more frictional sand as the filler for the passive flexible mold, a fluidization matrix is built at the bottom of the passive flexible mold, then the airflow in a certain area is controlled by switching each grid (Fig. 13). When the end effector is working in a certain area, the corresponding grid is supplied with compressed air upward to fluidize the sand around this area for forming, while the sand in other areas remains solid to constrain the deformation of the clay slab within a certain area. This assumption not only has the potential to improve the accuracy while forming, but also can control the airflow rate to get a similar drying speed on both sides, so it can improve the final accuracy and productivity.

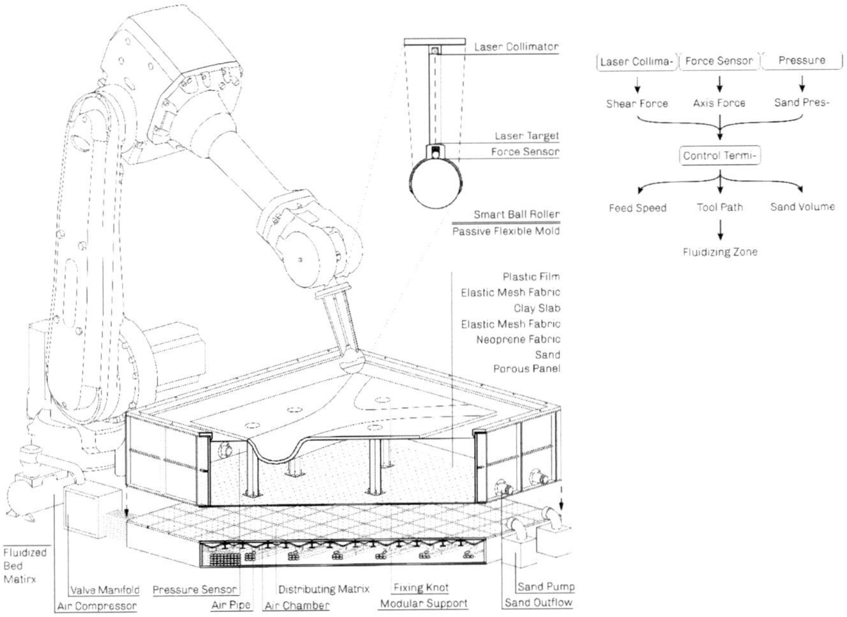

Fig. 13. Industrialized forming station proposal

5.3 Forming Strategy of Complicate Shape

As described above, the friction between the tool and clay slab can be used to redistribute the clay body and subsequently explore the potential of forming logic. For example, for a shape that has an undercut, it needs to be further researched whether it can be achieved by using friction to push the formed part horizontally.

5.4 Real-Time Feedback

For accuracy control, a tool as shown in Fig. 13 is envisioned. Real-time friction and shear forces can be collected by the laser collimator and force sensor for real-time toolpath adjustment. Another assumption is to use a closed-loop toolpath optimization by performing real-time 3D scanning [6]. A reasonable real-time adjustment algorithm can drastically improve accuracy, reduce the need for feedback forming, and avoid failure. Thus, this direction is worth studying.

5.5 Material Researches

As a study working with materials, the material itself should be further investigated [4]. A study of clay's micro and macro structures during the forming process would provide a useful reference for subsequent research.

References

1. Aguilar P, Borunda L, Pardal C (2020) Additive manufacturing of variable-density ceramics photocatalystic and filtering slats. eCAADe 2020(1):97–106
2. Bechthold M (2016) Ceramic Prototypes—design, computation, and digital fabrication. Inf Constr 68(544):e167. https://doi.org/10.3989/ic.15.170.m15
3. Celento D, Harrow D (2008) CeramiSKIN: Digital possibilities for ceramic cladding systems. ACADIA 2008:292–299
4. Centeno G, Bagudanch I, Martínez-Donaire A, García-Romeu M, Vallellano C (2014) Critical analysis of necking and fracture limit strains and forming forces in single-point incremental forming. Mater Des 63:20–29
5. Derby B, HändleF. (2007) Extrusion in ceramics. Heidelberg Springer, Berlin Heidelberg, Berlin
6. Duflou J, Habraken A, Cao J, Malhotra R, Bambach M, Adams D, Vanhove H, Mohammadi A, Jeswiet J (2017) Single point incremental forming: state-of-the-art and prospects. IntJ Mater Form 11(6):743–773
7. Emmens W, van den Boogaard A (2007) Strain in shear, and material behaviour in incremental forming. Key Eng Mater 344:519–526
8. Filice L, Fratini L, Micari F (2002) Analysis of material formability in incremental forming. CIRP Ann 51(1):199–202
9. Jeswiet J, Micari F, Hirt G, Bramley A, Duflou J, Allwood J (2005) Asymmetric single point incremental forming of sheet metal. CIRP Ann 54(2):88–114
10. Seibold Z, Hinz K, García del Castillo y López J, Alonso N, Mhatre S, Bechthold, M (2018) Ceramic morphologies. Precision and control in paste-based additive manufacturing. ACADIA:350-357
11. Tsuji Y, Kawaguchi T, Tanaka T (1993) Discrete particle simulation of two-dimensional fluidized bed. Powder Technol 77(1):79–87

Fabrication of Reinforced 3D Concrete Printing Formwork

Jiaxiang Luo[1], Tianyi Gao[1], and Philip F. Yuan[2(✉)]

[1] Fab-Union, 81 Gongqing Rd, Shanghai, China
ucbqj25@ucl.ac.uk
[2] College of Architecture and Urban Planning, Tongji University, 1239 Siping Rd, Shanghai, China
philipyuan007@tongji.edu.cn

Abstract. In recent years, the emerging 3D printing concrete technology has been proved to be an effective and intelligent strategy compared with conventional casting concrete construction. Due to the principle of additive manufacturing strategy, this concrete extrusion technique creates great opportunities for designing freeform geometries for surface decoration since this material has a promising performance of high compressive strength, low deformation, and excellent durability. However, the structure behavior is usually questioned, defined by the thickness and printing path. At the same time, the experiments for using 3D printing elements for structural and functional parts are still insufficient. Little investigation has been made into developing reinforcement strategies compatible with 3D printing concrete. In fact, conventional formwork and easy-to-install reinforcement support structures have various advantages in terms of labor costs but can hardly be reused. Thus, using 3D concrete printing as formwork for projects in different scales is an effective solution in the mass customized prefabrication era. Considering large-scale projects, the demand to provide concrete formwork with a proper reinforcement strategy for better toughness, flexibility, and strength is necessary. In this paper, we proposed different off-site reinforced 3D printing concrete strategies and evaluated them from time and material cost, deviation, and accessibility of fabrication.

Keywords: 3D concrete printing formwork Robotic fabrication · Structure behavior · Fiber reinforcement · Reinforcement strategy

1 Introduction

In recent years, 3D printing in the construction industry has generated a considerable interest, both in academic and industry, owing to significant benefits in terms of higher quality and productivity, faster construction processes, higher geometrical freedom, and cost-efficiency. Concrete 3D printing, as an essential part of digital construction, has been successfully applied in infrastructure due to its flexibility, speed, and economy, which can significantly save the material and time costs of engineering projects. Due to the principle of additive manufacturing strategy, this concrete extrusion technique creates

P. F. Yuan et al. (eds.), *Hybrid Intelligence*, Computational Design and Robotic Fabrication,
https://doi.org/10.1007/978-981-19-8637-6_44

great opportunities for designing freeform geometries for surface decoration since this material has a promising performance of high compressive strength, low deformation, and excellent durability.

In the era of mass customization, the manufacturing strategies for prefabricated free form architecture elements become popular and urgent. In the process of traditional complex concrete elements, to achieve high precision, it is necessary to make temporary timber formworks before construction. This traditional formwork installation system and structure reinforcement strategies have advantages for labor costs but often cause crucial material waste.

The erection of molds and the placement of reinforcement still requires physically demanding labor, when bespoke geometries are particularly needed. This results in personal health issues of construction workers that should be avoided as much as possible, particularly with an aging work force as in many developed countries [1]. The method of applying permanent 3D printing concrete formwork to build reinforced concrete prefabricated components, combining 3D printing with traditional construction techniques is a new field needed to be explored. In this paper, we investigate the possibility of introducing this technology in the real engineering project. Taking the geometry deformation, surface crack performance, curing time, usage of steel bars, printing time, weight, and accessibility of installation into consideration, two group of contrast experiments, using different types of steel rebars and fiber additive agent are represented for comprehensive results and analysis.

2 Background

2.1 Transition to Robotic Fabrication in the Digital Era

The construction industry is the pillar industry of the economy in China. The proportion of value reached 7.01%. As a labor-intensive industry, labor cost is one of the essential costs in the construction industry. With the continuous intensification of labor shortages, backward production technology, waste of production resources, and soft skills of workers in the industry in recent years, robotic fabrication as the core has become an important opportunity for upgrading the construction industry.

As one of the core parts of robotic fabrication, building robotics is a digital design and construction tool with multi-degree of freedom, high precision, and high efficiency. It surpasses the processing limitation of traditional technology and can more efficiently complete the customized processing production of large quantities of building components. At the same time, the robot can break through the limitations of arm span and load-bearing and enhance the realization of the limits that cannot be reached by manual operation, to complete large or complex multi-scale and multi-function construction tasks.

2.2 3D Printing Concrete as a New Normal

Due to the economy of concrete materials and its excellent structural performance, concrete has become the most significant amount of construction material today. 3D printing technology brings freedom of form and automated process to concrete construction that is unavailable in traditional construction modes. In the era of mass customized building components, the construction technology of 3D printing concrete has a great application prospect. However, due to the weak layer bonding, it is difficult to transmit shear and tensile forces, and can only be used for purely compressed structures [2]. The function of 3D printing concrete is still limited and restricted by structure performance. In fact, 3D printing permanent formwork technology is promising for promoting the application of 3DCP in the field of structural engineering because the reinforcement cage can be designed and fabricated according to existing design specifications [3].

2.3 Previous Attempts of Reinforcement Strategies in 3D Printing Concrete

Many previous attempts have been made to enhance the structural performance of 3D printing concrete. Due to lack of interlayer bonding and printing defects, the performance of 3D printed concrete is slightly lower than the mold cast concrete [5]. It is suggested that adding 13 mm PVA Fiber and combining it with primary concrete material is a practical way to avoid surface cracks and modify the bending strength of printed elements [4]. In this paper, we also use a fiber-reinforced high-strength cementitious mortar as printed material, and the large components of the mortar are shown in the following table. However, only adding fiber is not enough to ensure the quality, especially when large components are required in mass customization projects. Thus, it is still necessary to start a further investigation for optimal reinforcement placement, configuration, and reinforcement material (Fig. 1 and Table 1).

Material	**Quantity (kg/m³)**
White Portland Cement (P.W I 52.5)	654
Silver Sand (0-1mm)	785
Silica Fume	65
Metakaolin	65
Water	229
Superplasticizer	0.65
PVA Fiber (13mm)	2.5

Fig. 1. Material and mixture design

In fact, the existing reinforced solutions have many disadvantages and can hardly be used in the project. For instance, Bos uses 6 mm steel fiber to prepare 3D-printed concrete. The addition of fiber can significantly improve the flexural strength of concrete but cannot enhance the strain-softening behavior of concrete material. ETH Zurich has proposed a 3D printed concrete material reinforced with steel mesh, using a robotic arm system to build steel mesh and formwork simultaneously. This method can effectively improve the mechanical properties of 3D-printed concrete structures, but it is difficult to match the concrete printing process due to the local high temperature generated during the steel welding process. The toughening method of 3D printed reinforcement materials

Table 1. Relationship between time cost and reinforcement strategies

	Size A	Size B	Size C
Time consumption of printing	3 h	1.5 h	4 h
Time consumption of curing	12 h	8 h	15 h
Amount of reinforcement	30	15	42
Weight	0.6t	0.3t	0.8t

Size A: 1000 * 1000 * 100 mm
Size B: 1000 * 500 * 100 mm
Size C: Irregular Shape

is a method that uses 3D printing technology to print auxiliary materials and toughen concrete. It can significantly modify the shear and bending strength, but the time cost is massive, and it can barely work with concrete printing at the same time. In this way, only using robotic fabrication in construction is not a mature technology. In the current situation integrating necessary labor production and prefabricated reinforced steel bars is vital in engineering projects [5].

3 Methodology

We developed a new control group experiment program according to the existing research. Since the structural behavior and bond durability of the 3D printing concrete formwork are significant, this experiment aims to achieve a permanent use of external concrete printing formwork.

Firstly, the structural strength, including the compressive, shear strength, and the bonding strength between the printed formwork and the internal casting material, can be obtained by changing the fiber proportion in the gap between different layers of formwork. Secondly, by changing the shape of the steel reinforcement bar during the robotic fabrication, the effect of varying placement of reinforcement on the strength of the printed piece can be obtained. The experimental results have a decisive impact on the subsequent large-scale printing experiment.

The permanent concrete printed outer formwork and its own structural strength determines whether it can be used in specific engineering conditions. Due to the technical characteristics of 3D printing concrete formwork technology, the printing path design in the fabricating process is looped laminated printing. The internal cavity is bound to be generated, while the strength of the printed formwork produced by such a cavity printing mode is weaker than the printed mode without a cavity. Therefore, in the experiment, it is hoped to obtain the best material proportion that can improve the strength of the printed formwork by casting the cavity with different material proportions. At the same time, in the experiment, the internal casting material with different material ratios will also affect the bond strength between the casting concrete and the printed concrete. Therefore, the following control group was set up in the experimental process. The result was to explore the influence of different fiber adding ratios of casting concrete on the strength of the printing formworks under the condition of the same printing speed and material extrusion width.

Group A test plan:

Test the concrete structure strength of different material ratio results.

Group B test plan:

Test the bond strength of concrete with different material ratio.

Group C test plan:

Use different (L-shaped and C-shaped) steel bars to reinforce the externa formwork of 3D-printed concrete.

By combining the above three test plans, the experiment produced ten blocks of 3D printing concrete. (its size is 1000 * 500 * 100 mm) As shown in the figure, the processing of L-shaped steel bars is placed outwards with sharp corners. Each layer is set with 5 bars and 20 layers apart. In this way, the steel bars effectively reinforce the structure between the printing layers of concrete, and the gap formed can be used as an external formwork to add structural steel bars. However, the measurement results show that this method of placement is less efficient.

The sequence of steel bars in type C is staggered between 2 layers. Six steel bars are placed in the two adjacent layers and placed at an interval of 20 layers. In the end, the geometry deformation, surface crack performance, curing time, usage of steel bars, printing time, weight, and accessibility of installation (Figs. 2 and 3).

Fig. 2 L-shaped steel bars

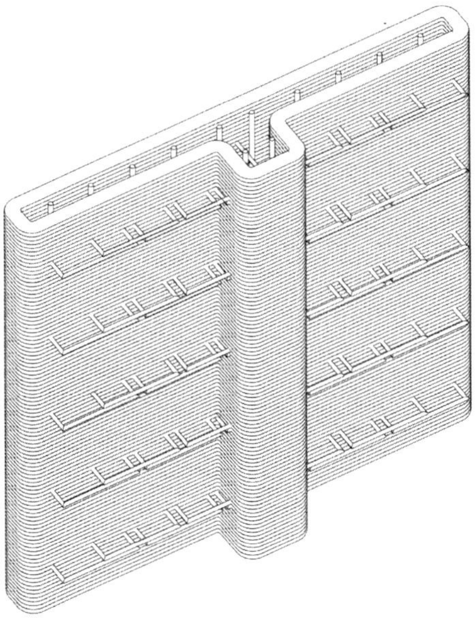

Fig. 3. C-shaped steel bars

4 Integrated Human and Robotic Fabrication

4.1 FU Robot as a Platform for 3D Printing Process

The digital design software of the toolpath generation and the stop point control are built in the Rhino/Grasshopper environment. After curvature and contour analysis of printing components, the toolpath is generated. Previous experience suggested that the foundation printing speed should be set to 0.1 m/s while the body part should be set around 0.2 m/s so that the extrusion width could be controlled at approximately 30 mm. Before traditional printing, the experiment is needed to check whether the toolpath has any problem, avoiding failure in the later printing process. The grasshopper parametric design program and complete digital analysis and design workflow are shown in Fig. 4.

4.2 Additive Concrete Formwork Manufacturing Process

In the construction of architectural concrete, concrete formwork and mold have played a critical role because of the construction characteristics of concrete. Concrete formwork and mold refer to a complete set of the structural system of formwork, mold, and supporting mold plate formed by pouring concrete. The process of formwork directly affects the final quality of concrete. In this concrete formwork manufacturing process, we first generate three layers of extrusions, serving as the foundation for a self-horizontal plane followed by a stopping point. Then, since each experimental material is one meter high and we have two different steel rebar types, one of the experimental groups has five more stop points. In contrast, the other group has ten more stop points averagely allocated in each layer for reinforced steel rebars. After setting the endpoint location of the whole

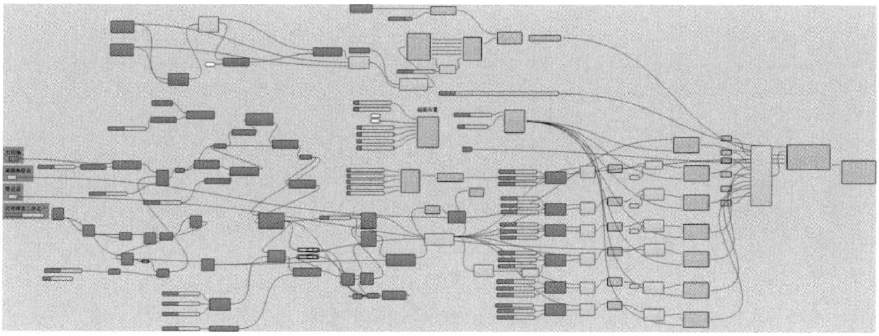

Fig. 4. Grasshopper and FU robot platform

printing file, the script is then generated by the entire environment and platforms so that the program can be copied into the KUKA demonstrator to start the fabrication process. When the printing process is finished, curing A specific flowchart of this design to the construction process is shown in Fig. 5.

4.3 Additive Concrete Formwork Manufacturing Process

Besides the robotic fabrication process, labor also costs a key role in determining the appearance and quality of printed concrete. In the entire process, mixing concrete with PVA fiber, placing reinforced steel rebars, and casting formworks with different mixed materials occupy most labor costs. Compared with traditional construction methods, it has enormous advantages, such as cost, safety, and construction speed. Moreover, when sustainable development is increasingly ethically significant, this approach can make construction operations environmentally friendly.

5 Result and Discussion

5.1 Time and Material Cost

The total processing time of concrete 3d printing outer formwork can be estimated using Eqs. (1–3) shown below. The total 3D-printing time depends on the volume of the concrete formwork, the height of the concrete printed layers, the horizontal cross-sectional area, the speed of the motor, the time it takes to manually place the steel rebar and the time of initial curing of printed concrete. The total time shown here is an estimate, however, the experimental time in real environment is often different, because the environmental factors (like temperature and humidity) also affect the time of 3D printing. However, the experimental time in the natural environment is often different because environmental factors (like temperature and humidity) also affect the time of 3D printing [6]

$$T_{total} = \frac{V_{Con}}{V_t \times N} t_1 + t_2 \tag{1}$$

$$V_t = S_{con} \times H_{con} \tag{2}$$

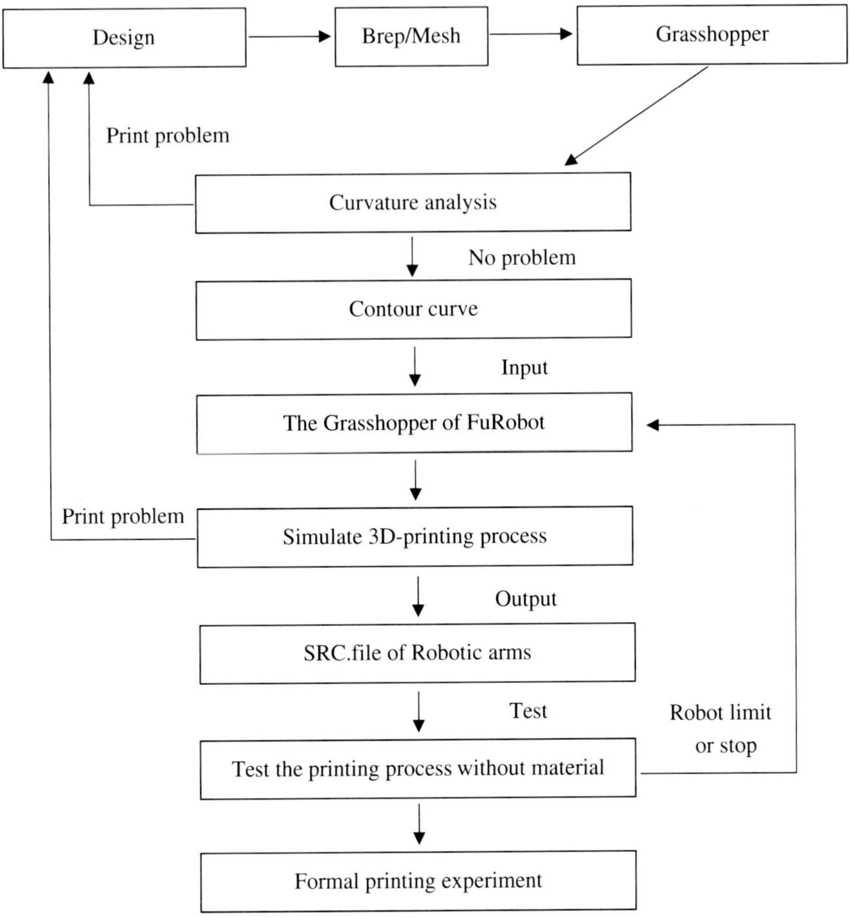

Fig. 5. Digital analysis and design workflow

$$t_2 = t_3 + t_4 \tag{3}$$

T_{total}–the estimated total time for 3D printing process, V_{Con}–the total volume of the design Brep, V_t–Concrete extrusion volume per unit time, N–the speed of the motor, S_{con}–The sectional area of the concrete printing formwork, H_{con}–the height of one-layer printed concrete, t_3–the estimated time for curing time of 3D printed concrete, t_4–the estimated time for manually place the steel rebar.

5.2 Fabrication Deviation

The current study shows that the proposed integrated reinforced methods have different structure behaviors and surface appearances. To conclude, the larger the components are, the more deformation and cracks will exist. Type C reinforced rebars have better deformation performance while it costs more labor in the printing process, optimizing bending strength significantly. However, compared with the Type L rebars, it can result in more flexural cracks (Figs. 6, 7 and Table 2).

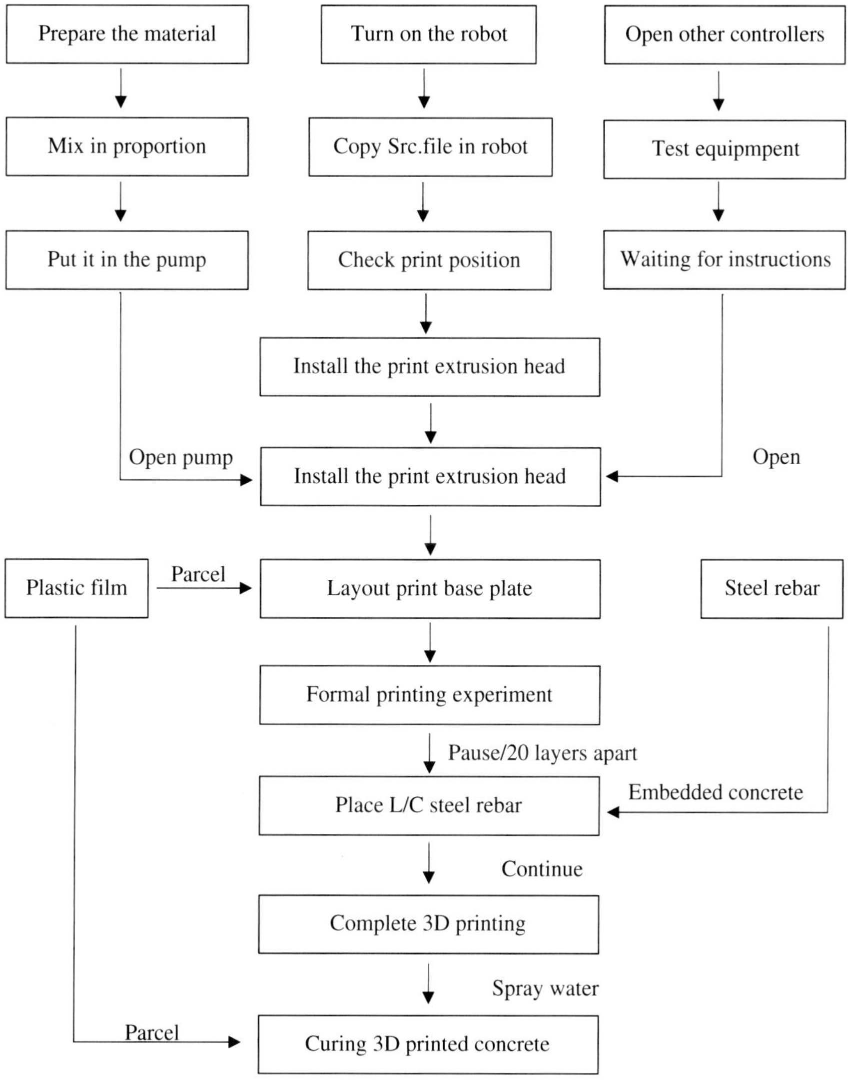

Fig. 6. Flowchart of design to construction process

Fig. 7. Printing process

Table 2. Surface and deformation performance

	Number of cracks	Length of cracks (cm)	Bending deformation (mm)
Size A 01+ L	2	30	10
Size A 02 + L	1	15	8
Size A 03 + L	0	0	3
Size A 04 + C	1	5	8
Size A 05 + C	0	0	3
Size B 01 + L	0	0	5
Size B 02 + L	1	10	5
Size B 03 + L	0	0	3
Size B 04 + C	0	0	3
Size B 05 + C	1	10	2

5.3 Different Types of Additive Agent

From the perspective of the additive agent, in the following table, it is essential to note that the time cost of preparing steel fiber takes twice as long as PVA fiber since we use the same proportion of PVA fiber in the casting part and printing part. Specifically, the steel fiber is much harder and stronger than PVA fiber so it is not allowed to be blended in the pump, which means another blender is required before printing. Furthermore, steel fiber will add additional weight in the formwork itself, occupying two percent of the volume fraction of casting concrete. The casting process and finished prototypes are shown in Figs. 8, 9 and Table 3.

Fig. 8. Casting process

6 Conclusion and Future Work

In this paper, we first compared different reinforcement strategies for 3D concrete printing and then proposed a new method that significantly improves the bending strength, toughness, and flexibility of permanent 3D concrete printing formworks. Integrating prefabricated and standardized rebar 3D printed concrete elements not only opens new function possibilities but also enhances the reliability of robotic fabrication. In addition, this system can be adopted for the construction of mass customization elements, where traditional concrete fabrication could cause more carbon footprint through the process of fabricating timber formworks and casting concrete. It is a potential approach that efficiently achieves complex 3D geometries through parametric design tools, promoting the specification related to 3D printing architecture integrity. However, more scientific experiments like four-point bending tests and toughness and ductility tests still need to be done to get specific bending strength and flexural characteristics of 3D printing concrete. Additionally, we can get more promising results and possibilities in engineering

Fig. 9. Finished surface

Table 3. Different additive agent and labor cost

	Casting material	Prepared time (h)	Casting wight (t)	Labor
Size A 01 + L	Non	0	0	2
Size A 02 +L	Modified PVA fiber	0.5	0.35	3
Size A 03 + L	Modified PVA fiber	0.5	0.35	2
Size A 04 + C	Modified PVA fiber	0.5	0.35	2
Size A 05 + C	Modified steel fiber	1	0.4	2
Size B 01 + L	Non	0	0	2
Size B 02 + L	Modified PVA fiber	0.3	0.18	1
Size B 03 + L	Modified PVA fiber	0.3	0.18	2
Size B 04 +C	Modified PVA fiber	0.3	0.18	1
Size B 05 +C	Modified steel fiber	0.7	0.23	1

projects with further modifications to the printing environment referring to temperature and humidity, more types of addition agents in the formworks, and more rational curing methods. Thus, more non-uniform geometries of large, prefabricated formwork components could be achieved.

References

1. Bos F, Wolfs R, Ahmed Z et al. (2016) Additive manufacturing of concrete in construction: Potentials and challenges of 3D concrete printing. Virtual Phys Prototyp 11(3):209–225
2. Zhan Q, Zhou X, Yuan PF (2021) Digital design and fabrication of a 3D concrete printed Prestressed bridge
3. Wang L, Yang Y, Yao L et al (2022) Interfacial bonding properties of 3D printed permanent formwork with the post-casted concrete. Cem Concr Compos 128:104457
4. Singh A, Liu Q, Xiao J et al (2022) Mechanical and macrostructural properties of 3D printed concrete dosed with steel fibers under different loading direction. Constr Build Mater 323:126616. https://doi.org/10.1016/j.conbuildmat.2022.126616
5. Marchment T, Sanjayan J (2021) Reinforcement method for 3D concrete printing using paste-coated bar penetrations. Autom Constr
6. Xiao J, Lv Z, Duan Z et al (2022) Study on preparation and mechanical properties of 3D printed concrete with different aggregate combinations. J Build Eng 2022(51):104282

Levelling Calibration and Intelligent Real-Time Monitoring of the Assembly Process of a DfD-Based Prefabricated Structure Using a Motion Capture System

Xiang Wang, Yang Li, Ziqi Zhou, Xueyuan Lv, Philip F. Yuan, and Lei Chen(✉)

China Construction First Group Cooporation Limited, Xisihuannanlu 52, Beijing 100161, China
18310021@tongji.edu.cn

Abstract. Conventional measuring techniques and equipment such as the level and total-station are commonly used in on-site construction to measure the position of building elements. However, a motion capture system can measure the dynamic 3D movements of markers attached to any target structure with high accuracy and high sampling rate. Considering the characteristics of prefabricated structures that is composed by lot of discrete building elements, advanced requirements for the on-site assembly monitoring is required. This paper introduces an innovative real-time monitoring technique for the DfD-based (Design for Disassembly) structure with the application of motion capture system and other hardware in an IoT-based BIM system. The design and construction method of the structure system, on-site setup of monitoring system and hardware, data acquisition and analysis method, calibration algorithm as well as the BIM system are further illustrated in the paper. The proposed method is finally applied in a real building project that is composed by thousand discrete building elements and covers a large area of 50*25 m. As demonstrator, such monitoring system is applied in the real construction of a DfD-based prefabricated steel structure in the "Water Cube" (Chinese National Aquatics Centre) in Beijing. The building process is successfully recorded and displayed on-site with the digital twin model in the BIM system. The construction states of the building elements are gathered with different kind of IoT techniques such as the RfID chips and QR-Codes. With the demand to control the flatness tolerance within 6 mm (within a 25*50 m area), a large area monitoring system was applied in the project and finally reduced the construction time within 20 days. The final tolerance is verified and further discussed2.

Keywords: Motion-capture system · Design for disassembly · Levelling calibration · Life-cycle monitoring · BIM system

1 Introduction

Nowadays, construction industry is facing a tremendous digital changing with the wide application of digital and intelligent building techniques. Prefabricated structure system is also greatly challenging the traditional building system and methods with the

P. F. Yuan et al. (eds.), *Hybrid Intelligence*, Computational Design and Robotic Fabrication,
https://doi.org/10.1007/978-981-19-8637-6_45

advantage that lot of building components and be standardized, and hence the design and fabrication process can be greatly improved by the application of BIM system and industrial manufacturing technique. Nevertheless, prefabricated structures also bring a restrict requirement of detail design and a series of building techniques that rely greatly on the assembly process and its accuracy. In this sense, apart from the digital fabrication technique, the digital or advanced assembly technique is of great importance in the present architectural researches and practice.

At the same time, prefabricated structure also brings a new view point of the relationship between architecture and its environment, and of adaptive and sustainable architecture. In this field, DfD (Design for Disassembly) is one of the most systematic theories, and focuses on the hierarchy and composition of the structure system, its components and their connection details, with the purpose to give full play to the performance of the material and to enhance the potential sustainability of a building [1]. Design for disassembly requires both the design process to split building into components system and the construction techniques to assemble the structure. On one hand, DfD system responds to the fast and repeatable installation and damage-free detachment. Therefore, it emphasizes the application of dry joint, including both structural members and enclosure members. On the other hand, due to the highly discretized structural system, the design and construction of such structure relies strongly on the building information model of the component system and the installation process itself. This also makes the on-site assembly procedure and its quality control (accuracy of the installation) very important.

In this research, an innovative on-site building process for the DfD structure system is introduced. The importance the possibility to use BIM models together with the IoT (Internet of Things) techniques is discussed. At the same time, a novel application of motion capture system in the levelling process is present. The main case in this research is a DfD-based quick assembly and disassembly steel structure system, which is mainly used for the construction of the competition platform of the Beijing 2022 Winter Olympics curling venues. For the first time in the history of the Olympic Games, the curling venue of the Beijing Winter Olympics has applied a DfD construction system that can be repeatedly disassembled and assembled (Fig. 1). The main swimming pool of the original 2008 Summer Olympics Swimming Centre- "the Water Cube"- is used for reconstruction, trying to achieve a repeatable "winter and summer" scene change for the different uses. Since the construction time of the structural conversion has greatly affected the daily operations of the venues, the efficiency of the structural conversion has become the core issue of this project. At the same time, because curling has extremely high requirements for the flatness of the venue platform (the vertical height difference of any two points of the venue platform within the entire 50*25 m range must be guaranteed within 6 mm), the overall stability of the prefabricated structure and the accuracy of the installation also determines the success of the entire project.

Fig. 1. Demonstrator DfD-based project of the Olympic curling hall in the Chinese National Aquatics Center ("Water-Cube"). The steel structure should be assembled and disassembled every year to accommodate the scenario changing requirement of the building

2 BIM System and the Motion Capture System

2.1 BIM-Based Construction Control System

The selected research project provides a high complexity of requirement that includes the status monitoring of each building component in the full construction cycle, the real-time dynamic monitoring of the high information of hundreds of plates, the high accurate calibration standard of levelling (± 3 mm). Therefore, a complex framework and a powerful construction data analysis platform is needed. The structural design and the component design consider both the possibilities to maintain the integrity of the structure and to make the component small enough for transportation. And this also makes the final discretized building structure composed of thousands of different structural elements.

To solve the problem, the BIM platform is introduced from the designing stage of the structure system. Firstly, a component definition system and its related building information model is built in REVIT. For each component, a customized family is created with several special defined fields to trace its condition, such as the transportation stages, assembly stages as well as the levelling information.

The tracing and monitoring of various components relies mainly on the sensors such as the RFID that gives the status information whether a component is on-site or still in transportation, the motion capture system that provide the height of each top plate as well as the QR Code information that the builders edited at any time. To integrate every different kind of information together, Autodesk REVIT is selected as the main software platform. With the use of the "Rhino-inside-REVIT" technique, a computational data analysis approach could be achieved by using Grasshopper Platform inside REVIT (Fig. 2). By developing corresponding interfaces in the system, one can get all the information at the same time and interact with the same data system.

2.2 Motion Capture System

A Motion Capture (MoCap) System comprises markers, cameras, a server, and a computer as the data analyzer and exchanger, as shown in Fig. 3. The infrared-reflective markers are attached to the target objects and the cameras can triangulate the location of the markers and record them with a high sampling rate.

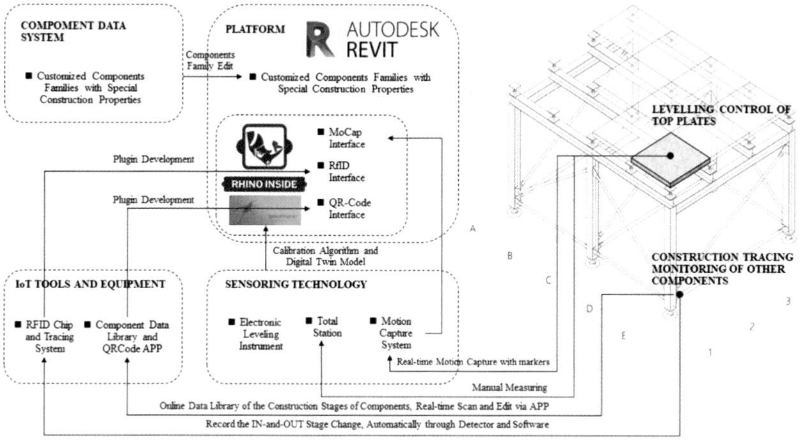

Fig. 2. On-site monitoring system and its workflow

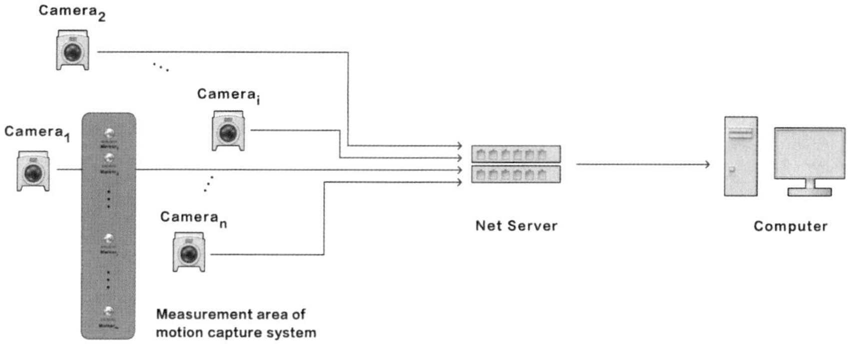

Fig. 3. Typical system configuration of a motion capture system

To obtain the 3D coordinates of a marker, at least 2 cameras must simultaneously track the positions of the target marker in the 2D images. At the same time, because the accuracy is decreased if the marker image in any camera(s) overlap or are closely positioned, a series of evenly distributed cameras are the common configuration in a MoCap system. All the cameras in a MoCap system must be synchronized to measure the dynamic motions of different individual markers. To achieve the synchronization, a server gets all the real-time 2D information of the cameras and send them to the computer. The software in the computer combines the 2D coordinates together and record all the history data for each marker.

To get the accurate 3D coordinate of the target markers, the real coordinates of the cameras should be calculated. Because there are often many cameras in one system, one simple static coordinate information of the markers is not enough. Therefore, a calibration procedure must be applied. The calibration uses a great amount of data, which is usually sampled by performing with a wand, as shown in Fig. 4. Because the wand provides the basic shape relationship and distances of the markers as a priori and the wand can be easily waved, the calibration method based on the wand can simplify the process and improve the calibration accuracy.

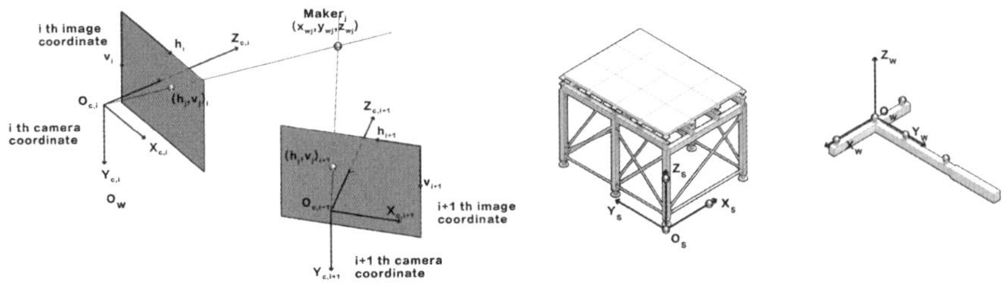

Fig. 4. Principles of the calibration of a motion capture system

The calibration process can be divided into two processes: static and dynamic. Firstly, static calibration is performed to establish the absolute spatial reference of the coordinate system. In this process, the T-shape of the wand will be used. As shown in Fig. 4, the wand is placed around the target, and the markers appended on the wand form two perpendicular axes (x_w, y_w). These axes are used to define a 3D space coordinated system (x_w, y_w, z_w). The relationship between the 2D image coordinates of the cameras (u, v) and the target 3D model coordinates (x_w, y_w, z_w) is defined by the algorithm as follows [2]:

$$\begin{pmatrix} u \\ v \end{pmatrix} = [C][I|0][R|T]\begin{pmatrix} x_w \\ y_w \\ z_w \\ 1 \end{pmatrix} = \begin{bmatrix} f_x & 0 & p_x \\ 0 & f_y & p_y \\ 0 & 0 & 1 \end{bmatrix}\begin{bmatrix} 1 & 0 & 0 & 0 \\ 0 & 1 & 0 & 0 \\ 0 & 0 & 1 & 0 \end{bmatrix}\begin{bmatrix} r_{11} & r_{12} & r_{13} & T_x \\ r_{21} & r_{22} & r_{23} & T_y \\ r_{31} & r_{32} & r_{33} & T_z \\ 0 & 0 & 0 & 1 \end{bmatrix}\begin{pmatrix} x_w \\ y_w \\ z_w \\ 1 \end{pmatrix} \tag{1}$$

where $[C]$ is the perspective projection matrix from the camera space to the image space, which is also an intrinsic matrix that consists of the focal length f_x and f_y and the principal point coordinates p_x and p_y. $[R|T]$ is the extrinsic matrix that shows the transformation from the wand space to the camera space. It contains the rotation matrix R and the translation vector T.

Three coordinate systems are used in the process to transform the 2D image coordinates to the 3D model coordinates. As shown in Fig. 4, the 2D image coordinates $(h_j, v_j)_\mathrm{i}$ of the j th marker from the i th camera are the result of the perspective projection of the camera coordinates. The camera coordinates $(x_c, y_c, z_c)_\mathrm{i}$ are the 3D coordinates of the ith camera corresponding to the marker. The wand coordinates (x_{wj}, y_{wj}, z_{wj}) are the virtual 3D coordinates of the jth marker relative to the location of the wand.

The camera matrix $[C]$ and the extrinsic matrix $[R|T]$ are optimized and finally determined via a dynamic calibration procedure, which is performed by waving the wand to record multiple coordinates data and to calculate the relative positions and directions of the camera. The accuracy of the reconstruction of the 3D coordinates is dependent on various factors such as the focal length, the lens distortion coefficients and the camera's resolution. During the calibration, direct linear transformation and nonlinear transformation algorithms are frequently employed [3].

3 Experiments and the Results

3.1 Design of the Structure and the BIM System

The final construction structure is 56.7 m long and 26.7 m wide. The supporting frame body is made of prefabricated steel columns and beams, covered with light-weight precast concrete slab to meet the ice-making conditions (Fig. 5). The structural beams and column members are all high-frequency welded thin-walled H-section steel. The steel used for the members is Q235B grade steel, and the connection is bolted. The connections used are M16 high-strength structural bolts. The surface layer adopts lightweight concrete prefabricated slabs, the main specification is 1*1*1 m, the strength grade is L40. Finally, a total of 1,568 prefabricated slabs and a steel structure of 140 t are used.

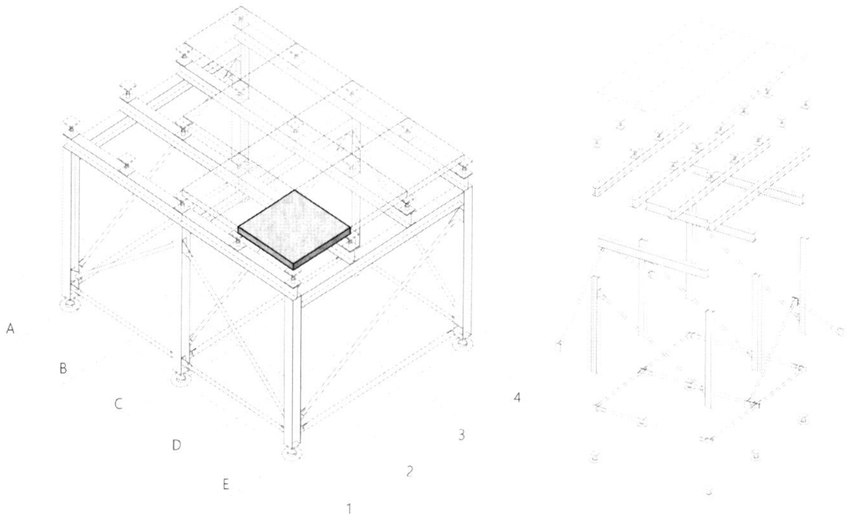

Fig. 5. Design of the DfD-based structure system

All components are defined parametrically in REVIT by means of customized family components through the use of computational design methods in the Grasshopper plugin, so the establishment of the component system database and custom modifications can be quickly realized. In order to realize real-time positioning and tracking and information visualization of the component system, all components in the database are named using unique coding rules. At the same time, in order to realize the collection and processing of information based on IoT theory, a plug-in suitable for reading RFID beacon information has been developed in the relevant Grasshopper platform. Meanwhile, a WECHAT-based Application which is suitable for QR code scanning and information filling has been created. Data crawling and real-time analysis of QR code database information is realized in Grasshopper (Fig. 6). Therefore, the repetitively disassembled steel structure system realizes the real-time perception and analysis of the construction information of the whole process at each component level, and is integrated accordingly through the BIM software platform.

Fig. 6. Data collection of the QR Code based APP

3.2 On-Site Levelling Calibration

The construction site has an area of 60*25 m, which is far beyond the sensing range of the sensors (such as the resolution range of the infrared cameras of the motion capture system). At the same time, due to the large height difference of the original swimming pool in the site, the motion capture cameras on site can only be placed on the side of the pool bank in a unilateral arrangement, and the single monitoring range can basically reach 12*15 m (Fig. 7). In previous researches, it is also shown that unevenly distributed cameras, different light environment and frequent change of camera visibility would also affect the final accuracy of the MoCap system [4, 5]. And it is helpful to use hybrid MoCap techniques and to work together with traditional surveying and measuring techniques to reduce the system tolerance [6]. Therefore, a calibration of the levelling analysis is required.

In order to solve this problem, a set of coordinate system compensation and correction algorithm based on ICP (Iterative Closest Points) algorithm was developed, and was finally developed as plugins inside Grasshopper. The basic principle is: after the main beams are installed, the top midpoints of the structural column are used to arrange the positioning markers, the relevant column top coordinates are then collected in the MoCap software to establish a grid system. At the same time, the electronic level and

Fig. 7. On-site placement of the motion capture system and the collected point cloud for each marker

the total station were used as reference data, and the total station was used to collect the horizontal XY coordinates, and the electronic level was used to collect the vertical Z coordinates to generate a second set of real coordinate grids for correcting errors. Finally, the corresponding regions of the two sets of grids were fitted, the global tolerance will be eliminated. Together with the BIM model, a real-time levelling support software is further developed (Fig. 8).

Fig. 8. On-site placement of the motion capture system and the collected point cloud for each marker

At the construction site, the BIM system and related levelling support system were connected to the central control screen, and the construction personnel were assisted to fine-tune the relevant structure through real-time playback. This construction method greatly saved the time required for on-site installation, especially the real-time display of the components' status and the levelling process, and provided a large amount of data support for the construction planning. Compared with the previous structural conversion construction using traditional level measurement, it reduced the construction time from 40 to 20 days and enhanced the quality of the assembled structure.

In the construction, a part of the construction area that did not use levelling monitoring was reserved as a comparison verification area in all areas. It was tested via scanning with the laser tracing instrument to test all the slab centre coordinates. The results showed that the area with levelling monitoring basically meet the requirements of the project for global and local elevation errors, and the overall accuracy was controlled between $-1.5-1.5$ mm (Fig. 9). Compared with the levelling monitoring area, the unmonitored area had a larger elevation error, and it was related to the experience of the construction personnel. A large number of measurement points in the area were lower than the global elevation, and the error was in the range of -2 to -10 mm.

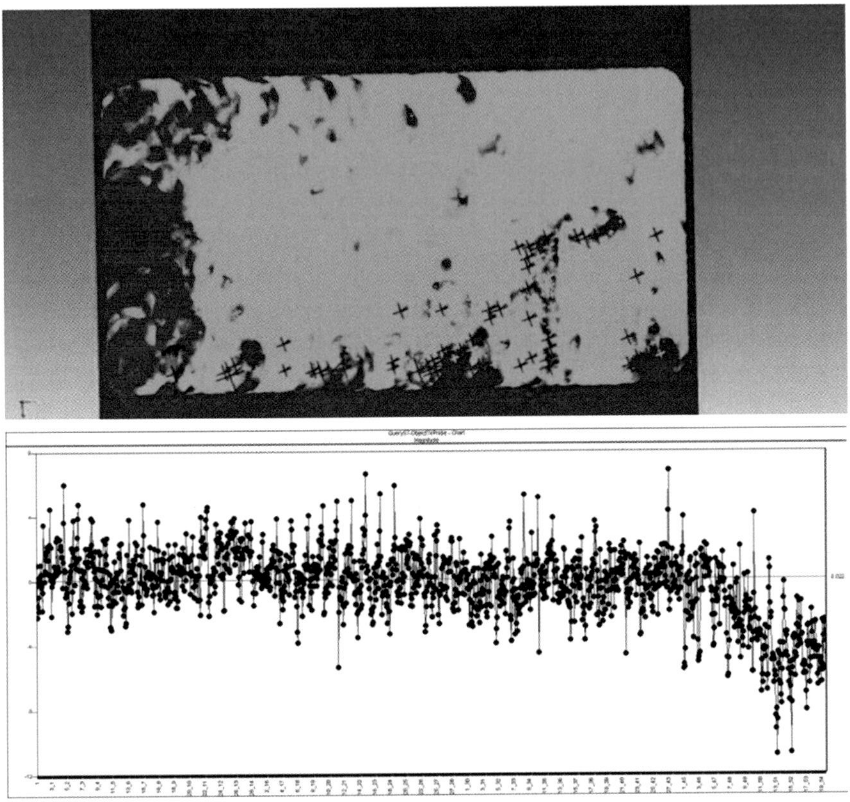

Fig. 9. Verification of the effect of the motion capture levelling system

4 Conclusion

This research presents and discusses the possibility to apply a BIM-based system for the intelligent and sustainable construction of some possible DfD-based structure systems. The system framework and key technologies are shown with an integrated BIM workflow, with corresponding IoT techniques, motion capture system technique and the software interface in the Grasshopper plugin. As demonstrator, such construction method is applied in the China National Aquatics Centre for the rapid and repeatable scene transition from the 2008 summer swimming venue to the 2022 winter curling venue, effectively reducing the required structural construction time and ensuring the structure itself the high standard of elevation control requirements. Under the current social development background, this construction method will provide a potential green, efficient and sustainable energy-saving construction technology, and provide an enlightening direction for the digital development, sustainable development, energy conservation and emission reduction of the future construction industry.

References

1. Guy B, Ciarimboli N (2003) Design for disassembly in the built environment. WA, Resource Venture, Inc. Pennsylvania State University, City of Seattle
2. Bailey B, Wolf A (2007) Real time 3D motion tracking for interactive computer simulations, vol 3. Imperial College, London, UK
3. Hinrichs RN, McLean SP (1995) NLT and extrapolated DLT: 3-D cinematography alternatives for enlarging the volume of calibration. J Biomech 28(10):1219–1223
4. Aurand AM, Dufour JS, Marras WS (2017) Accuracy map of an optical motion capture system with 42 or 21 cameras in a large measurement volume. J Biomech 58:237–240
5. Raghu SL, Kang C-k, Whitehead P, Takeyama A, Conners R (2019) Static accuracy analysis of Vicon T40s motion capture cameras arranged externally for motion capture in constrained aquatic environments. J Biomech 89:139–42
6. Nagymáté G, Tuchband T, Kiss RM (2018) A novel validation and calibration method for motion capture systems based on micro-triangulation. J Biomech 74:16–22

Author Index

P. F. Yuan et al. (eds.), *Hybrid Intelligence*, Computational Design and Robotic Fabrication,
https://doi.org/10.1007/978-981-19-8637-6

Printed in the United States
by Baker & Taylor Publisher Services